21ST CENTURY ASTRONOMY

SIXTH EDITION

21ST CENTURY ASTRONOMY

SIXTH EDITION

Laura Kay
Barnard College

Stacy Palen
Weber State University

George Blumenthal
University of California—Santa Cruz

W. W. NORTON & COMPANY

NEW YORK · LONDON

W. W. Norton & Company has been independent since its founding in 1923, when William Warder Norton and Mary D. Herter Norton first published lectures delivered at the People's Institute, the adult education division of New York City's Cooper Union. The firm soon expanded its program beyond the Institute, publishing books by celebrated academics from America and abroad. By midcentury, the two major pillars of Norton's publishing program—trade books and college texts—were firmly established. In the 1950s, the Norton family transferred control of the company to its employees, and today—with a staff of four hundred and a comparable number of trade, college, and professional titles published each year—W. W. Norton & Company stands as the largest and oldest publishing house owned wholly by its employees.

Editor: Erik Fahlgren
Project Editor: Diane Cipollone
Editorial Assistant: Sara Bonacum
Managing Editor, College: Marian Johnson
Managing Editor, College Digital Media: Kim Yi
Production Manager: Ashley Horna
Media Editor: Rob Bellinger
Associate Media Editor: Arielle Holstein
Media Project Editor: Danielle Belfiore
Media Editorial Assistant: Kelly Smith
Marketing Manager: Katie Sweeney
Design Director: Rubina Yeh
Photo Editor: Trish Marx
Director of College Permissions: Megan Schindel
Permissions Associate: Elizabeth Trammell
Composition: Graphic World
Manufacturing: Transcontinental

ISBN 978-0-393-64470-8 (pbk)

W. W. Norton & Company, Inc., 500 Fifth Avenue, New York, NY 10110-0017
wwnorton.com
W. W. Norton & Company Ltd., 15 Carlisle Street, London W1D 3BS
1 2 3 4 5 6 7 8 9 0

Aloha and Bon Voyage to our late co-author Brad Smith, who showed the world the outer solar system through his cameras on the Voyager spacecraft.

LAURA KAY thanks her wife, M.P.M. She dedicates this book to her late uncle, Lee Jacobi, for an early introduction to physics, and to her late colleagues at Barnard College, Tally Kampen and Sally Chapman.

STACY PALEN thanks her husband, John Armstrong, for his patient support during this project.

GEORGE BLUMENTHAL gratefully thanks his wife, Kelly Weisberg, and his children, Aaron and Sarah Blumenthal, for their support during this project. He also wants to thank Professor Robert Greenler for stimulating his interest in all things related to physics.

BRIEF CONTENTS

CONTENTS

1 Thinking Like an Astronomer   2

2 Patterns in the Sky—Motions of Earth and the Moon   22

3 Motion of Astronomical Bodies 58

4 Gravity and Orbits 82

5 Light 108

6 The Tools of the Astronomer 140

7 The Formation of Planetary Systems 170

8 The Terrestrial Planets and Earth's Moon 196

9 Atmospheres of the Terrestrial Planets 230

10 Worlds of Gas and Liquid—The Giant Planets 264

PART III STARS AND STELLAR EVOLUTION

Working It Out

AstroTours

AstroTour animations, Interactive Simulations, and Astronomy in Action Videos are all available from the free Student Site at the Digital Resources Site, and they are also integrated into assignable Smartwork5 exercises. **digital.wwnorton.com/astro6**

Interactive Simulations

Astronomy in Action Videos

Preface

Dear Student

Why is it a good idea to take a science course, and in particular, why is astronomy a course worth taking? Many people choose to learn about astronomy because they are curious about the universe. Your instructor likely has two basic goals in mind for you as you take this course. The first is to understand some basic physical concepts and how they apply to the universe around us. The second is to think like a scientist and learn to use the scientific method not only to answer questions in this course but also to be able to evaluate and understand new information you encounter in your life. We have written the sixth edition of *21st Century Astronomy* with these two goals in mind.

Throughout this book, we emphasize not only the content of astronomy (for example, the differences among the planets, the formation of chemical elements) but also *how* we know what we know. The scientific method is a valuable tool that you can carry with you and use for the rest of your life. One way we highlight the process of science is the **Process of Science Figures**. In each chapter, we have chosen one discovery and provided a visual representation illustrating the discovery or a principle of the process of science. In these figures, we try to illustrate that science is not a tidy process, and that discoveries are sometimes made by different groups, sometimes by accident, but always because people are trying to answer a question and show why or how we think something is the way it is.

The most effective way to learn something is to "do" it. Whether playing an instrument or a sport or becoming a good cook, reading "how" can only take you so far. The same is true of learning astronomy. We have written this book to help you "do" as you learn. Sometimes this means doing an activity to apply a concept and sometimes we provide a tool to make reading a more active process.

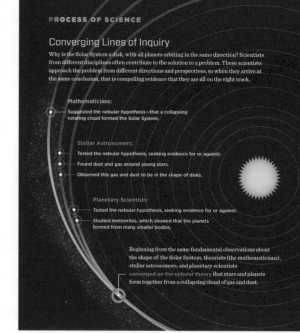

PROCESS OF SCIENCE

Converging Lines of Inquiry

Why is the Solar System a disk, with all planets orbiting in the same direction? Scientists from different disciplines often contribute to the solution to a problem. These scientists approach the problem from different directions and perspectives, so when they arrive at the same conclusion, that is compelling evidence that they are all on the right track.

Mathematicians:
- Suggested the nebular hypothesis—that a collapsing rotating cloud formed the Solar System.

Stellar Astronomers:
- Tested the nebular hypothesis, seeking evidence for or against.
- Found dust and gas around young stars.
- Observed this gas and dust to be in the shape of disks.

Planetary Scientists:
- Tested the nebular hypothesis, seeking evidence for or against.
- Studied meteorites, which showed that the planets formed from many smaller bodies.

Beginning from the same fundamental observations about the shape of the Solar System, theorists (the mathematicians), stellar astronomers, and planetary scientists converged on the nebular theory that stars and planets form together from a collapsing cloud of gas and dust.

- At the beginning of each chapter, we have provided a set of Learning Goals to guide you as you read. There is a lot of information in every chapter, and the Learning Goals should help you focus on the most important points.

- We present a big-picture question in association with the chapter-opening photo at the beginning of each chapter. For each of these, we have tried to pose a question that is not only relevant to its chapter but also something you may have

LEARNING GOALS

By the end of this chapter, you should be able to:

LG 1 Describe how our understanding of planetary system formation developed from the work of both planetary and stellar scientists.

LG 2 Discuss the role of gravity and angular momentum in explaining why planets orbit the Sun in a plane and why they revolve in the same direction that the Sun rotates.

LG 3 Explain how temperature at different locations in the protoplanetary disk affects the composition of planets, moons, and other bodies.

LG 4 Discuss the processes that resulted in the formation of planets and other objects in our Solar System.

LG 5 Describe how astronomers find planets around other stars and derive exoplanet properties.

How did our Solar System form?

wondered about. We hope that these questions, plus the photographs that accompany them, capture your attention as well as your imagination.

CHECK YOUR UNDERSTANDING 7.1

Which of the following pieces of evidence support the nebular hypothesis? (Choose all that apply.) (a) Planets orbit the Sun in the same direction. (b) The Solar System is relatively flat. (c) Earth has a large Moon. (d) We observe disks of gas and dust around other stars.

READING ASTRONOMY NEWS

ARTICLES QUESTIONS · A multiplanet system discovered around a relatively nearby star

Four Earth-sized planets detected orbiting the nearest Sun-like star

August 9, 2017, by TIM STEPHENS

A new study by an international team of astronomers reveals that four Earth-sized planets orbit the nearest Sun-like star, Tau Ceti, which is about 12 light years away and visible to the naked eye. These planets have masses as low as 1.7 Earth mass, making them among the smallest planets ever detected around nearby Sun-like stars. Two of them are super-Earths located in the habitable zone of the star, meaning they could support liquid surface water.

According to lead author Fabo Feng of the University of Hertfordshire, UK, the researchers are getting tantalizingly close to the 10-centimeter-per-second limit required for detecting Earth analogs. "Our detection of such weak wobbles is a milestone in the search for Earth analogs and the understanding of the Earth's habitability through comparison with these analogs," Feng said. "We have introduced new methods to remove the noise in the data in order to reveal the weak planetary signals."

The outer two planets around Tau Ceti

see how star's activity differed at different wavelengths and use that information to separate this activity from signals of planets," Tuomi said.

The researchers painstakingly improved the sensitivity of their techniques and were able to rule out two of the signals the team had identified in 2013 as planets. "But no matter how we look at the star, there seem to be at least four rocky planets orbiting it," Tuomi said. "We are slowly learning to tell the difference between wobbles caused by planets and those caused by stellar active surface."

ARTICLES QUESTIONS

1. How do planets cause their stars to "wobble"?
2. With the spectroscopic method, why is finding Earth-sized planets harder than finding giant planets?
3. How did the astronomers improve their ability to find small planets?
4. What is a debris disk?
5. What would be the disadvantage to a planet's being tidally locked to its star? (Hint: Refer to Chapter 2.)

- There are **Check Your Understanding** questions at the end of each chapter section. These questions are designed to be answered quickly if you have understood the previous section. The answers are provided in the back of the book so you can check your answer and decide if further review is necessary.

- As a citizen of the world, you make judgments about science, distinguishing between good science and pseudoscience. You use these judgments to make decisions in the grocery store, pharmacy, car dealership, and voting booth. You may base these decisions on the presentation of information you receive through the media, which is very different from the presentation in class. One important skill is the ability to recognize what is credible and to question what is not. To help you hone this skill, we have provided **Reading Astronomy News** sections at the end of every chapter. These features include a news article with questions to help you make sense of how science is presented to you. It is important that you learn to be critical of the information you receive, and these features will help you do that.

- While we know a lot about the universe, science is an ongoing process, and we continue to search for new answers. To give you a glimpse of what we don't know, we provide an **Unanswered Questions** feature near the end of each chapter. Most of these questions represent topics that scientists are currently studying.

? Unanswered Questions

- How typical is the Solar System? Only within the past decade have astronomers found other systems containing four or more planets, and so far the observed distributions of large and small planets in those multiplanet systems have looked different from those of the Solar System. Computer simulations of planetary system formation suggest that a system with stable orbits and a planetary distribution like those of the Solar System may develop only rarely. Improved supercomputers can run more complex simulations, which can be compared with the observations to better understand how solar systems are configured.

- How Earth-like must a planet be before scientists declare it to be "another Earth"? An editorial in the science journal *Nature* cautioned that scientists should define "Earth-like" in advance—before multiple discoveries of planets "similar" to Earth are announced and a media frenzy ensues. Must a planet be of similar size and mass, be located in the habitable zone, and have spectroscopic evidence of liquid water before we call it "Earth 2.0"?

- The language of science is mathematics, and it can be as challenging to learn as any other language. The choice to use mathematics as the language of science is not arbitrary; nature "speaks" math. To learn about nature, you will need to speak its language. We don't want the language of math to obscure the concepts, so we have placed this book's mathematics in **Working It Out** boxes to make it clear when we are beginning and ending a mathematical argument, so that you can spend time with the concepts in the chapter text and then revisit the mathematics of the concept to study the formal language of the argument. You will learn to work with data and identify when data aren't quite right. We want you to be comfortable reading, hearing, and speaking the language of science, and we will provide you with tools to make it easier.

7.1 Working It Out Angular Momentum

In its simplest form, the angular momentum (L) of a system is given by

$$L = m \times v \times r$$

where m is the mass, v is the speed at which the mass is moving, and r represents how spread out the mass is.

Let's apply that relationship to the angular momentum of Jupiter in its orbit about the Sun. The angular momentum from one body orbiting another is called *orbital* angular momentum, $L_{orbital}$. The mass (m) of Jupiter is 1.90×10^{27} kilograms (kg), the speed of Jupiter in orbit (v) is 1.31×10^4 meters per second (m/s), and the radius of Jupiter's orbit (r) is 7.79×10^{11} meters. Putting all that together gives

$$L_{orbital} = (1.90 \times 10^{27} \text{ kg}) \times (1.31 \times 10^4 \text{ m/s}) \times (7.79 \times 10^{11} \text{ m})$$

$$L_{orbital} = 1.94 \times 10^{43} \text{ kg m}^2/\text{s}$$

where R is the radius of the sphere and P is the rotation period of its spin.

Let's compare Jupiter's orbital angular momentum with the Sun's spin angular momentum to investigate the distribution of angular momentum in the Solar System. Appendix 2 provides the Sun's radius (6.96×10^8 meters) and mass (1.99×10^{30} kg), and its rotation period is 24.5 days = 2.12×10^6 seconds. If we assume that the Sun is a uniform sphere, the spin angular momentum of the Sun is

$$L_{spin} = \frac{4 \times \pi \times (1.99 \times 10^{30} \text{ kg}) \times (6.96 \times 10^8 \text{ m})^2}{5 \times (2.12 \times 10^6 \text{ s})}$$

$$L_{spin} = 1.14 \times 10^{42} \text{ kg m}^2/\text{s}$$

$L_{orbital}$ of Jupiter is about 17 times greater than L_{spin} of the Sun. Thus, most of the angular momentum of the Solar System now

- Each chapter concludes with an **Origins** section, which relates material or subjects found in the chapter to questions about the origin of the universe and the origin of life in the universe and on Earth. Astrobiologists have made much progress in recent years on understanding how conditions in the universe may have helped or hindered the origin of life, and in each Origins we explore an example from its chapter that relates to how the universe and life formed and evolved.

- At the end of each chapter, we have provided several types of questions, problems, and activities for you to practice your skills. The **Test Your Understanding** questions focus on more detailed facts and concepts from the chapter. **Thinking about the Concepts** questions ask you to synthesize information and explain the "how" or "why" of a situation. **Applying the Concepts** problems give you a chance to practice the quantitative skills you learned in the chapter and to work through a situation mathematically. The **Using the Web** questions represent other opportunities to "learn by doing." Using the Web sends you to websites of space missions, observatories, experiments, or archives to access recent observations, results, or press releases. Other sites are for "citizen science" projects in which you can contribute to the analysis of new data.

- **Explorations** show you how to use the concepts and skills you learned in an interactive way. Some of the book's Explorations ask you to use animations and simulations on the Student Site, while the others are hands-on, paper-and-pencil activities that use everyday objects such as ice cubes or balloons.

The resources outside of the book (at the Student Site) can help you understand and visualize many of the physical concepts described in the book. **AstroTours and Interactive Simulations** are represented by icons in the margins of the book. There is also a series of short **Astronomy in Action** videos that are represented by icons in the margins and available at the Student Site. These videos feature one of the authors (and several students) demonstrating physical concepts at work. Your instructor might assign these videos to you or you might choose to watch them on your own to create a better picture of each concept in your mind.

Astronomy gives you a sense of perspective that no other field of study offers. The universe is vast, fascinating, and beautiful, filled with a wealth of objects that, surprisingly, can be understood using only a handful of principles. By the end of this book, you will have gained a sense of your place in the universe.

USING THE WEB

46. Using the exoplanet catalogs:
    a. Go to the "Catalog" Web page (http://exoplanet.eu/catalog) of the Extrasolar Planets Encyclopedia. Look for a star that has multiple planets. Make a graph showing the distances of the planets from that star, and note the masses and sizes of the planets. Put the Solar System planets on the same axis. How does that extrasolar planet system compare with the Solar System?
    b. Go to the "Exoplanets Data Explorer" website (http://exoplanets.org) and click on "Table." That website lists planets that have detailed orbital data published in scientific journals, and it may have a smaller total count than the website in part (a). Pick a planet discovered recently, as specified in the "First Reference" column. What is the planet's minimum mass? What is its semimajor axis and the period of its orbit? What is the eccentricity of its orbit? Click on the star name in the first column to get more information. Does that planet have a radial velocity curve? Was it observed in transit, and if so, what is the planet's radius and density? Is it more like Jupiter or more like Earth?

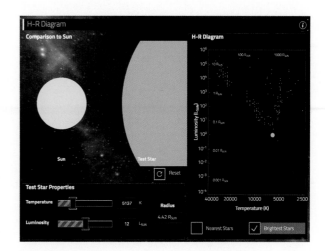

Dear Instructor

We wrote this book with a few overarching goals: to inspire students, to make the material interactive, and to create a useful and flexible tool that can support multiple learning styles.

As scientists and as teachers, we are passionate about the work we do. We hope to share that passion with students and inspire them to engage in science on their own. Through our own experience, familiarity with education research, and surveys of instructors, we have come to know a great deal about how students learn and what goals teachers have for their students. We have explicitly addressed many of these goals and learning styles in this book, sometimes in large, immediately visible ways such as the inclusion of features but also through less obvious efforts such as questions and problems that relate astronomical concepts to everyday situations or a fresh approach to organizing material.

For example, many teachers state that they would like their students to become "educated scientific consumers" and "critical thinkers" or that their students should "be able to read a news story about science and understand its significance." We have specifically addressed these goals in our **Reading Astronomy News** feature, which presents a news article and a series of questions that guide a student's critical thinking about the article, the data presented, and the sources.

Education research shows that the most effective way to learn is by doing. **Exploration** activities at the end of each chapter are hands-on, asking students to take the concepts they've learned in the chapter and apply them as they interact with animations and simulations on the Student Site or work through pencil-and-paper activities. Many of these Explorations incorporate everyday objects and can be used either in your classroom or as activities at home. The **Using the Web** problems direct students to "citizen science" projects, where they can contribute to the analysis of new astronomical data. Other problems send students to websites of space missions, observatories, collaborative projects, and catalogs to access the most current observations, results, and news releases. These Web problems can be used for homework, lab exercises, recitations, or "writing across the curriculum" projects.

We also believe students should be exposed to the more formal language of science—mathematics. We have placed the math in **Working It Out** boxes, so it does not interrupt the flow of the text or get in the way of students' understanding of conceptual material. But we've gone further by beginning with fundamental ideas in early Working It Out boxes and slowly building in complexity through the book. We've also worked to remove some of the stumbling blocks that affect student confidence by providing calculator hints, references to earlier Working It Out boxes, and detailed, fully worked examples. Many chapters include problems on reading and interpreting graphs. Appendix 1, "Mathematical Tools," summarizes some math concepts for students.

Discussion of basic physics is contained in Part I to accommodate courses that use the *Solar System* or *Stars and Galaxies* volumes. A "just-in-time" approach to introducing the physics is still possible by bringing in material from Chapters 2–6 as needed. For example, the sections on tidal forces in Chapter 4 can be taught along with the moons of the Solar System in Part II, or with mass transfer in binary stars in Part III, or with galaxy interactions in Part IV. Spectral lines in Chapter 5 can be taught with planetary atmospheres in Part II or with stellar spectral types in Part III, and so on.

In our overall organization, we have made several efforts to encourage students to engage with the material and build confidence in their scientific skills as they proceed through the book. For planets, stars and galaxies, we have organized the material to cover the general case first and then delve into more details with specific examples. Thus, you will find "planetary systems" before our own Solar System, "stars" before the Sun, and "galaxies" before the Milky Way. This allows us to avoid frustrating students by making assumptions about what they know about stars or galaxies or forward-referencing to basic definitions and overarching concepts. This organization also implicitly helps students understand their place in the universe: our galaxy and our star are each one of many. They are specific examples of a physical universe in which the same laws apply everywhere. Planets have been organized comparatively to emphasize that science is a process of studying individual examples that lead to collective conclusions. All of these organizational choices were made with the student perspective in mind and a clear sense of the logical hierarchy of the material.

We begin in Chapter 19 by introducing galaxies as a whole and our measurements of them, including recession velocities. Then we address the Milky Way in Chapter 20—a specific example of a galaxy that we can discuss in detail. This follows the repeating motif of moving from the general to the specific that exists throughout the text and gives students a basic grounding in the concepts of spiral galaxies, supermassive black holes, and dark matter before they need to apply those concepts to the specific example of our own galaxy. Chapter 21, "The Expanding Universe," covers the cosmological principle, the Hubble expansion, and the observational evidence for the Big Bang.

The **Origins** sections illustrate how astrobiologists and other scientists approach the study of a scientific question from the chapter related to the origin of the universe and of life. Material about exoplanets is introduced in Chapter 7, incorporated into other chapters when appropriate, and continued in Chapter 24. We revised each chapter, streamlining some topics, and updating the science to reflect the progress in the field. This includes:

- Updates on new telescopes and space missions
- New images and results from Solar System exploration missions to the Moon, Mars, Jupiter, Saturn, and Pluto
- Updated graphical data on climate change
- Results from Kepler and other missions on populations of exoplanets
- Results from the GAIA mission on stellar distances, the HR diagram, and the Milky Way
- Results from LIGO on the discovery of gravitational waves
- Updated cosmology from the Planck mission
- Many new articles for Reading Astronomy News, such as "2,500 Miles of Citizen Scientists", "NASA killed Cassini to Avoid Contaminating Saturn's Moons", and "LIGO Just Detected the Oldest Gravitational Waves Ever Discovered"
- Revised Process of Science figures for each chapter
- Revised simulations for some Explorations exercises

Many professors find themselves under pressure from accrediting bodies or internal assessment offices to assess their courses in terms of learning goals. Each chapter has Learning Goals and end-of-chapter Summary to correspond to the chapter's Learning Goals. In Smartwork5, questions and problems are tagged and can be sorted by Learning Goal. Smartwork5 contains more than 2,000 questions and problems that are tied directly to this text, including the Check Your Understanding questions and versions of the Reading Astronomy News and Exploration questions. Any of these could be used as a reading quiz to be completed before class or as homework. Every question in Smartwork5 has hints and answer-specific feedback so that students are coached to work toward the correct answer. An instructor can easily modify any of the provided questions, answers, and feedback or can create his or her own questions.

We've also created a series of 23 videos explaining and demonstrating concepts from the text, accompanied by questions in Smartwork5. You might assign these videos prior to lecture—either as part of a flipped modality or as a "reading quiz." In either case, you can use the diagnostic feedback from the questions in Smartwork5 to tailor your in-class discussions. Or you might show them in class, to stimulate discussion. Or you might simply use them as a jumping-off point—to get ideas for activities to do with your own students.

We continue to look for better ways to engage students, so please let us know how these features work for your students.

Digital Media for Students

digital.wwnorton.com/astro6

smartwork5 Online Homework

Steven Desch, Guilford Technical Community College

Diane Friend, University of Montana

Ana M. Larson, University of Washington

Violet Mager, Penn State Wilkes-Barre

Ryan Oelkers, Vanderbilt University

Todd Young, Wayne State College

William Younger, Tidewater Community College

More than 2,000 questions support *21st Century Astronomy, Sixth Edition*—all with answer-specific feedback, hints, and ebook links. Questions include Summary Self-Tests, versions of the Explorations (based on AstroTours and new Interactive Simulations), and questions based on the Reading Astronomy News features. Astronomy in Action video questions focus on getting students to come to class prepared and on overcoming common misconceptions. Process of Science Guided Inquiry Assignments help students apply the scientific method to important questions in astronomy, challenging them to think like scientists. Rounding out the Smartwork5 course are ranking, sorting, and labeling exercises—all classified according to the same Learning Goals used in the textbook and test bank.

Smartwork5 can be set up to work right in your LMS, with student scores flowing directly to your LMS gradebook. Your local Norton representative can help you set up LMS integration.

At Play in the Cosmos: The Videogame

Jeff Bary (Colgate University), Adam Frank (University of Rochester), Learning Games Network (University of Wisconsin and Massachusetts Institute of Technology), Gear Learning at UW-Madison's Wisconsin Center for Education Research

This ground-breaking and award-winning educational videogame is designed to engage introductory astronomy students by asking them to confront challenges using concepts that span the scope of the course. Using the process of science, students complete missions by flying their ship and choosing the right scientific tools for each task. The videogame can be downloaded to PC, Mac, and tablets and can be included with new copies of the Sixth Edition at no additional cost.

A complete videogame instructor's manual provides suggestions for meaningfully integrating the game into your course—for example, as a lab activity or as homework. The manual, written by lead game author Jeff Bary, provides mission overviews, plus multiple-choice and discussion questions for every mission.

Norton Ebook

The 21st Century Astronomy, Sixth Edition ebook provides students an enhanced reading experience at a fraction of the cost of a print textbook. Students are able to have an active reading experience and can take notes, bookmark, search, highlight, and even read offline. Instructors can add notes for students to see as they read the text. Norton Ebooks can be viewed on—and synced among—all computers and mobile devices. Every question in Smartwork5 provides a reference link to the ebook so that students have easy access to their textbook when completing homework assignments.

Student Site

W. W. Norton's student website features the following:

- Thirty AstroTour animations. These animations, some of which are interactive, use art from the text to help students visualize important physical and astronomical concepts. All are now tablet-compatible.

- Seven new Interactive Simulations, authored by Stacy Palen, pair with the Exploration activities in the text, allowing students to explore topics such as Moon phases, Kepler's laws, and the Hertzsprung-Russell diagram.

- Twenty-three Astronomy in Action videos that feature author Stacy Palen demonstrating the most important concepts in a visual, easy to understand, and memorable way.

Learning Astronomy by Doing Astronomy: Collaborative Lecture Activities

Stacy Palen, Weber State University

Ana Larson, University of Washington

Students learn best by doing. Devising, writing, testing, and revising suitable in-class activities that use real astronomical data, illuminate astronomical concepts, and pose probing questions that ask students to confront misconceptions can be challenging and time consuming. In this workbook, the authors draw on their experience teaching thousands of students in many different types of courses (large in-class, small in-class, hybrid, online, flipped, and so forth) to bring 30 field-tested activities that can be used in any classroom today. The activities have been designed to require no special software, materials, or equipment and to take no more than 50 minutes to do.

Starry Night Planetarium Software (College Version) and Workbook

Steven Desch, Guilford Technical Community College

Michael Marks, Bristol Community College

Starry Night is a realistic, user-friendly planetarium simulation program designed to allow students in urban areas to perform observational activities on a computer screen. Norton's unique accompanying workbook offers observation assignments that guide students' virtual explorations and help them apply what they've learned from the text reading assignments.

For Instructors

Instructor's Manual

Ana Larson, University of Washington

This resource includes brief chapter overviews, suggested discussion points, notes on the Process of Science figures, AstroTour animations, Astronomy in Action videos, Interactive Simulations, Reading Astronomy News, Explorations, and worked solutions to Check Your Understanding and end-of-chapter questions and problems. Also included are notes on teaching with *Learning Astronomy by Doing Astronomy: Collaborative Lecture Activities* and answers to the *Starry Night Workbook* exercises.

Norton Interactive Instructor's Guide (IIG)

This new and searchable online resource is designed to help instructors prepare for lecture in real time. It contains the following resources, each tagged by topic and chapter:

- Discussion Points
- Notes on the Process of Science figures
- AstroTour animations
- Astronomy in Action videos
- Interactive Simulations
- Notes on Reading Astronomy News and Explorations
- Notes on teaching with Learning Astronomy by Doing Astronomy: Collaborative Lecture Activites
- Solutions to *Check Your Understanding and end-of-chapter* questions and problems
- Answers to *Starry Night Workbook* exercises
- PowerPoint Lecture Slides
- All art and tables in JPEG and PowerPoint formats
- LMS Coursepacks, available in Blackboard, Canvas, Desire2Learn, and Moodle formats

All of the above content is also available on the traditional Instructor's Resource Site.

Teaching Astronomy by Doing Astronomy blog
tada101.com

This new resource features posts by Stacy Palen on teaching ideas, classroom activities, and current events. Instructors looking for additional ways to incorporate active learning will find this blog to be an invaluable resource.

PowerPoint Lecture Slides

Christa Speights, Northern Kentucky University

These ready-made lecture slides integrate selected textbook art, all Check Your Understanding and Working It Out questions from the text, class questions, and links to the AstroTour animations. These lecture slides are fully editable and are available in Microsoft PowerPoint format.

Test Bank

Matthew Newby, Temple University

The Test Bank has been revised using Bloom's Taxonomy and provides over 2,400 multiple-choice and short-answer problems. Each chapter of the Test Bank consists of five question levels classified according to Bloom's Taxonomy:

Remembering

Understanding

Applying

Analyzing

Evaluating

Problems are further classified by section and difficulty level, making it easy to construct tests and quizzes that are meaningful and diagnostic. The Test Bank assesses a common set of Learning Objectives consistent with the textbook and Smartwork5 online homework.

LMS Coursepacks

Norton's Coursepacks, available for use in various Learning Management Systems (LMSs), feature all Test Bank questions, links to the AstroTours, worksheets based on the Explorations and Astronomy in Action videos, and automatically graded versions of selected end-of-chapter Test Your Understanding multiple-choice questions. Coursepacks are available in BlackBoard, Canvas, Desire-2Learn, and Moodle formats.

Acknowledgments

The authors would like to acknowledge the extraordinary efforts of the staff at W. W. Norton: Sara Bonacum who kept things flowing smoothly; Jane Miller and Trish Marx for managing the numerous photos; and the copy editor, Gabe Waggoner, who made sure that all the grammar and punctuation survived the multiple rounds of the editing process. We would especially like to thank John Murdzek for the developmental editing process, and Diane Cipollone, who shepherded the manuscript through production.

Erik Fahlgren was the editor. Ashley Horna managed the production and Rubina Yeh was the design director. Rob Bellinger, Arielle Holstein, and Kelly Smith worked on the media and supplements, and Katie Sweeney will help get this book into the hands of people who can use it.

We gratefully acknowledge the contributions of the authors who worked on previous editions of *21st Century Astronomy*: Jeff Hester, Gary Wegner and the late Dave Burstein, Ron Greeley, Brad Smith, and Howard Voss, with special thanks to Dave for starting the project, to Jeff for leading the original authors through the first edition, and to Brad for leading the second and third editions.

Laura Kay

Stacy Palen

George Blumenthal

And we would like to thank the reviewers, whose input at every stage improved the book:

Sixth Edition Reviewers

Manuel Alvarado, El Paso Community College

Gagandeep Anand, Boston University

Simon Balm, Santa Monica College

Raymond Benge, Tarrant College

Ryan Bennett, University of North Texas

Brett Bochner, Hofstra University

Jack Brockway, Radford University

Shea Brown, University of Iowa

Spencer Buckner, Austin Peay State University

C. Austin Campbell, El Paso Community College

Christina Cavalli, Austin Community College

Gerald Cecil, University of North Carolina- Chapel Hill

Scott Cochran, Central Michigan University

Tara Cotton, University of Georgia

Robert Coyne, Texas Tech University

Steve Desche, Arizona State University

Michael Endl, Austin Community College

Duncan Farrah, University of Hawaii

Sunil Fernandes, University of Texas- San Antonio

Ken Gayley, University of Iowa

Christopher Gerardy, University of North Carolina- Charlotte

Guillermo Gonzalez, Ball State University

Ioannis Haranas, Wilfrid Laurier College

Javier Hasbun, University of West Georgia

Jim Higdon, Georgia Southern University

Scott Hildreth, Chabot College

Christopher Hodges, Sacramento State University

Mike Hood, Mt. San Antonio College

James Imamura, University of Oregon

Jennifer Jones, Arapahoe Community College

Bruno Jungweirt, North Carolina State University

Patrick Kelly, Dalhousie University

Bethuel Khamala, El Paso Community College

Keigo Fukumura, James Madison University

Rafael Lang, Purdue University

Kenneth Lanzetta, Stony Brook University

Lee LaRue, Paris Junior College

Denis Leahy, University of Calgary

Hector Leal, University of Texas- Rio Grande Valley

Hyun-chul Lee, University of Texas- Rio Grande Valley

Ludwik Lembryk, University of Toledo

Isaac Lopez, Boston University

Petrus Martens, Georgia State University

Justin Mason, Old Dominion University

Douglas McElroy, Chaffey College

Robert Morehead, Texas Tech University

Hon-kie Ng, Florida State University

Merav Opher, Boston University

Robert Parks, Louisiana State University

Jon Pedicino, College of the Redwoods

Nicolas Pereyra, University of Texas- Rio Grande Valley

Bob Powell, University of West Georgia

Barry Rice, Sierra College

Ruben Sandapen, Acadia University

Eric Schlegel, University of Texas- San Antonio

Michael Schwartz, Santa Monica College

Parampreet Singh, Louisiana State University

James Sy, El Paso Community College

Catherine Tabor, El Paso Community College

Benjamin Team, University of West Georgia

Fiorella Terenzi, Florida International University

Lennart Van Haaften, Texas Tech University

Jasper Wall, University of British Columbia

Colin Wallace, University of North Carolina- Chapel Hill

G. Scott Watson, Syracuse University

James Webb, Florida International University

Julia Wickett, Collin College

Kurt Williams, Texas A&M University

Fred Wilson, Collin College

David Wood, San Antonio College

Reviewers of Previous Editions

Scott Atkins, University of South Dakota

Timothy Barker, Wheaton College

Peter A. Becker, George Mason University

Timothy C. Beers, Michigan State University

David Bennum, University of Nevada–Reno

Edwin Bergin, University of Pittsburgh

William Blass, University of Tennessee

Steve Bloom, Hampden Sydney College

Daniel Boice, University of Texas at San Antonio

Bram Boroson, Clayton State University

David Branning, Trinity College

Julie Bray-Ali, Mt. San Antonio College

Suzanne Bushnell, McNeese State University

Paul Butterworth, George Washington University

Juan E. Cabanela, Minnesota State University–Moorhead

Amy Campbell, Louisiana State University

Michael Carini, West Kentucky University

Supriya Chakrabarti, Boston University

Robert Cicerone, Bridgewater State College

David Cinabro, Wayne State University

Judith Cohen, California Institute of Technology

Eric M. Collins, California State University–Northridge

John Cowan, University of Oklahoma–Norman

Debashis Dasgupta, University of Wisconsin–Milwaukee

Robert Dick, Carleton University

Gregory Dolise, Harrisburg Area Community College

Tom English, Guilford Technical Community College

David Ennis, The Ohio State University

John Finley, Purdue University

Matthew Francis, Lambuth University

Kevin Gannon, College of Saint Rose

Todd Gary, O'More College of Design

Christopher Gay, Santa Fe College

Parviz Ghavamian, Towson University

Martha Gilmore, Wesleyan University

Bill Gutsch, St. Peter's College

Karl Haisch, Utah Valley University

Charles Hawkins, Northern Kentucky University

Sebastian Heinz, University of Wisconsin–Madison

Barry Hillard, Baldwin Wallace College

Paul Hintzen, California State University–Long Beach

Paul Hodge, University of Washington

William A. Hollerman, University of Louisiana at Lafayette

Hal Hollingsworth, Florida International University

Olencka Hubickyj-Cabot, San Jose State University

Kevin M. Huffenberger, University of Miami

Adam Johnston, Weber State University

Steven Kawaler, Iowa State University

Monika Kress, San Jose State University

Jessica Lair, Eastern Kentucky University

Alex Lazarian, University of Wisconsin–Madison

Kevin Lee, University of Nebraska–Lincoln

Matthew Lister, Purdue University

M. A. K. Lodhi, Texas Tech University

Leslie Looney, University of Illinois at Urbana–Champaign

Jack MacConnell, Case Western Reserve University

Kevin Mackay, University of South Florida

Dale Mais, Indiana University–South Bend

Michael Marks, Bristol Community College

Norm Markworth, Stephen F. Austin State University

Kevin Marshall, Bucknell University

Stephan Martin, Bristol Community College

Amanda Maxham, University of Nevada–Las Vegas

Chris McCarthy, San Francisco State University

Ben McGimsey, Georgia State University

Charles McGruder, West Kentucky University

Janet E. McLarty-Schroeder, Cerritos College

Stanimir Metchev, Stony Brook University

Chris Mihos, Case Western University

Milan Mijic, California State University–Los Angeles

J. Scott Miller, University of Louisville

Scott Miller, Sam Houston State University

Kent Montgomery, Texas A&M University–Commerce

Andrew Morrison, Illinois Wesleyan University

Edward M. Murphy, University of Virginia

Kentaro Nagamine, University of Nevada–Las Vegas

Ylva Pihlström, University of New Mexico

Jascha Polet, California State Polytechnic University

Dora Preminger, California State University–Northridge

Daniel Proga, University of Nevada–Las Vegas

Laurie Reed, Saginaw Valley State University

Judit Györgyey Ries, University of Texas

Allen Rogel, Bowling Green State University

Kenneth Rumstay, Valdosta State University

Masao Sako, University of Pennsylvania

Samir Salim, Indiana University–Bloomington

Ata Sarajedini, University of Florida

Paul Schmidtke, Arizona State University

Ann Schmiedekamp, Pennsylvania State University

Jonathan Secaur, Kent State University

Ohad Shemmer, University of North Texas

Caroline Simpson, Florida International University

Paul P. Sipiera, William Rainey Harper College

Ian Skilling, University of Pittsburgh

Tammy Smecker-Hane, University of California–Irvine

Allyn Smith, Austin Peay State University

Jason Smolinski, State University of New York at Oneonta

Roger Stanley, San Antonio College

Ben Sugerman, Goucher College

Neal Sumerlin, Lynchburg College

Angelle Tanner, Mississippi State University

Christopher Taylor, California State University–Sacramento

Donald Terndrup, The Ohio State University

Todd Thompson, The Ohio State University

Glenn Tiede, Bowling Green State University

Frances Timmes, Arizona State University

Trina Van Ausdal, Salt Lake Community College

Walter Van Hamme, Florida International University

Karen Vanlandingham, West Chester University

Nilakshi Veerabathina, University of Texas at Arlington

Paul Voytas, Wittenberg University

Ezekiel Walker, University of North Texas

James Webb, Florida International University

Paul Wiita, Georgia State University

Richard Williamon, Emory University

David Wittman, University of California–Davis

About the Authors

Laura Kay is an award-winning professor and Chair of Physics and Astronomy at Barnard College, where she has taught since 1991. She received a BS degree in physics and an AB degree in feminist studies from Stanford University, and MS and PhD degrees in astronomy and astrophysics from the University of California, Santa Cruz. As a graduate student she spent 13 months at the Amundsen Scott station at the South Pole in Antarctica. She studies active galactic nuclei, using ground-based and space telescopes. She teaches courses in astronomy, astrobiology, women and science, and polar exploration. At Barnard she has served as chair of the Physics & Astronomy Department, chair of the Women's Studies Department, chair of Faculty Governance, and interim associate dean for Curriculum and Governance.

Stacy Palen is an award-winning professor in the physics department and the director of the Ott Planetarium at Weber State University. She received her BS in physics from Rutgers University and her PhD in physics from the University of Iowa. As a lecturer and postdoc at the University of Washington, she taught Introductory Astronomy more than 20 times over 4 years. Since joining Weber State, she has been very active in science outreach activities ranging from star parties to running the state Science Olympiad. Stacy does research in formal and informal astronomy education and the death of Sun-like stars. She spends much of her time thinking, teaching, and writing about the applications of science in everyday life. She then puts that science to use on her small farm in Ogden, Utah.

George Blumenthal has been a professor of astronomy and astrophysics at the University of California, Santa Cruz since 1972, and served as the university's Chancellor from 2006 to 2019. He received his BS degree from the University of Wisconsin–Milwaukee and his PhD in physics from the University of California, San Diego. As a theoretical astrophysicist, George's research encompasses several broad areas, including the nature of the dark matter that constitutes most of the mass in the universe, the origin of galaxies and other large structures in the universe, the earliest moments in the universe, astrophysical radiation processes, and the structure of active galactic nuclei such as quasars. Besides teaching and conducting research, he has served as Chair of the UC Santa Cruz Astronomy and Astrophysics Department, has chaired the Academic Senate for both the UC Santa Cruz campus and the entire University of California system, and has served as the faculty representative to the UC Board of Regents.

21ST CENTURY ASTRONOMY

SIXTH EDITION

1

Thinking Like an Astronomer

Now is a fascinating time to be studying this most ancient of the sciences. Loosely translated, the word **astronomy** means "patterns among the stars." But modern astronomy—the astronomy we talk about in this book—is about far more than looking at the sky and cataloging the visible stars. The contents of the universe, its origin and fate, and the nature of space and time have become the subjects of rigorous scientific investigation. Humans have long speculated about our *origins*. How and when did the Sun, Earth, and Moon form? Are other galaxies, stars, planets, and moons similar to our own? The answers that scientists are finding to those questions are changing our view of not only the cosmos but also ourselves.

LEARNING GOALS

In this chapter, we will begin studying astronomy by exploring our place in the universe and the methods of science. By the end of this chapter, you should be able to:

LG 1   Describe the size and age of the universe and Earth's place in it.

LG 2   Explain how astronomers use the scientific method to study the universe.

LG 3   Show how scientists use mathematics, including graphs, to find patterns in nature.

LG 4   Describe our astronomical origins.

The first view of Earth seen by humans from deep space. In December 1968, *Apollo 8* astronauts photographed Earth above the Moon's limb. ▶▶▶

Why is this considered one of the most famous photos ever taken?

3

Figure 1.1 Our cosmic address is Earth, Solar System, Milky Way Galaxy, Local Group, Virgo Supercluster, Laniakea Supercluster. We live on Earth, a planet in our Solar System, orbiting the Sun, a star in the Milky Way Galaxy. The Milky Way is a large galaxy within the Local Group of galaxies, located in the Virgo Supercluster, one of four superclusters in the Laniakea Supercluster.

1.1 Earth Occupies a Small Place in the Universe

Astronomers contemplate our place in the universe by studying Earth's position in space and time. Locating Earth in the larger universe is the first step in learning the science of astronomy. In this section, you will get a feel for the neighborhood in which Earth is located. You will also begin to explore the scale of the universe in space and time.

Our Place in the Universe

Most people receive their postal mail at an address—house or building number, street, city, state, and country. If we expand our view to include the enormously vast universe, however, our "cosmic address" might include our planet, star, galaxy, galaxy group or cluster, and galaxy supercluster.

We reside on a planet called Earth, which is orbiting under the influence of gravity around a star called the **Sun**. The Sun is an ordinary middle-aged star more massive and luminous than some stars but less massive and luminous than others. The Sun is extraordinary only because of its importance to us within our own **Solar System**. Our Solar System consists of eight planets (listed in order of distance from the Sun): Mercury, Venus, Earth, Mars, Jupiter, Saturn, Uranus, and Neptune. It also contains many smaller bodies, such as dwarf planets (for example, Pluto, Ceres, and Eris), asteroids (for example, Ida and Eros), and comets (for example, Halley). All those objects are gravitationally bound to the Sun.

The Sun is located about halfway out from the center of the **Milky Way Galaxy**, a flattened collection of stars, gas, and dust. Our Sun is just one among several hundred billion stars scattered throughout our galaxy, and planets orbit many of those stars.

The Milky Way is a member of a collection of a few dozen galaxies called the **Local Group**. Most galaxies in that group are much smaller than the Milky Way. As we look farther outward, the Local Group is part of a vastly larger collection of thousands of galaxies—a **supercluster**—called the Virgo Supercluster, which is part of an even larger grouping called the Laniakea Supercluster. The observable universe has millions of superclusters.

We can now define our cosmic address—Earth, Solar System, Milky Way Galaxy, Local Group, Virgo Supercluster, Laniakea Supercluster—as illustrated in **Figure 1.1**. Yet even that address is not complete because it encompasses only the *local universe*. The part of the universe that we can see—the *observable universe*—extends to many times the size of Laniakea in every direction. That volume contains about 1 trillion (thousand billion) to 2 trillion galaxies. The entire universe is much larger than the local universe and contains much more than the observed planets, stars, and galaxies. Astronomers estimate that about 95 percent of the universe is made up of matter that does not interact with light, known as *dark matter*, and a form of energy that permeates all space, known as *dark energy*. Dark matter and dark energy aren't well understood, and they are among the many exciting areas of research in astronomy.

The Scale of the Universe

As you saw in Figure 1.1, the size of the universe completely dwarfs our human experience. We can start by comparing astronomical sizes and distances to something more familiar. For example, the diameter of our Moon (3,474 kilometers [km]) is slightly greater than the distance between New York, New York, and Ogden, Utah (**Figure 1.2a**), where two of the authors of this book work. The distance from Earth to the Moon is about 100 times the Moon's diameter, and the planet Saturn with its majestic rings would fill much of that distance (**Figure 1.2b**). The distance from Earth to the Sun is about 400 times the Earth–Moon distance, and the distance to the planet Neptune is about 30 times the Earth–Sun distance.

As we move out from the Solar System to the stars, however, the enormous distances become difficult to comprehend. The nearest star is about 9,000 times farther from the Sun than the Sun's distance to Neptune. The diameter of our Milky Way Galaxy is 30,000 times the distance to that nearest star. The Andromeda Galaxy, the nearest large galaxy similar to the Milky Way, is about 600,000 times farther than that nearest star. The diameter of the Local Group of galaxies is about 4 times the distance to Andromeda, and the diameter of the Laniakea Supercluster, which includes the Local Group and many other galaxy groups and clusters, is 50 times larger than the Local Group. As noted earlier, that supercluster is just one of millions in the observable universe.

To get a better sense of those distances, imagine a model in which the objects and distances in the universe are 1 billion times smaller than they really are. In that model, Earth is about the size of a marble or a peanut M&M (about 1.3 centimeters [cm], or half an inch), the Moon is 38 cm away, and the Sun is 150 meters away. Neptune is 4.5 km from the Sun, and the nearest star to the Sun is about 40,000 km away (or about the length of the real Earth's circumference). The model Milky Way Galaxy would fill the real Solar System nearly to the orbit of Saturn. The distance between the model Milky Way and Andromeda galaxies would fill the real Solar System 20 times farther, beyond humanity's most distant space probe. The model Laniakea Supercluster would fill the real Solar System and go about one-eighth of the way to the nearest star.

When thinking about distances in the universe, discussing the time it takes to travel to various places can be helpful. If someone asks you how far it is to the nearest city, you might say 100 km or you might say 1 hour. In either case, you will have given that person an idea of how far the city is. In astronomy, the speed of a car on the highway is far too slow to be useful. Instead, we use the fastest speed in the universe—the speed of light. Light travels at 300,000 kilometers per second (km/s). **Figure 1.3** shows how distances that range from the size of Earth to the size of the observable universe can be measured in the time it takes light to travel that far. Light can circle Earth, for example—a distance of 40,000 km—in just under 1/7 of a second. The circumference of Earth, then, is 1/7 of a light-second. Even relatively small distances in astronomy are so vast that they are measured in units of **light-years (ly)**: the distance light travels in 1 year, about 9.5 trillion km, or 6 trillion miles.

Because light takes time to reach us, we see astronomical objects as they were in the past: how far back in time depends on the object's distance from us. Because light from the Moon takes 1 1/4 seconds to reach us, we see the Moon as it was 1 1/4 seconds ago. Because light from the Sun takes 8 1/3 minutes to reach us, we see the Sun as it was 8 1/3 minutes ago. We see the nearest star as it was more than 4 years ago and objects across the Milky Way as they were tens of

(a)

(b)

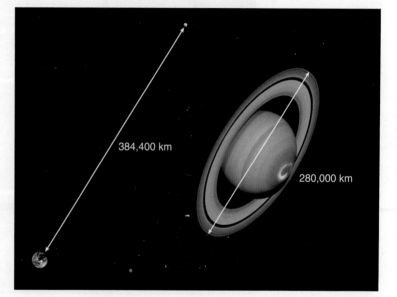

Figure 1.2 (a) The diameter of the Moon is about the same as the distance between New York, New York, and Ogden, Utah. (b) The size of Saturn, including the rings, is about 70 percent of the distance between Earth and the Moon.

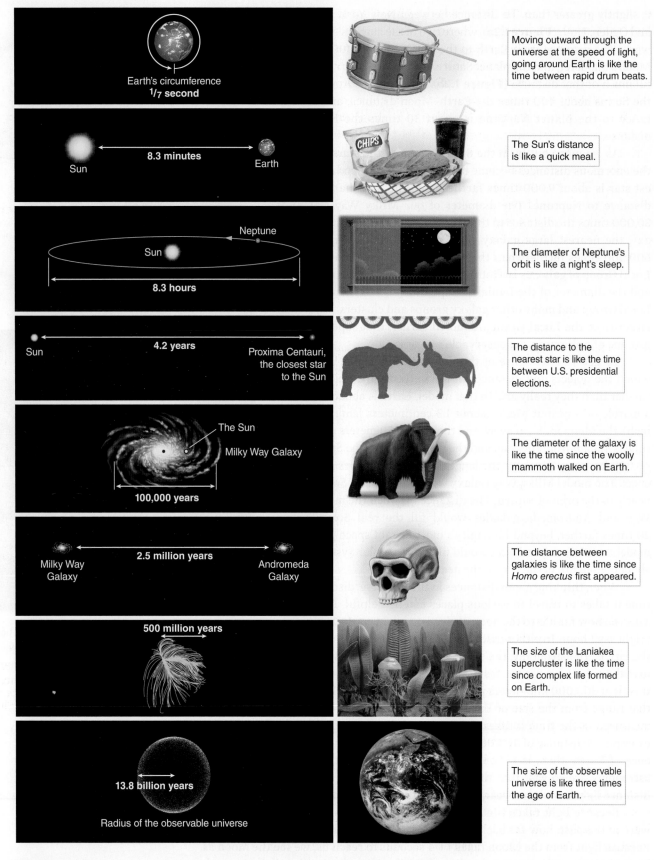

Earth's circumference
1/7 second

Moving outward through the universe at the speed of light, going around Earth is like the time between rapid drum beats.

Sun **8.3 minutes** Earth

The Sun's distance is like a quick meal.

Neptune
Sun
8.3 hours

The diameter of Neptune's orbit is like a night's sleep.

Sun **4.2 years** Proxima Centauri, the closest star to the Sun

The distance to the nearest star is like the time between U.S. presidential elections.

The Sun
Milky Way Galaxy
100,000 years

The diameter of the galaxy is like the time since the woolly mammoth walked on Earth.

Milky Way Galaxy **2.5 million years** Andromeda Galaxy

The distance between galaxies is like the time since *Homo erectus* first appeared.

500 million years

The size of the Laniakea supercluster is like the time since complex life formed on Earth.

13.8 billion years
Radius of the observable universe

The size of the observable universe is like three times the age of Earth.

Figure 1.3 Thinking about the time light takes to travel between objects helps us comprehend the vast distances in the observable universe.

thousands of years ago. The light from the Virgo Cluster of galaxies has travelled for 50 million years to reach us. The light from the most distant observable objects has been traveling for almost the age of the universe—nearly 13.8 billion years.

The vast distances from Earth to other objects in the universe tell us that we occupy a very small part of the space in the universe and a very small part of time. Earth and the Solar System are only about one-third as old as the universe. Animals have existed on Earth for even less time. Imagine the age of the universe and the important events in it as though they took place within a single day, as illustrated in **Figure 1.4**. In that timeline, the Big Bang begins the cosmic day at midnight, and the original light chemical elements are created within the first 2 seconds. The first stars and galaxies appear within the first 10 minutes. Our Solar System formed from recycled gas and dust left over from previous generations of stars, at about 4 P.M. on that cosmic clock. The first bacterial life appears on Earth at 5:20 P.M., the first land animals at 11:20 P.M., and modern humans at 11:59:59.8 P.M.—that is, with just 1/5 of a second left in the cosmic day. We humans appeared quite recently in the history of the universe.

CHECK YOUR UNDERSTANDING 1.1

Rank the following from smallest to largest: (a) a light-minute, (b) a light-year, (c) a light-hour, (d) the radius of Earth, (e) the distance from Earth to the Sun, (f) the radius of the Solar System.

1.2 Science Is a Way of Viewing the Universe

Humans have long paid attention to the sky and the stars, making new discoveries and contributing new ideas about the universe. Those discoveries and new ideas still occur often and evolve rapidly, making astronomy a dynamic science. To view the universe through the eyes of an astronomer, you need to understand how science itself works. Throughout this book, we emphasize not only scientific discoveries but also the *process* of science. We begin with this section, which examines the scientific method.

The Scientific Method

The **scientific method** is a systematic way of exploring the world on the basis of developing and then testing new ideas or explanations. Often, the method begins with a fact—an observation or a measurement. For example, you might observe that the weather changes predictably each year and wonder why that happens. You then create a **hypothesis**, a testable explanation of the observation: "I think that it is cold in the winter and warm in the summer because Earth is closer to the Sun in the summer." You come up with a test: if it is cold in the winter and warm in the summer because Earth is closer to the Sun in the summer, then everywhere on the planet will be cold in the winter—Australia should have winter at the same time of year as the United States. You can use that test to check your hypothesis. In January, you travel from the United States to Australia and find that it is summer in Australia. Your hypothesis has just been proved incorrect, so we say that it has been **falsified**. That is different from the word's meaning in common usage, in which "falsified" evidence has been manipulated to misrepresent the truth. Your test has two important elements that all scientific tests share. Your observation is reproducible: anyone who goes to Australia will find the same result. And

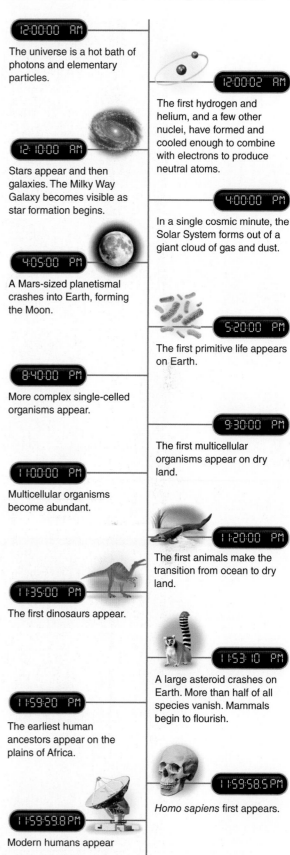

The universe is a hot bath of photons and elementary particles. — 12:00:00 AM

The first hydrogen and helium, and a few other nuclei, have formed and cooled enough to combine with electrons to produce neutral atoms. — 12:00:02 AM

Stars appear and then galaxies. The Milky Way Galaxy becomes visible as star formation begins. — 12:10:00 AM

In a single cosmic minute, the Solar System forms out of a giant cloud of gas and dust. — 4:00:00 PM

A Mars-sized planetismal crashes into Earth, forming the Moon. — 4:05:00 PM

The first primitive life appears on Earth. — 5:20:00 PM

More complex single-celled organisms appear. — 8:40:00 PM

The first multicellular organisms appear on dry land. — 9:30:00 PM

Multicellular organisms become abundant. — 11:00:00 PM

The first animals make the transition from ocean to dry land. — 11:20:00 PM

The first dinosaurs appear. — 11:35:00 PM

A large asteroid crashes on Earth. More than half of all species vanish. Mammals begin to flourish. — 11:53:10 PM

The earliest human ancestors appear on the plains of Africa. — 11:59:20 PM

Homo sapiens first appears. — 11:59:58.5 PM

Modern humans appear — 11:59:59.8 PM

Figure 1.4 This cosmic timeline presents the history of the universe as a 24-hour day.

your result is repeatable: if you conducted a similar test next year or the year after, you would get the same result. Because you have falsified your hypothesis, you must revise or replace it to be consistent with the new data.

Any idea that is not testable—that is, not falsifiable—must be accepted or rejected based on intuition alone, so it is not a scientific idea. A falsifiable hypothesis or idea does not have to be testable using current technology, but we must be able to imagine an experiment or observation that could prove the idea wrong if we could carry it out. As continuing tests support a hypothesis by failing to disprove it, scientists come to accept the hypothesis as a theory. A **theory** is a well-developed idea or group of ideas that must agree with known physical laws and make testable predictions of a given phenomenon or system. As with "falsified," the scientific meaning of "theory" is different from the meaning in common usage. In everyday language, theory may mean a guess: "Do you have a theory about who did it?" In everyday language, a theory can be something we don't take too seriously. "After all," people say, "it's only a theory."

In stark contrast, scientists use the word *theory* to mean a carefully constructed proposition that takes into account every piece of relevant data as well as our entire understanding of how the world works. A theory has been or will be used to make testable predictions, and all those predictions must come true. Every attempt to prove it false fails. A classic example is Einstein's theory of relativity, which Chapter 18 covers in some depth. For more than a century, scientists have tested the predictions of the theory of relativity and have not been able to falsify it. Even after 100 years of verification, if a prediction of the theory of relativity failed tomorrow, the theory would require revision or replacement. As Einstein himself noted, a theory that fails only one test is proved false. In that sense, all scientific knowledge is subject to challenge. In addition, scientific results should be reproducible so that other people get the same results.

In the loosely defined hierarchy of scientific knowledge, an *idea* is a notion about how something might be. Moving up the hierarchy, we come to a *fact*—an observation or measurement. The measured value of Earth's radius is a fact. A *hypothesis* is an idea that leads to testable predictions. A hypothesis may be the forerunner of a scientific theory, may be based on an existing theory, or both. At the top of the hierarchy is a *theory*: an idea that has been examined carefully, is consistent with all existing theoretical and observational knowledge, and makes testable predictions. Ultimately, the success of the predictions is the deciding factor between competing theories. A scientific *law* is a series of observations that enables making predictions but has no underlying explanation of why the phenomenon occurs. A *law* of daytime might say that the Sun rises and sets once each day, whereas a *theory* of daytime might say that the Sun rises and sets once each day because Earth spins on its axis. Scientists themselves can be sloppy about how they use those words, so you may sometimes see them used differently from how we have defined them here.

As the **Process of Science Figure** shows, the scientific method follows a specific sequence. Scientists begin with an observation or idea, followed by analysis, followed by a hypothesis, followed by prediction, followed by further observations or experiments to test the prediction. A hypothesis may lead to a scientific theory, may be based on an existing theory, or both. Ultimately, the basis for deciding among competing theories is the success of their predictions. Scientists can use theories to take their knowledge a step further by building theoretical models. A **theoretical model** describes the properties of a particular object or system in terms of known physical laws or theories, which are used to connect the theoretical model to the behavior of a complex system.

The Scientific Method

The scientific method is a formal procedure used to test the validity of scientific hypotheses and theories.

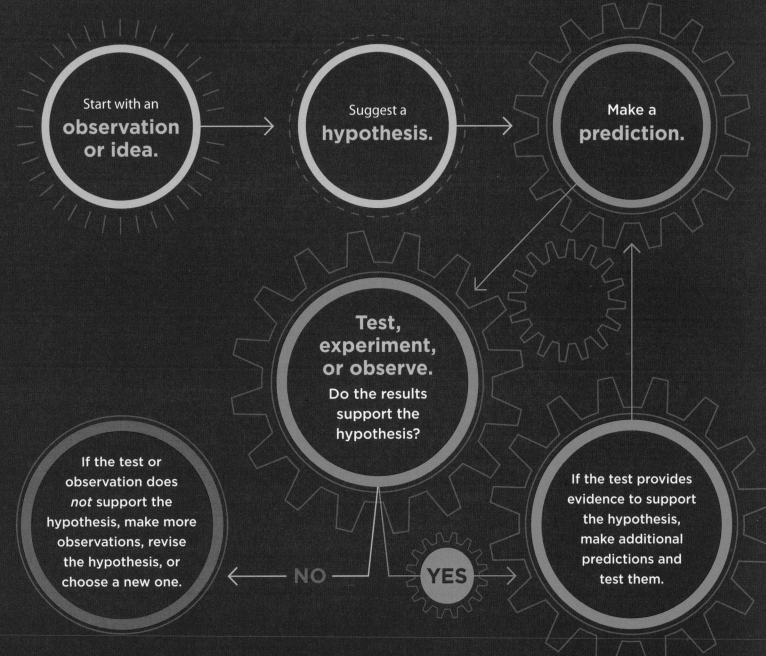

Start with an **observation or idea.**

Suggest a **hypothesis.**

Make a **prediction.**

Test, experiment, or observe.
Do the results support the hypothesis?

If the test or observation does *not* support the hypothesis, make more observations, revise the hypothesis, or choose a new one.

NO

YES

If the test provides evidence to support the hypothesis, make additional predictions and test them.

An idea or observation leads to a falsifiable hypothesis. This hypothesis is tested by observation or experiment, and is either accepted as a tested theory or rejected based on the results. The green loop repeats indefinitely as scientists continue to test the theory.

Scientific Principles

Scientific **principles** are general ideas or a sense about the universe that often guides the construction of new theories. Two principles used in studying astronomy are the *cosmological principle* and *Occam's razor.*

The **cosmological principle** is the testable assumption that the physical laws that apply here and now also apply everywhere and at all times. That principle also implies that the universe has no special locations or directions and that matter and energy obey the same physical laws throughout space and time as they do today on Earth. The principle means that the same physical laws that we observe and apply in laboratories on Earth can be used to understand what goes on in the centers of stars or in distant galaxies. Each new theory that comes from applying the cosmological principle to observations of the universe around us adds to our confidence in the validity of that principle, which we discuss in more detail in Chapter 21.

Occam's razor states that when facing two hypotheses that explain all the observations equally well, we should favor the one that requires the fewest assumptions until we have evidence to the contrary. For example, a hypothesis that atoms are constructed differently in the Andromeda Galaxy and in the Milky Way Galaxy would violate the cosmological principle. That hypothesis also would require many assumptions to explain how atoms in the Andromeda Galaxy are constructed differently and yet still appear to behave in the same way as atoms in the Milky Way. For example, suppose you assume that the center of an atom in Andromeda is negatively charged, whereas the center of an atom in the Milky Way is positively charged. That assumption would require you to make additional assumptions about the location of the boundary between Andromeda-like matter and Milky Way–like matter and about why atoms on the boundary between the two regions did not destroy each other. You also would need an assumption about why atoms in the two regions are constructed so differently. If reasonable experimental evidence ever challenges the validity of the cosmological principle, scientists will construct a new description of the universe that takes those new data into account. Until then, the cosmological principle is the hypothesis that has the fewest assumptions, thereby satisfying Occam's razor. To date, the cosmological principle has been repeatedly tested and remains valid.

In many sciences, researchers can conduct controlled experiments to test hypotheses. That experimental method is often unavailable to astronomers: we cannot change the tilt of Earth or the temperature of a star to see what happens. Instead, astronomers work from observations or models. Astronomers make multiple observations by using various methods and then create mathematical and physical models based on established science to explain the observations.

For example, when astronomers first discovered planets orbiting other stars, those new extrasolar planets, or *exoplanets*, were most often giant gaseous planets (similar to Jupiter) in short-period orbits very close to their star. Those planets are the easiest to discover because they have a strong pull on their star, as we discuss in Chapter 7. However, their proximity to their star is completely unlike the situation in our own Solar System, where the giant planets are far from the Sun. Those discoveries challenged existing ideas about how our Solar System formed. As different observers using multiple telescopes found more and more of those planets, astronomers realized they needed new explanations of planet formation to explain how such large planets could wind up so close to their star. Astronomers could not build different solar systems—but they could use the known laws of physics to create computer simulations of planetary systems. When researchers did that, they found that planets can migrate, moving to orbits closer to or farther from their star. Planetary scientists are searching for evidence to test that idea in our own Solar System, where that process may have occurred early in the life of the Sun. In that example, the theory was motivated by the earlier observations.

Alternatively, an astronomer might make predictions from an existing successful mathematical or physical model and then conduct observations and analyze data to test the predictions. One example is the discovery of black holes. In the late 18th century, two scientists hypothesized the existence of "dark stars": massive objects having such strong gravity that light could not escape. At that time, the scientists had no way to test that hypothesis. More than 100 years later, in the early 20th century, Karl Schwarzschild (1873–1916) studied Einstein's relativity equations and calculated that those collapsed dark stars would be very small, with a radius of only a few kilometers. Fifty years later, those objects were named *black holes*. No evidence of their existence was available until the 1970s and 1980s, when the new technology of space-based X-ray telescopes made possible the observations needed to test the hypothesis. Einstein's existing theory made the discovery of black holes possible.

The scientific method provides the rules for testing whether an idea is false, but it offers no insight into where the idea came from in the first place or how an experiment was designed. Scientists discussing their work use words such as *insight*, *intuition*, and *creativity*. Scientists speak of a beautiful theory in the same way that an artist speaks of a beautiful painting or a musician speaks of a beautiful song. Science has an aesthetic that is as human and as profound as any found in the arts.

In one important respect, though, science is different from art or music. Whereas art and music are judged by a human jury alone, the validity of a scientific hypothesis or theory is subject to the natural world. Nature alone is the judge of which theories can be kept and which theories must be discarded. What we want to be true does not matter. In the history of science, many a beautiful and beloved theory has been abandoned.

Scientific Revolutions

Scientific inquiry is necessarily dynamic. Scientists do not have all the answers and must constantly refine their ideas in response to new data and new insights. The vulnerability of knowledge may seem like a weakness. "Gee, you really don't know anything," the cynical person might say. But that vulnerability is actually the scientific process's greatest strength: it means that science self-corrects. New information eventually overturns incorrect ideas. In science, even our most cherished ideas about the nature of the physical world remain fair game, subject to challenge by new evidence. Many of history's best scientists earned their status by falsifying a universally accepted idea. That is a powerful motivation for scientists to challenge old ideas constantly—to formulate and test new explanations for their observations.

For example, the classical physics that Sir Isaac Newton developed in the 17th century to explain motion, forces, and gravity withstood the scrutiny of scientists for more than 200 years. During the late 19th and early 20th centuries, however, a series of scientific revolutions completely changed our understanding of the nature of reality. The work of Albert Einstein (**Figure 1.5**) is representative of those scientific revolutions. Einstein's special and general theories of relativity replaced Newton's mechanics. Einstein did not prove Newton wrong but rather

Figure 1.5 Albert Einstein is perhaps the most famous scientist of the 20th century, and he was *Time* magazine's selection for Person of the Century. Einstein helped usher in two scientific revolutions.

showed that Newton's theories were a special case of a far more general and powerful set of physical theories. Einstein's new ideas unified the concepts of mass and energy and destroyed the conventional notion of space and time as separate concepts.

Throughout this text, you will encounter many other discoveries that forced scientists to abandon accepted theories. Einstein himself never embraced the view of the world offered by *quantum mechanics*—a second revolution he helped start. Yet quantum mechanics, a statistical description of the behavior of particles smaller than atoms, has held up for more than 100 years. In science, all authorities are subject to challenge, even Einstein.

Science is a way of thinking about the world. It is a search for the relationships that make our world what it is. Scientific knowledge is an accumulated collection of ideas about how the universe works, yet scientists are always aware that what is known today may be superseded tomorrow. A scientist assumes that order exists in the universe and that the human mind can grasp the essence of the rules underlying that order. Scientists build on those assumptions to make and then test predictions, finding the underlying rules that allow humanity to solve problems, invent new technologies, or find a new appreciation for the natural world. In the final analysis, science has found such a central place in our civilization because *science works*.

CHECK YOUR UNDERSTANDING 1.2

The scientific method is a process by which scientists: (a) prove theories to be known facts; (b) gain confidence in theories by failing to prove them wrong; (c) show all theories to be wrong; (d) survey what most scientists think about a theory.

1.3 Astronomers Use Mathematics to Find Patterns

Scientific thinking allows scientists to make predictions. Once a pattern has been observed, such as the daily rising and setting of the Sun, scientists can predict what will happen next.

Imagine that the patterns in your life became disrupted, so that the world became entirely unpredictable. For example, what would life be like if you could not predict whether an object you dropped would fall up or down? Or what if one morning the Sun rose in the east and set in the west, and the next day it rose in the west and set in the east? In fact, objects do fall toward the ground. The Sun rises, sets, and then rises again at predictable times and in predictable locations. Spring turns into summer, summer turns into autumn, autumn turns into

Figure 1.6 Since ancient times, people recognized that patterns in the sky change with the seasons. Those and other patterns shape our lives. These star maps show the sky in the Northern Hemisphere during each season.

Summer

Spring

Fall

Winter

winter, and winter turns into spring again. The rhythms of nature produce patterns in our lives, and those patterns give us clues about the nature of the physical world. Astronomers identify and characterize those patterns and use them to understand the world around us. Some of the most easily identified patterns in nature are those we see in the sky. What in the sky will look different or the same a week from now? A month from now? A year from now? As you can see in **Figure 1.6**, patterns in the sky mark the changing of the seasons. Because the seasons determine planting and harvesting times, astronomy—which studies those patterns—is the oldest of all sciences. We describe many other examples of patterns in the sky in Chapter 2.

Astronomers use mathematics to analyze patterns and to communicate complex material compactly and accurately. Because studying patterns in nature is so important to science, it should come as no surprise that mathematics is the language of science. Many people find mathematics to be a major obstacle that prevents them from appreciating the beauty and elegance of the world as seen through the eyes of a scientist. In this book, we strive to explain any necessary math in everyday language. We will describe what equations mean and help you use them in a way that allows you to connect scientific concepts to the world. Your responsibility is to accept the challenge and make an honest effort to understand the material. **Working It Out 1.1** and **Working It Out 1.2** review some basics of mathematical tools and graphs. At the back of the book, Appendix 1 explains some essential mathematical tools, and Appendix 2 contains physical constants of nature. Other appendixes contain data tables with key information about planets, moons, and stars.

CHECK YOUR UNDERSTANDING 1.3

When you see a pattern in nature, it is usually evidence of: (a) a theory's being displayed; (b) a breakdown of random clustering; (c) an underlying physical law; (d) a coincidence.

1.1 Working It Out Mathematical Tools

Scientists use mathematics to understand the patterns they observe and to communicate that understanding to others. The following mathematical tools will be useful in our study of astronomy:

Scientific notation. Scientific notation is how we handle numbers of vastly different sizes. Writing out 7,540,000,000,000,000,000,000 in standard notation is inefficient. Scientific notation uses the first few digits (the *significant* ones) and counts the number of decimal places to create the condensed form 7.54×10^{21}. Similarly, instead of writing out 0.000000000005, we write 5×10^{-12}. The exponent on the 10 is positive or negative depending on whether the decimal point moves right or left, respectively. For example, the average distance to the Sun is 149,600,000 km, but astronomers usually express that value as 1.496×10^8 km.

Ratios. Ratios are a useful way to compare objects. A star may be "10 times as massive as the Sun" or "10,000 times as luminous as the Sun." Those expressions are ratios.

Proportionality. Often, understanding a concept amounts to understanding the *sense* of the relationships that it predicts or describes. "If you have twice as far to go, getting there will take you twice as long." "If you have half as much money, you will be able to buy only half as much gas." Those are examples of proportionalities.

Appendix 1 further explains the mathematical tools used in this book.

1.2   Working It Out Reading a Graph

Scientists often convey complex information and mathematical patterns in graphical form. Reading graphs is an important skill not only in astronomy but also in life. Economists, social scientists, mortgage brokers, financial analysts, urban planners, historians, medical professionals, and scientists all use graphs to evaluate and communicate important information.

Graphs typically have two axes: a horizontal axis (the x-axis) and a vertical axis (the y-axis). The x-axis usually shows an independent variable, the one a researcher might have control over in an experiment, whereas the y-axis shows the dependent variable, which—in many experiments—is the variable a researcher is studying.

Graphs can take different shapes. Suppose we plot the distance a car travels over time, as shown in **Figure 1.7a**. In a linear graph, each interval on an axis represents the same-sized step. Each step on the horizontal axis of the graph in Figure 1.7a represents 5 minutes. Each step on the vertical axis represents a distance of 5 km traveled by the car. Data are plotted on the graph, with one dot for each observation; for example, after 20 minutes, the car has traveled 20 km.

Drawing a line through those data indicates the trend (or relationship) of the data. To understand what the trend means, scientists often find the slope of the line, which is the relationship of the line's rise along the y-axis to its movement along the x-axis. To find the slope, we look at the change between two points on the vertical axis divided by the change between the same two points on the horizontal axis. For example, finding the slope of the line gives

$$\text{Slope} = \frac{\text{Change in vertical axis}}{\text{Change in horizontal axis}}$$

$$\text{Slope} = \frac{(15 - 10) \text{ km}}{(15 - 10) \text{ min}}$$

$$= 1 \text{ km/min}$$

There, the trend tells us that the car is traveling at 1 kilometer per minute (km/min), or 60 kilometers per hour (km/h).

Many observations of natural processes do not yield a straight line on a graph—for example, an exponential process. Think about what happens when you catch a cold. You feel fine when you get up in the morning at 7 A.M. At 9 A.M. you feel a little tired. By 11 A.M. you have a bit of a sore throat or a sniffle and think, "I wonder whether I'm getting sick," and by 1 P.M., you have a runny nose, congestion, fever, and chills. That process is exponential because the virus that has infected you reproduces exponentially.

For the sake of this discussion, suppose that the virus produces one copy of itself each time it invades a cell. (Viruses actually produce 1,000–10,000 copies each time they invade a cell, so the exponential curve is much steeper.) One virus infects a cell and multiplies, so now two viruses exist—the original and a copy. Those viruses invade two new cells, and each one produces a copy. Now four viruses are there. After the next cell invasion, we have eight. Then 16, 32, 64, 128, 256, 512, 1,024, 2,048, and so on. That behavior is plotted in **Figure 1.7b**.

Seeing what's happening in the early stages of an exponential curve can be difficult because the later numbers are so much larger than the earlier ones. That's why we sometimes plot that type of data *logarithmically*, by putting the logarithm (the power of 10) of the data on the vertical axis, as shown in **Figure 1.7c**. Now each step on the axis represents 10 times as many viruses as the previous step. Even though we draw all the steps the same size on the page, they represent different-sized steps in the data—the number of viruses, for example. We often use that technique in astronomy because it has a second, related advantage: very large variations in the data can easily fit on the same graph.

Each time you see a graph, you should first understand the axes—what data are plotted on the graph? Then you should check whether the axes are linear or logarithmic. Finally, you can look at the actual data or lines in the graph to understand how the system behaves.

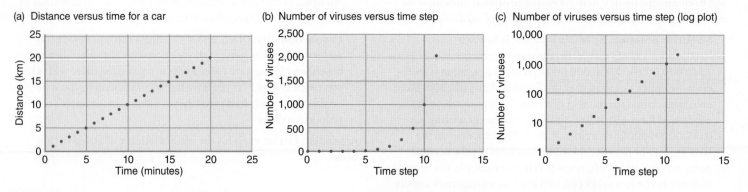

Figure 1.7 Graphs such as these show relationships between quantities. (a) The relationship between time and distance traveled. (b) The relationship between time and number of viruses. (c) The data in part (b) plotted logarithmically.

Origins An Introduction

How and when did the universe begin? What combination of events led to the existence of humans as sentient beings living on a small, rocky planet orbiting a typical middle-aged star? Are others like us scattered throughout the galaxy?

Earlier in the chapter we mentioned the theme of origins. Throughout this book, that theme involves much more than how humans came to be on Earth. In these Origins sections, which conclude each chapter, we look into the origin of the universe and the origin of life on Earth. We also examine the possibilities of life elsewhere in the Solar System and beyond—a subject called **astrobiology**. Our origins theme includes the discovery of planets around other stars and how they compare with the planets of our own Solar System.

Later in the book, we present observational evidence that supports the **Big Bang** theory, which states that the universe started expanding from an infinitesimal size about 13.8 billion years ago. In the early universe, only the lightest chemical elements were found in substantial amounts: hydrogen and helium, and tiny amounts of lithium and beryllium. However, we live on a planet with a central core consisting mostly of heavy elements such as iron and nickel, surrounded by outer layers made up of rocks containing large amounts of silicon and various other elements—all heavier than the original elements. The human body contains carbon, nitrogen, oxygen, calcium, phosphorus, and a host of other chemical elements—except for hydrogen itself, all are heavier than hydrogen and helium. If those heavier elements that make up Earth and our bodies were not present in the early universe, where did they come from?

The answer to that question lies within the stars themselves (**Figure 1.8**). In the core of a star, light elements, such as hydrogen, combine to form more massive atoms, which eventually leads to atoms such as carbon. When a star nears the end of its life, it often loses much of its material—including some of the new atoms formed in its interior—by blasting it back into interstellar space. That material combines with material lost from other stars—some of which produced even more massive atoms as they exploded—to form large clouds of dust and gas. Those clouds go on to make new stars and planets, similar to our Sun and Solar System. Earlier "generations" of stars supplied the building blocks for the chemical processes that we see in the universe, including life. The atoms that make up much of what we see were formed in the cores of stars. The phrase "we are stardust" is not just poetry. We are actually made of recycled stardust.

Studying origins also gives examples of the process of science. Many physical processes in chemistry, planetary science, physics, and astronomy that are seen on Earth or in the Solar System are observed across the Milky Way and throughout the universe. As of this writing, though, the only biology we know about is that which exists on Earth. At this point in human history, then, much of what scientists can say about the origin of life on Earth and the possibility of life elsewhere is reasoned extrapolation and educated speculation. In these Origins sections, we address some of those hypotheses and try to be clear about which are speculative and which have been tested.

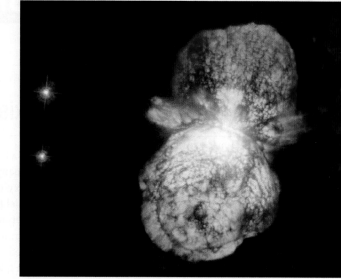

Figure 1.8 You and everything around you are composed of atoms forged in the interiors of stars that lived and died before the Sun and Earth formed. The supermassive star Eta Carinae, shown here, is ejecting a cloud of chemically enriched material just as earlier generations of stars once did to enrich the gas that would become our Solar System.

ARTICLES   QUESTIONS

Astronomers think there are 10 times more galaxies than previously estimated.

A Universe of 2 Trillion Galaxies

Royal Astronomical Society

An international team of astronomers, led by Christopher Conselice, professor of astrophysics at the University of Nottingham, have found that the universe contains at least 2 trillion galaxies, 10 times more than previously thought. The team's work, which began with seedcorn funding from the Royal Astronomical Society, appears in the *Astrophysical Journal.*

Astronomers have long sought to determine how many galaxies there are in the observable universe, the part of the cosmos where light from distant objects has had time to reach us. Over the last 20 years, scientists have used images from the Hubble Space Telescope to estimate that the universe we can see contains around 100–200 billion galaxies. Current astronomical technology allows us to study just 10 percent of these galaxies, and the remaining 90 percent will be only seen once bigger and better telescopes are developed.

Professor Conselice's research is the culmination of 15 years' work, part-funded by a research grant from the Royal Astronomical Society awarded to Aaron Wilkinson, an undergraduate student at the time. Aaron, now a PhD student at the University of Nottingham, began by performing the initial galaxy-counting analysis, work that was crucial for establishing the feasibility of the larger-scale study.

Professor Conselice's team then converted pencil beam images of deep space from telescopes around the world, and especially from the Hubble telescope, into three-dimensional maps. These allowed them to calculate the density of galaxies as well as the volume of one small region of space after another. This painstaking research enabled the team to establish how many galaxies we have missed—much like an intergalactic archaeological dig.

The results of this study are based on the measurements of the number of observed galaxies at different epochs—different instances in time—through the universe's history. When Professor Conselice and his team at Nottingham, in collaboration with scientists from the Leiden Observatory at Leiden University in the Netherlands and the Institute for Astronomy at the University of Edinburgh, examined how many galaxies there were at a given epoch, they found that there were significantly more at earlier times.

It appears that when the universe was only a few billion years old, there were 10 times as many galaxies in a given volume of space as there are within a similar volume today. Most of these galaxies were low-mass systems with masses similar to those of the satellite galaxies surrounding the Milky Way.

Professor Conselice said, "This is very surprising as we know that, over the 13.7 billion years of cosmic evolution since the Big Bang, galaxies have been growing through star formation and mergers with other galaxies. Finding more galaxies in the past implies that significant evolution must have occurred to reduce their number through extensive merging of systems."

He continued, "We are missing the vast majority of galaxies because they are very faint and far away. The number of galaxies in the universe is a fundamental question in astronomy, and it boggles the mind that over 90 percent of the galaxies in the cosmos have yet to be studied. Who knows what interesting properties we will find when we study these galaxies with the next generation of telescopes?"

1. Explain how observing galaxies at different distances tells us about how galaxies existed in different periods.
2. Why does that result suggest that galaxies started small?
3. What is meant by "observable universe"?
4. Watch Hubblecast 96 based on this news story (http://spacetelescope.org/videos/heic1620a/). What else did you learn from the video?
5. Do a search to see if there is a new estimate of the number of observable galaxies.

Source: "A universe of two trillion galaxies," Royal Astronomical Society News Archive, October 13, 2016. Reprinted with permission.

Summary

Astronomy seeks answers to many compelling questions about the universe. Astronomy uses all available tools to follow the scientific method. The process of science is based on objective reality, physical evidence, and testable hypotheses. Scientists continually strive to improve their understanding of the natural world and must be willing to challenge accepted truths as new information becomes available.

LG 1 **Describe the size and age of the universe and Earth's place in it.** We reside on a planet orbiting a star at the center of a solar system in a vast galaxy that is one of many in the universe. We occupy a very tiny part of the universe in space and time.

LG 2 **Explain how astronomers use the scientific method to study the universe.** The scientific method is an approach to learning about the physical universe. The method includes observations, forming hypotheses, making predic-

tions to test and refine those hypotheses, and repeated testing of theories. All scientific knowledge is provisional. Like art, literature, and music, science is a creative human activity; it is also a remarkably powerful, successful, and aesthetically beautiful way of viewing the world.

LG 3 **Show how scientists use mathematics, including graphs, to find patterns in nature.** Mathematics provides many of the tools that astronomers need to understand the patterns we see and to communicate that understanding to others.

LG 4 **Describe our astronomical origins.** We are a product of the universe: The very atoms we're made of formed in stars that died long before the Sun and Earth formed. The atoms of the chemical elements that make up matter come from particles that have been around since the Big Bang, processed through stars, and collected into the Sun and the Solar System.

? Unanswered Questions

- What makes up the universe? We have listed planets, stars, and galaxies as making up the cosmos, but astronomers now have evidence that 95 percent of the universe is composed of dark matter and dark energy, which we do not understand. Scientists are using the largest telescopes and particle colliders on Earth, as well as telescopes and experiments in space, to explore what makes up dark matter and what constitutes dark energy.

- Does life as we know it exist elsewhere in the universe? At the time of this writing, no scientific evidence indicates that life exists on any other planet. Our universe is enormously large and has existed for a great length of time. What if life is too far away or existed too long ago for us ever to "meet"?

Questions and Problems

Test Your Understanding

1. Rank the following in order from smallest to largest.
   a. Local Group
   b. Milky Way
   c. Solar System
   d. universe
   e. Sun
   f. Earth
   g. Laniakea Supercluster
   h. Virgo Supercluster

2. If an event were to take place on the Sun, approximately how long would the light it generates take to reach us?
   a. 8 minutes
   b. 11 hours
   c. 1 second
   d. 1 day
   e. It would reach us instantaneously.

3. *Understanding* in science means that
   a. we have accumulated lots of facts.
   b. we can connect facts through an underlying idea.
   c. we can predict events on the basis of accumulated facts.
   d. we can predict events on the basis of an underlying idea.

4. The cosmological principle states that
   a. on a large scale, the universe is the same everywhere at a given time.
   b. the universe is the same at all times.
   c. our location is special.
   d. all of the above

5. The Sun is part of
   a. the Solar System.
   b. the Milky Way Galaxy.
   c. the universe.
   d. all of the above

6. A light-year is a measure of
   a. distance.
   b. time.
   c. speed.
   d. mass.

7. Occam's razor states that
   a. the universe is expanding in all directions.
   b. the laws of nature are the same everywhere in the universe.
   c. if two hypotheses fit the facts equally well, the simpler one is more likely to apply.
   d. patterns in nature are really manifestations of random occurrences.

8. The circumference of Earth is 1/7 of a light-second. Therefore,
   a. if you were traveling at the speed of light, you would travel around Earth 7 times in 1 second.
   b. light travels a distance equal to Earth's circumference in 1/7 of a second.
   c. neither a nor b
   d. both a and b

9. According to the graphs in Figures 1.7b and c, by how much did the number of viruses increase in four time steps?
   a. It doubled.
   b. It tripled.
   c. It quadrupled.
   d. It went up more than 10 times.

10. Any explanation of a phenomenon that includes a supernatural influence is not scientific because
    a. it is not based on a hypothesis.
    b. it is wrong.
    c. people who believe in the supernatural are not credible.
    d. science is the study of the natural world.

11. "All scientific knowledge is provisional." In that context, *provisional* means
    a. "wrong."
    b. "relative."
    c. "temporary."
    d. "incomplete."

12. When we observe a star that is 10 light-years away, we are seeing that star
    a. as it is today.
    b. as it was 10 days ago.
    c. as it was 10 years ago.
    d. as it was 20 years ago.

13. Which of the following was *not* made in the Big Bang?
    a. hydrogen
    b. lithium
    c. beryllium
    d. carbon

14. "We are stardust" means that
    a. Earth exists because two stars collided.
    b. the atoms in our bodies have passed through (and often formed in) stars.
    c. Earth is formed primarily of material that used to be in the Sun.
    d. Earth and the other planets will eventually form a star.

15. According to our current understanding of the universe, the following astronomical events led to the formation of you. Place them in order of their occurrence over astronomical time.
    a. Stars die and distribute heavy elements into the space between the stars.
    b. Hydrogen and helium are made in the Big Bang.
    c. Enriched dust and gas gather into clouds in interstellar space.
    d. Stars are born and process light elements into heavier ones.
    e. The Sun and planets form from a cloud of interstellar dust and gas.

Thinking about the Concepts

16. Suppose you lived on the planet named "Tau Ceti e" that orbits Tau Ceti, a nearby star in our galaxy. How would you write your cosmic address?

17. Imagine yourself living on a planet orbiting a star in a very distant galaxy. What does the cosmological principle tell you about the physical laws at that distant location?

18. If the Sun suddenly exploded, how soon after the explosion would we know about it?

19. If a star exploded in the Andromeda Galaxy, how long would that information take to reach Earth?

20. Give an example of a scientific theory that a newer theory has superseded. As scientists developed that new theory, where on the Process of Science Figure did a change occur so that the old theory became invalid and the new theory was accepted?

21. Some people have proposed the theory that extraterrestrials (aliens) visited Earth in the remote past. Can you think of any tests that could support or refute that theory? Is it falsifiable?

22. What does the word *falsifiable* mean? Give an example of an idea that is not falsifiable. Give an example of an idea that is falsifiable.

23. Explain how the word *theory* is used differently in the context of science than it is in everyday language.

24. What is the difference between a *hypothesis* and a *theory* in science?

25. Suppose the tabloid newspaper at your local supermarket claimed that children born under a full Moon become better students than children born at other times.
    a. Is that theory falsifiable?
    b. If so, how could it be tested?

26. A textbook published in 1945 stated that light from the Andromeda Galaxy takes 800,000 years to reach Earth. In this book, we assert that it takes 2,500,000 years. What does that difference tell you about a scientific "fact" and how our knowledge evolves?

27. Astrology makes testable predictions. For example, it predicts that the horoscope for your star sign on any day should fit you better than horoscopes for other star signs. Read the daily horoscopes for all the astrological signs in a newspaper or online. How many of them might fit the day you had yesterday? Repeat the experiment every day for a week and record which horoscopes fit your day each day. Did your horoscope sign consistently best describe your experiences?

28. A scientist on television states that it is a known fact that life does not exist beyond Earth. Would you consider that scientist's statement reputable? Explain your answer.

29. Some astrologers use elaborate mathematical formulas and procedures to predict the future. Does doing so show that astrology is a science? Why or why not?

30. Why can it be said that we are made of stardust? Explain why current theory supports that statement.

Applying the Concepts

31. Review Working It Out 1.1., and then convert the following numbers to scientific notation:
    a. 7,000,000,000
    b. 0.00346
    c. 1,238

32. Review Working It Out 1.1, and then convert the following numbers to standard notation:
    a. 5.34×10^8
    b. 4.1×10^3
    c. 6.24×10^{-5}

33. If a car is traveling at 35 km/h, how far does it travel in
    a. 1 hour?
    b. half an hour?
    c. 1 minute?

34. Review Appendix 1. The surface area of a sphere is proportional to the square of its radius. How many times larger is the surface area if the radius is
    a. doubled?
    b. tripled?
    c. halved (divided by 2)?
    d. divided by 3?

35. The average distance from Earth to the Moon is 384,400 km. How many days would it take you, traveling at 800 km/h—the typical speed of jet aircraft—to reach the Moon?

36. The average distance from Earth to the Moon is 384,400 km. In the late 1960s, astronauts reached the Moon in about 3 days. How fast (on average) must they have been traveling (in kilometers per hour) to cover that distance in that time? Compare that speed with the speed of a jet aircraft (800 km/h).

37. (a) If light takes about 8 1/3 minutes to travel from the Sun to Earth, and Neptune is 30 times farther from Earth than the Sun is, how long does light from Neptune take to reach Earth? (b) Radio waves travel at the speed of light. What does that fact imply about the problems you would have if you tried to conduct a two-way conversation between Earth and a spacecraft orbiting Neptune?

38. The distance from Earth to Mars varies from 56 million km to 400 million km. How long does a radio signal traveling at the speed of light take to reach a spacecraft on Mars when Mars is closest and when Mars is farthest away?

39. The surface area of a sphere is proportional to the square of its radius. The radius of the Moon is only about one-quarter that of Earth. How does the surface area of the Moon compare with that of Earth?

40. A remote Web page may sometimes reach your computer by going through a satellite orbiting approximately 3.6×10^4 km above Earth's surface. What is the minimum delay, in seconds, before the Web page shows up on your computer?

41. Imagine that you have become a biologist, studying rats in Indonesia. Usually, Indonesian rats maintain a constant population. Every half century, however, those rats suddenly begin to multiply exponentially. Then the population crashes back to the constant level. Sketch a graph that shows the rat population over two of those episodes.

42. New York is 2,444 miles from Los Angeles. What is that distance in car-hours? In car-days? (Assume a travel speed of 60 mph.)

43. Some people theorize that a tray of hot water will freeze more quickly than a tray of cold water when both are placed in a freezer.
    a. Does that theory make sense to you?
    b. Is the theory falsifiable?
    c. Do the experiment yourself. Note the results. Was your intuition borne out?

44. A pizzeria offers a 9-inch-diameter pizza for $12 and an 18-inch-diameter pizza for $24. Are both offerings equally economical? If not, which is the better deal? Explain your reasoning.

45. The circumference of a circle is given by $C = 2\pi r$, where r is the radius of the circle.
    a. Calculate the approximate circumference of Earth's orbit around the Sun, assuming that the orbit is a circle with a radius of 1.5×10^8 km.
    b. A year has 8,766 hours, so how fast, in kilometers per hour, does Earth move in its orbit?
    c. How far along in its orbit does Earth move in 1 day?

USING THE WEB

46. Go to the interactive "The Scale of the Universe 2" (Web page at https://apod.nasa.gov/apod/ap140112.html, or download the app or find on YouTube). Start at 10^0 (human size) and scale upward; clicking on an object gives you its exact size. What astronomical bodies are about the size of the United States? What objects are about the size of Earth? What stars are larger than the distance from Earth to the Sun? How many light-days is the distance of *Voyager 1* to Earth? What objects are about the size of the distance from the Sun to the nearest star? How many times larger is the Milky Way than the Solar System? How many times larger is the Local Group than the Milky Way? How many times larger is the observable universe than the Local Group?

47. a. For a video representation of the scale of the universe, view the short video *The Known Universe* at the Hayden Planetarium website (https://amnh.org/our-research/hayden-planetarium/digital-universe), which takes the viewer on a journey from the Himalayan mountains to the most distant galaxies. How far have broadcast radio programs from Earth traveled? Is the Sun a particularly luminous star in comparison with others? Do you think the video is effective for showing the size and scale of the universe?

    b. A similar film produced in 1996 in IMAX format, *Cosmic Voyage*, can be found online on YouTube. Watch the "powers of 10" zoom out to the cosmos, starting at the 7-minute mark, for about 5 minutes. Do the "powers of 10" circles add to your understanding of the size and scale of the universe?

48. Go to the Astronomy Picture of the Day (APOD) app or website (https://apod.nasa.gov/apod) and click on "Archive" to look at the recent pictures and videos. Submissions to the website come from all around the world. Pick one and read the explanation. Was the image or video taken from Earth or from space? Is it a combination of several images? Does it show Earth, our Solar System, objects in our Milky Way Galaxy, more distant galaxies, or something else? Would someone who has not studied astronomy understand the explanation? Do you think the website promotes a general interest in astronomy?

49. Throughout this book, we examine how the media cover discoveries in astronomy and space. On your favorite news website (or one assigned by your instructor), find a recent article about astronomy or space. Does the website have a separate section for science? Is the article you selected based on a press release, on interviews with scientists, or on an article in a scientific journal? Use Google News or the equivalent to see how widespread the coverage of the story is. Have many newspapers carried it? Has it been picked up internationally? Has it been discussed in blogs? Do you think the story was interesting enough to be covered?

50. Go to a blog about astronomy or space. Is the blogger a scientist, a science writer, a student, or an enthusiastic amateur astronomer? What is the current topic of interest? Is it controversial? Are readers making many comments? Is the blog something you would want to read again?

digital.wwnorton/astro6

Logic is fundamental to studying science and to scientific thinking. A logical fallacy is an error in reasoning, which good scientific thinking avoids. For example, "because Einstein said so" is not an adequate argument. No matter how famous the scientist is (even if he is Einstein), he or she must still supply a logical argument and evidence to support a claim. Anyone who claims that something must be true because Einstein said it has committed the logical fallacy known as an *appeal to authority*. Many types of logical fallacies exist, but a few crop up often enough in discussions about science that you should be aware of them.

Ad hominem. In that fallacy, you attack the person making the argument instead of the argument itself. For example, "A well-known actor says dogs are better pets than cats, but I think that actor is an idiot, so cats must be better."

Appeal to belief. That fallacy has the general pattern "Most people believe X is true; therefore, X is true." For example, "Most people believe Earth orbits the Sun. Therefore, Earth orbits the Sun." Even if your conclusion is correct, you may still have committed a logical fallacy in your argument.

Begging the question. In that fallacy, also known as circular reasoning, you assume that the claim is true and then use that assumption as evidence to prove the claim is true. For example, "I am trustworthy; therefore, I must be telling the truth." No real evidence is presented for the conclusion.

Biased sample. If a sample drawn from a smaller pool has a bias, conclusions about the sample cannot be applied to a larger pool. For example, imagine you poll students at your university and find that 30 percent of them visit the library at least once per week. Then you conclude that 30 percent of Americans visit the library at least once per week. You have committed the biased sample fallacy because university students are not a representative sample of the American public.

Post hoc ergo propter hoc. That term is Latin for "after this, therefore because of this." Just because one thing follows another doesn't mean that one caused the other. For example, "An eclipse occurred, and then the king died. Therefore, the eclipse killed the king." That fallacy is often connected to related inverse reasoning: "If we can prevent an eclipse, the king won't die."

Slippery slope. In that fallacy, you claim that a chain reaction of events will take place, inevitably leading to a conclusion that no one could want. For example, "If I don't get A's in all my classes, I will not ever be able to get into graduate school, and then I won't ever be able to get a good job, and then I will be living in a van down by the river until I'm old and starve to death." None of those steps actually follows inevitably from the one before.

The following are examples of logical fallacies. Identify the fallacy represented. Each fallacy just discussed is represented once.

1 You get a chain email threatening terrible consequences if you break the chain. You move the message to your spam box. Later that day you get in a car accident. The next morning, you retrieve the chain email and send it along.

_____

_____

2 If I get question 1 on the assignment wrong, then I'll get question 2 wrong as well, and before you know it, I will never catch up in the class.

_____

_____

3 All my friends love the band Degenerate Electrons. Therefore, all people my age love that band.

_____

_____

4 Eighty percent of Americans believe in the tooth fairy. Therefore, the tooth fairy exists.

_____

5 My professor says that the universe is expanding. But my professor is a geek, and I don't like geeks, so the universe can't be expanding.

_____

6 When applying for a job, you use a friend as a reference. Your prospective employer asks you how she can be sure your friend is trustworthy, and you say, "I can vouch for him."

_____

_____

2

Patterns in the Sky—Motions of Earth and the Moon

Ancient peoples learned to use the patterns they observed in the sky to predict how the length of day, the seasons, and the appearance of the Moon would change. Some people understood those patterns well enough to create complicated calendars and predict rare eclipses. But now we can see those patterns with the perspective of centuries of modern science, and we can explain those changes as being a consequence of the motions of Earth and the Moon. Discovering what causes those patterns has shown us the way outward into the universe.

....

LEARNING GOALS

Here in Chapter 2, we examine the patterns in the sky and on Earth and the underlying motions that cause them. By the end of this chapter, you should be able to:

LG 1 Describe how Earth's rotation about its axis and its revolution around the Sun affect how we perceive celestial motions from different places on Earth.

LG 2 Explain why seasons change throughout the year.

LG 3 Describe the factors that create the phases of the Moon.

LG 4 Sketch the alignment of Earth, the Moon, and the Sun during eclipses of the Sun and the Moon.

What causes a
solar eclipse?

(a)
(b)
(c)

Figure 2.1 (a) One suspected use of Stonehenge 4,000 years ago was to keep track of celestial events. (b) The Mayan El Caracol at Chichén Itzá in Mexico (906 CE) is believed to have been designed to align with Venus. (c) The Beijing Ancient Observatory in China (1442 CE) includes ancient astronomical instruments as well as some brought by European Jesuits in the 17th and 18th centuries.

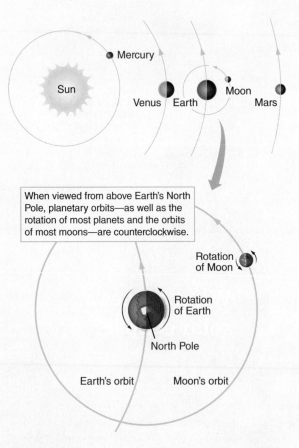

When viewed from above Earth's North Pole, planetary orbits—as well as the rotation of most planets and the orbits of most moons—are counterclockwise.

Figure 2.2 Motions in the Solar System, as viewed from above Earth's North Pole. (Not drawn to scale.)

2.1 Earth Spins on Its Axis

Ancient humans may not have known that they were "stardust," but they did sense a connection between their lives on Earth and the sky above. Before our modern technological civilization, people's lives were more attuned to the ebb and flow of nature, which includes the patterns in the sky. By watching the repeating patterns of the Sun, Moon, and stars in the sky, people could predict when the seasons would change and when the rains would come and the crops would grow. Ancient astronomers with knowledge of the sky—priests and priestesses, natural philosophers, and explorers—all had knowledge of the world that others did not, and knowledge was power. Some of those early observations and ideas live on today in the names of stars and the apparent grouping of stars we call **constellations**, in calendars based on the Moon and Sun that many cultures still use, and in the astronomical names of the days of the week.

From Mesopotamia to Africa, from Europe to Asia, from the Americas to the British Isles, the archaeological record holds evidence that ancient cultures built structures to study astronomical positions and events. **Figure 2.1** shows some examples. From the 8th through 17th centuries, people used pre-telescopic astronomical observatories to study the sky for timekeeping and navigation. Many of those structures and observatories are now national historical or UNESCO World Heritage sites.

The Celestial Sphere

Long before Christopher Columbus sailed west from Spain, Aristotle and other Greek philosophers knew that Earth was a sphere. However, because Earth seems stationary, they did not realize that its motion caused the changes they observed in the sky from day to day and year to year. As we explain here, Earth's rotation on its axis determines the rising and setting of the Sun, Moon, and stars—one of the rhythms that governs life on Earth (**Figure 2.2**).

One reason the ancients did not suspect that Earth rotates was that they could not perceive its spinning motion. As Earth rotates about its axis, the planet's surface is moving quite fast—about 1,674 kilometers per hour (km/h) at the equator. We do not feel that motion any more than we would feel the motion of a car with a perfectly smooth ride cruising down a straight highway. Nor do we feel the direction of Earth's spin, although the hourly motion of the Sun, Moon, and stars across the sky reveals that direction. Earth's **North Pole** is at the north end of Earth's rotation axis. Imagine that you are in space far above Earth's North Pole. From there you would see Earth complete a counterclockwise rotation, once each 24-hour period, as Figure 2.2 shows. As the rotating Earth carries an observer on the surface from west to east, objects in the sky appear to move in the opposite direction, from east to west. As seen from Earth's surface, each object seems to move across the sky in a path called its **apparent daily motion**.

To help visualize the apparent daily motions of the Sun and stars, think of the sky as a huge sphere with the stars painted on its surface and Earth at its center. From ancient Greek times to the Renaissance, many people believed that to be a true representation of the heavens. Astronomers call that imaginary sphere the **celestial sphere** (**Figure 2.3**). The celestial sphere is a useful concept because it is easy to visualize, but never forget that it is, in fact, imaginary.

Each point on the celestial sphere indicates a direction in space. Directly above Earth's North Pole is the **north celestial pole** (**NCP**). Directly above Earth's **South Pole**, which is at the south end of Earth's rotation axis, is the **south celestial pole** (**SCP**). Directly above Earth's **equator** is the **celestial equator**, an imaginary circle that divides the sky into a northern half and a southern half. Just as the north celestial pole is the projection of the direction of Earth's North Pole into the sky, the celestial equator is the projection of the plane of Earth's equator into the sky. Just as Earth's North Pole is 90° away from Earth's equator, the north celestial pole is 90° away from the celestial equator. If you are in the Northern Hemisphere and you point one arm toward the celestial equator and one arm toward the north celestial pole, your arms will always form a right angle, so the north celestial pole is 90° away from the celestial equator. The same holds true for the Southern Hemisphere: the angle between the celestial equator and the south celestial pole is always 90° as well.

Between the celestial poles and the equator, objects have positions on the celestial sphere with coordinates analogous to latitude and longitude on Earth. **Latitude** indicates distance north or south from Earth's equator. On the celestial sphere, **declination** similarly indicates the distance of an object north or south of the celestial equator (from 0° to ±90°). On Earth, **longitude** measures how far east or west you are from the Royal Observatory in Greenwich, England. On the celestial sphere, **right ascension** is similar to longitude on Earth. Right ascension measures a celestial body's angular distance eastward along the celestial equator from the point where the Sun's path crosses the celestial equator from south to north. Those coordinates are used to quickly locate objects in the sky. The **ecliptic** is the Sun's path in the sky throughout the year (Figure 2.3). In Appendix 7, we discuss detailed descriptions and illustrations of latitude and longitude as well as coordinates used with the celestial sphere.

The **zenith** is the point in the sky directly above you, wherever you are, as **Figure 2.4a** shows. You can find the **horizon** by standing up and pointing your right hand at the zenith and your left hand straight out from your side. Turn in a complete circle. Your left hand has traced out the entire horizon. You can divide the sky into an eastern half and a western half with a line that runs from the horizon at due north through the zenith to the horizon at due south. That imaginary north–south line is the **meridian**, shown as a dashed line in Figure 2.4a. **Figure 2.4b** shows those locations on the celestial sphere. The meridian line continues around the far side of the celestial sphere, through the **nadir** (the point directly below you), and back to the starting point due north.

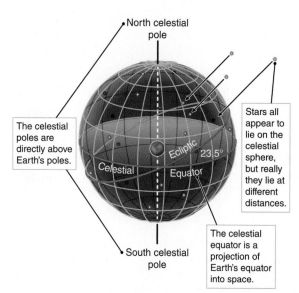

The celestial poles are directly above Earth's poles.

North celestial pole

South celestial pole

Stars all appear to lie on the celestial sphere, but really they lie at different distances.

The celestial equator is a projection of Earth's equator into space.

Ecliptic   23.5°

Celestial   Equator

Figure 2.3 The celestial sphere is a useful fiction in thinking about the stars' appearance and apparent motion in the sky.

AstroTour: The Celestial Sphere and the Ecliptic

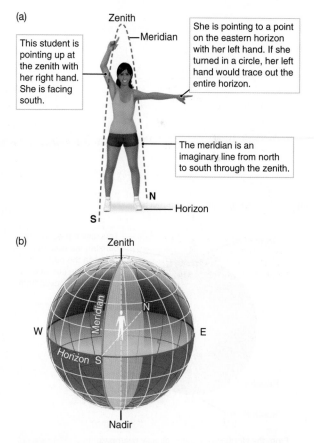

(a)

Zenith

Meridian

This student is pointing up at the zenith with her right hand. She is facing south.

She is pointing to a point on the eastern horizon with her left hand. If she turned in a circle, her left hand would trace out the entire horizon.

The meridian is an imaginary line from north to south through the zenith.

Horizon

N

S

(b)

Zenith

Meridian

W   N   E

Horizon S

Nadir

Figure 2.4 (a) The meridian is a line on the celestial sphere that runs from north to south, dividing the sky into an east half and a west half. (b) At any location on Earth, the sky is divided into an east half and a west half by an imaginary meridian projected onto the celestial sphere.

Take a moment to visualize all those locations in space. To see how to use the celestial sphere, consider the Sun at noon and at midnight. Local noon occurs when the Sun crosses the meridian at your location. That is the highest point above the horizon that the Sun will reach on any given day. The highest point is almost never the zenith. You have to be in a specific place on a specific day for the Sun to be directly overhead at noon. One such place and time is at latitude 23.5° north of the equator on June 20. From our perspective on Earth, the celestial sphere appears to rotate, carrying the Sun across the sky to its highest point at noon, over toward the west to set in the evening. In *reality*, the Sun remains in the same place in space through the entire 24-hour period, and Earth rotates so that any given location on Earth faces a different direction at every moment. When it is noon where you live, Earth has rotated so that you face most directly toward the Sun. Half a day later, at midnight, your location on Earth has rotated to face most directly away from the Sun. Local midnight occurs when the Sun is precisely opposite from its position at local noon.

The View from the Poles

The apparent daily motions of the stars and the Sun depend on where you live. For example, the apparent daily motions of celestial objects in a northern place such as Alaska are different from the apparent daily motions seen from a tropical island such as Hawaii. To understand why your location matters, let's examine the view of the stars from the poles—and then use them to guide our thinking about the view of the stars from other latitudes.

Imagine that you are the person in **Figure 2.5a** who is standing on the North Pole watching the sky. There, the north celestial pole is directly overhead at the zenith. Ignore the Sun for the moment and pretend that you can always see stars in the sky. You are standing where Earth's axis of rotation intersects its surface, which is like standing at the center of a rotating carousel. As Earth rotates, the spot directly above you remains fixed over your head while everything else in the sky appears to revolve counterclockwise around that spot. **Figure 2.5b** depicts that overhead view.

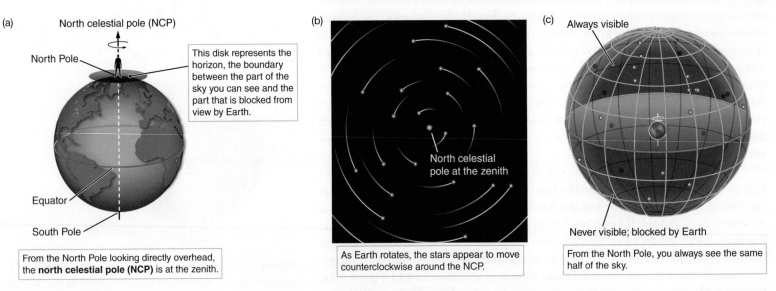

(a)

North celestial pole (NCP)

North Pole

This disk represents the horizon, the boundary between the part of the sky you can see and the part that is blocked from view by Earth.

Equator

South Pole

From the North Pole looking directly overhead, the **north celestial pole (NCP)** is at the zenith.

(b)

North celestial pole at the zenith

As Earth rotates, the stars appear to move counterclockwise around the NCP.

(c)

Always visible

Never visible; blocked by Earth

From the North Pole, you always see the same half of the sky.

Figure 2.5 As viewed from (a) Earth's North Pole, (b) stars move throughout the night in counterclockwise, circular paths about the zenith. (c) The same half of the sky is always visible from the North Pole.

Wherever you are on Earth, you can see only half the sky. The horizon is the boundary between the part of the sky you can see and the other half of the sky that is blocked by Earth. Except at the poles, the visible half of the sky changes constantly as Earth rotates because the zenith points to different locations in the sky as Earth carries you around. If you are standing at the North Pole, however, the zenith is always in the same location in space, so objects visible from the North Pole follow circular paths that always have the same **altitude**, or angle above the horizon. Objects close to the zenith appear to follow small circles, whereas objects near the horizon follow the largest circles (Figure 2.5b). The view from the North Pole is special: you will always see the *same* half of the celestial sphere from there because nothing rises or sets each day as Earth turns (**Figure 2.5c**).

The view from Earth's South Pole is much the same—with two major differences. First, the South Pole is on the opposite side of Earth from the North Pole, so the visible half of the sky at the South Pole is precisely the half hidden from view at the North Pole. Second, stars appear to move clockwise around the south celestial pole rather than counterclockwise as they do at the north celestial pole. To visualize why those motions are different, stand up and spin around from right to left. As you look at the ceiling, things appear to move counterclockwise, but as you look at the floor, they appear to be moving clockwise.

CHECK YOUR UNDERSTANDING 2.1A

No matter where you are on Earth, stars appear to rotate about a point called the:
(a) zenith; (b) celestial pole; (c) nadir; (d) meridian.

....

The View Away from the Poles

Suppose that you leave the North Pole to travel south to lower latitudes. Imagine a line from the center of Earth to your location on the surface, as in **Figure 2.6**. Now imagine a second line from Earth's center to the point on the equator closest to you. The angle between those two lines is your latitude. At the North Pole, for example, those two imaginary lines form a 90° angle. At the equator, they form a 0° angle. So the latitude of the North Pole is 90° north, and the latitude of the equator is 0°. The South Pole is at latitude 90° south.

Your latitude determines the part of the sky that you can see throughout the year. As you move south from the North Pole, your zenith moves away from the north celestial pole, and so the horizon moves as well. At the North Pole, the horizon makes a 90° angle with the north celestial pole, which is at the zenith. At a latitude of 60° north, as in Figure 2.6, your horizon is tilted 60° from the north celestial pole. No matter where you are on Earth, the angle between your horizon and the north celestial pole is equal to your latitude. The situation is the same in the Southern Hemisphere—your latitude is the altitude of the south celestial pole. At the equator, at a latitude of 0°, the north and south celestial poles would be at the northern and southern horizons, respectively.

One way to solidify your understanding of the view of the sky at different latitudes is to draw pictures like the one in Figure 2.6. If you can draw a picture like that for any latitude—filling in the values for each angle in the drawing and imagining what the sky looks like from there—you are developing a working knowledge of the appearance of the sky. That knowledge will prove useful later when we discuss a variety of phenomena, such as the changing of the seasons.

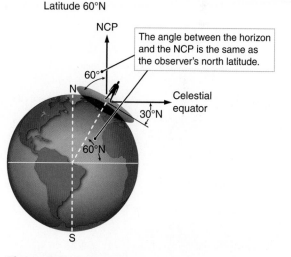

Figure 2.6 Your perspective on the sky depends on your location on Earth. The locations of the celestial poles and the celestial equator in an observer's sky depend on the observer's latitude. Here, an observer at latitude 60° north sees the north celestial pole at an altitude of 60° above the northern horizon and the celestial equator 30° above the southern horizon.

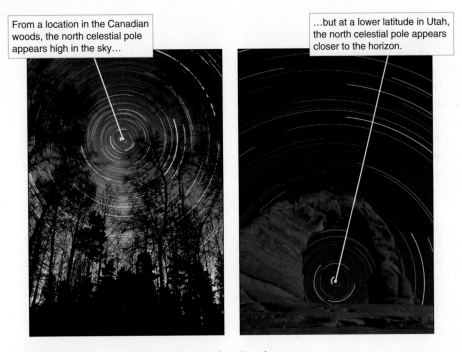

From a location in the Canadian woods, the north celestial pole appears high in the sky…

…but at a lower latitude in Utah, the north celestial pole appears closer to the horizon.

Figure 2.7 Time exposures of the sky showing the apparent motions of stars through the night. Note the difference in the circumpolar portion of the sky as seen from the two northern latitudes.

MOTIONS OF THE STARS AND THE CELESTIAL POLES **Figure 2.7** shows two time-lapse views of the sky from different latitudes. The stars' apparent motions about the celestial poles also differ from latitude to latitude. The visible part of the sky constantly changes as stars rise and set with Earth's rotation. From that perspective the horizon appears fixed, whereas the stars appear to move. From those latitudes, if we focus our attention on the north celestial pole, we see much the same thing we saw from Earth's North Pole. The north celestial pole remains fixed in the sky, and all the stars appear to move throughout the night in counterclockwise, circular paths around that point. The north celestial pole is no longer directly overhead as it was at the North Pole, however, so the apparent circular paths of the stars are now tipped with respect to the horizon. (More correctly, your horizon is now tipped with respect to the apparent circular paths of the stars.)

From the vantage point of an observer in the Northern Hemisphere, stars located near the north celestial pole in the sky never dip below the horizon and thus stay visible all night. Recall from Figure 2.6 that the latitude is equal to the altitude of the north celestial pole. Stars closer to the north celestial pole than that angle never dip below the horizon as they complete their apparent paths around the pole. Those stars are **circumpolar** ("around the pole"). Another group of stars, near the south celestial pole, never rises above the horizon in the Northern Hemisphere. Stars between those that never rise and those in the circumpolar region (that is, the ones that never set) can be seen for *only part* of each 24-hour day. Those stars appear to rise and set as Earth turns. The equator is the only place on Earth where you can see the entire sky over the course of 24 hours. From there, the north and south celestial poles sit on the northern and southern horizons, respectively, and all the stars move through the sky each 24-hour day. (Even though the Sun lights the sky for roughly half that time, the stars are still there.)

Figure 2.8 shows the orientation of the sky as seen by observers at four latitudes. For an observer at the North Pole (Figure 2.8a), the celestial equator lies exactly along the horizon. The north celestial pole is at the zenith, and the southern half of the sky is never visible. Stars neither rise nor set; their paths form circles parallel to the horizon. For an observer at the equator (Figure 2.8b), the celestial poles are both at the horizon, and all the stars are visible in a 24-hour period, rising straight up and setting straight down each day.

At other latitudes, the celestial equator intersects the horizon due east and due west. Therefore, a star on the celestial equator rises due east and sets due west. Stars located north of the celestial equator rise north of east and set north of west. Stars located south of the celestial equator rise south of east and set south of west.

From everywhere else on Earth (except at the poles), half of the celestial equator is always visible above the horizon. Therefore, any object located on the celestial equator is visible half the time—above the horizon for 12 hours each day. Objects not on the celestial equator are above the horizon for different amounts of time. Figures 2.8c and d show that stars in the observer's hemisphere are visible

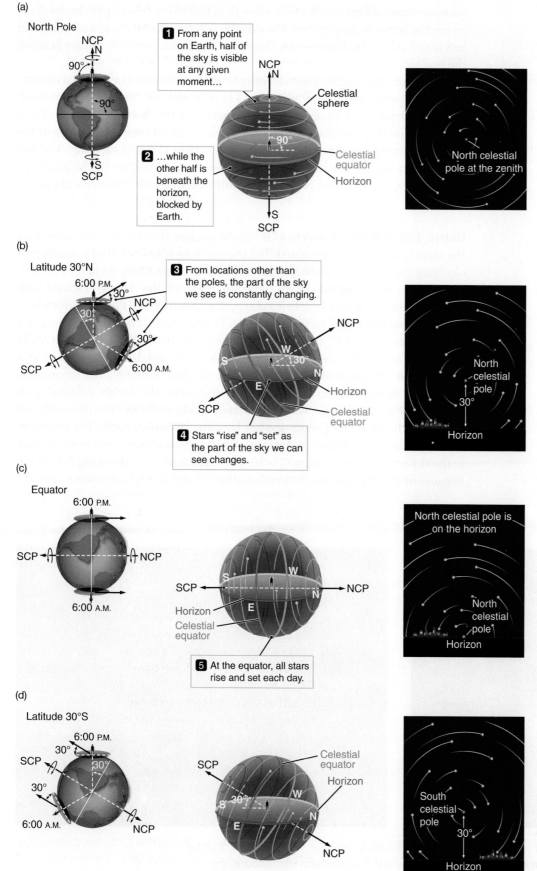

(a)

North Pole

1 From any point on Earth, half of the sky is visible at any given moment…

2 …while the other half is beneath the horizon, blocked by Earth.

North celestial pole at the zenith

(b)

Latitude 30°N

3 From locations other than the poles, the part of the sky we see is constantly changing.

4 Stars "rise" and "set" as the part of the sky we can see changes.

North celestial pole

Horizon

(c)

Equator

5 At the equator, all stars rise and set each day.

North celestial pole is on the horizon

North celestial pole

Horizon

(d)

Latitude 30°S

South celestial pole

Horizon

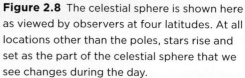

Figure 2.8 The celestial sphere is shown here as viewed by observers at four latitudes. At all locations other than the poles, stars rise and set as the part of the celestial sphere that we see changes during the day.

for more than half the day because more than half of each star's path in the sky is above the horizon. In contrast, stars in the opposite hemisphere are visible for less than half the day because less than half of each star's path in the sky is above the horizon.

As seen from the Northern Hemisphere, stars north of the celestial equator remain above the horizon for more than 12 hours each day. The farther north the star is, the longer it stays up. Circumpolar stars are the extreme example of that phenomenon; they are always above the horizon. In contrast, stars south of the celestial equator are above the horizon for less than 12 hours each day. The farther south a star is, the less time it is visible. For an observer in the Northern Hemisphere, stars located close to the south celestial pole never rise above the horizon.

USING THE STARS TO NAVIGATE   Since ancient times, travelers have used the stars to navigate. They would find the north or south celestial poles by recognizing the stars that surround them. In the Northern Hemisphere, a moderately bright star happens to be located close to the north celestial pole (**Figure 2.9a**). That star is called Polaris, the "North Star." Polaris's altitude in the sky is nearly equal to your latitude. If you are in Phoenix, Arizona, for example (latitude 33.5° north), Polaris has an altitude very close to 33.5°. In Fairbanks, Alaska (latitude 64.6° north), Polaris sits much higher, with an altitude of 64.6°. In the Southern Hemisphere, the constellation Crux (commonly called the Southern Cross) points to a star near the south celestial pole (**Figure 2.9b**). A navigator who has located a pole star can identify north and south and therefore east, west, and her latitude. Doing so enables the navigator to determine which direction to travel. The location of the north celestial pole in the sky was used to measure Earth's size, as described in **Working It Out 2.1**. Because of Earth's rotation, determining your longitude by astronomical meth-

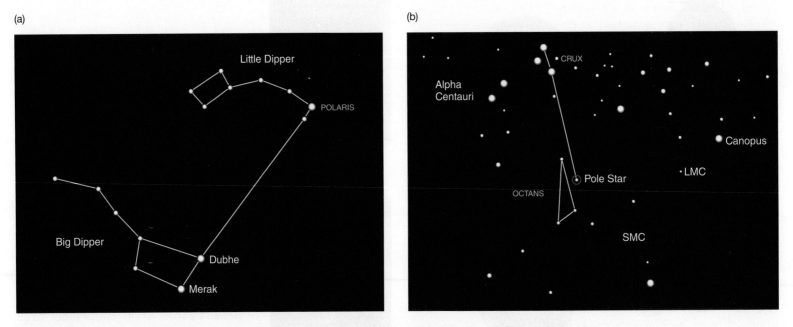

Figure 2.9 Groups of stars near the pole stars in the sky can be used to locate a pole star. (a) Two bright "pointer stars" in the cup of the Big Dipper point toward Polaris, the "North Star." (b) In the Southern Hemisphere, the constellation Crux and two of its bright pointer stars can be used to locate the relatively faint southern pole star.

2.1) Working It Out How to Estimate Earth's Size

We can use the location of the north celestial pole in the sky to estimate Earth's size. Suppose we start out in Phoenix, Arizona, and we observe the north celestial pole to be 33.5° above the horizon. If we head north, by the time we reach the Grand Canyon, about 290 km from Phoenix, we notice that the north celestial pole has risen to about 36° above the horizon. That difference between 33.5° and 36° (2.5°) is 1/144 of the way around a circle. (A circle is 360°, and 2.5°/360° = 1/144.)

That means we must have traveled 1/144 of the way around Earth's circumference, C, by traveling the 290 km between Phoenix and the Grand Canyon. In other words,

$$\frac{1}{144} \times C = 290 \text{ km}$$

When we rearrange the expression, Earth's circumference is given by

$$C = 144 \times 290 \text{ km} \approx 42{,}000 \text{ km}$$

Earth's actual circumference is just over 40,000 km, so our simple calculation was close. The circumference of a circle is equal to $2\pi r$, where r is its radius. Earth's radius, then, is given by

$$\text{Radius} = \frac{C}{2\pi} = \frac{40{,}000 \text{ km}}{2\pi} = 6{,}400 \text{ km}$$

In about 230 BCE, using much the same method, the Greek astronomer Eratosthenes (276–194 BCE) was the first to accurately measure Earth's size. As **Figure 2.10** illustrates, Eratosthenes used the distance between his home city of Alexandria, Egypt, and the city of Syene (now Aswân, also in Egypt), which was 5,000 "stadia." He noticed that on the first day of summer in Syene, the sunlight reflected directly off the water in a deep well, so the Sun must have been nearly at the zenith. By measuring the shadow of the Sun from an upright stick in Alexandria, he saw that the Sun was about 7.2° south of the zenith on the same date. Assuming that Earth was spherical and that Syene was directly south of Alexandria, he determined the distance between the two cities to be 7.2° divided by 360, or 1/50 of Earth's circumference.

Accurately estimating distances in those days was hard. Although historians know Eratosthenes concluded that the cities were 5,000 stadia apart, they are still not at all sure of the value of his stadion unit. If the stadion was 185 meters, the size of Greek stadia, Eratosthenes would have worked the math similarly:

$$\frac{1}{50} \times C = 5{,}000 \text{ stadia} \times 185 \text{ meters/stadion}$$

$$= 925{,}000 \text{ meters} = 925 \text{ km}$$

Eratosthenes would have found Earth's circumference to be

$$C = 50 \times 925 \text{ km} = 46{,}250 \text{ km}$$

That result is only about 16 percent higher than the modern value.

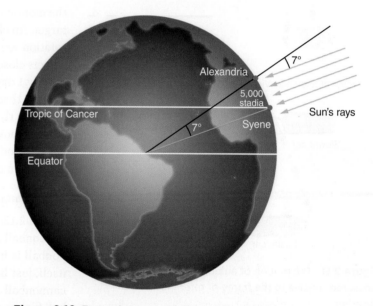

Figure 2.10 Eratosthenes used observations and basic calculations to estimate Earth's size.

ods is much more complicated. Longitude cannot easily be determined from astronomical observation alone.

Relative Motions and Frame of Reference

Why don't we feel motion as Earth spins on its axis and moves through space in its orbit around the Sun? The answer lies in the concept of a **frame of reference**, a coordinate system within which an observer measures positions and motions. The difference in motion between two individual frames of reference is called the

(a) Frame of reference: Viewer on the street

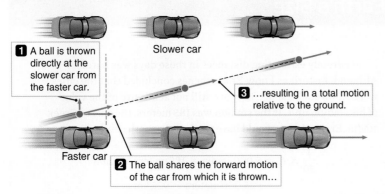

1 A ball is thrown directly at the slower car from the faster car.

Slower car

3 ...resulting in a total motion relative to the ground.

Faster car

2 The ball shares the forward motion of the car from which it is thrown...

(b) Frame of reference: Viewer in faster car

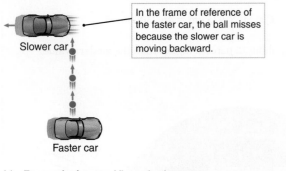

In the frame of reference of the faster car, the ball misses because the slower car is moving backward.

Slower car

Faster car

(c) Frame of reference: Viewer in slower car

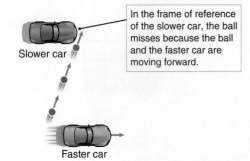

In the frame of reference of the slower car, the ball misses because the ball and the faster car are moving forward.

Slower car

Faster car

Figure 2.11 The motion of an object is always measured relative to the frame of reference of the observer.

relative motion. For example, imagine that you are riding in a car traveling down a straight section of highway at a constant speed. Under those conditions, without looking out the window or feeling road vibrations, you cannot easily tell the difference between being in the moving car and sitting in a parked car. Because everything in the car is moving together, you can measure only the relative motions between objects in the car, and they are all zero—no motion is observed. Similarly, the resulting relative motions between objects that are near each other on Earth are zero because they have the same rotational motion. That is why we do not notice Earth's motion.

Now imagine two cars driving down the road at different speeds, as **Figure 2.11a** shows. For the moment, ignore any real-world complications, such as wind resistance. If you were to throw a ball from the faster-moving car directly out the side window at the slower-moving car as the two cars passed, you would miss. The ball shares the forward motion of the faster car, so the ball outruns the forward motion of the slower car. From your perspective in the faster car (**Figure 2.11b**), the slower car lagged behind the ball. From the slower car's perspective (**Figure 2.11c**), your car and the ball sped on ahead.

Although we cannot feel Earth's rotation, it influences things as diverse as the motion of weather patterns and how an artillery gunner must aim at a distant target. An object on Earth's surface moves in a circle each day around the planet's rotation axis. That circle is larger for objects near the equator and smaller for objects closer to one of the poles, but all objects complete their circular motion in exactly 1 day. As you can see in **Figure 2.12**, Earth's surface moves faster at the equator than at higher latitudes. An object closer to the equator has farther to go each day than one nearer a pole does. Therefore, the object nearer the equator must be moving faster than the one at higher latitude. If an object starts out at one latitude and then moves to another, that difference in speed influences its apparent motion over the surface.

Now consider two locations at different latitudes. Imagine a cannonball launched directly north from a point in the Northern Hemisphere, as in Figure 2.12a. Because the cannon is closer to the equator than its target is, the cannonball is moving toward the east faster than its target. Even though the cannonball is fired toward the north, it shares the eastward velocity of the cannon itself, just like the ball thrown from the faster car in Figure 2.11. Therefore, the cannonball is *also* moving toward the east faster than its target. As the cannonball flies north, it too moves toward the east faster than the ground underneath it does. To an observer on the ground, the cannonball appears to curve toward the east as it outruns the eastward motion of the ground it is crossing. The farther north the cannonball flies, the greater the difference between its eastward velocity and the ground's. Thus, the cannonball follows a path that appears to curve more and more to the east the farther north it goes. If you are located in the Northern Hemisphere and fire a cannonball *south* toward the equator, as in Figure 2.12b, the opposite effect will occur. Now the cannon is moving toward the east more slowly than its target. As the cannonball flies toward the south, its eastward motion lags behind that of the ground underneath it, and the cannonball appears to curve toward the west.

That curving motion of objects as a result of the difference in Earth's rotation speeds at different latitudes is called the **Coriolis effect**. In the Northern Hemisphere, the Coriolis effect causes a cannonball fired north to drift to the east as seen from the ground. In other words, the cannonball appears to curve to the

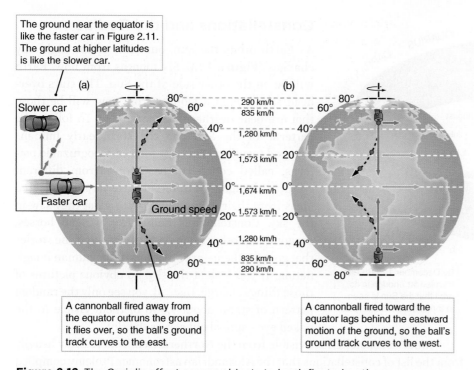

The ground near the equator is like the faster car in Figure 2.11. The ground at higher latitudes is like the slower car.

(a)

Slower car

Faster car

Ground speed

80°
60°
40°
20°
0°
20°
40°
60°
80°

290 km/h
835 km/h
1,280 km/h
1,573 km/h
1,674 km/h
1,573 km/h
1,280 km/h
835 km/h
290 km/h

(b)

80°
60°
40°
20°
0°
20°
40°
60°
80°

A cannonball fired away from the equator outruns the ground it flies over, so the ball's ground track curves to the east.

A cannonball fired toward the equator lags behind the eastward motion of the ground, so the ball's ground track curves to the west.

Figure 2.12 The Coriolis effect causes objects to be deflected as they move across Earth's surface. The green dashed curve shows the cannonball's path as measured by a local observer.

right. A cannonball fired south appears to curve to the west, which also gives it the appearance of curving to the right. In the Northern Hemisphere, the Coriolis effect seems to deflect things to the *right*. In the Southern Hemisphere, it seems to deflect things to the *left*. In between, at the equator itself, the Coriolis effect vanishes.

On Earth, the effect is enough to deflect a fly ball hit north or south into deep left field in a stadium in the northern United States by about a half a centimeter. At some time or other, the Coriolis effect has probably determined the outcome of a baseball game.

CHECK YOUR UNDERSTANDING 2.1B

If the star Polaris has an altitude of 35°, we know that: (a) our longitude is 55° east; (b) our latitude is 55° north; (c) our longitude is 35° west; (d) our latitude is 35° north.

. . . .

2.2 Revolution about the Sun Leads to Changes during the Year

Earth orbits (or **revolves**) around the Sun in the same direction that Earth spins about its axis—counterclockwise as viewed from above Earth's North Pole (see Figure 2.2). A **year** is the time Earth takes to complete one revolution around the Sun. Earth's motion around the Sun is responsible for many of the patterns of change we see in the sky and on Earth, including changes in the stars we see overhead. Because of that motion, the stars in the night sky change throughout the year, and Earth experiences seasons.

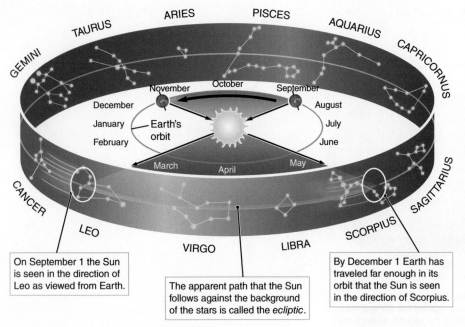

On September 1 the Sun is seen in the direction of Leo as viewed from Earth.

The apparent path that the Sun follows against the background of the stars is called the *ecliptic*.

By December 1 Earth has traveled far enough in its orbit that the Sun is seen in the direction of Scorpius.

Figure 2.13 As Earth orbits the Sun, different stars appear in the night sky, and the Sun's apparent position against the background of stars changes. The imaginary circle that the Sun's annual path traces is called the ecliptic. Constellations along the ecliptic form the zodiac.

Constellations and the Zodiac

As Earth orbits the Sun, our view of the night sky changes (**Figure 2.13**). Six months from now, Earth will be on the other side of the Sun. The stars overhead at midnight 6 months from now will be those that are near overhead at noon today. To follow the patterns of the Sun and the stars, early humans grouped together stars that formed recognizable patterns, called constellations. People from different cultures, though, saw different patterns and projected ideas from their own cultures onto what they saw in the sky. Constellations named for winged horses, dragons, and other imaginary things—and the stories that go with them—are creations of the human imagination. If you look at the sky, no obvious pictures of those things emerge. Instead, you see only the random pattern of stars—about 9,000 of them visible to the naked eye—spread out across the sky.

Modern constellations visible from the Northern Hemisphere draw heavily from the list of constellations that the Alexandrian astronomer Ptolemy compiled 2,000 years ago. Modern constellation names in the southern sky come from European explorers who visited the Southern Hemisphere during the 17th and 18th centuries. Today, astronomers use an officially sanctioned set of 88 constellations to serve as a road map of the sky. Every star in the sky lies within the borders of a single constellation, and the names of constellations are used in naming their component stars. For example, Sirius, the brightest star in the sky, lies within the constellation Canis Major (meaning "big dog"). Following the Greek alphabet, astronomers call the brightest star in a constellation *alpha*; the second brightest, *beta*; and so forth. The official name of Sirius is Alpha Canis Majoris, indicating that it is the brightest star in Canis Major. Appendix 7 includes sky maps showing the constellations.

If you noted the position of the Sun in relation to the stars each day for a year, you would find that the Sun traces out a great circle against the background of the stars. On September 1, the Sun appears to be in the direction of the constellation Leo. Six months later, on March 1, Earth is on the other side of the Sun, and the Sun appears from our perspective on Earth to be in the direction of the constellation Aquarius. Recall that the Sun's apparent path against the background of the stars is called the ecliptic (the blue band in Figure 2.13). The constellations that lie along the ecliptic through which the Sun appears to move are called the constellations of the **zodiac**.

Earth's Axial Tilt and the Seasons

To understand why the seasons change, we need to consider the combined effects of Earth's rotation on its axis and revolution around the Sun. Many people believe that seasons change because Earth is closer to the Sun in summer and farther away in winter. How can we falsify that hypothesis? We can predict that if the distance between Earth and the Sun caused the seasons, then all of Earth should experience summer at the same time of year. But the United States has summer in June, whereas Australia has summer in December. In modern times, we can directly measure the distance: Earth is actually closest to the Sun at the beginning of January. We have just falsified the hypothesis tying seasonal change to distance, and we need to look for another explanation that explains all the available facts.

As Earth orbits the Sun, the Sun appears to move along the ecliptic, which is tilted 23.5° with respect to the celestial equator. That tilt occurs because Earth's axis of rotation is tilted 23.5° from the perpendicular to Earth's orbital plane. The days are longer in summer than in winter, and the Sun is higher in the sky as it crosses the meridian in summer than in winter.

Figure 2.14 shows that as Earth moves around the Sun, its axis always points in the same direction in space (toward Polaris). As it orbits, Earth is sometimes on one side of the Sun and sometimes on the other. As a result, Earth's North Pole is sometimes tilted more toward the Sun, and other times the South Pole is tilted more toward the Sun. When Earth's North Pole is tilted toward the Sun, an observer on Earth views the Sun north of the celestial equator. For observers in the Northern Hemisphere, the Sun is above the horizon more than 12 hours each day, thus making the days longer than 12 hours. Six months later, when Earth's North Pole is tilted away from the Sun, an observer in the same place views the Sun south of the celestial equator and the days are shorter.

AstroTour: The Earth Spins and Revolves

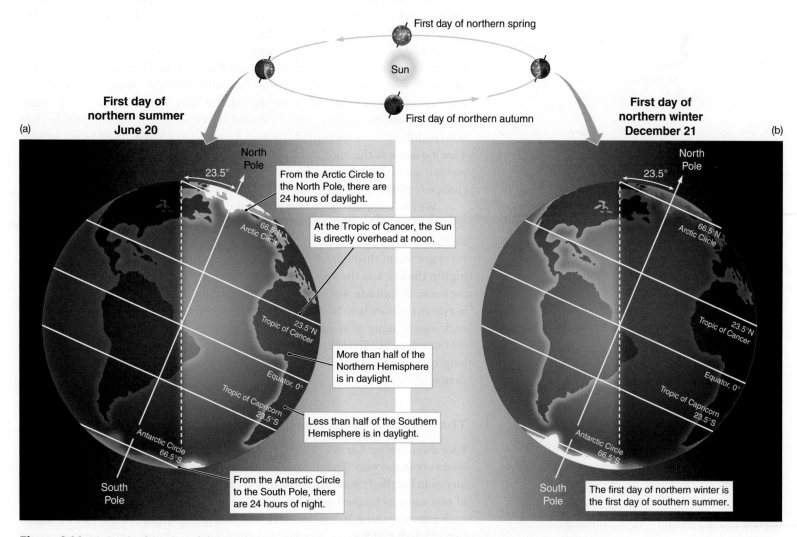

Figure 2.14 (a) On the first day of the northern summer (around June 20, the summer solstice), the northern end of Earth's axis is tilted most nearly toward the Sun, whereas the Southern Hemisphere is tipped away. (b) Six months later, on the first day of the northern winter (around December 21, the winter solstice), the situation is reversed. Seasons are opposite in the two hemispheres.

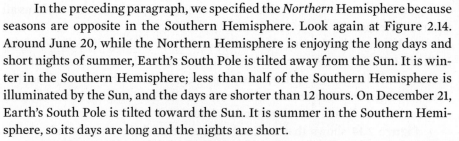

Astronomy in Action: The Cause of Earth's Seasons

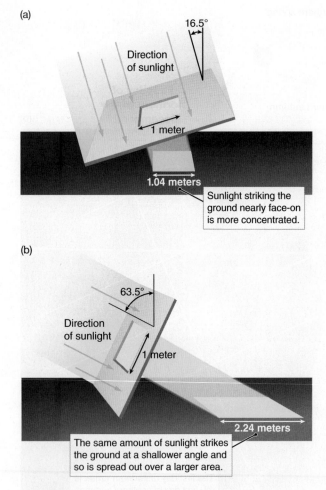

Figure 2.15 Local noon at latitude 40° north. (a) On the first day of northern summer, sunlight strikes the ground almost face-on. (b) On the first day of northern winter, sunlight strikes the ground more obliquely, and less than half as much sunlight falls on each square meter of ground each second.

In the preceding paragraph, we specified the *Northern* Hemisphere because seasons are opposite in the Southern Hemisphere. Look again at Figure 2.14. Around June 20, while the Northern Hemisphere is enjoying the long days and short nights of summer, Earth's South Pole is tilted away from the Sun. It is winter in the Southern Hemisphere; less than half of the Southern Hemisphere is illuminated by the Sun, and the days are shorter than 12 hours. On December 21, Earth's South Pole is tilted toward the Sun. It is summer in the Southern Hemisphere, so its days are long and the nights are short.

To understand how the combination of Earth's axial tilt and its path around the Sun creates seasons, consider a limiting case. If Earth's spin axis were exactly perpendicular to the plane of Earth's orbit (the **ecliptic plane**), the Sun would always be on the celestial equator. At every latitude (except the poles), the Sun would follow the same path through the sky every day, rising due east each morning and setting due west each evening. The Sun would be above the horizon exactly half the time, and days and nights would always be exactly 12 hours long everywhere on Earth. In short, each day would be just like the last, and Earth would have no seasons.

The changing length of days through the year only partly explains seasonal temperature changes. Another important effect relates to the angle at which the Sun's rays strike Earth. The Sun is higher in the sky during summer than during winter, and sunlight strikes the ground *more directly* during summer than during winter. To see why that is important, study **Figure 2.15**. During summer, Earth's surface is more nearly face-on to the incoming sunlight. More energy falls on each square meter of ground each second; the light is concentrated and bright. During winter, the surface is more tilted with respect to the sunlight, so the light is more diffuse. Less energy falls on each square meter of the ground each second. That variation is the main reason why summer is hotter and winter is colder. As you can see in the **Process of Science Figure**, determining the causes of seasonal change requires accounting for all the known facts.

We can compare the average temperatures at different latitudes on Earth to see the effect of the Sun's height in the sky. Near the equator, the Sun passes high overhead every day, regardless of the season. As a result, the average temperatures are warm throughout the year. At high latitudes, however, the Sun is *never* high in the sky, and the average temperatures can be cold even during summer. In between, at latitude 40° north, which stretches across the United States from northern California to New Jersey, more than twice as much solar energy falls on each square meter of ground per second at noon on June 20 as falls there at noon on December 21. Those two effects—the directness of sunlight and the different lengths of the night—mean that the Sun heats a hemisphere more during summer than winter.

The Solstices and the Equinoxes

Four days during Earth's orbit (two solstices and two equinoxes) mark unique moments in the year. The day when the Sun is highest in the sky as it crosses the meridian is called the **summer solstice**. On that day, the Sun rises farthest north of east and sets farthest north of west. In the Northern Hemisphere, that occurs each year near June 20, the first day of summer. Figure 2.14a shows that orientation of Earth and Sun.

Six months after the summer solstice, around December 21, the North Pole is tilted away from the Sun (Figure 2.14b). That day is the **winter solstice** in the Northern Hemisphere—the shortest day of the year and the first day of winter in

Theories Must Fit All the Known Facts

Many people misunderstand the phenomenon of changing seasons
because they do not account for all the relevant facts.

HYPOTHESIS 1

*We have seasons because Earth is closer to
the Sun in the summer and farther away in winter.*

THE TEST If this were true, both the Northern and Southern
Hemispheres would have summer in July. However, the Northern
and Southern Hemispheres experience opposite seasons.

THE CONCLUSION The hypothesis is falsified.

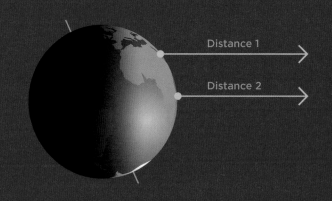

HYPOTHESIS 2

*We have seasons because the tilt of Earth's axis
causes one hemisphere to be significantly closer
to the Sun than the other.*

THE TEST If this were true, the distances would have to be
very different to cause such a large effect. Earth is tiny compared
to its distance from the Sun: The distances between each
hemisphere and the Sun differ by less than 0.004 percent.

THE CONCLUSION The hypothesis is falsified.

HYPOTHESIS 3

*We have seasons because Earth's tilt changes
the distribution of energy—one hemisphere
receives more intense light than the other.*

THE TEST If this were true, the amount of sunlight striking the
ground in the summer would have to be more than in the winter,
and the days would have to be longer in summer.

THE CONCLUSION Seasons are caused primarily by a change in
illumination due to Earth's tilt. During winter, less energy falls on
each square meter of ground per second.

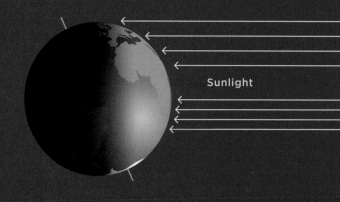

New information often challenges misconceptions.
Incorporating this new information may alter current theories
to improve the explanation of observed phenomena.

(a)

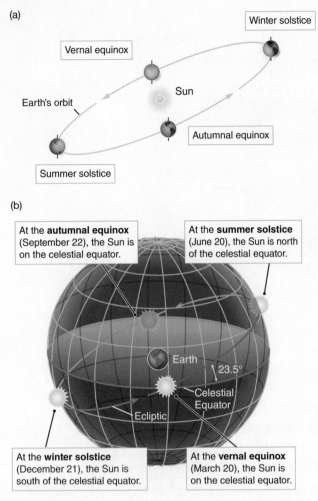

(b)

Figure 2.16 Earth's motion around the Sun from the frame of reference of (a) the Sun and (b) Earth.

Astronomy in Action: The Earth-Moon-Sun System

the Northern Hemisphere. Almost all cultural traditions in the Northern Hemisphere include some sort of major celebration in late December. Those winter festivals celebrate the return of the source of Earth's light and warmth. The days have stopped growing shorter and are beginning to get longer, and spring will come again.

Between the two solstices, the ecliptic crosses the celestial equator on two days. On those days, the Sun lies directly above Earth's equator. We call those days *equinoxes*, which means "equal night," because the entire Earth experiences 12 hours of daylight and 12 hours of darkness (except at the poles). Halfway between summer solstice and winter solstice, the **autumnal equinox** marks the beginning of fall in the Northern Hemisphere; it occurs around September 22. Halfway between winter solstice and summer solstice, the **vernal equinox** marks the beginning of spring in the Northern Hemisphere; it occurs around March 20. In the Southern Hemisphere, the dates for the summer and winter solstices, and the autumnal and vernal equinoxes, are reversed from those in the Northern Hemisphere.

Figure 2.16 shows the solstices and equinoxes from two perspectives. Figure 2.16a shows Earth in orbit around a stationary Sun, whereas Figure 2.16b shows the Sun's apparent motion along the celestial sphere, which is how it appears to observers on Earth. In both cases, we are looking at the plane of Earth's orbit from the side, so that it is shown in perspective and looks quite flattened. We also have tilted the images so that Earth's North Pole points straight up. The equinoxes correspond to points in the sky where the celestial equator meets the ecliptic. Practice shifting between those perspectives. You will know that you understand them when you can look at a position in either panel and predict the corresponding positions of the Sun and Earth in the other panel.

Just as a pot of water on a stove takes time to heat up when the burner is turned on and to cool when the burner is turned off, Earth takes time to respond to changes in heating from the Sun. The hottest months of northern summer are usually July and August, which come after the summer solstice, when days are growing shorter. Similarly, the coldest months of northern winter are usually January and February, which occur after the winter solstice, when days are growing longer. Temperature changes on Earth lag behind changes in the amount of heating we receive from the Sun.

That picture of the seasons must be modified somewhat near Earth's poles. At latitudes north of 66.5° north and south of 66.5° south, the Sun is circumpolar

Figure 2.17 This composite photo shows the midnight Sun, which is visible in latitudes above 66.5° north (or south). In the 360-degree panoramic view, the Sun moves 15° each hour.

for a part of the year surrounding the first day of summer. Those lines of latitude are the **Arctic Circle** and the **Antarctic Circle**, respectively (see Figure 2.14). When the Sun is circumpolar, it is above the horizon 24 hours per day, earning the polar regions the nickname "land of the midnight Sun" (**Figure 2.17**). An equally long period surrounds the first day of winter, when the Sun never rises and the nights are 24 hours long. The Sun never rises high in the Arctic or Antarctic sky, so the sunlight is never very direct. Even with the long days at the height of summer, the Arctic and Antarctic regions remain relatively cool.

In contrast, on the equator, *all* stars, including the Sun, are above the horizon approximately 12 hours per day. Days and nights there are 12 hours long throughout the year. The Sun passes directly overhead on the first day of spring and the first day of autumn because on those days the Sun is on the celestial equator. Sunlight is most direct, perpendicular to the ground, at the equator on those days. On the summer solstice, the Sun is at its northernmost point along the ecliptic. On that day, and on the winter solstice, the Sun is farthest from the zenith at noon, and therefore sunlight is least direct at the equator.

Latitude 23.5° north is called the Tropic of Cancer, and latitude 23.5° south is called the Tropic of Capricorn (Figure 2.14). The band between those latitudes is called the **Tropics**. If you live in the tropics—in Rio de Janeiro or Honolulu, for example—the Sun will be directly overhead at noon twice during the year.

Precession of the Equinoxes

Two thousand years ago, when Ptolemy and his associates were formalizing their knowledge of the positions and motions of objects in the sky, the Sun appeared in the constellation Cancer on the first day of northern summer and in the constellation Capricornus on the first day of northern winter. Today, the Sun is in Taurus on the first day of northern summer and in Sagittarius on the first day of northern winter. Why have the constellations in which solstices appear changed? Two motions are actually associated with Earth and its axis: Earth spins on its axis, but its axis also wobbles like the axis of a spinning top slowing down (**Figure 2.18a**). The wobble is very slow: the north celestial pole takes about 26,000 years to complete one trip around a large circle centered on the north ecliptic pole. Polaris is the star we now see near the north celestial pole. However, if you could travel several thousand years into the past or the future, the point about which the northern sky appears to rotate would no longer be near Polaris. Instead, the stars would rotate about another point on the path shown in Figure 2.18b. That figure shows the path of the north celestial pole through the sky during one cycle of that wobble.

The celestial equator is perpendicular to Earth's axis. Therefore, as Earth's axis wobbles, the celestial equator also must wobble. As it does so, the locations where the celestial equator crosses the ecliptic—the equinoxes—change as well. During each 26,000-year wobble of Earth's axis, the locations of the equinoxes make one complete circuit around the celestial equator. That change of the position of the equinox, due to the wobble of Earth's axis, is called the **precession of the equinoxes**.

CHECK YOUR UNDERSTANDING 2.2

If Earth's axis were tilted by 45°, instead of its actual tilt of 23.5°, how (if at all) would the seasons be different from what they are now? (a) The seasons would remain the same. (b) Summers would be colder. (c) Winters would be shorter. (d) Winters would be colder.

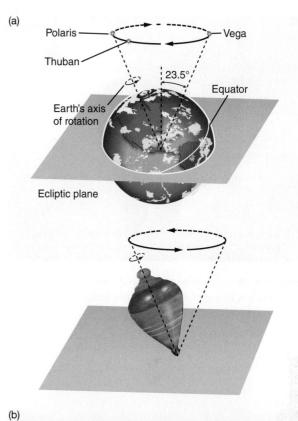

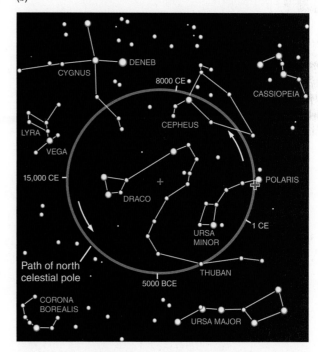

Figure 2.18 (a) Earth's axis of rotation changes orientation in the same way that the axis of a spinning top changes orientation. Polaris is the current North Pole star, Thuban was the North Pole star in 3000 BCE, and Vega will be the Pole star in 14,000 CE. (b) That precession causes the projection of Earth's rotation axis to move in a circle, centered on the north ecliptic pole (orange cross in the center). The red cross on the circle marks the projection of Earth's axis on the sky in the early 21st century.

Figure 2.19 An orange and a lamp can help you visualize the changing phases of the Moon.

2.3 The Moon's Appearance Changes as It Orbits Earth

After the Sun, the most prominent object in our sky is the Moon. Just as Earth orbits around the Sun, the Moon orbits around Earth once every 27.32 days. In this section, we discuss the phases of the Moon as seen from Earth.

The Changing Phases of the Moon

The Moon and its changing aspects have been the frequent subject of mythology, art, literature, and music. In mythology, the Moon was the Greek goddess Artemis, the Roman goddess Diana, and the Inuit god Igaluk. We speak of the "man in the Moon," the "harvest Moon," and sometimes a "blue Moon."

Unlike the Sun, the Moon has no light source of its own; instead, the Moon shines by reflected sunlight. As the Moon orbits Earth, our view of the illuminated portion of the Moon constantly changes. Those different appearances of the Moon are called **phases**. During a new Moon, when the Moon is between Earth and the Sun, the side facing away from us is illuminated, and during a full Moon, when Earth is between the Sun and the Moon, the side facing toward us is illuminated. The rest of the time, we can see only part of the illuminated portion from Earth. Sometimes the Moon appears as a circular disk in the sky. Other times it is nothing more than a thin sliver, or its face appears dark.

To help you visualize the changing phases of the Moon, use an orange, a lamp, and your head (**Figure 2.19**). Your head is Earth, the orange is the Moon, and the lamp is the Sun. Turn off all the other lights in the room, and step back as far from the lamp as you can. Hold up the orange slightly above your head so that the lamp illuminates it from one side. Move the orange clockwise around your head and watch how the appearance of the orange changes. When you are between the orange and the lamp, the face of the orange that is toward you is fully illuminated. The orange appears to be a bright, circular disk. As the orange moves around its circle, you will see a progression of lighted shapes, depending on how much of the bright side and how much of the dark side of the orange you can see. That progression of shapes mimics the changing phases of the Moon.

Figure 2.20 shows the changing phases of the Moon. The **new Moon** occurs when the Moon is between Earth and the Sun. The far side is illuminated, but the near side is in darkness and we cannot see it. The new Moon appears close to the Sun in the sky, so it is up in the daytime with the Sun: it rises in the east at sunrise, crosses the meridian near noon, and sets in the west near sunset. A new Moon is never above the horizon in the nighttime sky.

A few days after a new Moon, as the Moon orbits Earth, a sliver of its illuminated half, called a **waxing crescent Moon**, becomes visible. *Waxing* here means "growing in size and brilliance"; the name refers to the fact that the Moon appears to be "filling out" from night to night at that time. From our perspective, the Moon has also moved away from the Sun in the sky. Because the Moon travels around Earth in the same direction in which Earth rotates, we now see the Moon trailing the Sun, so the Moon is east of the Sun in the sky. A waxing crescent Moon is visible in the western sky in the evening, near the setting Sun but remaining above the horizon after the Sun sets. The "horns" of the crescent always point directly away from the Sun.

Astronomy in Action: Phases of the Moon

AstroTour: The Moon's Orbit: Eclipses and Phases

Simulation: Phases of the Moon

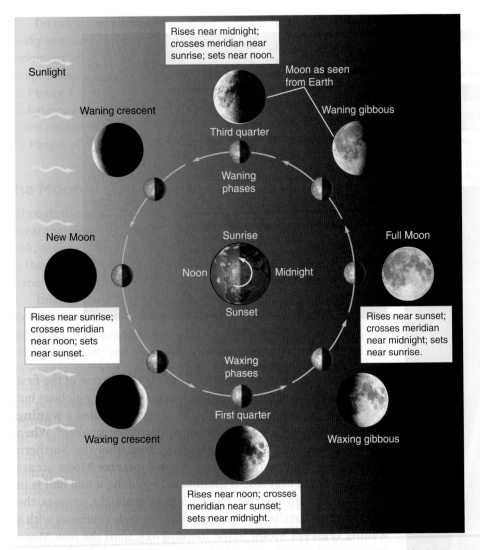

Rises near midnight; crosses meridian near sunrise; sets near noon.

Sunlight

Waning crescent

Moon as seen from Earth

Waning gibbous

Third quarter

Waning phases

New Moon

Sunrise

Noon   Midnight

Full Moon

Sunset

Rises near sunrise; crosses meridian near noon; sets near sunset.

Rises near sunset; crosses meridian near midnight; sets near sunrise.

Waxing phases

First quarter

Waxing crescent

Waxing gibbous

Rises near noon; crosses meridian near sunset; sets near midnight.

Figure 2.20 The inner circle of images (connected by blue arrows) shows the Moon as it orbits Earth, as seen by an observer far above Earth's North Pole. The Sun is on the left. The outer ring of images shows the corresponding phases of the Moon as seen from the northern hemisphere of Earth.

As the Moon moves farther along in its orbit, and the angle between the Sun and Moon grows larger, more and more of its near side becomes illuminated. About a week after the new Moon, half of the near side of the Moon is illuminated and half is in darkness. That phase is called a **first quarter Moon** because the Moon has moved a quarter of the way around Earth and has completed the first quarter of its cycle from new Moon to new Moon (Figure 2.20). The first quarter Moon rises at noon, crosses the meridian at sunset, and sets at midnight.

As the Moon moves beyond first quarter, more than half of its near side is illuminated. That phase is called a **waxing gibbous Moon**, from the Latin *gibbus*, meaning "hump." The waxing gibbous Moon continues nightly to "grow" until finally we see the entire near side of the Moon illuminated—a **full Moon**. Earth is now between the Sun and the Moon, which appear opposite each other in the sky when viewed from Earth. The full Moon rises as the Sun sets, crosses the meridian at midnight, and sets in the morning as the Sun rises.

3.1 Working It Out How Copernicus Computed Orbital Periods and Scaled the Solar System

Orbital Periods

Copernicus calculated the sidereal period by observing the synodic period. He didn't know the actual Earth–Sun distance in miles or kilometers, so he let the Earth–Sun distance equal 1. We call that distance 1 **astronomical unit** (**AU**). Let P be the sidereal period of a planet and S, its synodic period. E is Earth's sidereal period, which equals 1 year, or 365.25 days. By thinking about the distance that Earth and the planet move in one synodic period, and noting that an inferior planet orbits the Sun faster than Earth does, we can show that

$$\frac{1}{P} = \frac{1}{E} + \frac{1}{S}$$

for an inferior planet, with P, E, and S all in the same units of days or years. Similarly, Earth orbits the Sun faster than a superior planet does, so the planet has traveled only part of its orbit around the Sun after 1 Earth year. The equation for a superior planet is

$$\frac{1}{P} = \frac{1}{E} - \frac{1}{S}$$

For Saturn, the time that passes between oppositions—the date of maximum brightness—shows that Saturn's synodic period (S) is 378 days, or 378 ÷ 365.25 = 1.035 years. Then, to compute Saturn's sidereal period (P) in years, we use $S = 1.035$ yr and $E = 1$ yr in the equation for a superior planet:

$$\frac{1}{P} = \frac{1}{1\,\text{yr}} - \frac{1}{1.035\,\text{yr}} = 1 - 0.966\,\text{yr}^{-1} = 0.034\,\text{yr}^{-1}$$

Thus,

$$P = \frac{1}{0.034\,\text{yr}^{-1}} = 29.4\,\text{yr}$$

Saturn's sidereal period is 29.4 years, which means that Saturn takes 29.4 years to travel around the Sun and return to where it started in space.

Scaling the Solar System

Copernicus used the configurations of the planets shown in Figure 3.7 along with their sidereal periods to compute the relative distances of the planets. For the superior planets, he measured the fraction of the circular orbit that the planet completed in the time between opposition and quadrature, and then he used trigonometry to solve for the planet–Sun distance in astronomical units (see Figure 3.7a). For the inferior planets, he had a right triangle at the point of greatest elongation, and then he used right-triangle trigonometry to solve for the planet–Sun distance in astronomical units (see Figure 3.7b). Copernicus's values are impressively similar to modern values (see Table 3.1). Copernicus still did not know the actual value of the astronomical unit in miles, but he was the first to accurately compute the relative distances of the planets from the Sun.

3.2 Kepler's Laws Describe Planetary Motion

Copernicus did not understand *why* the planets move about the Sun, but he realized that his heliocentric picture offered a way to compute the planets' relative distances. His theory is an example of **empirical science**, which seeks to describe patterns in nature as accurately as possible. Copernicus's work was revolutionary because he challenged the accepted geocentric model and proposed that Earth is one planet among many. His conclusions paved the way for other great empiricists, including Tycho Brahe and Johannes Kepler.

Tycho Brahe's Observations

Tycho Brahe (1546–1601—**Figure 3.9**) was a Danish astronomer of noble birth who entered university at age 13 to study philosophy and law. After seeing a partial solar eclipse in 1560, Tycho (conventionally referred to by his first name) became interested in astronomy. A few years later, he observed Jupiter and Saturn near each other in the sky, though not exactly where one would expect

Figure 3.9
Tycho Brahe, known commonly as Tycho, was one of the greatest astronomical observers before the invention of the telescope.

from the astronomical tables based on Ptolemy's model. Tycho gave up studying law and devoted himself to making better tables of the planets' positions in the sky.

The king of Denmark granted Tycho the island of Hven, between Sweden and Denmark, to build an observatory. Tycho designed and built new instruments, operated a printing press, and taught students and others how to conduct observations. With the help of his sister Sophie, Tycho carefully measured the precise positions of planets in the sky over several decades, developing the most comprehensive set of planetary data then available. He created his own geo-heliocentric model, shown in **Figure 3.10**. In Tycho's model, the planets orbit the Sun, and the Sun and planets orbit Earth. His model gained some acceptance among people who preferred to keep Earth at the center for philosophical or religious reasons. Tycho lost his financial support when the king died, and in 1600 he relocated to Prague.

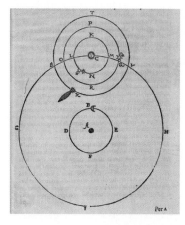

Figure 3.10
Tycho's geo-heliocentric model, showing the Moon and Sun orbiting Earth, with the other planets orbiting the Sun.

Kepler's Laws

In 1600, Tycho hired a more mathematically inclined astronomer, Johannes Kepler (1571–1630—**Figure 3.11**), as his assistant. Kepler, who had studied Copernicus's ideas, was responsible for the next major step toward understanding the planets' motions. Kepler inherited the records of Tycho's observations when he died. Working first with Tycho's observations of Mars, Kepler deduced three rules, now generally referred to as **Kepler's laws**, which accurately describe how the planets move. Kepler's laws are empirical: they use existing data to make predictions about future behavior but do not include an underlying theory of why the objects behave as they do.

Figure 3.11
Johannes Kepler explained the motions of the planets with three empirically determined laws.

KEPLER'S FIRST LAW Comparing Tycho's extensive planetary observations with predictions from Copernicus's heliocentric model, Kepler expected the data to confirm circular orbits for planets orbiting the Sun. Instead, he found disagreements between his predictions and Tycho's observations. He was not the first to notice such discrepancies. Rather than discard Copernicus's model, Kepler revised it.

By replacing Copernicus's circular orbits with *elliptical* orbits, Kepler could predict the positions of planets for any day, and his predictions fit Tycho's observations almost perfectly. Looking like an elongated circle, an **ellipse** is symmetric from right to left and from top to bottom. As shown in **Figure 3.12a**, you can draw an ellipse by attaching the two ends of a piece of string to a piece of paper, stretching the string tight with the tip of a pencil, and then drawing around those two points while keeping the string tight. Each point at which the string is attached is a **focus** (plural: **foci**) of the ellipse. Notice in **Figure 3.12b** that the ellipse becomes more circular as the two foci move closer together. As the two foci move farther apart, the ellipse becomes more and more elongated. The **eccentricity** (*e*) of an ellipse measures that elongation; it is determined by the separation between the

Figure 3.12 (a) We can draw an ellipse by attaching a length of string to a piece of paper at two points (called foci) and then pulling the string around as shown. The long axis is called the major axis. (b) Ellipses range from circles (e = 0) to elongated eccentric shapes. e = eccentricity.

(a)

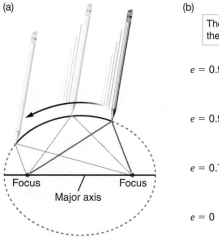

Focus Focus
Major axis

(b)

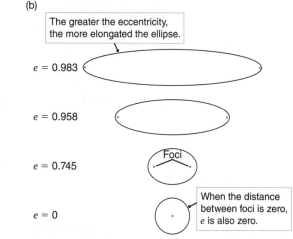

The greater the eccentricity, the more elongated the ellipse.

e = 0.983

e = 0.958

e = 0.745

Foci

e = 0

When the distance between foci is zero, e is also zero.

Figure 3.13 According to Kepler's first law, planets move in elliptical orbits with the Sun at one focus. (Nothing is at the other focus.) The orbit's eccentricity is the center-to-focus distance divided by the semimajor axis.

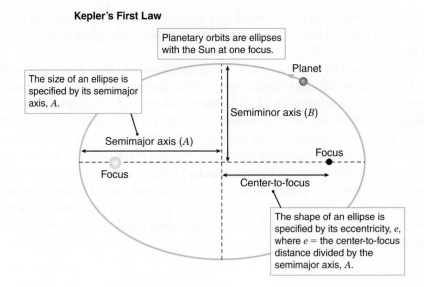

Kepler's First Law

Planetary orbits are ellipses with the Sun at one focus.

The size of an ellipse is specified by its semimajor axis, A.

Planet

Semiminor axis (B)

Semimajor axis (A)

Focus

Focus

Center-to-focus

The shape of an ellipse is specified by its eccentricity, e, where e = the center-to-focus distance divided by the semimajor axis, A.

AstroTour: Kepler's Laws

Simulation: Planetary Orbits

two foci divided by the length of the long axis, called the **major axis**. A circle has an eccentricity of 0 because the two foci coincide at the center. The more elongated the ellipse becomes, the closer its eccentricity gets to 1.

Kepler's first law of planetary motion states that the orbit of each planet is an ellipse with the Sun located at one focus (see the **Process of Science Figure**). Nothing but empty space is at the other focus. **Figure 3.13** illustrates Kepler's first law and shows the features of an ellipse that can be measured. The dashed lines in Figure 3.13 represent the major (long) and minor (short) axes of the ellipse. Half the length of the major axis of the ellipse is called the **semimajor axis**, A. The average distance between the Sun and a planet equals the semimajor axis of the planet's orbit.

The eccentricities of planetary orbits vary widely, but most planetary objects in our Solar System have nearly circular orbits. As **Figure 3.14a** shows, Earth's orbit, with an eccentricity of 0.017, is very nearly a circle centered on the Sun; therefore the distance variation is small. In contrast, dwarf planet Pluto's orbit, depicted in **Figure 3.14b**, has an eccentricity of 0.249. The orbit is noticeably elongated, with the Sun offset from center.

KEPLER'S SECOND LAW   By analyzing Tycho's observational data of changes in the planets' positions, Kepler found that a planet moves fastest when closest to the Sun and slowest when farthest from the Sun. For example, we now measure Earth's average speed in its orbit around the Sun at 29.8 kilometers per second

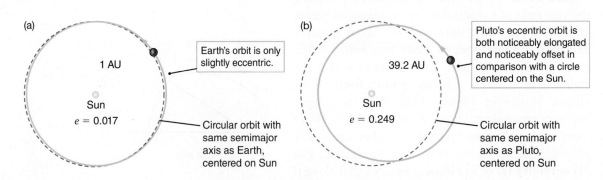

(a)

1 AU

Sun

$e = 0.017$

Earth's orbit is only slightly eccentric.

Circular orbit with same semimajor axis as Earth, centered on Sun

(b)

39.2 AU

Sun

$e = 0.249$

Pluto's eccentric orbit is both noticeably elongated and noticeably offset in comparison with a circle centered on the Sun.

Circular orbit with same semimajor axis as Pluto, centered on Sun

Figure 3.14 The orbits of (a) Earth and (b) Pluto in comparison with circles around the Sun. e = eccentricity.

Theories Are Falsifiable

Early astronomers studied the motions of the planets,
but did not understand why the planets behaved as they do.

Copernicus
1473–1543

The Hypothesis
Copernicus proposed that planets moved in
circular orbits (with epicycles) around the Sun.

Tycho
1546–1601

The Observation
Tycho, born three years after Copernicus died, observed and
collected enormous amounts of data about planet positions.

Kepler
1571–1630

The Prediction
Kepler, 25 years younger than Tycho, used Copernicus's model
and Tycho's data to predict where the planets should be.

The Test
When Kepler compared the predictions with actual data,
they disagreed. Copernicus's idea was falsified!

The New Hypothesis
Planet orbits are not circular. They are elliptical.

Newton
1643–1726

This new hypothesis gained acceptance a century later
when Newton showed that planetary orbits are an
inevitable consequence of gravitational attraction.

In order for a theory to be scientific, it must be falsifiable, even if the test
can't be carried out until decades or centuries later. Disproving an old theory is part of
the self-correcting nature of science: It always leads to deeper understanding.

Kepler's Second Law

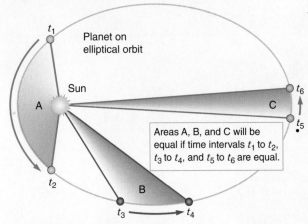

Figure 3.15 An imaginary line between a planet and the Sun sweeps out an area as the planet orbits. According to Kepler's second law, if the three time intervals shown are equal, the three areas A, B, and C will be the same.

(km/s). When closest to the Sun, Earth travels at 30.3 km/s. At its farthest from the Sun, Earth travels at 29.3 km/s.

Kepler found a way to elegantly describe the changing speed of a planet in its orbit around the Sun. **Figure 3.15** shows a planet at six points along the ellipse of its orbit (t_1 to t_6). Imagine a straight line connecting the Sun with that planet. We can think of the line as "sweeping out" an area as it moves with the planet from one point to another. Area A (in orange) is swept out between times t_1 and t_2, area B (in blue) is swept out between times t_3 and t_4, and area C (in green) is swept out between times t_5 and t_6. When the planet is closest to the Sun (area A), it is moving the fastest, but the distance between the planet and the Sun is small. Kepler realized that changes in the distance from the Sun to a planet and changes in a planet's speed work together so that the area a planet sweeps out in the same amount of time is always the same, regardless of where the planet is in its orbit. Therefore, if the three time intervals in the figure are equal (that is, $t_1{\rightarrow}t_2 = t_3{\rightarrow}t_4 = t_5{\rightarrow}t_6$), the three areas A, B, and C must be equal as well.

Kepler's second law, also called Kepler's **law of equal areas**, states that the imaginary line connecting a planet to the Sun sweeps out equal areas in equal times, regardless of where the planet is along the ellipse of its orbit. That law applies to only one planet at a time. The area swept out by Earth in a given time interval is always the same. Likewise, the area swept out by Mars in a given time is always the same. But the area swept out by Earth and the area swept out by Mars in a given time are *not* the same. That law can be used to find the speed of a planet anywhere in its orbit.

KEPLER'S THIRD LAW Kepler looked for patterns in the planets' orbital periods. Compared with planets closer to the Sun, he found that planets farther from the Sun have longer orbits and orbit the Sun more slowly. Kepler discovered a mathematical relationship between a planet's sidereal period—how many years the planet takes to go around the Sun and return to the same position in the Solar System—and its average distance from the Sun in astronomical units. **Kepler's third law** states that in those units, the square of the sidereal

---

3.2 Working It Out Kepler's Third Law

Kepler's third law states that the square of the period of a planet's orbit measured in years, P_{years}, is equal to the cube of the semimajor axis of the planet's orbit measured in astronomical units, A_{AU}. As an equation, the law says

$$(P_{\text{years}})^2 = (A_{\text{AU}})^3$$

For mathematical convenience, Kepler used units based on Earth's orbit—astronomical units and years. If other units were used, such as kilometers and days, P^2 would still be proportional to A^3, but the constant of proportionality would not be 1.

To calculate the average size of Neptune's orbit in astronomical units, you first need to find out how long Neptune's period is in

Earth years, which you can determine by observing the synodic period and from that computing its sidereal period (using Working It Out 3.1). Neptune's sidereal period is 165 years. Plugging that number into Kepler's third law gives the following result:

$$(P_{\text{years}})^2 = (165)^2 = 27{,}225 = (A_{\text{AU}})^3$$

To solve that equation, you must first square 165 to get 27,225 and then take its cube root (see Appendix 1 for calculator hints). Then

$$A_{\text{AU}} = \sqrt[3]{27{,}225} = 30.1$$

The semimajor axis of Neptune's orbit—that is, the average distance between Neptune and the Sun—is 30.1 AU.

TABLE 3.2 Kepler's Third Law: $P^2 = A^3$

Planet	Period P (years)	Semimajor Axis A (AU)	$\dfrac{P^2}{A^3}$
Mercury	0.241	0.387	$\dfrac{0.241^2}{0.387^3} = 1.00$
Venus	0.615	0.723	$\dfrac{0.615^2}{0.723^3} = 1.00$
Earth	1.000	1.000	$\dfrac{1.000^2}{1.000^3} = 1.00$
Mars	1.881	1.524	$\dfrac{1.881^2}{1.524^3} = 1.00$
Ceres	4.599	2.765	$\dfrac{4.599^2}{2.765^3} = 1.00$
Jupiter	11.86	5.204	$\dfrac{11.86^2}{5.204^3} = 1.00$
Saturn	29.46	9.582	$\dfrac{29.46^2}{9.582^3} = 0.99*$
Uranus	84.01	19.201	$\dfrac{84.01^2}{19.201^3} = 1.00$
Neptune	164.79	30.047	$\dfrac{164.79^2}{30.047^3} = 1.00$
Pluto	247.92	39.482	$\dfrac{247.92^2}{39.482^3} = 1.00$
Eris	557.00	67.696	$\dfrac{557.00^2}{67.696^3} = 1.00$

*Ratios are not exactly 1.00 because of slight perturbations from the gravity of other planets.

period (P) of a planet's orbit is equal to the cube of the semimajor axis (A) of the planet's orbit.

Table 3.2 lists the periods and semimajor axes of the orbits of the eight classical planets and three of the dwarf planets, along with the values of the ratio P^2 divided by A^3. Those data are also plotted in **Figure 3.16**. Kepler referred to that relationship as his **harmonic law** or, more poetically, as the "Harmony of the Worlds." Kepler's third law is explored further in **Working It Out 3.2**. Kepler's laws enhanced the heliocentric mathematical model of Copernicus and led to its greater acceptance.

CHECK YOUR UNDERSTANDING 3.2

Order the following from largest to smallest semimajor axis: (a) a planet with a period of 84 Earth days; (b) a planet with a period of 1 Earth year; (c) a planet with a period of 2 Earth years; (d) a planet with a period of 0.5 Earth years.

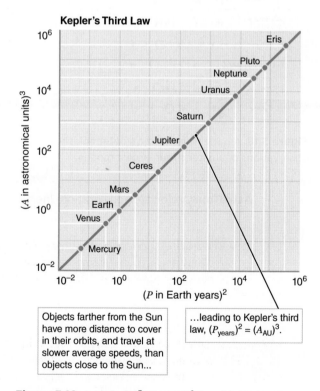

Kepler's Third Law

Objects farther from the Sun have more distance to cover in their orbits, and travel at slower average speeds, than objects close to the Sun...

...leading to Kepler's third law, $(P_{\text{years}})^2 = (A_{\text{AU}})^3$.

Figure 3.16 A plot of A^3 versus P^2 for objects in our Solar System shows that they obey Kepler's third law. (By plotting powers of 10 on each axis, we can fit both large and small values on the same plot. We will use that approach often.)

Figure 3.17
Galileo Galilei laid the physical framework for Newton's laws.

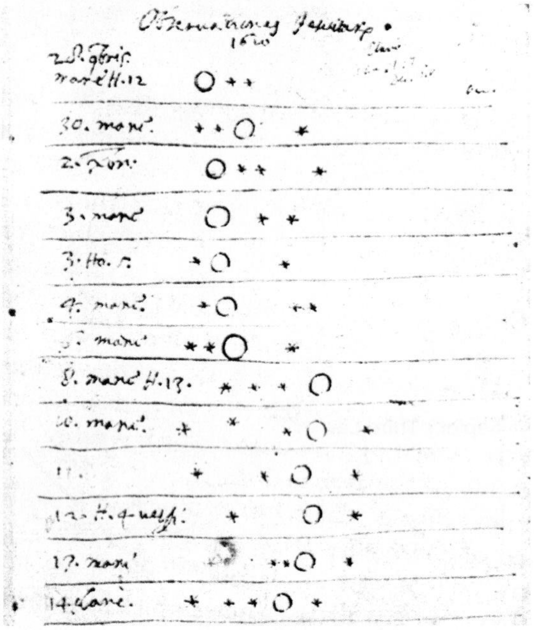

Figure 3.18 A page from Galileo's notebook shows his observations of Jupiter's four largest moons.

AstroTour: Velocity, Acceleration, Inertia

3.3 Galileo's Observations Supported the Heliocentric Model

Galileo Galilei (1564–1642—**Figure 3.17**) was the first to use a telescope to make and report significant discoveries about astronomical objects. Galileo's telescopes were relatively small, yet sufficient for him to observe spots on the Sun, the uneven surface and craters of the Moon, and the many stars in the band of light in the sky called the Milky Way.

Galileo's Observations

Galileo provided the first observational evidence that some objects in the sky do not orbit Earth. When Galileo turned his telescope on Jupiter, he observed four "stars" in a line near the planet. Over time he observed that the objects changed position from night to night (**Figure 3.18**). Galileo hypothesized that those objects were moons orbiting Jupiter. Those four moons are the largest of Jupiter's many moons and are still called the Galilean moons. Galileo also estimated the relative distance of each moon from Jupiter and the periods of their orbits, and he showed that they followed Kepler's third law.

Galileo also observed that Venus went through phases like the Moon. He noticed that the phases of Venus were correlated with how large Venus appeared in his telescope. In a geocentric model in which Venus orbits Earth like the Moon does, the apparent size of Venus would change only slightly, when it looped an epicycle. In the heliocentric model, however, the Earth–Venus distance varies by a lot, and Venus's size changes accordingly. When Venus is in its gibbous to full phases, it is farther away, on the other side of the Sun from Earth, and smaller in the sky. When Venus is in its crescent to new phases, it is closer, on the same side of the Sun as Earth, and larger in the sky (**Figure 3.19**). His observations of Jupiter's moons and the phases of Venus in particular convinced Galileo that Copernicus was correct to place the Sun at the center of the Solar System.

In addition to his astronomical observations, Galileo did important work on the motion of objects. Unlike natural philosophers, who thought about objects in motion but did not actually experiment with them, Galileo conducted experiments with falling and rolling objects. As with his telescopes, Galileo improved or developed new technology to enable him to conduct those experiments. For example, by carefully rolling balls down an inclined plane and by dropping various objects from a height, he found that a falling object travels a distance proportional to the square of the time it has been falling. If he simultaneously dropped two objects of different masses, they reached the ground at the same time, showing that all objects falling to Earth accelerate at the same rate, independent of their mass.

Galileo's observations and experiments with many types of moving objects, such as carts and balls, led him to disagree with the Greek philosophers about when and why objects continue to move or come to rest. Before Galileo, it was thought that an object's natural state was to be at rest. But he found that an object naturally does what it was doing until a force acts on it. That is, an object in motion continues moving along a straight line with a constant speed until a force acts on it to change that state of motion. That idea of *inertia*, which Newton later adopted as his first law of motion, has implications for not only the motion of carts and balls but also the orbits of planets.

Dialogue Concerning the Two Chief World Systems

Galileo faced considerable danger because of his work. His later life was consumed by conflict with the Catholic Church over his support of the Copernican system. In 1632, Galileo published his best-selling book, *Dialogo sopra i due massimi sistemi del mondo* ("Dialogue Concerning the Two Chief World Systems"). The *Dialogo* presents a brilliant philosopher named Salviati as the champion of the Copernican heliocentric view of the universe. The defender of an Earth-centered universe, Simplicio—who uses arguments made by the classical Greek philosophers and the pope—sounds silly and ignorant.

Galileo, a religious man with two daughters in a convent, thought he had the Catholic Church's tacit approval for his book. But when he placed several of the pope's geocentric arguments in the unflattering mouth of Simplicio, the perceived attack on the pope got the church's attention. Galileo was put on trial for heresy, sentenced to prison, and eventually placed under house arrest. To escape a harsher sentence, Galileo was forced to publicly recant his belief in the Copernican theory that Earth moves around the Sun. According to one story, as he left the courtroom, Galileo stamped his foot on the ground and muttered, "And yet it moves!"

The *Dialogo* was placed on the pope's Index of Prohibited Books, along with Copernicus's *De Revolutionibus*. Yet Galileo's work traveled across Europe, was translated into other languages, and was read by other scientists. Galileo spent his final years compiling his research on inertia and other ideas into the book *Discourses and Mathematical Demonstrations Relating to Two New Sciences*, which was published in 1638 in Holland, outside the Catholic Church's jurisdiction. (In 1992, Pope John Paul II apologized for the "Galileo Case.")

Figure 3.19 Modern photographs of the phases of Venus show that when we see Venus more illuminated, it also appears smaller, implying that Venus is farther away then.

CHECK YOUR UNDERSTANDING 3.3

Which of Galileo's astronomical observations did Copernicus's model explain better than Ptolemy's? (a) sunspots; (b) craters on the Moon; (c) moons of Jupiter; (d) apparent size and phases of Venus

3.4 Newton's Three Laws Help Explain How Celestial Bodies Move

Empirical laws, such as Kepler's laws, describe what happens, but they do not explain why. Kepler described the orbits of planets as ellipses, but he did not explain why they should be so. To take that next step in the scientific process, scientists use basic physical principles and the tools of mathematics to derive the empirically determined laws. Alternatively, a scientist might start with physical laws and predict relationships, which are then verified or falsified through experiment and observation. If those predictions are verified, the scientist may have determined something fundamental about how the universe works.

Sir Isaac Newton (1642–1727—**Figure 3.20**) took that next step in explaining the nature of motion. Newton was a student of mathematics at Cambridge University when it closed because of the Great Plague and students were sent home to the safer countryside. Over the next 2 years, he studied on his own, and at the age of 23 he invented calculus, which would become crucial to his development of the physics of motion. (The German mathematician Gottfried Leibniz independently developed calculus around the same time.)

Figure 3.20
Sir Isaac Newton formulated three laws of motion.

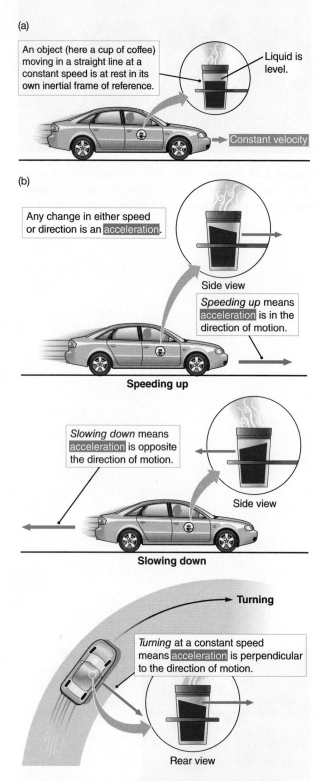

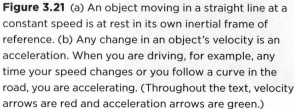

Figure 3.21 (a) An object moving in a straight line at a constant speed is at rest in its own inertial frame of reference. (b) Any change in an object's velocity is an acceleration. When you are driving, for example, any time your speed changes or you follow a curve in the road, you are accelerating. (Throughout the text, velocity arrows are red and acceleration arrows are green.)

Building on the work of Kepler, Galileo, and others, Newton proposed three physical laws that govern the motions of all objects in the sky and on Earth. To understand how the planets and all other celestial bodies move, you must understand these three laws.

Newton's First Law: Objects at Rest Stay at Rest; Objects in Motion Stay in Motion

A **force** (**F**) is a push or a pull on an object. Two or more forces can oppose one another such that they are perfectly balanced and cancel out. For example, gravity pulls down on you as you sit in your chair, but the chair pushes up on you with an exactly equal and opposite force. As a result, you stay motionless. Forces that cancel out do not affect an object's motion. When forces combine to produce an effect, we often use the term *net force*, or sometimes just *force*.

Imagine that you are driving a car, and your book is on the seat next to you. A rabbit runs across the road in front of you, and you hit the brakes hard. You feel the seat belt tighten to restrain you. At the same time, your book flies off the seat and hits the dashboard. You have just experienced what Newton describes in his first law of motion. **Inertia** is an object's tendency to maintain its state—either of uniform motion or of rest—until a net force pushes or pulls it. In the stopping car, you did not hit the dashboard because the force of the seat belt slowed you down. The book, however, hit the dashboard because no such force acted upon it.

Newton's first law of motion describes inertia and states that an object in motion tends to stay in motion, in the same direction, until a net force acts upon it; and an object at rest tends to stay at rest until a net force acts upon it. Galileo's law of inertia became the cornerstone of physics as Newton's first law.

Recall from Section 2.1 the concept of a frame of reference. Within a frame of reference, only the relative motions between objects have any meaning. Without external clues, you cannot tell the difference between sitting still and traveling at constant speed in a straight line. For example, if you close your eyes while riding in the passenger seat of a quiet car on a smooth road, you would feel as though you were sitting still. In the earlier example in this chapter, your book was "at rest" beside you on the front seat of your car, but a person standing by the side of the road would see the book moving past at the same speed as the car. People in a car approaching you would see the book moving quite fast—at the speed they are traveling plus the speed you are traveling! All those perspectives are equally valid, and all those speeds of the book are correct when measured in the appropriate reference frame.

A reference frame moving in a straight line at a constant speed is called an **inertial frame of reference**. Any inertial frame of reference is as good as another. In the inertial frame of reference of a cup of coffee, for example, **Figure 3.21a** shows that the cup is at rest in its own frame even if the car is moving quickly down the road.

Newton's Second Law: Motion Is Changed by Forces

What if a net force does act? In the earlier example, you were traveling in the car, and your motion slowed when the force of the seat belt acted upon you. Forces change an object's motion—by changing either the speed or the direction. That effect reflects **Newton's second law of motion**: if a net force acts on an object, the object's motion changes.

In the driver's seat of a car, you have several controls, including an accelerator and a brake pedal, which you use to make the car speed up or slow down. A *change in speed* is one way the car's motion can change. But you also have the steering wheel in your hands. When you are moving down the road and you turn the wheel, your speed does not necessarily change, but the direction of your motion does. A *change in direction* also is a kind of change in motion.

Together, an object's speed and direction are called **velocity** (**v**). "Traveling at 50 kilometers per hour (km/h)" indicates speed; "traveling north at 50 km/h" indicates velocity. **Acceleration** (**a**) describes the changes in an object's velocity. For example, if you go from 0 to 100 km/h in 4 seconds, you feel a strong push from the seat back as it shoves your body forward, causing you to accelerate along with the car. However, if you take 2 minutes to go from 0 to 100 km/h, the acceleration is so slight that you hardly notice it.

Partly because a car's gas pedal is often called the accelerator, some people think *acceleration* always means that an object is speeding up. But in physics, *any change in speed or direction is an acceleration*. **Figure 3.21b** illustrates that point by showing what happens to the coffee in a cup as the car speeds up, slows down, or turns. Slamming on your brakes and going from 100 to 0 km/h in 4 seconds is just as much an acceleration as going from 0 to 100 km/h in 4 seconds. Similarly, the acceleration you experience as you go through a fast, tight turn at a constant speed is every bit as real as the acceleration you feel when you slam your foot on the accelerator or the brake. Whether speeding up, slowing down, turning left, or turning right—if you are not moving in a straight line at a constant speed, you are accelerating.

Newton's second law of motion says that a net force causes acceleration. An object's acceleration depends on two things. First, as shown in **Figure 3.22**, the acceleration depends on the strength of the net force acting on the object to change its motion. If the forces acting on the object do *not* add up to zero, a net force is present and the object accelerates (Figure 3.22a). The stronger the net force, the greater the acceleration (Figure 3.22b). Push on something twice as hard and it experiences twice as much acceleration. Push on something 3 times as hard and its acceleration is 3 times as great. The resulting change in motion occurs in the direction the net force points. Push an object away from you, and it will accelerate away from you.

An object's acceleration also depends on its inertia. You can push some objects easily, such as the empty box from a new refrigerator. But you can't easily shove the actual refrigerator around, however, even though it is about the same size as the box. Figure 3.22c shows that the greater the mass, the greater the inertia, and the *less* acceleration that will occur in response to the same net force. That relationship among acceleration, force, and mass is expressed mathematically in **Working It Out 3.3**.

Newton's Third Law: Whatever Gets Pushed Pushes Back

Imagine that you are standing on a skateboard and pushing yourself along with your foot. Each shove of your foot against the ground sends you faster along your way. But why? You accelerate because as you push on the ground, the ground pushes back on you.

Part of Newton's genius was his ability to see patterns in such everyday events. Newton realized that *every* time one object exerts a force on another, the second object exerts a matching force on the first. That second force is as strong as the first force but is in the opposite direction. When you accelerate yourself on the skateboard, you push backward on Earth, and Earth pushes you forward. As

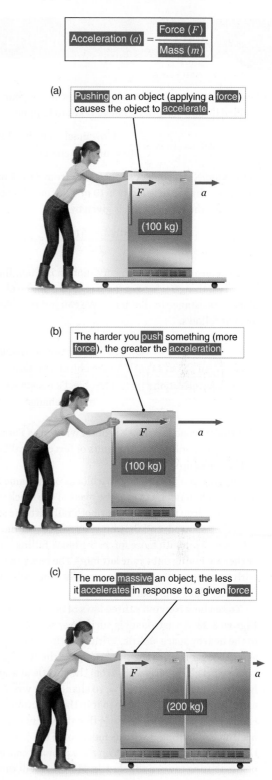

$$\text{Acceleration } (a) = \frac{\text{Force } (F)}{\text{Mass } (m)}$$

(a) Pushing on an object (applying a force) causes the object to accelerate.

F a

(100 kg)

(b) The harder you push something (more force), the greater the acceleration.

F a

(100 kg)

(c) The more massive an object, the less it accelerates in response to a given force.

F a

(200 kg)

Figure 3.22 According to Newton's second law of motion, an object's acceleration is the force acting on the object divided by the object's mass. (Throughout the text, force arrows are blue.)

3.3 Working It Out Using Newton's Laws

Your acceleration is determined by how much your velocity changes, divided by how long that change takes to happen:

$$\text{Acceleration} = \frac{\text{How much velocity changes}}{\text{How long the change takes to happen}}$$

For example, if an object's speed goes from 5 to 15 meters per second (m/s), the change in velocity is 10 m/s. If that change happens in 2 seconds, the acceleration is given by

$$a = \frac{15\,\text{m/s} - 5\,\text{m/s}}{2\,\text{s}} = 5\,\text{m/s}^2$$

To determine how an object's motion is changing, we need to know two things: what net force is acting on the object and the object's resistance to that force. We can put that idea into equation form as follows:

$$\left(\begin{array}{c}\text{An object's}\\\text{Acceleration}\end{array}\right) = \frac{\begin{array}{c}\text{The force acting to change}\\\text{the object's motion}\end{array}}{\begin{array}{c}\text{The object's resistance}\\\text{to that change}\end{array}} = \frac{\text{Force}}{\text{Mass}}$$

Newton's second law above is often written as Force = mass × acceleration, or $F = ma$. The units of force are called **newtons** (**N**), so $1\,\text{N} = 1\,\text{kg m/s}^2$.

Suppose you are holding two blocks of the same size, the block in your right hand has twice the mass of the block in your left hand. When you drop the blocks, they both fall with the same acceleration, as Galileo showed, and they hit your two feet at the same time. Which will hit with more force: the block falling onto your right foot or the one falling onto your left foot? The block in your right hand, with twice the mass, will hit your right foot with twice the force that the other block hits your left foot.

To see how Newton's three laws of motion work together, study **Figure 3.23**. An astronaut is adrift in space, motionless with respect to the nearby space shuttle. With no tether to pull on, how can the astronaut get back to the ship? Suppose the 100-kg astronaut throws a 1-kg wrench directly away from the shuttle at a speed of 10 m/s. Newton's second law says that to change the wrench's motion, the astronaut must apply a force to it in the direction away from the shuttle. Newton's third law says that the wrench must therefore push back on the astronaut with as much force but in the opposite direction. The force of the wrench on the astronaut causes the astronaut to begin drifting toward the shuttle. How fast will the astronaut move? Turn to Newton's second law again. Because the astronaut has more mass, he will accelerate less than the wrench will. A force that causes the 1-kg wrench to accelerate to 10 m/s will have much less effect

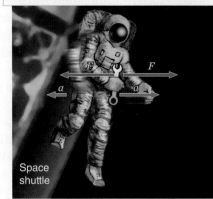

An astronaut adrift in space pushes on a wrench, which, according to Newton's third law, pushes back on the astronaut.

Space shuttle

While in contact with each other, the wrench and the astronaut experience accelerations proportional to the inverse of their masses…

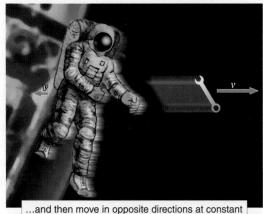

…and then move in opposite directions at constant velocities, in accordance with Newton's first law.

Figure 3.23 According to Newton's laws, if an astronaut adrift in space throws a wrench, the two will move in opposite directions. Their speeds will depend on their masses: the same force will produce a smaller acceleration of a more massive object than of a less massive object. (Acceleration and velocity arrows not drawn to scale.)

on the 100-kg astronaut. Because acceleration equals force divided by mass, the 100-kg astronaut will experience only 1/100 as much acceleration as the 1-kg wrench. The astronaut will drift toward the shuttle, but only at the leisurely rate of 1/100 × 10 m/s, or 0.1 m/s.

shown in **Figure 3.24**, when a woman moves a load on a cart by pulling a rope, the rope pulls back, and when a car tire pushes back on the road, the road pushes forward on the tire. Similarly, when Earth pulls on the Moon, the Moon pulls on Earth, and when a rocket engine pushes hot gases out of its nozzle, those hot gases push back on the rocket, propelling it into space.

All those force pairs are examples of **Newton's third law of motion**, which says that forces always come in pairs, with those two forces always equal in strength but opposite in direction. The forces in those pairs always act on two objects. Your weight pushes down on the floor, and the floor pushes back up on your feet with the same amount of force. For every force, an equal force in the opposite direction is *always* there.

CHECK YOUR UNDERSTANDING 3.4

Imagine a planet moving in a perfectly circular orbit around the Sun. Is that planet accelerating? (a) Yes, because it is constantly changing its speed. (b) Yes, because it is constantly changing its direction. (c) No, because its speed is not constantly changing. (d) No, because planets do not accelerate.

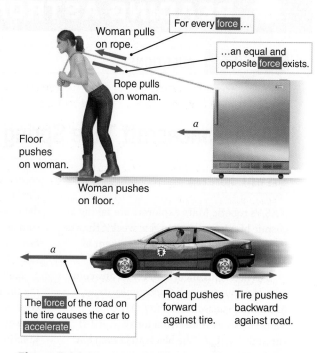

Figure 3.24 Newton's third law states that for every force, an equal and opposite force is always present. Those opposing forces always act on the two objects in the same pair.

Astronomy in Action: Velocity, Force, and Acceleration

AstroTour: Velocity, Acceleration, Inertia

Origins Planets and Orbits

In addition to the planets in our own Solar System, researchers have detected thousands of planets orbiting stars other than our own Sun. Those planets' orbits can be calculated and understood by applying the same three Kepler's laws that we have discussed here. Astrobiologists think that a planet's orbit around its star affects the chances of life developing.

Consider the average distance of the planet from its star. A planet close to its star receives more energy than a planet farther out. If Earth were closer to the Sun, it would be hotter throughout the year—probably so hot that water would evaporate and no longer exist as a liquid. If Earth were farther from the Sun, it would be colder and all surface water would probably freeze. Liquid water was crucial for life to form on Earth, so some astronomers look for planets at a distance from their star such that liquid water can exist (that distance varies depending on the star's temperature and size).

Next consider the eccentricity of a planet's orbit. Recall from Figure 3.14a that Earth's orbit differs from a circle by less than 2 percent. Thus, Earth's distance from the Sun varies by only about 5 million km throughout the year; and as we saw in Chapter 2, seasons change on Earth because of the tilt of Earth's axis, not because of slight changes in distance from the Sun. Mars, however, has about the same axial tilt as Earth. But because the orbit of Mars is more eccentric, Mars has greater seasonal variation. The distance between Mars and the Sun varies from 1.38 AU (207 million km) to 1.67 AU (249 million km)—an eccentricity of 9 percent (or ~40 million km). As a result, the seasons on Mars are unequal. They are shorter when Mars is closer to the Sun and moving faster, and longer when Mars is farther from the Sun and moving slower. The inequality of the martian seasons affects the overall stability of the planet's temperature and climate. When we look at planets orbiting other stars, we see that many have orbital eccentricities even higher than that of Mars and therefore large variations in temperature.

Earth is at the right distance from the Sun to have temperatures that permit water to be liquid, and its orbital eccentricity is low enough that the average planetary temperature does not change much during its orbit. Those orbital characteristics have contributed to making the conditions on Earth suitable for life to develop.

NASA Spacecraft Take Spring Break at Mars

MIKE WALL, Space.com

NASA's robotic Mars explorers are taking a cosmic break for the next few weeks, thanks to an unfavorable planetary alignment of Mars, the Earth, and the Sun.

Mission controllers won't send any commands to the agency's *Opportunity* rover, *Mars Reconnaissance Orbiter* (*MRO*), or *Mars Odyssey* orbiter from today (April 9) through April 26. The blackout is even longer for NASA's car-size *Curiosity* rover, which is slated to go solo from April 4 through May 1.

The cause of the communications moratorium is a phenomenon called a Mars solar conjunction, during which the Sun comes between Earth and the Red Planet (**Figure 3.25**). Our star can disrupt and degrade interplanetary signals in this formation, so mission teams won't be taking any chances.

"Receiving a partial command could confuse the spacecraft, putting them in grave danger," NASA officials explain in a video posted last month by the agency's Jet Propulsion Laboratory (JPL) in Pasadena, California.

Opportunity and *Curiosity* will continue performing stationary science work, using commands already beamed to the rovers. *Curiosity* will focus on gathering weather data, assessing the martian radiation environment, and searching for signs of subsurface water and hydrated minerals, officials said Monday (April 8).

MRO and *Odyssey* will also keep studying the Red Planet from above, and they'll continue to serve as communications links between the rovers and Earth. The conjunction will also affect the European Space Agency's *Mars Express* orbiter, officials have said.

Odyssey will send rover data home as usual during conjunction, though the orbiter may have to relay information multiple times due to dropouts. *MRO*, on the other hand, entered record-only mode on April 4. The spacecraft will probably have about 52 gigabits of data to relay when it's ready to start transmitting again on May 1, *MRO* officials have said.

Mars solar conjunctions occur every 26 months, so NASA's Red Planet veterans have dealt with them before. This is the fifth conjunction for *Opportunity*, in fact, and the sixth for *Odyssey*, which began orbiting Mars in 2001.

But it'll be the first for *Curiosity*, which touched down on August 5, kicking off a

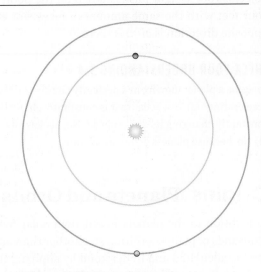

Figure 3.25 Every 26 months, Mars and Earth are on opposite sides of the Sun, preventing communication between the planets.

2-year surface mission to determine if the Red Planet has ever been capable of supporting microbial life.

"The biggest difference for this 2013 conjunction is having *Curiosity* on Mars," *Odyssey* mission manager Chris Potts, of NASA's Jet Propulsion Laboratory in Pasadena, California, said in a statement last month.

1. How often do Mars solar conjunctions occur? Does that interval correspond to the sidereal period or to the synodic period of Mars?
2. Give two reasons that conjunction is a bad time to view Mars in the sky.
3. Make a sketch of the Mars solar conjunction, showing Mars, Earth, and the Sun. Using the orbital period data in Table 3.2, add the positions of Mars and Earth at the next Mars conjunction. Will it take place at the same location in space?
4. View this short video from NASA: https://www.jpl.nasa.gov/video/details.php?id=1204. Do you think it explains the concepts well to someone who is not studying astronomy?
5. If people were on a mission on Mars, losing contact would be troubling. How might NASA plan to avoid losing contact with astronauts on Mars during conjunctions?

Summary

Early astronomers hypothesized that Earth was stationary at the center of the Solar System. Later astronomers realized that a Sun-centered Solar System was much simpler and could explain what people observed in the sky. Planets, such as Jupiter, orbit the Sun, not Earth. Kepler's laws describe the planets' elliptical orbits around the Sun, including details about how fast a planet travels at various points in its orbit. Those laws helped Newton develop his laws of motion, which govern how all objects move (not just orbiting ones). Orbital semimajor axis, eccentricity, and stability may affect a planet's suitability to foster life.

LG 1 **Compare the geocentric and heliocentric models of the Solar System.** Earth's motion is hard to detect, so before the Copernican Revolution, most people accepted a geocentric model of the Solar System, in which all objects orbit Earth. But in that model, the planets' apparent retrograde motion was hard to understand. Copernicus created the first comprehensive mathematical model of the Solar System with the Sun at the center, called a heliocentric model. His model explained apparent retrograde motion as a visual illusion seen when an inner planet passes an outer planet in their orbits.

LG 2 **Use Kepler's laws to describe how objects in the Solar System move.** Using Tycho's observational data, Kepler developed empirical rules to describe the motions of the planets. Kepler's three laws state that (1) planets move in elliptical orbits around the Sun; (2) planets move fastest when closest to the Sun and slowest when farthest from the Sun, so that the planets sweep out equal areas in equal times; and (3) the orbital period (P) of a planet squared equals the semimajor axis (A) of its orbit cubed: $P^2 = A^3$.

LG 3 **Explain how Galileo's astronomical discoveries convinced him that the heliocentric model was correct.** Galileo invented astronomical telescopes and used them to observe sunspots, craters on the Moon, and moons orbiting Jupiter. He also saw Venus going through phases, like the Moon, but changing its apparent size in each phase. The observations of Venus were hard to explain with Ptolemy's model.

LG 4 **Describe the physical laws of motion discovered by Galileo and Newton.** Galileo studied the physics of falling objects and discovered the principle of inertia. Newton's laws state that (1) objects do not change their motion unless they experience a net force, (2) force = mass × acceleration, and (3) every force has an equal and opposite force. Net forces cause accelerations (changes in motion). Inertia resists changes in motion.

Unanswered Questions

- Would the history of scientific discoveries in physics and astronomy have been different if the Inquisition had not prosecuted Galileo? Galileo wrote the *Dialogo* after the Catholic Church in 1616 ordered him not to "hold or defend" the idea that Earth moves and the Sun is still. And he wrote his equally famous *Discorsi e Dimostrazioni Matematiche* (often shortened in English to "Two New Sciences") while under house arrest after his trial. However undeterred Galileo appeared to be, the effects of the decrees, prohibitions, and prosecutions might have dissuaded other scientists in Catholic countries from pursuing that type of work. Indeed, after Galileo's experiences, the center of the scientific revolution moved north to Protestant Europe.

- What percentage of planets are in unstable orbits? In younger planetary systems, planets might migrate in their orbits because of the presence of other massive planets nearby. We explain in Chapter 7 that Uranus and Neptune might have migrated in that way. Some planets have been discovered moving through the galaxy without any obvious orbit around a star—and therefore are not in the stable orbits we see in our own Solar System.

Questions and Problems

Test Your Understanding

1. An *empirical science* is one based on
   a. hypothesis.
   b. calculus.
   c. computer models.
   d. observed data.

2. When Earth catches up to a slower-moving outer planet and passes it in its orbit, the outer planet
   a. exhibits retrograde motion.
   b. slows down because it feels Earth's gravitational pull.
   c. becomes dimmer as it passes through Earth's shadow.
   d. moves into a more elliptical orbit.

3. Copernicus's model of the Solar System was superior to Ptolemy's because
   a. it had a mathematical basis and computed the spacing of the planets.
   b. it was much more accurate.
   c. it did not require epicycles.
   d. it fit the telescopic data better.

4. An inferior planet is one that is
   a. smaller than Earth.
   b. larger than Earth.
   c. closer to the Sun than Earth.
   d. farther from the Sun than Earth.

5. The time a planet takes to come back to the same position in space in relation to the Sun is called its _____ period.
   a. synodic
   b. sidereal
   c. heliocentric
   d. geocentric

6. Suppose a planet is discovered orbiting a star in a highly elliptical orbit. While the planet is close to the star, it moves _____, but while it is far away, it moves _____.
   a. faster; slower
   b. slower; faster
   c. retrograde; prograde
   d. prograde; retrograde

7. If a superior planet is observed from Earth to have a synodic period of 1.2 years, what is its sidereal period?
   a. 0.54 years
   b. 1.8 years
   c. 4.0 years
   d. 6.0 years

8. A net force must be acting when an object
   a. accelerates.
   b. changes direction but not speed.
   c. changes speed but not direction.
   d. all of the above

9. For Earth, $P^2/A^3 = 1.0$ (in appropriate units). Suppose a new dwarf planet is discovered that is 14 times farther from the Sun than Earth is. For that planet,
   a. $P^2/A^3 = 1.0$.
   b. $P^2/A^3 > 1.0$.
   c. $P^2/A^3 < 1.0$.
   d. we can't know the value of P^2/A^3 without more information.

10. Galileo observed that Venus had phases that correlated with the size of its image in his telescope. From that information, you may conclude that Venus
    a. is the center of the Solar System.
    b. orbits the Sun.
    c. orbits Earth.
    d. orbits the Moon.

11. Kepler's second law says that
    a. planetary orbits are ellipses with the Sun at one focus.
    b. the square of a planet's orbital period equals the cube of its semimajor axis.
    c. net forces cause changes in motion.
    d. planets move fastest when closest to the Sun.

12. The average speed of a new planet in orbit has been reported to be 33 km/s. When it is closest to its star it moves at 31 km/s, and when it is farthest from its star it moves at 35 km/s. Those data must be in error because
    a. the average speed is far too fast.
    b. Kepler's third law says the planet has to sweep out equal areas in equal times, so the speed of the planet cannot change.
    c. Kepler's second law says the planet must move fastest when closest to its star, not when farthest away.
    d. with those numbers, the square of the orbital period will not be equal to the cube of the semimajor axis.

13. Galileo observed that Jupiter has moons. From that information, you may conclude that
    a. Jupiter is the center of the Solar System.
    b. Jupiter orbits the Sun.
    c. Jupiter orbits Earth.
    d. some things do not orbit Earth.

14. If you start from rest and accelerate at 10 mph/s and end up traveling at 60 mph, how long did it take?
    a. 1 second
    b. 6 seconds
    c. 60 seconds
    d. 0.6 seconds

15. Planets with high eccentricity may be unlikely candidates for life because
    a. the speed varies too much.
    b. the period varies too much.
    c. the temperature varies too much.
    d. the orbit varies too much.

Thinking about the Concepts

16. Study Figure 3.2. During normal motion, does Mars move toward the east or west? Which direction does it travel when moving retrogradely? For how many days did Mars move retrogradely? If one of the martian missions were photographing *Earth* in the sky during those days, what would it have observed?

17. Copernicus and Kepler engaged in what is called empirical science. What do we mean by *empirical*?

18. Explain why Saturn's synodic period is very close to 1 Earth year (a sketch may help).

19. Make a sketch of Earth, Venus, and the Sun in a geocentric model and in a heliocentric model. Label the Sun and Earth, and then show the changing phases of Venus during one of its orbits. What would we observe in each model? Why was the invention of the telescope necessary to distinguish between those models?

20. Experiment with falling objects as Galileo did. When you drop pairs of objects with different masses, do they reach the ground at the same time? Do they hit the ground with the same force? Does that approach work with a sheet of paper or a tissue—why or why not?

21. A planet's orbital speed around the Sun varies. When is the planet moving fastest in its orbit? Slowest?

22. The Moon's orbit around Earth is elliptical, too, with an eccentricity of 0.05. How does that value compare with the eccentricity of Earth's orbit? How do those elliptical orbits explain the types of solar eclipses discussed in Chapter 2?

23. Galileo came up with the concept of inertia. What do we mean by *inertia*?

24. If Kepler had lived on Mars, would he have deduced the same empirical laws for the motion of the planets? Explain.

25. What is the difference between speed and velocity? Between velocity and acceleration?

26. When involved in an automobile collision, a person not wearing a seat belt will move through the car and often strike the windshield directly. Which of Newton's laws explains why the person continues forward, even though the car stopped?

27. When riding in a car, we can sense changes in speed or direction through the forces that the car applies on us. Do we wear seat belts in cars and airplanes to protect us from speed or from acceleration? Explain your answer.

28. An astronaut standing on Earth can easily lift a wrench having a mass of 1 kg, but not a scientific instrument with a mass of 100 kg. On the International Space Station, he can manipulate both, although the scientific instrument responds much more slowly than the wrench. Explain why.

29. The Process of Science Figure illustrates that scientific ideas are always open to challenge. Construct an argument that the constant process of challenging and falsifying ideas is a strength of science, rather than a weakness.

30. How might you expect conditions on Earth to be different if the eccentricity of its orbit were 0.17 instead of 0.017?

Applying the Concepts

31. Study the graph in Figure 3.16. Is that graph linear or logarithmic? From the data on the graph, find the approximate semimajor axis and period of Saturn. Show your work.

32. Study Figure 3.19, which shows that Venus's apparent size changes as it goes through phases. Approximately how many times larger is Venus in the sky at the thinnest crescent than at the gibbous phase shown? Approximately how many times closer is Venus to us at the phase of that thinnest crescent than at the gibbous phase?

33. Suppose a dwarf planet is discovered orbiting the Sun with a semimajor axis of 50 AU. What would be the planet's orbital period?

34. The orbital period of Uranus is 84 years. Compute the semimajor axis of its orbit. How much time passes between oppositions of Uranus?

35. Dwarf planet Ceres is 2.77 AU from the Sun. Its synodic period is 1.278 years.
    a. Use Working It Out 3.1 to find the sidereal period in years.
    b. Use Kepler's law to find the sidereal period in years.
    c. Compare your results for (a) and (b).

36. Suppose you read online that "experts have discovered a new planet with a distance from the Sun of 2 AU and a period of 3 years." Use Kepler's third law to argue that this is impossible.

37. Show, as Galileo did, that Kepler's third law applies to the four Galilean moons of Jupiter by calculating P^2 divided by A^3 for each moon. (Data on the moons can be found in Appendix 4.)

38. In a period of 3 months, a planet travels 30,000 km with an average speed of 3.8 m/s. Some time later, the same planet travels 65,000 km in 3 months. How fast is the planet traveling at that later time? During which period is the planet closer to the Sun?

39. If you were on Mars, how often would you see retrograde motion of *Earth* in the martian night sky? (You can view a picture at https://mars.jpl.nasa.gov/allaboutmars/nightsky/retrograde.)

40. The elliptical orbit of a comet that a spacecraft recently visited is 1.24 AU from the Sun at its closest approach and 5.68 AU from the Sun at its farthest.
    a. Sketch the comet's orbit. When is it moving fastest? Slowest?
    b. What is the semimajor axis of its orbit? How long does the comet take to go around the Sun?
    c. What is the distance from the Sun to the "center" of the ellipse? What is the eccentricity of the comet's orbit?

41. You are driving down a straight road at a speed of 90 km/h, and you see another car approaching you at a speed of 110 km/h along the road.
    a. With respect to your own frame of reference, how fast is the other car approaching you?
    b. With respect to the other driver's frame of reference, how fast are you approaching the other driver's car?

42. During the latter half of the 19th century, a few astronomers thought a planet might be circling the Sun inside Mercury's orbit. They even gave it a name: Vulcan. We now know that Vulcan does not exist. If a planet with an orbit one-fourth the size of Mercury's actually existed, what would its orbital period be in comparison with that of Mercury?

43. Suppose you are pushing a small refrigerator of mass 50 kg on wheels. You push with a force of 100 N.
    a. What is the refrigerator's acceleration?
    b. Assume the refrigerator starts at rest. How long will the refrigerator accelerate at that rate before it gets away from you (that is, before it is moving faster than you can run—on the order of 10 m/s)?

44. If a 100-kg astronaut pushes on a 5,000-kg satellite and the satellite accelerates at 0.1 m/s^2, what acceleration does the astronaut experience in the opposite direction?

45. Sketch the orbit of Mars by using the information provided in the "Origins: Planets and Orbits" section of the chapter for the closest and farthest distances of Mars from the Sun.
    a. What is the orbit's major axis? Semimajor axis?
    b. What is the distance from the "center" of the orbit to the Sun? Compute the orbit's eccentricity. Compare that value with the eccentricity of Earth's orbit.

USING THE WEB

46. Go to the Web page "This Week's Sky at a Glance" (https://www.skyandtelescope.com/observing/sky-at-a-glance/) at the *Sky & Telescope* magazine website. Which planets are visible in your sky this week? Why are Mercury and Venus visible only in the morning before sunrise or in the evening just after sunset? Before telescopes, how did people know the planets were different from the stars?

47. Look up the dates for the next opposition of Mars, Jupiter, or Saturn. One source is the NASA "Sky Events Calendar" (https://eclipse.gsfc.nasa.gov/SKYCAL/SKYCAL.html). Check only the "Planet Events" box in "Section 2: Sky Events"; and in Section 3, generate a Sky Events Calendar or Table for the year. If you are coming up on an opposition (Opp), take pictures of the planet over the next few weeks. Can you see its position move in retrograde motion with respect to the background stars?

48. Refer to the Web page from question 47 to find the current observational positions of all the planets.
    a. Which ones are in or near to conjunction, opposition, or greatest elongation?
    b. Which are visible in the morning sky? In the evening sky?
    c. To connect the time we can see the planets on Earth with their physical alignments with Earth and the Sun in space, sketch the Solar System with Earth, the Sun, and the planets as it looks today from "above." Check your result with NASA's "Solar System Simulator" (https://space.jpl.nasa.gov): set it for "Show Me Solar System" as seen from above, and set the field of view at 2°, then 20°, and then 45° to see the inner and then the outer planets. Does your sketch agree with the simulator?

49. Go to the Museo Galileo website and view the Room VII video and exhibit at https://catalogue.museogalileo.it/room/RoomVII.html. What did his telescope look like? What other instruments did he use? From the museum page you can link to short videos (in English) on his science and his trial (http://catalogue.museogalileo.it/index/VideoIndexByThematicArea.html#s7). For example, click on "Galileo's micrometer." How did he measure the separation of the moons from Jupiter? How did that measurement allow him to show that the moons obeyed Kepler's third law? Why is Galileo often considered the first modern scientist? Why is his middle finger on display in the museum?

50. Go to the online "Extrasolar Planets Encyclopedia" (http://exoplanet.eu/catalog).
    a. Find a planet with an orbital period similar to that of Earth (click on "Period" twice and it will show decreasing values in days). What is the semimajor axis of the exoplanet's orbit? If it is very different from 1 AU, the star's mass is different from the Sun's. Click on the star name in the first column to see the star's mass. What is the planet's orbital eccentricity?
    b. Click on "Planet" to sort by name, and select a star with multiple planets. Verify that Kepler's third law applies by showing that the value of P^2/A^3 is about the same for each planet of that star. How eccentric are the orbits of the multiple planets?

digital.wwnorton.com/astro6

In this Exploration, we examine how Kepler's laws apply to the orbit of Mercury. Visit the Digital Resources Page and on the Student Site open the "Planetary Orbits" Interactive Simulation in Chapter 3. That simulator animates the orbits of the planets, enabling you to control the simulation speed, as well as several other parameters. Here we focus on exploring Mercury's orbit, but you may wish to spend some time examining the orbits of other planets as well.

Kepler's First Law

In the "Orbit Settings" panel, use the drop-down menu next to "set parameters for" to select "Mercury" and then click "OK." Click the "Kepler's 1st Law" tab at the bottom of the control panel. Use the radio buttons to select "show empty focus" and "show center."

1   How would you describe the shape of Mercury's orbit?

_____

_____

Deselect "show empty focus" and "show center," and select "show semiminor axis" and "show semimajor axis." Under "Visualization Options," select "show grid."

2   Use the grid markings to estimate the ratio of the semiminor axis to the semimajor axis.

_____

_____

3   Calculate the eccentricity of Mercury's orbit from that ratio by using $e = [1 - (\text{Ratio})^2]^{1/2}$.

_____

_____

Kepler's Second Law

Click on "reset" near the top of the control panel, set parameters for Mercury, and click "OK." Then click on the "Kepler's 2nd Law" tab at the bottom of the control panel. Slide the "adjust size" slider to the right until the fractional sweep size is 1/8.

Click on "start sweeping." The planet moves around its orbit, and the simulation fills in area until one-eighth of the ellipse is filled. Click on "start sweeping" again as the planet arrives at the rightmost point in its orbit (that is, at the point in its orbit farthest from the Sun). You may need to slow the animation rate by using

the slider under "Animation Controls." Click on "show grid" under the visualization options. (If the moving planet annoys you, pause the animation.) One easy way to estimate an area is to count the squares.

4   Count the squares in the yellow area and in the red area. You will need to decide what to do with fractional squares. Are the areas the same? Should they be?

_____

_____

Kepler's Third Law

Click on "reset" near the top of the control panel, set parameters for Mercury, and then click on the "Kepler's 3rd Law" tab at the bottom of the control panel. Select "show solar system orbits" in the "Visualization Options" panel. Study the graph. Use the eccentricity slider to change the simulated planet's eccentricity. Make the eccentricity first smaller and then larger.

5   Did anything in the graph change?

_____

6   What do your observations of the graph tell you about the dependence of the period on the eccentricity?

_____

Set the parameters back to those for Mercury. Now use the semi-major axis slider to change the semimajor axis of the simulated planet.

7   What happens to the period when you make the semimajor axis smaller?

_____

_____

8   What happens when you make it larger?

_____

_____

9   What do those results tell you about the dependence of the period on the semimajor axis?

_____

_____

4 Gravity and Orbits

Here in Chapter 4, we explore the physical laws that explain the regular patterns in the motions of the planets. Because the Sun is far more massive than all the other parts of the Solar System combined, its gravity shapes the motions of every object in its vicinity, from the almost circular orbits of some planets to the extremely elongated orbits of comets.

. . . .

LEARNING GOALS

By the end of this chapter, you should be able to:

LG 1 Explain the elements of Newton's universal law of gravitation.

LG 2 Use the laws of motion and gravitation to explain planetary orbits.

LG 3 Explain how tidal forces from the Sun and Moon create Earth's tides.

LG 4 Describe how tidal forces affect solid bodies.

The International Space Station in orbit around Earth. ▶ ▶ ▶

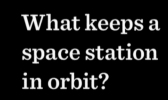

What keeps a
space station
in orbit?

interstellar cloud of gas and dust to collapse 4.5 billion years ago to form the Sun, Earth, and the rest of the Solar System. Gravity binds colossal groups of stars into galaxies. Gravity shapes space and time, and it can affect the ultimate fate of the universe. We return often to the concept of gravity because it is central to an understanding of the universe.

CHECK YOUR UNDERSTANDING 4.1A

If the distance between Earth and the Sun were cut in half, the gravitational force between those two objects would: (a) decrease by a factor of 4; (b) decrease by a factor of 2; (c) increase by a factor of 2; (d) increase by a factor of 4.

Gravity Differs from Place to Place within an Object

You can think of Earth as a collection of small masses, each of which feels a gravitational attraction toward every other small part of Earth. The mutual gravitational attraction that occurs among all parts of the same object is called *self-gravity*.

As you sit reading this book, you are exerting a gravitational attraction on every other fragment of Earth, and every other fragment of Earth is exerting a gravitational attraction on you. Your gravitational interaction is strongest with the parts of Earth closest to you. The parts of Earth on the other side of our planet are much farther from you, so their pull on you is correspondingly less.

The net effect of all those forces is to pull you (or any other object) toward Earth's center. If you drop a hammer, it falls directly toward the ground. Because Earth is nearly spherical, for every piece of Earth pulling you toward your right, a corresponding piece of Earth is pulling you toward your left with just as much force. For every piece of Earth pulling you forward, a corresponding piece of Earth is pulling you backward. Because Earth is almost spherically symmetric, all those "sideways" forces (see the blue arrows in **Figure 4.4a**) cancel out, leaving behind an overall force that points toward Earth's center (**Figure 4.4b**).

Some parts of Earth are closer to you and others are farther away, but an average distance exists between you and all the small fragments of Earth pulling on you. That average distance is the distance between you and the center of Earth. As illustrated in Figure 4.4b, the overall pull that you feel is the same as it would be if all the mass of Earth were concentrated at a single point located at the very center of the planet.

That relationship is true for any spherically symmetric object. Outside the object, its gravity behaves as though all the mass of that object were concentrated at a point at its center. That relationship is important in many applications. For example, when you estimate your weight on another planet, you are calculating the force of gravity between you and the planet. The "distance" in the gravitational equation will be the distance between you and the center of the planet, which is just the radius of the planet plus your altitude.

CHECK YOUR UNDERSTANDING 4.1B

If Earth shrank to a smaller radius but kept the same mass, would the gravitational force between Earth and the Moon: (a) decrease; (b) increase; or (c) stay the same? Would everyone's weight at Earth's surface: (a) increase; (b) decrease; or (c) stay the same?

(a)

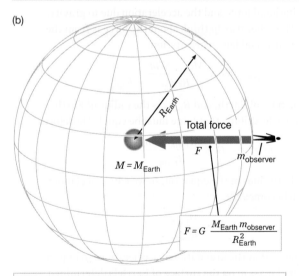

Center of Earth
m

$m_{observer}$

Total mass = M_{Earth}

An object on the surface of a spherical mass (such as Earth) feels a gravitational attraction toward each small part of the sphere.

(b)

R_{Earth}

Total force

F

$m_{observer}$

$M = M_{Earth}$

$$F = G \frac{M_{Earth}\, m_{observer}}{R_{Earth}^2}$$

The net force is the same as if we scooped up the mass of the entire sphere and concentrated it at a point at the center.

Figure 4.4 Outside a sphere, the net gravitational force due to a spherical mass is the same as the gravitational force from the same mass concentrated at a point at the center of the sphere.

4.2 An Orbit Is One Body "Falling Around" Another

Kepler's laws on the motions of planets enabled astronomers to predict the positions of the planets accurately, but those laws did not explain why the planets behave as they do. Newton's work explained why planets orbit the Sun.

Newton used his laws of motion and his proposed law of gravity to predict the paths of planetary orbits. His calculations showed that those orbits should be ellipses with the Sun at one focus, that planets should move faster when closer to the Sun, and that the square of the period of a planet's orbit should vary as the cube of the semimajor axis of that elliptical orbit. Newton's universal law of gravitation *predicted* that planets should orbit the Sun in just the way that Kepler's empirical laws described. By explaining Kepler's laws, Newton found important corroboration for his law of gravitation.

Gravity and Orbits

Newton's laws tell us how forces change an object's motion and how objects interact with one another through gravity. To know where an object will be at any given time, we have to "add up" the object's motion over time. Newton invented calculus to do that, but here we aim just for a conceptual understanding.

In Newton's time, the closest thing to making a heavy object fly was shooting cannonballs out of a cannon, so he used cannonballs in "thought experiments" about planetary motions. If you drop a cannonball, it falls directly to the ground, like any other mass does. A cannonball fired out of a cannon that is level with the ground behaves differently, however, as shown in **Figure 4.5a**. The cannonball still falls to the ground in the same amount of time as it does when it is dropped, but while falling it also travels over the ground, following a curved path that carries it a horizontal distance before it finally lands. The faster the cannonball moves when it is fired from the cannon, the farther it will go before it hits the ground (**Figure 4.5b**).

In the real world, that experiment reaches a natural limit. To travel through air, the cannonball must push air out of its way—an effect normally referred to as *air resistance*—which slows it down. But we can ignore such real-world complications in this thought experiment. Instead imagine that, having inertia, the cannonball continues along its course until it runs into something. The faster the cannonball moves when it is fired, the farther it goes before hitting the ground. If the cannonball flies far enough, Earth's surface curves out from under it, as shown in **Figure 4.5c**. Eventually a point is reached (**Figure 4.5d**) at which the cannonball is flying so fast that the surface of Earth curves away from the cannonball at the same rate at which the cannonball is falling toward Earth. When that occurs, the cannonball, which always falls toward the center of Earth, is, in a sense, falling around the world. An **orbit** is the path of one object that freely falls around another.

Why do astronauts appear to float freely about the cabin of a spacecraft? It is not because they have escaped Earth's gravity; Earth's gravity is what holds them in their orbit. Instead, the answer lies in Galileo's early observation that all objects fall with the same acceleration, regardless of their mass. The astronauts and the spacecraft are both in orbit around Earth, moving in the same direction, at the same speed, and experiencing the same gravitational acceleration, so they fall around Earth together. **Figure 4.6** demonstrates that point. The astronaut is orbiting Earth just as the spacecraft is orbiting Earth. On the surface of Earth,

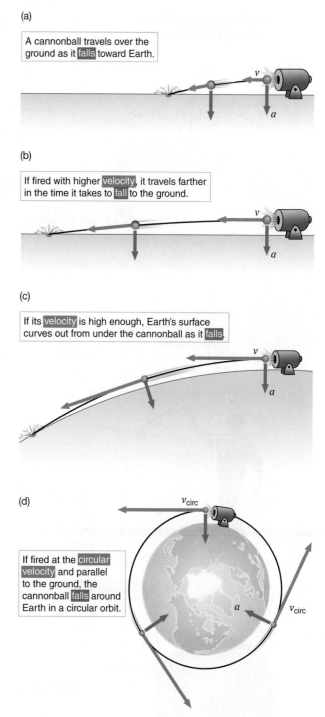

(a) A cannonball travels over the ground as it **falls** toward Earth.

(b) If fired with higher **velocity**, it travels farther in the time it takes to **fall** to the ground.

(c) If its **velocity** is high enough, Earth's surface curves out from under the cannonball as it **falls**.

(d) If fired at the **circular velocity** and parallel to the ground, the cannonball **falls** around Earth in a circular orbit.

Figure 4.5 Newton realized that a cannonball fired at the right speed would fall around Earth in a circle. Velocity (*v*) is indicated by a red arrow and acceleration (*a*), by a green arrow.

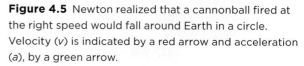

AstroTour: Newton's Laws and Universal Gravitation

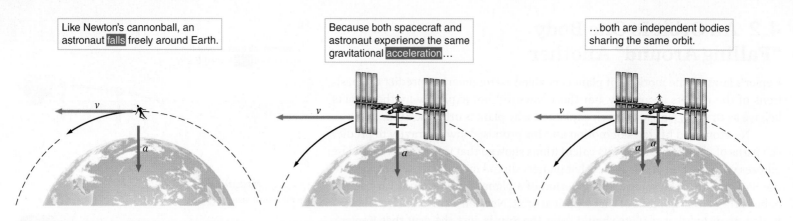

Like Newton's cannonball, an astronaut falls freely around Earth.

Because both spacecraft and astronaut experience the same gravitational acceleration...

...both are independent bodies sharing the same orbit.

Figure 4.6 A "weightless" astronaut has not escaped Earth's gravity. Rather, an astronaut and a spacecraft share the same orbit as they fall around Earth together.

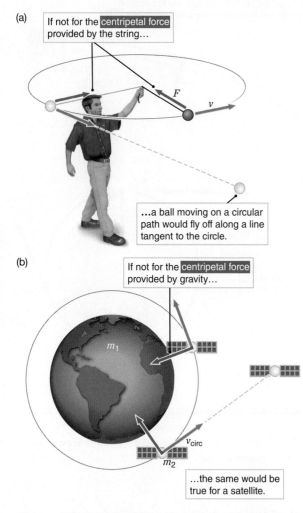

(a) If not for the centripetal force provided by the string...

...a ball moving on a circular path would fly off along a line tangent to the circle.

(b) If not for the centripetal force provided by gravity...

m_1

m_2

v_{circ}

...the same would be true for a satellite.

Figure 4.7 (a) A string provides the centripetal force that keeps a ball moving in a circle. (We are ignoring the smaller force of gravity that also acts on the ball.) (b) Gravity provides the centripetal force that holds a satellite in a circular orbit about Earth.

your body tries to fall toward the center of Earth, but the ground gets in the way. You feel your weight when you are standing on Earth because the ground pushes on you to oppose the force of gravity, which pulls you downward. In the spacecraft, however, nothing interrupts the astronaut's fall because the spacecraft is falling around Earth in the same orbit as the astronaut. The astronaut is in **free fall**, falling freely in Earth's gravity. The **Process of Science Figure** illustrates the universality of Newton's law of gravitation.

What Velocity Is Needed to Reach Orbit?

How fast must Newton's cannonball be fired for it to fall around the world? The cannonball would be in **uniform circular motion**, meaning it moves along a circular path at constant speed. That type of motion is discussed in more depth in Appendix 8. Another example of uniform circular motion is a ball whirling around your head on a string, as shown in **Figure 4.7a**. If you let go of the string, the ball will fly off in a straight line in whatever direction it is traveling at the time, just as Newton's first law predicts for an object in motion. The string prevents the ball from flying off by constantly changing the direction the ball is traveling. The central force of the string on the ball is called a **centripetal force**: a force toward the center of the circle. Using a more massive ball, speeding up its motion, and making the string shorter so that the turn is tighter all are ways to increase the force needed to keep a ball moving in a circle.

For Newton's cannonball (or a satellite), no string holds the ball in its circular motion. Instead, gravity supplies the force, as illustrated in **Figure 4.7b**. The force of gravity must be just enough to keep the satellite moving on its circular path. Because that force has a specific strength, the satellite must therefore be moving at a particular speed around the circle, which we call its **circular velocity** (v_{circ}). If the satellite were moving at any other velocity, it would not be moving in a circular orbit. Remember the cannonball: If it is moving too slowly, it will drop below the circular path and hit the ground. Similarly, if the cannonball is moving too fast, its motion will carry it above the circular orbit. Only a cannonball moving at just the right velocity—the circular velocity—will fall around Earth on a circular path (see Figure 4.5d).

Newton's thought experiment became a reality in 1957, when the Soviet Union launched the first artificial object to orbit Earth. They used a rocket to lift Sputnik 1, an object about the size of a basketball, high enough above Earth's upper atmosphere that air resistance wasn't an issue. Sputnik 1 achieved a high enough speed that it fell around Earth, just like Newton's imaginary cannonball.

Universality

The laws of physics are the same everywhere and at all times.
The principle of universality underlies our understanding of the natural world.

1604

Galileo

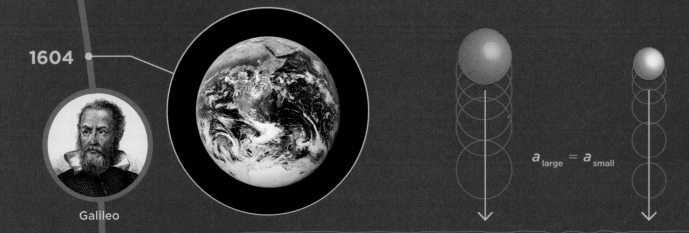

$a_{large} = a_{small}$

In 1604, Galileo determined that all objects on Earth have the same gravitational acceleration, *a*. Universality means that all objects on any world share the same gravitational acceleration with each other.

1971

David Scott
and *Apollo 15*

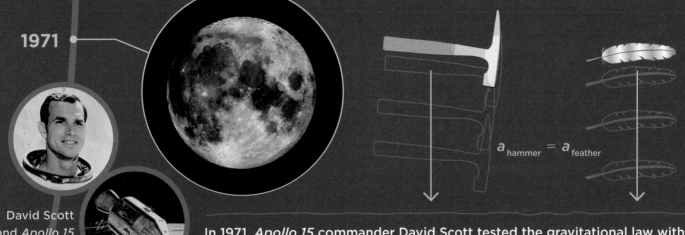

$a_{hammer} = a_{feather}$

In 1971, *Apollo 15* commander David Scott tested the gravitational law with a hammer and a falcon feather on the Moon. With no air resistance, the objects had the same gravitational acceleration.

The physical laws that apply to falling objects also apply to planets orbiting the Sun, to stars orbiting within the galaxy, and to galaxies orbiting each other. This universality allows us to draw conclusions about very distant objects.

When one object is falling around a much more massive object, we say that the less massive object is a **satellite** of the more massive object. Planets are satellites of the Sun, and moons are natural satellites of planets. Newton's imaginary cannonball and Sputnik 1 were satellites (*sputnik* means "satellite" in Russian). An Earth-orbiting spacecraft and the astronauts inside it are independent satellites of Earth that conveniently happen to share the same orbit.

The Shape of Orbits

Some Earth satellites travel a circular path at constant speed. Just like the ball on the string, satellites traveling at the circular velocity always stay the same distance from Earth, neither speeding up nor slowing down in orbit. But what if the satellite then fired its rockets and started traveling *faster* than the circular velocity? Earth's pull is as strong as ever, but because the satellite has greater speed, Earth's gravity does not bend the satellite's path sharply enough to hold it in a circle. When that happens, the satellite begins to climb above a circular orbit.

AstroTour: Elliptical Orbit

As the distance between the satellite and Earth increases, the satellite slows down. Think about a ball thrown upward into the air, as shown in **Figure 4.8a**. As the ball climbs higher, the pull of Earth's gravity opposes its motion, slowing the ball down. The ball climbs more and more slowly until its vertical motion stops for an instant and then is reversed; the ball then begins to fall back toward Earth, speeding up along the way. A satellite does the same thing. As the satellite climbs above a circular orbit and begins to move away from Earth, Earth's gravity opposes the satellite's outward motion, slowing the satellite down. The farther the satellite is from Earth, the more slowly the satellite moves—just like the ball thrown into the air. Also just like the ball, the satellite reaches a maximum height on its curving path and then begins falling back toward Earth. When it does, Earth's gravity speeds the satellite up as it gets closer and closer to Earth. The satellite's orbit has changed from circular to elliptical.

Any object in an elliptical orbit, including a planet orbiting the Sun (**Figure 4.8b**), will therefore lose speed as it pulls away from what it is orbiting

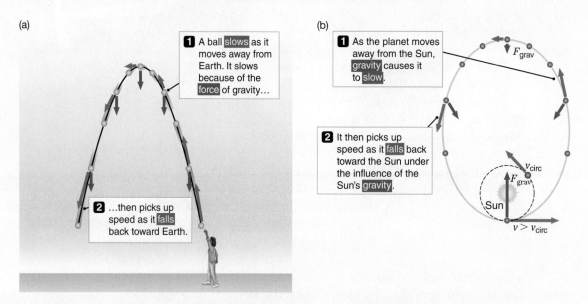

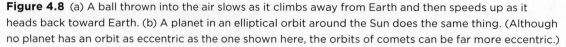

Figure 4.8 (a) A ball thrown into the air slows as it climbs away from Earth and then speeds up as it heads back toward Earth. (b) A planet in an elliptical orbit around the Sun does the same thing. (Although no planet has an orbit as eccentric as the one shown here, the orbits of comets can be far more eccentric.)

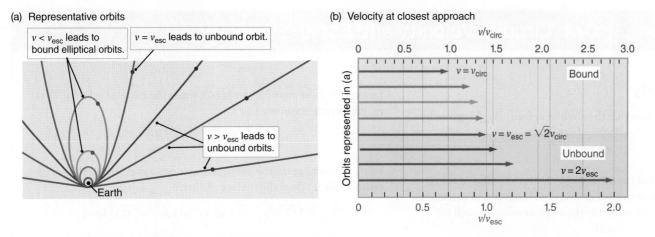

Figure 4.9 (a) Various orbits that share the same point of closest approach but differ in velocity at that point. (b) Closest-approach velocities for the orbits in (a). An object's velocity at closest approach determines the orbit's shape and whether the orbit is bound or unbound. v_{circ} = circular velocity; v_{esc} = escape velocity.

and then gain that speed back again as gravity causes it to fall inward toward what it is orbiting. Gravity, therefore, helps explain Kepler's second law from Section 3.2—namely, that a planet moves fastest when closest to the Sun and slowest when farthest from the Sun.

Newton's laws do more than explain Kepler's laws: they predict different types of orbits beyond Kepler's empirical experience. **Figure 4.9a** shows a series of satellite orbits, each with the same point of closest approach to Earth but with different velocities at that point, as indicated in **Figure 4.9b**. The more speed a satellite has at its closest approach to Earth, the farther the satellite can pull away from Earth, and the more eccentric its orbit becomes. As long as it remains elliptical, no matter how eccentric, the orbit is called a **bound orbit** because the satellite is gravitationally bound to the object it is orbiting.

In that sequence of faster and faster satellites, a point comes at which the satellite is moving so fast that gravity cannot reverse its outward motion, so the object travels away from Earth, never to return. The lowest speed at which an object can permanently leave the gravitational grasp of another mass is called the **escape velocity**, v_{esc}. Once a satellite's velocity at closest approach equals or exceeds v_{esc}, and it is no longer gravitationally bound to the object it was orbiting, we say it is in an **unbound orbit**.

As Figure 4.9a shows, an object with a velocity *less* than the escape velocity (v_{esc}) has an elliptically shaped orbit and follows the same path over and over again, whereas unbound orbits do not close like an ellipse. An object such as a comet on an unbound orbit makes only a single pass around the Sun and then continues away from the Sun into deep space, never to return. Circular velocity and escape velocity are further explored in **Working It Out 4.2**.

Estimating Mass by Using Newton's Version of Kepler's Laws

Newton's calculations opened up a new way to investigate the universe. He showed that the same physical laws that describe the flight of a cannonball on Earth—or the legendary apple falling from a tree—also describe the motions of the planets through the heavens. His laws of motion and gravitation predict all three of Kepler's empirical laws of planetary motion. Newton's version of Kepler's

4.2 Working It Out Circular Velocity and Escape Velocity

Circular Velocity

In Appendix 8, we show that the circular velocity (v_{circ}) is given by

$$v_{circ} = \sqrt{\frac{GM}{r}}$$

where M is the mass of the orbited object, r is the radius of the circular orbit, and G is the universal gravitational constant. A cannonball moving at just the right velocity—the circular velocity—will fall around Earth on a circular path.

We can use that equation to show how fast Newton's cannonball would have to travel to stay in its circular orbit. The average radius of Earth is 6,370 km, the mass of Earth is 5.97×10^{24} kg, and the gravitational constant is 6.67×10^{-20} km³/(kg s²). Inserting those values into the expression for v_{circ}, we get

$$v_{circ} = \sqrt{\frac{[6.67 \times 10^{-20}\,\text{km}^3/(\text{kg s}^2)] \times (5.97 \times 10^{24}\,\text{kg})}{6,370\,\text{km}}} = 7.91\,\text{km/s}$$

To stay in its circular orbit, Newton's cannonball would have to be traveling about 8 kilometers per second (km/s)—more than 28,000 kilometers per hour (km/h). That's well beyond the range of a typical cannon, but rockets routinely reach those speeds.

Now let's compare that speed with that needed to launch a satellite from the Moon into orbit just above the lunar surface. The radius of the Moon is 1,740 km, and its mass is 7.35×10^{22} kg. Those values give a considerably lower circular velocity than the one for Earth:

$$v_{circ} = \sqrt{\frac{(6.67 \times 10^{-20}\,\text{km}^3/\text{kg s}^2) \times (7.35 \times 10^{22}\,\text{kg})}{1,740\,\text{km}}} = 1.68\,\text{km/s}$$

Escape Velocity

Sending a spacecraft to another planet requires launching it with a velocity greater than Earth's escape velocity. The escape velocity is a factor of $\sqrt{2}$, or approximately 1.41, times the circular velocity. That relation can be expressed as

$$v_{esc} = \sqrt{2} \times v_{circ} = \sqrt{\frac{2GM}{R}}$$

Using the numbers in the above example, we can calculate the escape velocity from the surface of Earth:

$$v_{esc} = \sqrt{2} \times v_{circ} = 1.41 \times 7.91\,\text{km/s} = 11.15\,\text{km/s}$$

To leave Earth, a rocket must have a speed of at least 11.15 km/s, or 40,140 km/h.

As with weight, the escape velocity from other astronomical objects is different from the escape velocity from Earth. Ida is a small asteroid orbiting the Sun between the orbits of Mars and Jupiter. Ida has an average radius of 15.7 km and a mass of 4.2×10^{16} kg. Therefore,

$$v_{esc} = \sqrt{\frac{2 \times [6.67 \times 10^{-20}\,\text{km}^3/(\text{kg s}^2)] \times (4.2 \times 10^{16}\,\text{kg})}{15.7\,\text{km}}}$$

$$v_{esc} = 0.019\,\text{km/s} = 68\,\text{km/h}$$

A baseball thrown at about 130 km/h (81 mph) would easily escape from Ida's surface and fly off into interplanetary space.

When two objects are closer to having the same mass, however—such as dwarf planet Pluto and its moon Charon, or a large planet and a star, or two stars—both objects experience significant accelerations in response to their mutual gravitational attraction. The two objects are both orbiting about a common point located between them, called the **center of mass**, so we now must think of them as falling around each other. Each mass is moving on its own elliptical orbit around the two objects' mutual center of mass. From measuring the size and period of any orbit, we can calculate the *sum* of the masses of the orbiting objects. Almost all knowledge about the masses of astronomical objects comes directly from applying Newton's version of Kepler's third law.

laws can be used to estimate the mass of the Sun from the orbit of Earth, as seen in **Working It Out 4.3**.

When Earth, a much less massive object, is orbiting the Sun, a much more massive object, the Sun's gravity has a strong influence on Earth, but Earth's gravity has little effect on the Sun. Therefore, the Sun remains nearly motionless from Earth's orbit around it.

Astronomy in Action: Center of Mass

CHECK YOUR UNDERSTANDING 4.2

Rank the circular velocities of the following objects from smallest to largest: (a) a 5-kg object orbiting Earth halfway to the Moon; (b) a 10-kg object orbiting Earth just above Earth's surface; (c) a 15-kg object orbiting Earth at the same distance as the Moon.

4.3 Working It Out Calculating Mass from Orbital Periods

Newton's Version of Kepler's Third Law

The time a planet takes to complete one orbit around the Sun (the orbital period, P) equals the distance traveled divided by the planet's speed. For simplicity, let's assume the orbit is circular. Thus, the time an object takes to make one trip around the Sun is the circumference of the circle ($2\pi r$) divided by the object's speed (v). The speed of the planet must be equal to the circular velocity discussed in Working It Out 4.2. Putting those relationships together, we have

$$\text{Orbital period } (P) = \frac{\text{Circumference of orbit}}{\text{Circular velocity}} = \frac{2\pi r}{\sqrt{\dfrac{GM_{\text{Sun}}}{r}}}$$

Squaring both sides of the equation gives

$$P^2 = \frac{4\pi^2 r^2}{\dfrac{GM_{\text{Sun}}}{r}} = \frac{4\pi^2}{GM_{\text{Sun}}} \times r^3$$

The square of the period of an orbit is equal to a constant ($4\pi^2/GM_{\text{Sun}}$) multiplied by the cube of the radius of the orbit. That is Kepler's third law ($P^2 = \text{constants} \times A^3$) applied to circular orbits. When Kepler used Earth units of years and astronomical units for the planets, he was taking a ratio of their periods and orbital radii with those for Earth, so the constants canceled out. Using calculus, Newton similarly could derive Kepler's third law for elliptical orbits with semimajor axis A instead of radius r:

$$P^2 = \frac{4\pi^2}{GM_{\text{Sun}}} \times A^3$$

Mass of the Sun

If we can measure the size and period of any orbit, we can use Newton's universal law of gravitation to calculate the mass of the object being orbited. To do so, we rearrange Newton's form of Kepler's third law above to read

$$M = \frac{4\pi^2}{G} \times \frac{A^3}{P^2}$$

Everything on the right side of that equation is either a constant (4, π, and G) or a measurable quantity (the semimajor axis A and period P of an orbit). The left side of the equation is the mass of the object at the focus of the ellipse. For example, we can find the mass of the Sun by noting the period and semimajor axis of the orbit of a planet around the Sun. Let's use the numbers for Earth. Whenever we have an equation with G, putting everything else into the same units as G (km, kg, s) is best. So first we must compute the number of seconds in 1 year: $P = 1\,\text{yr} = 365.24\,\text{days/yr} \times 24\,\text{h/day} \times 60\,\text{min/h} \times 60\,\text{s/min} = 3.16 \times 10^7\,\text{s}$. The semimajor axis $A = 1\,\text{AU} = 1.5 \times 10^8\,\text{km}$. Then the mass of the Sun can be computed:

$$M_{\text{Sun}} = \frac{4\pi^2}{G} \times \frac{A^3}{P^2} = \frac{4\pi^2}{6.67 \times 10^{-20}\,\text{km}^3/(\text{kg s}^2)} \times \frac{(1.5 \times 10^8\,\text{km})^3}{(3.16 \times 10^7\,\text{s})^2}$$

$$M_{\text{Sun}} = 2.0 \times 10^{30}\,\text{kg}$$

We could have used the period and semimajor axis of another Solar System planet to get the same result for the mass of the Sun.

4.3 Tidal Forces Are Caused by Gravity

The rise and the fall of the oceans are called Earth's **tides**. Coastal dwellers long ago noted that the strength of the tides varies with the phase of the Moon. Tides are strongest during a new or a full Moon and are weakest during first quarter or third quarter Moon. Here in Section 4.3, we see how tides result from differences between the strength of the gravitational pull of the Moon and Sun on one part of Earth in comparison with their pull on other parts of Earth.

Astronomy in Action: Tides

Tides and the Moon

Recall from Figure 4.4 that each small part of an object feels a gravitational attraction toward every other small part of the object, and that self-gravity differs from place to place. In addition, each small part of an object feels a gravitational attraction toward every other mass in the universe, and those external forces also differ from place to place within the object.

The Moon's gravity pulls on Earth as though the Moon's mass were concentrated at its center. The side of Earth that faces the Moon is closer to the Moon

Simulation: Tides on Earth

AstroTour: Tides and the Moon

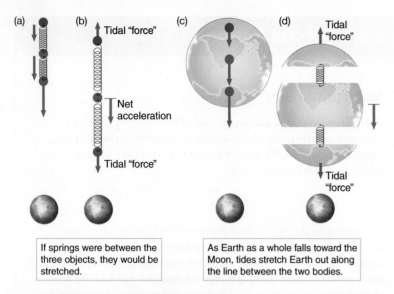

If springs were between the three objects, they would be stretched.

As Earth as a whole falls toward the Moon, tides stretch Earth out along the line between the two bodies.

Figure 4.10 (a) Imagine three objects connected by springs. (b) The springs are stretched as though forces were pulling outward on each end of the chain. (c) Similarly, three locations on Earth experience different gravitational attractions toward the Moon. (d) The difference in the Moon's gravitational attraction across Earth causes Earth's tides.

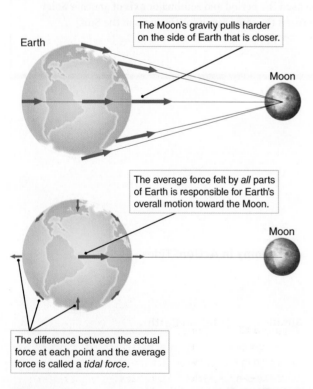

Earth

The Moon's gravity pulls harder on the side of Earth that is closer.

Moon

The average force felt by *all* parts of Earth is responsible for Earth's overall motion toward the Moon.

Moon

The difference between the actual force at each point and the average force is called a *tidal force*.

Figure 4.11 Tidal forces stretch Earth along the line between Earth and the Moon but compress Earth perpendicular to that line.

than the rest of Earth is, so it feels a stronger-than-average gravitational attraction toward the Moon. In contrast, the side of Earth facing away from the Moon is farther than average from the Moon, so it feels a weaker-than-average attraction toward the Moon. The pull of the Moon on the near side of Earth is about 7 percent greater than its pull on the far side of Earth.

To understand the consequence of that variation in the pull of the Moon, imagine three rocks being pulled by gravity toward the Moon. A rock closer to the Moon feels a stronger force than a rock farther from the Moon. Now suppose the three rocks are connected by springs (**Figure 4.10a**). As the rocks are pulled toward the Moon, the purple rock pulls away from the red rock, and the red rock pulls away from the blue rock. Therefore, the differences in the gravitational forces they feel will stretch *both* springs (**Figure 4.10b**). Now, instead of springs, imagine that the rocks are at different places on Earth (**Figure 4.10c**). On the side of Earth away from the Moon, the force is smaller (as indicated by the shorter arrow), so that part gets left behind (**Figure 4.10d**). Those differences in the Moon's gravitational attraction on different parts of Earth are called **tidal forces**.

Figure 4.11 shows how Earth is stretched, causing a tidal bulge. The Moon is not pushing the far side of Earth away; rather, the Moon simply is not pulling on the far side of Earth as hard as it is pulling on the planet as a whole. The far side of Earth is "left behind" as the rest of the planet is pulled more strongly toward the Moon. Figure 4.11 shows that a net force also is squeezing inward on Earth in the direction perpendicular to the line between Earth and the Moon. Together, the stretching by tidal forces along the line between Earth and the Moon and the squeezing by tidal forces perpendicular to that line distort Earth's shape, like a rubber ball caught in the middle of a tug-of-war.

If the surface of Earth were perfectly smooth and covered with an ocean of uniform depth and Earth did not rotate, the Moon would pull our oceans into an elongated **tidal bulge** like that shown in **Figure 4.12a**. The water would be at its deepest on the side toward the Moon and on the side away from the Moon, and the water would be at its shallowest midway between. But Earth is *not* covered with perfectly uniform oceans, and Earth *does* rotate. As any point on Earth rotates through the ocean's tidal bulges, that point experiences the ebb and flow of the tides. In addition, friction between the spinning Earth and its tidal bulge drags the oceanic tidal bulge around in the direction of Earth's rotation, as illustrated in **Figure 4.12b**.

Figure 4.12c shows the tides you would experience throughout a day. Begin as the rotating Earth carries you through the tidal bulge on the Moonward side of the planet. Because Earth's rotation drags the tidal bulge, the Moon is not exactly overhead but is instead high in the western sky. When you are at the high point in the tidal bulge, the ocean around you has risen higher than average—called a *high tide*. The Moon rises 50 minutes later than it did on the previous day, so the time between the Moon's return to the same position in the sky from the previous day is 24h 50m. Each high and low tide is 1/4 of that, or about 6 1/4 hours later. So, 6 1/4 hours after the high tide, somewhat after the Moon has settled beneath the western horizon, the rotation of Earth carries you through a point where the ocean is lower than average—called a *low tide*. If you wait another 6 1/4 hours, it is again high tide. You are now passing through the region where ocean water is "left behind" (with respect to Earth as a whole) in the tidal bulge on the side of Earth that is away from the Moon. The Moon at that time is hidden from view on

(a)

The Moon's tidal forces stretch Earth and its oceans into an elongated shape. The departure from spherical is called Earth's *tidal bulge*.

(b)

Because of friction, Earth's rotation drags its tidal bulge around, out of perfect alignment with the Moon.

(c)

Ocean tides rise and fall as the rotation of Earth carries us through the ocean's tidal bulges.

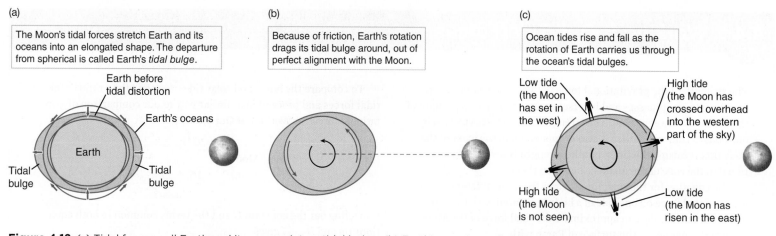

Earth before tidal distortion

Earth's oceans

Earth

Tidal bulge

Tidal bulge

Low tide (the Moon has set in the west)

High tide (the Moon has crossed overhead into the western part of the sky)

High tide (the Moon is not seen)

Low tide (the Moon has risen in the east)

Figure 4.12 (a) Tidal forces pull Earth and its oceans into a tidal bulge. (b) Earth's rotation pulls its tidal bulge slightly out of alignment with the Moon. (c) As Earth's rotation carries us through those bulges, we experience the ocean tides. The magnitude of the tides has been exaggerated here for clarity. In these figures, the observer is looking down from above Earth's North Pole. Sizes and distances are not to scale.

the far side of Earth. About 6 1/4 hours later, sometime after the Moon has risen above the eastern horizon, it is low tide. About 25 hours after you started this journey—the amount of time the Moon takes to return to the same point in the sky from which it started—you again pass through the tidal bulge on the near side of the planet. That is the age-old pattern by which mariners have lived their lives for millennia: the twice-daily coming and going of high tide, shifting through the day in lockstep with the passing of the Moon.

Ocean currents and local geography, such as the shapes of Earth's shorelines and ocean basins, complicates the simple picture of tides. In addition, oceanwide oscillations occur, similar to water sloshing around in a basin. As the oceans respond to the tidal forces from the Moon, they flow around the various landmasses that break up the water covering our planet. Some places, such as the Mediterranean Sea and the Baltic Sea, are protected from tides by their relatively small sizes and the narrow passages connecting those bodies of water to the larger ocean. In other places, the shape of the land funnels the tidal surge from a large region of ocean into a relatively small area, concentrating its effect, as at eastern Canada's Bay of Fundy (**Figure 4.13**).

Solar Tides

Tides resulting from the pull of the Moon are called **lunar tides**, but the Sun also influences Earth's tides. The gravitational pull of the Sun causes Earth to stretch along a line pointing approximately in the direction of the Sun. The side of Earth closer to the Sun is pulled toward the Sun more strongly than the side of Earth away from the Sun, just as the side of Earth closest to the Moon is pulled more strongly toward the Moon. Tides on Earth resulting from differences in the gravitational pull of the Sun are called **solar tides**. The absolute strength of the Sun's pull on Earth is nearly 200 times greater than the strength of the Moon's pull on Earth. However, the Sun's gravitational attraction does not change by much from one side of Earth to the other because the Sun is much farther away than the Moon. As a result, solar tides are only about half as strong as lunar tides (**Working It Out 4.4**).

(a)

(b)

Figure 4.13 The world's most extreme tides are found in the Bay of Fundy in eastern Canada. Water rocks back and forth in the bay with a period of about 13 hours, close to the 12.5-hour period of the tides. The shape of the basin amplifies the tides so that the difference in water depth between low tide (a) and high tide (b) is extreme—typically about 14.5 meters and as much as 16.6 meters.

4.4 Working It Out Tidal Forces

The strength of the gravitational force between two bodies is proportional to their masses and inversely proportional to the square of the distance between them. The strength of tidal forces caused by one body acting on another is also proportional to the mass of the body that is raising the tides, but the strength is inversely proportional to the *cube* of the distance between them.

The equation for the tidal force comes from the difference of the gravitational force on one side of a body compared with the force on the other side. We can approximate the tidal force of the Moon acting on a mass m on the surface of Earth with

$$F_{tidal}(\text{Moon}) = \frac{2GM_{\text{Moon}}R_{\text{Earth}}m}{d_{\text{Earth−Moon}}^3}$$

where R_{Earth} is Earth's radius and $d_{\text{Earth-Moon}}$ is the distance between Earth and the Moon.

Let's compare the Moon's tidal force acting on Earth with the Sun's tidal force acting on Earth. F_{tidal} from the Moon is given in the preceding equation; F_{tidal} from the Sun is given by

$$F_{tidal}(\text{Sun}) = \frac{2GM_{\text{Sun}}R_{\text{Earth}}m}{d_{\text{Earth−Sun}}^3}$$

To compare the lunar and solar tides, we can take a ratio of the tidal forces and proceed in a similar way to our comparison of gravitational forces in Working It Out 4.1:

$$\frac{F_{tidal}(\text{Moon})}{F_{tidal}(\text{Sun})} = \frac{\dfrac{2GM_{\text{Moon}}R_{\text{Earth}}m}{d_{\text{Earth−Moon}}^3}}{\dfrac{2GM_{\text{Sun}}R_{\text{Earth}}m}{d_{\text{Earth−Sun}}^3}}$$

Canceling out the constant G and the terms common in both equations (m and R_{Earth}) gives

$$\frac{F_{tidal}(\text{Moon})}{F_{tidal}(\text{Sun})} = \frac{M_{\text{Moon}}}{M_{\text{Sun}}} \times \frac{d_{\text{Earth−Sun}}^3}{d_{\text{Earth−Moon}}^3} = \frac{M_{\text{Moon}}}{M_{\text{Sun}}} \times \left(\frac{d_{\text{Earth−Sun}}}{d_{\text{Earth−Moon}}}\right)^3$$

Using the values from Appendix 4, $M_{\text{Moon}} = 7.35 \times 10^{22}$ kg, $M_{\text{Sun}} = 2 \times 10^{30}$ kg, $d_{\text{Earth−Moon}} = 384{,}400$ km, and $d_{\text{Earth−Sun}} = 1.5 \times 10^8$ km, gives

$$\frac{F_{tidal}(\text{Moon})}{F_{tidal}(\text{Sun})} = \frac{7.35 \times 10^{22}\ \text{kg}}{2 \times 10^{30}\ \text{kg}} \times \left(\frac{1.5 \times 10^8\ \text{km}}{384{,}400\ \text{km}}\right)^3 = 2.2$$

Thus, the tidal force between Earth and the Moon is 2.2 times stronger than the tidal force between Earth and the Sun. The Sun is an important factor, too, though, and that's why the tides change depending on the alignment of the Moon and the Sun with Earth.

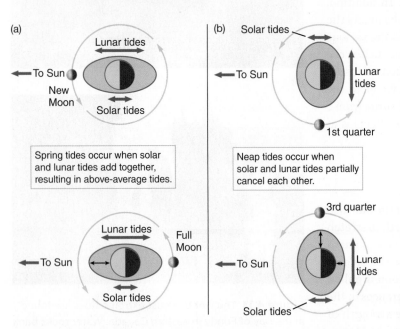

Spring tides occur when solar and lunar tides add together, resulting in above-average tides.

Neap tides occur when solar and lunar tides partially cancel each other.

Figure 4.14 Solar tides are about half as strong as lunar tides. The interactions of solar and lunar tides result in either (a) spring tides when they are added together or (b) neap tides when they partially cancel each other.

Solar and lunar tides interact. When the Moon and the Sun are lined up with Earth (**Figure 4.14a**), at either new or full Moon, the lunar and solar tides on Earth overlap. That alignment creates more extreme tides that range from higher high tides to lower low tides. The extreme tides near the new or full Moon are called **spring tides**—not because of the season but because the water appears to spring out of the sea. Conversely, when the Moon, Earth, and Sun make a right angle (**Figure 4.14b**), at the Moon's first and third quarters, the lunar and solar tidal forces stretch Earth in different directions, creating less extreme tides known as **neap tides**. The word *neap* is derived from the Saxon word *neafte*, which means "scarcity": at those times of the month, shellfish and other food gathered in the tidal region are less accessible because the low tide is higher than at other times. Neap tides are only about half as strong as average tides and only a third as strong as spring tides.

CHECK YOUR UNDERSTANDING 4.3

Rank the following in order of weakest to strongest total tides: (a) new Moon in July; (b) first quarter Moon in July; (c) full Moon in January; (d) third quarter Moon in January.

4.4 Tidal Forces Affect Solid Bodies

In Section 4.3, we focused on how the liquid of Earth's oceans moves in response to the tidal forces from the Moon and Sun. But those tidal forces also affect the solid body of Earth. As Earth rotates through its tidal bulge, the solid body of the planet is constantly being deformed by tidal forces. Earth is somewhat elastic (like a rubber ball), and tidal stresses cause a vertical displacement of about 30 centimeters (cm) between high tide and low tide, or roughly a third of the displacement of the oceans. Deforming the shape of a solid object requires energy. (For a practical demonstration of that fact, hold a rubber ball in your hand and squeeze and release it a few dozen times.) That energy from the deformation is converted into thermal energy by friction in Earth's interior. That friction opposes and takes energy from the rotation of Earth, causing Earth to gradually slow. Earth's internal friction adds to the slowing caused by friction between Earth and its oceans as the planet rotates through the tidal bulge of the oceans. As a result, Earth's days are currently getting about 1.7 milliseconds (ms) longer every century. That sounds insignificant, but it adds up: when dinosaurs ruled 200 million years ago, the day was closer to 23 hours long, and 200 million years into the future, the day will be close to 25 hours.

Tidal Locking

Other solid bodies besides Earth experience tidal forces. For example, the Moon has no bodies of liquid to make tidal forces obvious, but its shape is distorted in the same manner as Earth's. Because of Earth's much greater mass and the Moon's smaller radius, Earth's tidal effects on the Moon are stronger than the Moon's tidal effects on Earth. Given that the average tidal deformation of Earth is about 30 cm, the average tidal deformation of the Moon should be about 6 meters. However, what we actually observe on the Moon is a tidal bulge of about 20 meters. That unexpectedly large displacement exists because the Moon's tidal bulge was "frozen" into its relatively rigid crust at an earlier time, when the Moon was closer to Earth and tidal forces were much stronger than they are today. Planetary scientists sometimes call that deformation the Moon's *fossil tidal bulge.*

Recall that the Moon's rotation period equals its orbital period (see Figure 2.22). That synchronous rotation of the Moon is a result of **tidal locking**. Early in the Moon's history, the period of its rotation was almost certainly different from its orbital period. As the Moon rotated through its extreme tidal bulge, however, friction within the Moon's crust was tremendous, rapidly slowing the Moon's rotation. After a fairly short time, the period of the Moon's rotation equaled the period of its orbit. When its orbital and rotation periods became equalized, the Moon no longer rotated with respect to its tidal bulge. Instead, the Moon and its tidal bulge rotated *together*, in lockstep with the Moon's orbit around Earth. As illustrated in **Figure 4.15**, that scenario continues today as the tidally distorted Moon orbits Earth, always keeping the same face and the long axis of its tidal bulge toward Earth.

Tidal forces affect not only the rotations of the Moon and Earth but also their orbits. Because of its tidal bulge, Earth is not perfectly spherical. Therefore, the material in Earth's tidal bulge on the side nearer the Moon pulls on the Moon more strongly than does material in the tidal bulge on the back side of Earth. Because the tidal bulge on the Moonward side of Earth "leads" the Moon

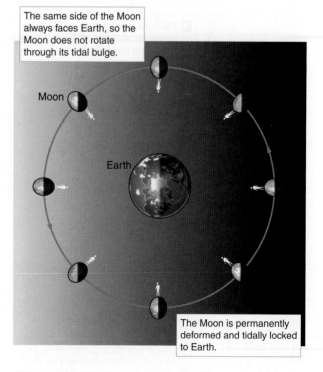

The same side of the Moon always faces Earth, so the Moon does not rotate through its tidal bulge.

Moon

Earth

The Moon is permanently deformed and tidally locked to Earth.

Figure 4.15 Tidal forces due to Earth's gravity lock the Moon's rotation to its orbital period.

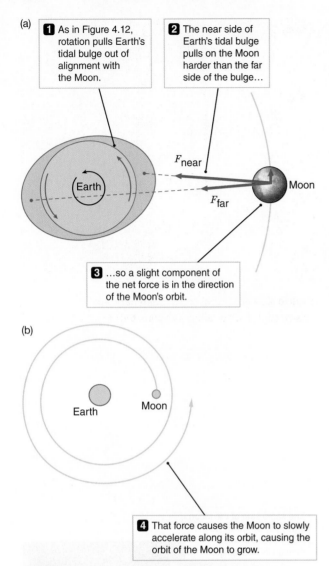

(a)

1 As in Figure 4.12, rotation pulls Earth's tidal bulge out of alignment with the Moon.

2 The near side of Earth's tidal bulge pulls on the Moon harder than the far side of the bulge…

F_{near}

Earth

Moon

F_{far}

3 …so a slight component of the net force is in the direction of the Moon's orbit.

(b)

Earth   Moon

4 That force causes the Moon to slowly accelerate along its orbit, causing the orbit of the Moon to grow.

Figure 4.16 Interaction between Earth's tidal bulge and the Moon causes the Moon to accelerate in its orbit and the Moon's orbit to grow.

somewhat, as shown in **Figure 4.16**, the gravitational attraction of the bulge causes the Moon to accelerate slightly along the direction of its orbit around Earth. The rotation of Earth is dragging the Moon along with it. The acceleration of the Moon in the direction of its orbital motion causes the orbit of the Moon to grow larger. At present, the Moon is drifting away from Earth at a rate of 3.83 cm per year.

As the Moon grows more distant, the lunar month gets about 0.014 second longer each century. If that increase in the radius of the Moon's orbit were to continue long enough (about 50 billion years), Earth would become tidally locked to the Moon, just as the Moon is now tidally locked to Earth. At that point, the Earth's period of rotation, the Moon's period of rotation, and the Moon's orbital period would all be the same—about 47 of our current days—and the Moon would be about 43 percent farther from Earth than it is today. That situation will never actually occur, however, because the Sun will have burned itself out long before then.

The effects of tidal forces are apparent throughout the Solar System. Most of the moons in the Solar System are tidally locked to their parent planets, and for dwarf planet Pluto and its largest moon, Charon, each is tidally locked to the other.

Tidal locking is only one way that orbits and rotations can be coupled. Tidal forces have coupled the planet Mercury's rotation to its very elliptical orbit around the Sun. Unlike the Moon's synchronous rotation, however, Mercury spins on its axis three times for every two trips around the Sun. The period of Mercury's orbit—87.97 Earth days—is exactly 1 1/2 times the 58.64 days that Mercury takes to spin once on its axis. When Mercury comes to the point in its orbit that is closest to the Sun, one hemisphere faces the Sun, and then in the next orbit, the other hemisphere faces the Sun.

Tidal Forces on Many Scales

We normally think of the effects of tidal forces as small in comparison with the force of gravity holding an object together, yet tidal effects can be extremely destructive. Consider for a moment the fate of a small moon, asteroid, or comet that wanders too close to a massive planet such as Jupiter or Saturn. All objects in the Solar System larger than about a kilometer in diameter are held together by their self-gravity. However, the self-gravity of a small object such as an asteroid, a comet, or a small moon is feeble. In contrast, the tidal forces close to a massive object such as Jupiter can be very strong. If the tidal forces trying to tear an object apart become greater than the self-gravity trying to hold the object together, the object will break into pieces.

The **Roche limit** is the distance at which a planet's tidal forces become greater than the self-gravity of a smaller object—such as a moon, asteroid, or comet—causing the object to break apart. For a smaller object having the same density as the planet, the Roche limit is about 2.45 times the planet's radius. Such an object bound together solely by its own gravity can remain intact when it is outside a planet's Roche limit, but not when it is inside the limit. Objects such as the International Space Station and other Earth satellites are not torn apart, even though they orbit well within Earth's Roche limit, because chemical and physical bonds hold them together, not just self-gravity.

We have concentrated on the role that tidal forces play on Earth and the Moon, but tidal forces exist throughout the Solar System and the universe.

Anytime two objects of significant size or two collections of objects interact gravitationally, the gravitational forces will differ from one place to another within the objects, giving rise to tidal effects. Tidal disruption of small bodies is the source of the particles that make up the rings of the giant planets. Tidal interactions can cause material from one star in a binary pair to be pulled onto the other star. Tidal effects can strip stars from clusters consisting of thousands of stars. Galaxies can pass close enough together to strongly interact gravitationally. When that happens, as in **Figure 4.17**, tidal effects can grossly distort both galaxies. Tidal forces even play a role in shaping huge collections of galaxies—the largest known structures in the universe.

CHECK YOUR UNDERSTANDING 4.4

The Moon always keeps the same face toward Earth because of: (a) tidal locking; (b) tidal forces from the Sun; (c) tidal forces from the Earth and Sun.

Figure 4.17 The tidal "tails" seen here are characteristic of tidal interactions between galaxies.

Origins Tidal Forces and Life

Earth's rotation is slowing down as the Moon slowly moves away into a larger orbit, so in the distant past the Moon was closer and Earth rotated faster. In that time, tides would have been stronger and the interval between high tides would have been shorter. Tides are also affected by the configuration of the continents and oceans on Earth, so precisely how much faster Earth rotated billions of years ago is unknown. What is known, though, is that the stronger and more frequent tides would have provided additional energy to the oceans of the young Earth.

Scientists debate whether life on Earth originated deep in the ocean, on the surface of the ocean, or in hot springs on land (see Chapter 24). The tides shaped the regions in the margins between land and ocean, such as tide pools and coastal flats. Some researchers think that those border regions, which alternate between wet and dry with the tides, could have been places where concentrations of biochemicals periodically became high enough for more complex reactions to take place. Those complex reactions were important to the earliest life. Later, those border regions may have been important as advanced life moved from the sea to the land.

Elsewhere in the Solar System, the giant planets Jupiter and Saturn are far from the Sun and thus very cold. Both Jupiter and Saturn have many moons, and the closest moons would experience strong tidal forces from their respective planets. Several of those moons are thought to have a liquid ocean underneath an icy surface. Tidal forces from Jupiter or Saturn supply the heat to keep the water in a liquid state. Astrobiologists think that those subsurface liquid oceans are perhaps the most probable location for life elsewhere in the Solar System.

On Earth, the solar tides are about half as strong as lunar tides. A planet with a closer orbit would experience much stronger tidal forces from its star. Many planets detected outside the Solar System have orbits very close to their stars (see Chapter 7), so those exoplanets experience strong tidal forces. The planets might be tidally locked, too, so they have a synchronous rotation like the Moon does around Earth, with one side of the planet always facing the star and one side facing away. How might life on Earth have evolved differently if half the planet was in perpetual night and half in perpetual day?

ARTICLES   QUESTIONS

Astronomers and physicists have wondered whether the value of Newton's gravitational constant, G, has changed over time. This press release reports on one experiment to measure the value of G.

Exploding Stars Prove Newton's Law of Gravity Unchanged over Cosmic Time

By Swinburne Media Centre

Australian astronomers have combined all observations of supernovae ever made to determine that the strength of gravity has remained unchanged over the last 9 billion years.

Newton's gravitational constant, known as G, describes the attractive force between two objects, together with the separation between them and their masses. It had previously been suggested that G could have been slowly changing over the 13.8 billion years since the Big Bang.

If G had been decreasing over time, for example, this would mean the Earth's distance to the Sun was slightly smaller in the past, meaning that we would experience longer seasons now compared to much earlier points in the Earth's history.

But researchers at Swinburne University of Technology in Melbourne have now analysed the light given off by 580 supernova explosions in the nearby and far universe and have shown that the strength of gravity has not changed.

"Looking back in cosmic time to find out how the laws of physics may have changed is not new," said Professor Jeremy Mould. "But supernova cosmology now allows us to do this with gravity."

A Type Ia supernova marks the violent death of a star called a white dwarf, which is as massive as our Sun but packed into a ball the size of our Earth.

Telescopes can detect the light from this explosion and use its brightness as a "standard candle" to measure distances in the universe, a tool that helped Australian astronomer Professor Brian Schmidt in his 2011 Nobel Prize–winning work discovering the mysterious force Dark Energy.

Professor Mould and his PhD student Syed Uddin at the Swinburne Centre for Astrophysics and Supercomputing and the ARC Centre of Excellence for All-sky Astrophysics (CAASTRO) assumed that these supernova explosions happen when a white dwarf reaches a critical mass or after colliding with other stars to "tip it over the edge."

"This critical mass depends on Newton's gravitational constant G and allows us to

monitor it over billions of years of cosmic time—instead of only decades as was the case in previous studies," Professor Mould said.

Despite these vastly different time spans, their results agree with findings from the Lunar Laser Ranging Experiment that has been measuring the distance between the Earth and the Moon since NASA's *Apollo* missions in the 1960s and has been able to monitor possible variations in G at very high precision.

"Our cosmological analysis complements experimental efforts to describe and constrain the laws of physics in a new way and over cosmic time," Mr. Uddin said.

In their current publication, the Swinburne researchers were able to set an upper limit on the change in Newton's gravitational constant of 1 part in 10 billion per year over the past 9 billion years.

ARTICLES   QUESTIONS

1. Explain how observing distant supernovae means observing back in time.
2. Why are physicists still measuring G, centuries after Newton described that constant?
3. What quantities that you studied here in Chapter 4 would change if the value of G changed over time?
4. Why is it important to the process of science that this result agrees with a result from a totally different experiment?
5. If the average distance to the Sun were slightly less in the past, would solar tides have been weaker or stronger? Would you expect that change to have been as important to tides on Earth as changes in the distance to the Moon?

Source: "Exploding stars prove Newton's law of gravity unchanged over cosmic time," Swinburne University of Technology Media Center News Release, March 24, 2014. Reprinted by permission of Swinburne University of Technology.

Summary

Objects stay in orbit because of gravity. Newton's laws of motion and his proposed law of gravity predict the paths of planetary orbits and explain Kepler's laws. Newton's calculations showed that those orbits should be ellipses with the Sun at one focus, that planets should move faster when closer to the Sun, and that the square of the period of a planet's orbit should vary as the cube of the semimajor axis of that elliptical orbit. Newton also mathematically confirmed Galileo's conclusion that falling objects have an accelerated motion independent of their mass. Tidal forces provide energy to Earth's oceans. Tide pools on Earth may have been a site of early biochemical reactions. Some moons in the outer Solar System might have liquid water because of tidal heating from their respective planets.

LG 1 **Explain the elements of Newton's universal law of gravitation.** Gravity is a force between any two objects with mass. Gravity is one of the fundamental forces of nature—it binds the universe together. The force of gravity is proportional to the mass of each object and inversely proportional to the square of the distance between them.

LG 2 **Use the laws of motion and gravitation to explain planetary orbits.** An orbit is one body "falling around" another.

Planets orbit the Sun in elliptical orbits. All objects affected by gravity move either in bound elliptical orbits or unbound paths. Orbits are ultimately given their shape by the gravitational attraction of the objects involved, which in turn reflects the masses of those objects.

LG 3 **Explain how tidal forces from the Sun and Moon create Earth's tides.** Tides on Earth are the result of differences between how hard the Moon and Sun pull on one part of Earth in comparison with their pull on other parts of Earth. The primary cause of tides is the Moon, which stretches out Earth. The tides are the strongest when the Sun, Moon, and Earth are aligned. As Earth rotates, tides rise and fall twice each day.

LG 4 **Describe how tidal forces affect solid bodies.** Tidal forces lock the Moon's rotation to its orbit around Earth. Tidal forces can break up an object if its gets too close to a more massive object. Tidal forces are observed throughout the universe—in planets and moons, pairs of stars, and interacting galaxies.

? Unanswered Question

- What range of gravities will support human life? Humans have evolved to live on Earth's surface, but what happens when humans go elsewhere? What are the limits for our hearts, lungs, eyes, and bones? At the higher end of human tolerance, fighter pilots have been trained to experience about 10 times the normal surface gravity on Earth for very short periods (too long and they black out). Astronauts who spend several months in near-weightless conditions experience medical problems such as bone loss. On the Moon or Mars, humans will weigh much less than on Earth. Many science fiction tales have been written about what happens to children born on a space station or on another planet or moon with low surface gravity: would their hearts and bodies ever be able to adjust to the higher surface gravity of Earth, or must they stay in space forever?

Questions and Problems

Test Your Understanding

1. In Newton's universal law of gravitation, the force is
   a. proportional to both masses.
   b. proportional to the radius.
   c. proportional to the radius squared.
   d. inversely proportional to the orbiting mass.

2. If we wanted to increase a satellite's altitude above Earth and keep it in a stable orbit, we also would need to
   a. increase its orbital speed.
   b. increase its weight.
   c. decrease its weight.
   d. decrease its orbital speed.

3. An object in a(n) _____ orbit in the Solar System will remain in its orbit forever. An object in a(n) _____ orbit will escape from the Solar System.
   a. unbound; bound
   b. circular; elliptical
   c. bound; unbound
   d. elliptical; circular

4. Compared with your mass on Earth, your mass on the Moon would be
   a. lower because the Moon is smaller than Earth.
   b. lower because the Moon has less mass than Earth.
   c. higher because of the combination of the Moon's mass and size.
   d. the same—mass doesn't change.

5. If you went to Mars, your weight would be
   a. higher because you are closer to the center of the planet.
   b. lower because Mars has two small moons instead of one big moon, so less tidal force is present.
   c. lower because the lower mass and smaller radius of Mars combine to a lower surface gravity.
   d. the same as on Earth.

6. Venus has about 80 percent of Earth's mass and about 95 percent of Earth's radius. Your weight on Venus will be
   a. 20 percent more than on Earth.
   b. 20 percent less than on Earth.
   c. 10 percent more than on Earth.
   d. 10 percent less than on Earth.

7. The connection between gravity and orbits enables astronomers to measure the _____ of stars and planets.
   a. distances
   b. sizes
   c. masses
   d. compositions

8. If the Moon were twice as massive, how would the strength of lunar tides change?
   a. The highs would be higher, and the lows would be lower.
   b. Both the highs and the lows would be higher.
   c. The highs would be lower, and the lows would be higher.
   d. Nothing would change.

9. If Earth had half its current radius, how would the strength of lunar tides change?
   a. The highs would be higher, and the lows would be lower.
   b. Both the highs and the lows would be higher.
   c. The highs would be lower, and the lows would be higher.
   d. Both the highs and the lows would be lower.

10. If the Moon were 2 times closer to Earth than it is now, the gravitational force between Earth and the Moon would be
    a. 2 times stronger.
    b. 4 times stronger.
    c. 8 times stronger.
    d. 16 times stronger.

11. If the Moon were 2 times closer to Earth than it is now, the tides would be
    a. 2 times stronger.
    b. 4 times stronger.
    c. 8 times stronger.
    d. 16 times stronger.

12. If two objects are tidally locked to each other,
    a. the tides always stay on the same place on each object.
    b. the objects always remain in the same place in each other's sky.
    c. the objects are falling together.
    d. both a and b

13. Spring tides occur only when
    a. the Sun is near the vernal equinox.
    b. the Moon's phase is new or full.
    c. the Moon's phase is first quarter or third quarter.
    d. it is either spring or fall.

14. If an object crosses from farther to closer than the Roche limit, it
    a. can no longer be seen.
    b. begins to accelerate very quickly.
    c. slows down.
    d. may be torn apart.

15. Self-gravity is
    a. the gravitational pull of a person.
    b. the force that holds objects such as people and lamps together.
    c. the gravitational interaction of all the parts of a body.
    d. the force that holds objects on Earth.

Thinking about the Concepts

16. Both Kepler's laws and Newton's laws tell us something about the motion of the planets, but they have a fundamental difference. What is that difference?

17. Explain the difference between circular velocity and escape velocity. Which of those must be larger? Why?

18. Explain the difference between weight and mass.

19. Weight on Earth is proportional to mass. On the Moon, too, weight is proportional to mass, but the constant of proportionality is different on the Moon from what it is on Earth. Why? Explain why that difference does not violate the universality of physical law, as described in the Process of Science Figure.

20. Two comets are leaving the vicinity of the Sun, one traveling in an elliptical orbit and the other in an unbound orbit. What can you say about the future of those two comets? Would you expect either of them to return eventually?

21. What is the advantage of launching satellites from spaceports located near the equator? Would you expect satellites to be launched to the east or to the west? Why?

22. Explain how to use celestial orbits to estimate an object's mass. What observational quantities do you need to make that estimation?

23. Suppose astronomers discovered an object approaching the Sun in an unbound orbit. What would that say about the origin of the object?

24. What determines the strength of gravity at various radii between Earth's center and its surface?

25. The best time to dig for clams along the seashore is when the ocean tide is at its lowest. What phases of the Moon would be best for clam digging? What would be the best times of day during those phases?

26. The Moon is on the meridian at your seaside home, but your tide calendar does not show that it is high tide. What might explain that apparent discrepancy?

27. We may have an intuitive feeling for why lunar tides raise sea level on the side of Earth facing the Moon, but why is sea level also raised on the side facing away from the Moon?

28. Tides raise and lower the level of Earth's oceans. Can they do the same for Earth's landmasses? Explain your answer.

29. Lunar tides raise the ocean surface by less than 1 meter. How can tides as large as 5–10 meters occur?

30. Most commercial satellites are well inside the Roche limit as they orbit Earth. Why are they not torn apart?

Applying the Concepts

31. Mars has about one-tenth the mass of Earth and about half of Earth's radius. What is the value of gravitational acceleration on the surface of Mars in comparison with that on Earth? Compare your estimated mass and weight on Mars with your mass and weight on Earth. Do Hollywood movies showing people on Mars accurately portray that difference in weight?

32. Earth speeds along at 29.8 km/s in its orbit. Neptune's nearly circular orbit has a radius of 4.5×10^9 km, and the planet takes 164.8 years to make one trip around the Sun. Calculate how fast Neptune moves along in its orbit.

33. Venus's circular velocity is 35.03 km/s, and its orbital radius is 1.082×10^8 km. Use that information to calculate the mass of the Sun.

34. At the surface of Earth, the escape velocity is 11.2 km/s. What would be the escape velocity at the surface of a very small asteroid having a radius 10^{-4} that of Earth and a mass 10^{-12} that of Earth?

35. How long does Newton's cannonball, moving at 7.9 km/s just above Earth's surface, take to complete one orbit around Earth?

36. When a spacecraft is sent to Mars, it is first launched into an Earth orbit with circular velocity.
    a. Describe the shape of that orbit.
    b. What minimum velocity must we give the spacecraft to send it on its way to Mars?

37. Earth's average radius is 6,370 km and its mass is 5.97×10^{24} kg. Show that the acceleration of gravity at the surface of Earth is 9.81 m/s^2.

38. Using 6,370 km for Earth's radius, compare the gravitational force acting on a NASA rocket when it is sitting on its launchpad with the gravitational force acting on it when it is orbiting 350 km above Earth's surface.

39. The International Space Station travels on a nearly circular orbit 350 km above Earth's surface. What is its orbital speed?

40. Using the last equation in Working It Out 4.1, calculate the value of g on Earth. Compare that result with the measured value.

41. As described in Working It Out 4.4, tidal force is proportional to the masses of the two objects and is inversely proportional to the cube of the distance between them. Some astrologers claim that your destiny is determined by the "influence" of the planets that are rising above the horizon at the moment of your birth. Compare the tidal force of Jupiter (mass, 1.9×10^{27} kg; distance, 7.8×10^8 km) with that of the doctor in attendance at your birth (mass, 80 kg; distance, 1 meter).

42. The asteroid Ida (mass, 4.2×10^{16} kg) is attended by a tiny asteroidal moon, Dactyl, which orbits Ida at an average distance of 90 km. If you ignore the mass of the tiny moon, what is Dactyl's orbital period in hours?

43. Suppose you go skydiving.
    a. Just as you fall out of the airplane, what is your gravitational acceleration?
    b. Would that acceleration be bigger, smaller, or the same if you were strapped to a flight instructor and so had twice the mass?
    c. Just as you fall out of the airplane, what is the gravitational force on you? (Assume that your mass is 70 kg.)
    d. Would that force be bigger, smaller, or the same if you were strapped to a flight instructor and so had twice the mass?

44. Assume that a planet just like Earth is orbiting the bright star Vega at a distance of 1 astronomical unit (AU). Vega has twice the mass of the Sun.
    a. How long in Earth years will that planet take to complete one orbit around Vega?
    b. How fast is the Earth-like planet traveling in its orbit around Vega?

45. Suppose that in the past the Moon was 80 percent of the distance from Earth that it is now. Calculate how much stronger the lunar tides would have been. How would the neap and spring tides be different from now?

USING THE WEB

46. Go to NASA's "Apollo 15 Hammer-Feather Drop" Web page (https://nssdc.gsfc.nasa.gov/planetary/lunar/apollo_15_feather_drop.html or https://youtube.com/watch?v=oYEgdZ3iEKA) and watch the video from *Apollo 15* of astronaut David Scott dropping the hammer and falcon feather on the Moon. What did that experiment show? What would happen if you tried that on Earth with a feather and a hammer? Would it work? Suppose instead you dropped the hammer and a big nail. How would they fall? How does the acceleration of falling objects on the Moon compare with the acceleration of falling objects on Earth?

47. Go to the Exploratorium's "Your Weight on Other Worlds" Web page (http://exploratorium.edu/ronh/weight), which will calculate your weight on other planets and moons in our Solar System. On which objects would your weight be higher than it is on Earth? What difficulties would human bodies have in a higher-gravity environment? For example, could you easily get up out of bed and walk? What are the possible short-term and long-term effects of lower gravity on the human body? What types of life on Earth might adapt well to a different gravity?

48. a. Watch the first 12 minutes of the episode "Gravity" in season 2 of the series *The Universe* (http://history.com/shows/the-universe/season-2/episode-17 or http://dailymotion.com/video/x1p90qs or on YouTube) to see several illustrated examples of gravity. Why does the tennis ball appear to float when dropped at the top of the roller coaster? How could Newton imagine a satellite orbiting Earth centuries before it was possible? Why does the speed of the cannonball determine whether it goes into orbit? What was the technical difficulty in launching a satellite?

   b. In the same video as in part (a), watch the trip on the zero-G plane (the "vomit comet"), at 23–29 minutes. How does the plane simulate zero-G? Why does it last for only 20–30 seconds? How is that similar to the roller coaster in part (a)?

49. Go to a website that will show you the times for high and low tides—for example, http://saltwatertides.com. Pick a location and bring up the tide table for today and the next 14 days. Why do two high tides and two low tides occur every day? What is the difference in the height of the water between high and low tides? In the last few columns of the table, the times of moonrise and moonset are indicated, as well as the percentage of lunar illumination. Does the time of the high tide lead or follow the highest position of the Moon in the sky? Compare with Figure 4.12c: what phases of the Moon have the greatest differences in the height of high and low tides?

50. Figure 4.17 shows two galaxies pulling at each other, most likely after they have already collided. Watch "The Mice—Interacting Galaxies" on YouTube (https://youtube.com/watch?v=pR4pzthwqkA) to see a simulation of that interaction. Galaxies are not solid objects: they contain stars and gas and dust and a lot of empty space. Explain how those tidal tails can result from that type of interaction.

EXPLORATION

digital.wwnorton.com/astro6

In the Exploration of Chapter 3, we used the "Planetary Orbits" Interactive Simulation to explore Kepler's laws for Mercury. Now that we know how Newton's laws explain why Kepler's laws describe orbits, we will revisit the simulator to explore the Newtonian features of Mercury's orbit. Visit the Digital Resources Page and on the Student Site open the "Orbits" Interactive Simulation in Chapter 3.

Acceleration

To begin exploring the simulation, set parameters for "Mercury" in the "Orbit Settings" panel and then click "OK." Click the "Newtonian Features" tab at the bottom of the control panel. Select "show solar system orbits" and "show grid" under "Visualization Options." Change the animation rate to 0.01, and select the "start animation" button.

Examine the graph at the bottom of the panel.

1 Where is Mercury in its orbit when the acceleration is smallest?

_____

_____

2 Where is Mercury in its orbit when the acceleration is largest?

_____

_____

3 What are the values of the largest and smallest accelerations?

_____

_____

In the "Newtonian Features" graph, mark the boxes for "vector" and "line" that correspond to the acceleration. Specifying those parameters will insert an arrow that shows the direction of the acceleration and a line that extends the arrow.

4 To what Solar System object does the arrow point?

_____

_____

5 In what direction is the force on the planet?

_____

_____

Velocity

Examine the graph at the bottom of the panel again.

6 Where is Mercury in its orbit when the velocity is smallest?

_____

7 Where is Mercury in its orbit when the velocity is largest?

_____

_____

8 What are the values of the largest and smallest velocities?

_____

_____

Add the velocity vector and line to the simulation by clicking on the boxes in the graph window. Study the resulting arrows carefully.

9 Are the velocity and the acceleration always perpendicular (is the angle between them always 90°)?

_____

10 If the orbit were a perfect circle, what would be the angle between the velocity and the acceleration?

_____

_____

Hypothetical Planet

In the "Orbit Settings" panel, change the semimajor axis to 0.8 AU.

11 How does that imaginary planet's orbital period now compare with Mercury's?

_____

_____

Now change the semimajor axis to 0.1 AU.

12 How does the planet's orbital period now compare with Mercury's?

_____

_____

13 Summarize your observations of the relationship between the speed of an orbiting object and the semimajor axis.

_____

_____

5 Light

Our knowledge of the universe beyond Earth comes from light emitted, absorbed, or reflected by astronomical objects. Light carries information about the temperature, composition, and speed of the objects. Light also tells us about the nature of the material that the light passed through on its way to Earth. Yet light plays a far larger role in astronomy than that of being a messenger. Light is one of the primary means by which energy is transported throughout the universe. Stars, planets, and vast clouds of gas and dust filling the space between the stars heat up as they absorb light and cool off as they emit light. Light carries energy generated in the heart of a star outward through the star and off into space. Light transports energy from the Sun outward through the Solar System, heating the planets; and light carries energy away from each planet, allowing each one to cool. The balance between those two processes establishes each planet's temperature and therefore a planet's possible suitability for life.

LEARNING GOALS

An astronomer must try to understand the universe by studying the light and other particles that reach Earth from distant objects. By the end of this chapter, you should be able to:

LG 1 Describe the wave and particle properties of light, and describe the electromagnetic spectrum.

LG 2 Describe how to measure the chemical composition of distant objects by using the unique spectral lines of different types of atoms.

LG 3 Describe the Doppler effect and how it can be used to measure the motion of distant objects.

LG 4 Explain how the spectrum of light that an object emits depends on its temperature.

LG 5 Differentiate luminosity from brightness, and illustrate how distance affects each.

The visible part of the electromagnetic spectrum is laid out in the colors of this rainbow. ▶ ▶ ▶

What is light?

5.1 Light Brings Us the News of the Universe

Is light a **wave**—a disturbance that travels from one point to another—or is it made up of particles? Scientists have come to understand that light sometimes acts like a wave and sometimes acts like a particle. We begin with a discussion of how fast light travels. We then discuss its wavelike and particle-like properties.

The Speed of Light

In the 1670s, Danish astronomer Ole Rømer (1644–1710) studied the movement of the moons of Jupiter, measuring the times when each moon disappeared behind the planet. To his amazement, the observed times did not follow the regular schedule that he predicted from Kepler's laws. Sometimes the moons disappeared behind Jupiter sooner than expected, whereas at other times they disappeared behind Jupiter later than expected. Rømer realized that the difference depended on where Earth was in its orbit. If he began tracking the moons when Earth was closest to Jupiter, the moons were almost 17 minutes "late" by the time Earth was farthest from Jupiter. When Earth was once again closest to Jupiter, the moons again passed behind Jupiter at the predicted times.

Rømer correctly concluded that his observations did not represent a failure of Kepler's laws. Instead, he was seeing the first clear evidence that light travels at a finite speed. As shown in **Figure 5.1**, the moons appeared "late" when Earth was farther from Jupiter because of the time needed for light to travel the extra distance between the two planets. Over the course of Earth's yearly trip around the Sun, the distance between Earth and Jupiter changes by 2 astronomical units (AU). The speed of light equals that distance divided by Rømer's 16.7-minute delay, or about 3×10^5 kilometers per second (km/s). The value that Rømer actually announced in 1676 was a bit on the low side—2.25×10^5 km/s—because the size of Earth's orbit was not well known. Modern measurements of the speed of light give a value of 2.99792458×10^5 km/s in a **vacuum** (a region of space devoid of matter). The speed of light in a vacuum is one of nature's fundamental constants, usually written as c (lowercase). The speed of light through any medium, such as air or glass, is always less than c.

The International Space Station moves around Earth at a speed of about 28,000 kilometers per hour (km/h), taking 91 minutes to complete one orbit. Light travels almost 40,000 times faster than that and can circle Earth in only

Figure 5.1 Danish astronomer Ole Rømer realized that apparent differences between the predicted and observed orbital motions of Jupiter's moons depend on the distance between Earth and Jupiter. He used those observations to measure the speed of light. (The superscript letters in "$16^m 40^s$" stand for minutes and seconds of time, respectively.)

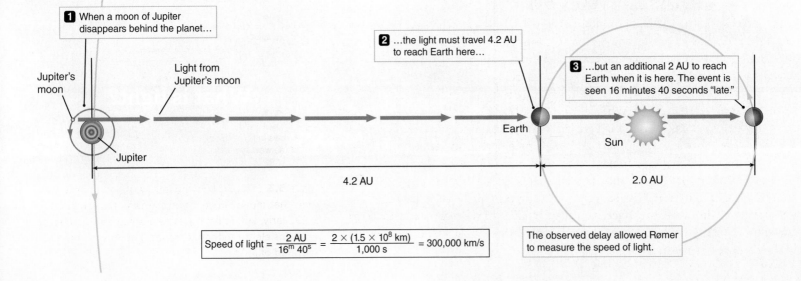

1 When a moon of Jupiter disappears behind the planet…

Jupiter's moon

Light from Jupiter's moon

Jupiter

2 …the light must travel 4.2 AU to reach Earth here…

3 …but an additional 2 AU to reach Earth when it is here. The event is seen 16 minutes 40 seconds "late."

Earth

Sun

4.2 AU

2.0 AU

$$\text{Speed of light} = \frac{2\ \text{AU}}{16^m\ 40^s} = \frac{2 \times (1.5 \times 10^8\ \text{km})}{1{,}000\ \text{s}} = 300{,}000\ \text{km/s}$$

The observed delay allowed Rømer to measure the speed of light.

1/7 of a second. Because light is so fast, its travel time is a convenient way to express cosmic distances. Light takes 1 1/4 seconds to travel between Earth and the Moon, so we say that the Moon is 1 1/4 light-*seconds* from Earth. The Sun is 8 1/3 light-*minutes* away, and the next-nearest star is 4 1/3 light-*years* distant. Thus, a *light-year* is defined as the distance light travels in 1 year, or about 9.5 trillion km. The light-year is a measure of distance.

While traveling at that high speed, light carries energy from place to place. **Energy** is the ability to do work, and it comes in many forms. **Kinetic energy** is the energy of moving objects. **Thermal energy** is closely related to kinetic energy and is the sum of all the random motion of atoms, molecules, and particles, by which we measure their temperature. For example, when light from the Sun strikes the pavement, the pavement heats up. Light carried that energy from the Sun to the pavement. Rømer knew how long light took to travel a given distance, but physicists would take more than 200 years to figure out what light actually is.

Light as an Electromagnetic Wave

In the late 19th century, the Scottish physicist James Clerk Maxwell (1831–1879) introduced the concept that electricity and magnetism are two components of the same physical phenomenon. An **electric force** is the push or pull between electrically charged particles that make up atoms, such as protons and electrons, arising from their electric charges. Particles with opposite charges attract, and those with like charges repel. A **magnetic force** is a force between electrically charged particles arising from their motion.

To describe those electric and magnetic forces, Maxwell considered what happens when charged particles move in *electric fields* and *magnetic fields*. An **electric field** is a measure of the electric force on a charge at any point in space. Similarly, a **magnetic field** is a measure of the magnetic force acting on a small magnet at any point in space.

Maxwell summarized the behavior of electric fields and magnetic fields in four elegant equations. Among other things, those equations say that a changing electric field causes a magnetic field and that a changing magnetic field causes an electric field. A change in the motion of a charged particle causes a changing electric field, which causes a changing magnetic field, which causes a changing electric field, and so on. You can see that interaction in **Figure 5.2**. Once the process starts, a self-sustaining procession of oscillating electric and magnetic fields moves out in all directions through space. In other words, an accelerating charged particle gives rise to an **electromagnetic wave**. Those electromagnetic waves, and the accelerating charges that generate them, are the sources of electromagnetic radiation. Maxwell's equations also predict the speed at which an electromagnetic wave should travel, which agrees with the measured speed of light (*c*).

Maxwell's wave description of light also suggests how light originates and how it interacts with matter. When a drop of water falls from the faucet into a sink full of water, it causes a disturbance, or wave, like the one shown in **Figure 5.3a**. The wave moves outward as a ripple on the surface of the water. **Figure 5.3b** shows that electromagnetic waves resulting from periodic changes in the strength of the electric and magnetic fields move out through space away from their source in much the same way. The ripples in the sink, however, are distortions of the water's surface, and they require a **medium**: a substance to travel through. Light waves move through empty space—what we call a vacuum— in the absence of a medium.

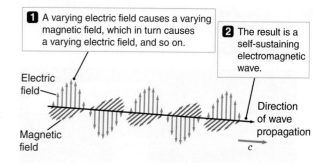

1 A varying electric field causes a varying magnetic field, which in turn causes a varying electric field, and so on.

2 The result is a self-sustaining electromagnetic wave.

Electric field

Magnetic field

Direction of wave propagation

c

Figure 5.2 An electromagnetic wave consists of oscillating electric and magnetic fields that are perpendicular both to each other and to the direction in which the wave travels.

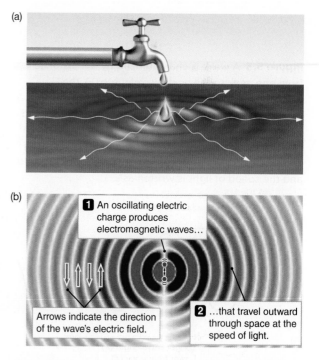

(a)

(b)

1 An oscillating electric charge produces electromagnetic waves...

Arrows indicate the direction of the wave's electric field.

2 ...that travel outward through space at the speed of light.

Figure 5.3 (a) A drop falling into water generates waves that move outward across the water's surface. (b) Similarly, an oscillating (accelerated) electric charge generates electromagnetic waves that move away at the speed of light.

(a)

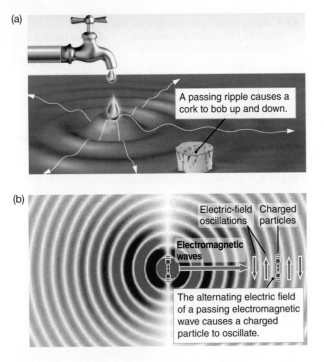

A passing ripple causes a cork to bob up and down.

(b)

Electric-field oscillations | Charged particles

Electromagnetic waves

The alternating electric field of a passing electromagnetic wave causes a charged particle to oscillate.

Figure 5.4 (a) When waves moving across the surface of water reach a cork, they cause the cork to bob up and down. (b) Similarly, a passing electromagnetic wave causes an electric charge to oscillate in response to the wave.

Figure 5.5 A wave is characterized by the distance from one peak to the next (wavelength, λ), the frequency of the peaks (f), the maximum deviations from the medium's undisturbed state (amplitude), and the velocity (v) at which the wave pattern travels from one place to another. In an electromagnetic wave, the amplitude is the maximum strength of the electric field, and the speed of light is written as c.

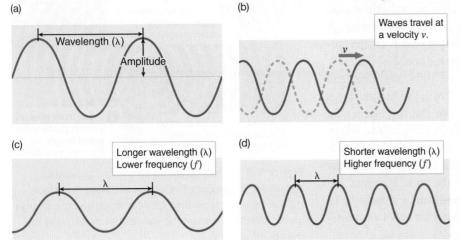

(a)

Wavelength (λ)

Amplitude

(b)

Waves travel at a velocity v.

v

(c)

Longer wavelength (λ)
Lower frequency (f)

λ

(d)

Shorter wavelength (λ)
Higher frequency (f)

λ

Now imagine that a cork is floating in the sink (**Figure 5.4a**). The cork remains stationary until the ripple from the dripping faucet reaches it. As the ripple passes by, the rising and falling water causes the cork to rise and fall. That can happen only if the wave is carrying energy—a conserved quantity that gives objects and particles the ability to do work. Light waves similarly carry energy through space and cause electrically charged particles to vibrate, as in **Figure 5.4b**.

Characterizing Waves

We discuss several kinds of waves in this book, including electromagnetic waves crossing the vast expanse of the universe and seismic waves traveling through Earth. Waves are generally characterized by four quantities: *amplitude, speed, frequency,* and *wavelength*. Each of those quantities is illustrated in **Figure 5.5**. The **amplitude** of a wave is the height of the wave above the undisturbed position (Figure 5.5a). For water waves, the amplitude is how far the water is lifted up by the wave. For light waves, the amplitude is related to the brightness of the light. A water wave travels at a particular speed, v (Figure 5.5b), through the water. The water itself doesn't travel; it just moves up and down at the same location. For waves such as those in water, that speed varies and depends on the density of the substance the wave moves through, among other things. Light, in contrast, always moves through a vacuum at the same speed, $c \approx 300{,}000$ km/s.

The distance from one crest of a wave to the next is the **wavelength**, usually denoted by the Greek letter lambda, λ (Figure 5.5c). The number of wave crests passing a point in space each second is called the wave's **frequency**, f. The unit of frequency is cycles per second, which is called **hertz** (**Hz**) after the 19th-century physicist Heinrich Hertz (1857–1894), who was the first to experimentally confirm Maxwell's predictions about electromagnetic radiation.

Figures 5.5c and 5.5d show that waves with longer (larger) wavelengths have lower (smaller) frequencies, whereas waves with shorter (smaller) wavelengths have higher (larger) frequencies. Higher-frequency waves carry more energy. Imagine standing on an ocean beach with the waves lapping up against you: the amount of energy you feel from the waves increases as the frequency of the ocean waves increases.

Waves travel a distance of one wavelength each cycle, so the speed of a wave can be found by multiplying the frequency and the wavelength. Translating that idea into math, we have $v = \lambda f$. The speed of light in a vacuum is always c, so once the wavelength of a wave of light is known, its frequency is known, and vice versa. Because light travels at constant speed, its wavelength and frequency are inversely proportional to each other: if the wavelength increases, the frequency decreases. Waves can carry a tremendous amount of information, such as complex and beautiful music, which travels by sound waves. As you continue your study of the universe, time and time again you will find that the information we receive, whether about the interior of Earth or about a distant star or galaxy, rides in on a wave.

The Electromagnetic Spectrum

Most light signals are made up of many wavelengths. A rainbow, such as the one shown in the chapter-opening figure, is created when white light interacts with water droplets and is spread out into its component colors. Light spread out by wavelength is called a **spectrum**. At the long-wavelength (and therefore low-frequency) end of the visible spectrum is red light. At the short-wavelength (high-frequency) end is violet light. A commonly used unit for the wavelength of visible light is the **nanometer (nm)**. A nanometer is one-billionth (10^{-9}) of a meter. Human eyes can see light between violet (about 380 nm) and red (750 nm). Stretched out between the two, in a rainbow, is the rest of the visible spectrum.

The light-sensitive cells in our eyes respond to visible light, but that is only a small sample of the range of possible wavelengths for electromagnetic radiation. Light can have wavelengths much shorter or much longer than your eyes can perceive. The whole range of wavelengths of light—collectively called the **electromagnetic spectrum**—is shown in **Figure 5.6**. Most of the electromagnetic spectrum—and therefore most of the information in the universe—is invisible to the human eye. To detect light outside the visible range, we must use specialized detectors of various kinds, as we explain in Chapter 6.

Refer to Figure 5.6 as we tour the electromagnetic spectrum, beginning with the shortest wavelengths and working our way to the longest ones. The very shortest wavelengths of light are called **gamma rays**, or sometimes gamma radiation. Because that light has the shortest wavelengths, it has the highest frequency and the highest energy, so it penetrates matter easily. Wavelengths between 0.1 and 40 nm are called **X-rays**. You have probably encountered X-rays at the dentist's office or in a hospital's emergency room—X-ray light has enough energy to penetrate skin and muscle but is stopped by denser bone. **Ultraviolet (UV) radiation** has wavelengths between 40 and about 380 nm—longer than X-rays but shorter than visible light. You are familiar with that type of light from sunburns: UV light has enough energy to penetrate your skin, but not much deeper.

Infrared (IR) radiation has longer wavelengths than the reddest wavelengths in the visible range. You are familiar with infrared radiation because you often feel it as heat. When you hold your hand next to a hot stove, some of the heat you feel is carried to your hand by infrared radiation emitted from the stove. In that sense, you could think of your skin as being a giant infrared eyeball—it is

Figure 5.6 By convention, the electromagnetic spectrum is divided into loosely defined regions ranging from gamma rays to radio waves. Throughout the book, we use the following labels to indicate the form of radiation used to produce astronomical images, with an icon to remind you: G = gamma rays; X = X-rays; U = ultraviolet; V = visible; I = infrared; R = radio. If more than one region is represented, multiple labels are highlighted.

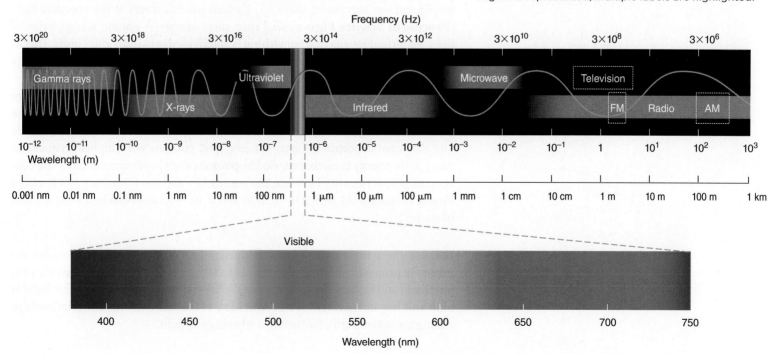

sensitive to infrared wavelengths. Infrared radiation also is used in television remote controls, and night vision goggles detect infrared radiation from warm objects such as animals. A useful unit for infrared light is the **micron** (**μm**, where μ is the Greek letter mu). One micron is 1,000 nm, or one-millionth (10^{-6}) of a meter. Infrared wavelengths are longer than red light and shorter than 500 microns.

Microwave radiation has even longer wavelengths than those of infrared radiation. The microwave in your kitchen heats the water in food by using light of those wavelengths. The longest-wavelength light, which has wavelengths longer than a few centimeters, is called **radio waves**. Light of those wavelengths in the form of FM, AM, television, and cell phone signals is used to transmit information.

CHECK YOUR UNDERSTANDING 5.1A

Rank the following in order from longest wavelength to shortest wavelength: (a) gamma rays; (b) visible light; (c) infrared light; (d) ultraviolet light; (e) radio waves

· ·

Light as a Particle

Although the wave theory of light describes many observations, it does not provide a complete picture of the properties of light. Many of the difficulties with the wave model of light have to do with how light interacts with small particles such as atoms and molecules.

The work of Albert Einstein and other scientists modified our understanding of light to show that light both acts like a wave and acts like a particle. In 1905, Einstein explained the *photoelectric effect*, in which subatomic particles called electrons are emitted when surfaces are illuminated by electromagnetic radiation greater than a certain frequency. He showed that the rate at which electrons are emitted depends only on the amount of incoming light and that the speed of the electrons depends only on the frequency of the incoming radiation. That work earned Einstein the Nobel Prize in Physics in 1921. In the particle model, light is made up of massless particles called **photons** (*phot-* means "light," as in *photograph*; and *-on* signifies a particle). Photons always travel at the speed of light (**Process of Science Figure**), and they carry energy. A photon is a *quantum* of light; *quantized* means that something is subdivided into individual units. **Quantum mechanics** is a branch of physics that deals with the quantization of energy and of other properties of matter.

The energy of a photon and the frequency of the electromagnetic wave are directly proportional to each other: the higher the frequency of the light, the greater the energy each photon carries. That relationship connects the particle and the wave concepts of light. For example, photons with higher frequencies carry more energy than that carried by photons with lower frequencies. The constant of proportionality between the energy (E) and the frequency (f) is called Planck's constant (h), which is equal to 6.63×10^{-34} joule-seconds (a joule is a unit of energy):

$$E = hf$$

Because the wavelength (λ) and frequency (f) of electromagnetic waves are inversely proportional, that also means that the photon energy is inversely proportional to the wavelength: $E = hc/\lambda$. In visible light, high-energy light is blue and low-energy light is red. **Working It Out 5.1** explores the relationships among the wavelength, frequency, and energy of light.

Agreement between Fields

Scientists working on very different problems in different fields all found the same results: light has a speed that can be measured.

Ole Rømer (1644–1710) studied eclipses of Jupiter's moons.

Rømer calculated the speed of light from the eclipse delays of Jupiter's moons.

In 1727, James Bradley (1693–1762) studied the apparent motions of stars, which appeared to make small circles because of the relative motion of Earth.

A half century after Rømer's measurement, Bradley's motion studies led to a more accurate measurement of the speed of light.

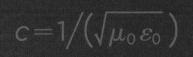

$$c = 1/\left(\sqrt{\mu_0 \varepsilon_0}\right)$$

James Clerk Maxwell (1831–1879) studied electricity and magnetism.

Maxwell determined that the speed of electromagnetic waves must be a constant, and is equal to the speed of light.

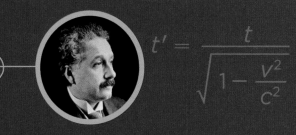

$$t' = \frac{t}{\sqrt{1 - \dfrac{v^2}{c^2}}}$$

Einstein (1879–1955) explored the consequences of the fact that the speed of light is the same for all observers.

Predictions from this theory have been tested repeatedly since 1905.

Astronomers and physicists converged on an understanding that photons travel at the "speed of light"—a fundamental constant of the universe that is the same for all observers.

5.1)) Working It Out Working with Electromagnetic Radiation

Wavelength and Frequency

When you tune to a radio station to, say, 770 AM, you are receiving an electromagnetic signal that travels at the speed of light and is broadcast at a frequency of 770 kilohertz (kHz), or 7.7×10^5 Hz. We can use the relationship between wavelength and frequency, $c = \lambda f$, to calculate the wavelength of the AM signal:

$$\lambda = \frac{c}{f} = \left(\frac{3 \times 10^8 \, \text{m/s}}{7.7 \times 10^5 \text{/s}} \right) = 390 \, \text{m}$$

FM wavelengths are much shorter than AM wavelengths. For example, a station at 89.5 FM broadcasts signals with a wavelength of 3.4 m.

The human eye is most sensitive to light in green and yellow wavelengths, about 500–590 nm. If we examine green light with a wavelength of 530 nm, we can compute its frequency:

$$f = \frac{c}{\lambda} = \left(\frac{3 \times 10^8 \, \text{m/s}}{530 \times 10^{-9} \, \text{m}} \right) = 5.66 \times 10^{14} \text{/s} = 5.66 \times 10^{14} \, \text{Hz}$$

That frequency corresponds to 566 *trillion* wave crests passing by each second.

Photon Energy

How does the energy of an X-ray photon with a wavelength of 1 nm compare with the energy of a visible-light photon with a wavelength of 530 nm (the same green light used in the previous calculation)? The equation for the energy of a photon is $E = hf$. Because $f = c/\lambda$, substituting c/λ for f yields the inverse relationship, $E = hc$. Because we are making a *comparison*, we can take a ratio, and then the constants h and c cancel out:

$$\frac{E_{\text{X-ray photon}}}{E_{\text{visible photon}}} = \frac{hc/\lambda_{1\,\text{nm}}}{hc/\lambda_{530\,\text{nm}}} = \frac{h\!\!\!/c\!\!\!/}{h\!\!\!/c\!\!\!/} \times \frac{\lambda_{500\,\text{nm}}}{\lambda_{1\,\text{nm}}} = \frac{530\,\text{nm}}{1\,\text{nm}} = 530$$

The X-ray photon has 530 times the energy of the visible-light photon.

AstroTour: Light as a Wave, Light as a Photon

In the particle description of light, the electromagnetic spectrum is a spectrum of photon energies. The higher the frequency of the electromagnetic wave, the greater the energy carried by each photon. Photons of shorter wavelength (higher frequency) carry more energy than that carried by photons of longer wavelength (lower frequency). For example, photons of blue light carry more energy than that carried by photons of longer-wavelength red light. Ultraviolet photons carry more energy than that carried by photons of visible light, and X-ray photons carry more energy than that carried by ultraviolet photons. The lowest-energy photons are radio wave photons. A beam of red light can carry just as much energy as a beam of blue light, but the red beam will have more photons than the blue beam.

CHECK YOUR UNDERSTANDING 5.1B

As wavelength increases, the energy of a photon _____ and its frequency _____. (a) increases; decreases (b) increases; increases (c) decreases; decreases (d) decreases; increases

---

5.2 The Quantum View of Matter Explains Spectral Lines

Light and matter interact, and that interaction allows us to detect matter even at great distances in space. To understand that interaction, we must first understand the building blocks of matter. Here in Section 5.2, we review atomic structure and the process by which astronomers identify the chemical elements in astronomical objects.

Atomic Structure

Matter is anything that occupies space and has mass. Atoms are composed of a central massive **nucleus**, which contains **protons** with a positive charge and **neutrons**, which have no charge. A cloud of negatively charged **electrons** surrounds the nucleus. Atoms with the same number of protons are all of the same type of chemical **element**. For example, an atom with two protons, shown in **Figure 5.7a**, is the element helium. An atom with six protons is the element carbon, one with eight protons is the element oxygen, and so forth. An element may have many **isotopes**: atoms with the same number of protons but different numbers of neutrons. **Molecules** are groups of atoms bound together by shared electrons. A single teaspoon of water contains about 10^{23} atoms—about as many atoms as the number of stars in the observable universe.

For an atom to be electrically neutral, it must have the same number of electrons as protons. Electrons have much less mass than protons or neutrons, so almost all the mass of an atom is found in its nucleus. The Danish physicist Niels Bohr (1885–1962) proposed a model of the atom in 1913 in which a massive nucleus sits in the center and the smaller electrons orbit around it, much as planets orbit around the Sun (**Figure 5.7b**).

The **Bohr model**, however, is an incomplete description of the atom. Just as waves of light have particle-like properties, particles of matter also have wavelike properties. Once that was understood, the Bohr model was modified so that the positively charged nucleus is surrounded by electron "clouds" or "waves," as shown in **Figure 5.7c**. In that quantum mechanical model, it is not possible to know precisely where the electron is in its orbit. The wave characteristics of particles make it impossible to pin down simultaneously their exact location and their exact velocity; some uncertainty will always exist. That is why a featureless cloud is used to represent electrons in orbit around an atomic nucleus.

Atomic Energy Levels

Because of their wavelike properties, electrons in an atom can take on only certain specific energies that depend on the energy states of the atom. The form that the electron waves take depends on the possible energy states of atoms. The energy states of atoms are like the shelves of the bookcase depicted in **Figure 5.8a**. The energy of an atom might correspond to the energy of one state or to the energy of the next state, but the energy of the atom is never found between the two states, just as a book can be on only one shelf at a time and cannot be partly on one shelf and partly on another. A given atom may have many energy states available to it, but those stable states are *discrete*.

Astronomers keep track of the allowed states of an atom by using energy level diagrams, as shown in **Figure 5.8b**. Both the bookcase and the energy level diagram are simplifications of the

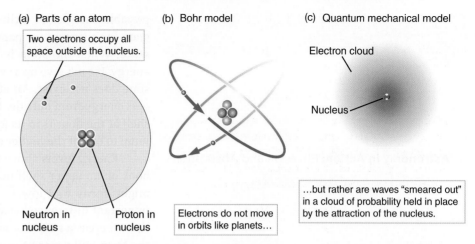

(a) Parts of an atom

Two electrons occupy all space outside the nucleus.

Neutron in nucleus

Proton in nucleus

(b) Bohr model

Electrons do not move in orbits like planets…

(c) Quantum mechanical model

Electron cloud

Nucleus

…but rather are waves "smeared out" in a cloud of probability held in place by the attraction of the nucleus.

Figure 5.7 (a) An atom (here, helium) is made up of a nucleus consisting of positively charged protons and electrically neutral neutrons and is surrounded by much less massive negatively charged electrons. (b) Atoms are often drawn as miniature "solar systems," but this Bohr model is incorrect. (c) Electrons are actually smeared out around the nucleus in quantum mechanical clouds of probability.

AstroTour: Atomic Energy Levels and the Bohr Model

Figure 5.8 (a) Energy states of an atom are analogous to shelves in a bookcase. You can move a book from one shelf to another, but books can never be placed between shelves. (b) Atoms exist in one allowed energy state or another but never in between. The ground state is the lowest possible level.

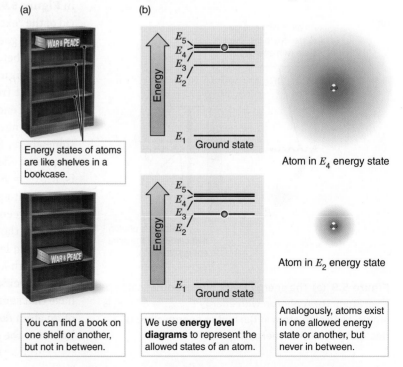

(a)

WAR & PEACE

Energy states of atoms are like shelves in a bookcase.

You can find a book on one shelf or another, but not in between.

(b)

Energy

E_5
E_4
E_3
E_2

E_1 Ground state

Atom in E_4 energy state

Energy

E_5
E_4
E_3
E_2

E_1 Ground state

Atom in E_2 energy state

We use **energy level diagrams** to represent the allowed states of an atom.

Analogously, atoms exist in one allowed energy state or another, but never in between.

possible energies of a three-dimensional system. The lowest possible energy state for a system (or part of a system) such as an atom is called the **ground state**. When the atom is in the ground state, the electron has its minimum energy. It can't give up any more energy to move to a lower state because a lower state does not exist. An atom will remain in its ground state forever unless it gets energy from outside. In the bookcase analogy, a book sitting on the bottom shelf at the floor is in its ground state. It has nowhere left to fall, and it cannot jump to one of the higher shelves on its own.

Energy levels above the ground state are called **excited states**. Just as a book on an upper shelf might fall to a lower shelf, an atom in an excited state might **decay** to a lower state by getting rid of some of its extra energy. An important difference between the atom and the book on the shelf, however, is that whereas a snapshot might catch the book falling between the two shelves, the atom will never be caught between two energy states. When the transition occurs from a higher state to a lower one, the difference in energy between the two states is carried off all at once. A common way for an atom to do that is to emit a photon. The photon emitted by the atom carries away exactly the amount of energy that atom loses as it goes from the higher energy state to the lower energy state. Similarly, atoms moving from a lower energy state to a higher energy state can absorb only certain specific energies.

To make an analogy with money, suppose you have a penny (1 cent), a nickel (5 cents), and a dime (10 cents), totaling 16 cents. Now imagine that you give away the nickel and are left with 11 cents. You never had exactly 13 cents or 13.6 cents. You had 16 cents and then 11 cents. Atoms don't accept and give away money to change energy states, but they do accept and give away photons with well-defined energies.

EMISSION SPECTRA   Imagine a hypothetical atom that has only two available energy states. The energy of the lower energy state (the ground state) is E_1, and the energy of the higher energy state (the excited state) is E_2. The energy levels of that atom can be represented in an energy level diagram such as the one in **Figure 5.9a**. An atom in the excited state moves to the ground state by getting rid of the "extra" energy all at once. In that **emission** process, a photon is released. The atom goes from one energy state to another, but it never has an amount of energy in between.

In **Figure 5.9b**, the green arrow pointing straight down indicates that the atom went from the higher state (E_2) to the lower state (E_1). The atom lost an amount of energy equal to $E_2 - E_1$, the difference between the two states. Because energy is never truly lost or created, the energy the atom lost has to show up somewhere. Here, the energy shows up in the form of a photon emitted by the atom. The energy of the emitted photon matches the energy lost by the atom; that is, $E_{\text{photon}} = E_2 - E_1$.

An atom can emit photons with energies corresponding *only* to the difference between two of its allowed energy states. Because the energy of a photon is related to the frequency or wavelength of electromagnetic radiation, in the transition from level 2 to level 1, a photon of energy $E_{\text{photon}} = E_2 - E_1$ has a specific wavelength $\lambda_{2 \to 1}$ and a specific frequency $f_{2 \to 1}$. Therefore, those emitted photons have a specific color, and every photon emitted in any transition from E_2 to E_1 will have that same color defined by a specific wavelength. The energy level structure of an atom determines the wavelengths of the photons it emits—the color of the light that the atom gives off.

Astronomy in Action: Emission and Absorption

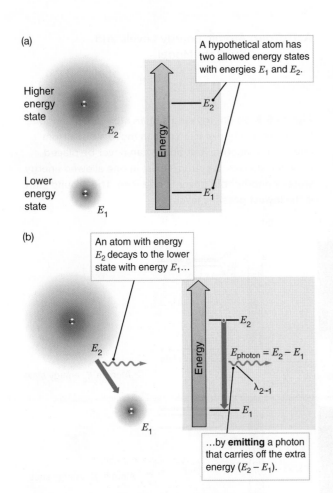

(a)

Higher energy state

E_2

Lower energy state

E_1

A hypothetical atom has two allowed energy states with energies E_1 and E_2.

Energy

E_2

E_1

(b)

An atom with energy E_2 decays to the lower state with energy E_1...

E_2

E_1

Energy

E_2

$E_{\text{photon}} = E_2 - E_1$

$\lambda_{2 \to 1}$

E_1

...by **emitting** a photon that carries off the extra energy ($E_2 - E_1$).

Figure 5.9 (a) The energy levels of a hypothetical two-level atom. (b) A photon with energy $E_{\text{photon}} = E_2 - E_1$ is emitted when an atom in the higher energy state (E_2) decays to the lower energy state (E_1).

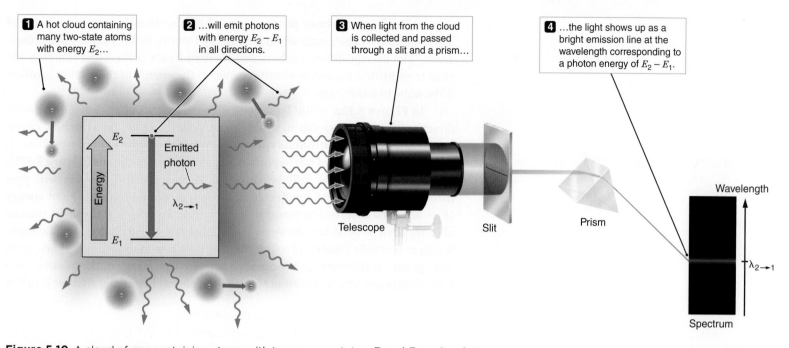

1 A hot cloud containing many two-state atoms with energy E_2...

2 ...will emit photons with energy $E_2 - E_1$ in all directions.

3 When light from the cloud is collected and passed through a slit and a prism...

4 ...the light shows up as a bright emission line at the wavelength corresponding to a photon energy of $E_2 - E_1$.

Energy

E_2

Emitted photon

$\lambda_{2\to1}$

E_1

Telescope

Slit

Prism

Wavelength

Spectrum

$\lambda_{2\to1}$

Figure 5.10 A cloud of gas containing atoms with two energy states, E_1 and E_2, emits photons with an energy $E_{\text{photon}} = E_2 - E_1$, which appear in the spectrum (right) as a single bright *emission line*.

Figure 5.10 illustrates the light coming from a cloud of gas consisting of the hypothetical two-state atoms in Figure 5.9. Any atom in the higher energy state (E_2) quickly decays and emits a photon in a random direction, and an enormous number of photons come pouring out of the cloud of gas. Instead of containing photons of all different energies—light of all different colors—that light contains only photons with the specific energy $E_2 - E_1$ and wavelength $\lambda_{2\to1}$. In other words, all the light coming from those atoms in the cloud is the same color. If you spread the light out into its component colors, only one color would be present—a single bright line called an **emission line**.

Why was the atom in the excited state E_2 in the first place? An atom sitting in its ground state will remain there unless it absorbs just the right amount of energy to kick it up to an excited state. In general, the atom either absorbs the energy of a photon or it collides with another atom or an unattached electron and absorbs some of the other particle's energy. In a neon sign, an alternating electric field inside the glass tube pushes electrons in the gas back and forth through the neon gas inside the tube. Some of those electrons crash into atoms of the gas, knocking them into excited states. The atoms then drop back down to their ground states by emitting photons, causing the gas inside the tube to glow a color characteristic of the element neon.

ABSORPTION SPECTRA An atom in a low energy state can absorb the energy of a passing photon and jump up to a higher energy state, as shown in **Figure 5.11**. Once again, the energy required to go from E_1 to E_2 is the difference in energy between the two states, $E_2 - E_1$. For a photon to cause an atom to jump from E_1 to E_2, it must provide exactly that much energy. The only photons that can excite atoms from E_1 to E_2 are photons with $E_{\text{photon}} = E_2 - E_1$. As with emission, those photons have a corresponding frequency ($f_{1\to2} = E_{\text{photon}}/h$) and wavelength

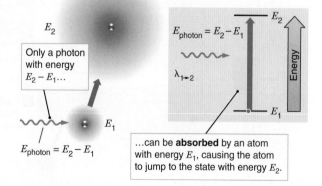

E_2

Only a photon with energy $E_2 - E_1$...

E_1

$E_{\text{photon}} = E_2 - E_1$

$E_{\text{photon}} = E_2 - E_1$

$\lambda_{1\to2}$

E_2

E_1

Energy

...can be **absorbed** by an atom with energy E_1, causing the atom to jump to the state with energy E_2.

Figure 5.11 An atom in a lower energy state (E_1) may absorb a photon of energy $E_{\text{photon}} = E_2 - E_1$, leaving the atom in a higher energy state (E_2).

($\lambda_{1\rightarrow2} = hc/E_{photon}$). Those photons have the same energy—the same color of light—emitted by the atoms when they decay from E_2 to E_1. That is not a coincidence. The energy difference between the two levels is the same whether the atom is emitting a photon or absorbing one, so the energy of the photon involved is the same in either case.

In **Figure 5.12a**, white light (with all wavelengths of photons in it) passes directly through a glass prism, which spreads out the light into a rainbow of colors. That is called a **continuous spectrum**. When the white light passes through a cool cloud composed of our hypothetical gas of two-state atoms (**Figure 5.12b**), however, some photons will be absorbed. Almost all the photons will pass through the cloud of gas unaffected because they do not have the right energy ($E_2 - E_1$) to be absorbed by atoms of the gas. Photons with just the right amount of energy can be absorbed, however, and as a result, those photons will be missing from the light passing through the prism. When that occurs, a sharp, dark line appears at the wavelength corresponding to the energy of the missing photons. That process by which atoms capture the energy of passing photons is called

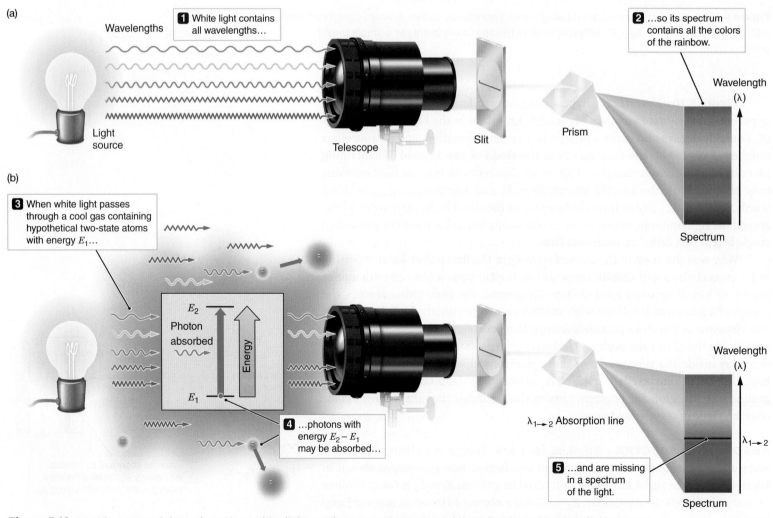

Figure 5.12 (a) When passed through a prism, white light produces a spectrum containing all colors. (b) When light of all colors passes through a cloud of hypothetical two-state atoms, photons with energy $E_{photon} = E_2 - E_1$ may be absorbed, leading to the dark absorption line in the spectrum.

absorption, and the dark line seen in the spectrum is called an **absorption line**. **Figure 5.13a** shows the absorption lines in the spectrum of a star. The same spectrum is shown again in **Figure 5.13b** as a graph of the brightness at every wavelength. The brightness in Figure 5.13b drops abruptly at the wavelengths corresponding to the dark lines in Figure 5.13a. Places between the dark lines are brighter and therefore higher on the graph than the absorption lines.

When an atom absorbs a photon, it may quickly decay to its previous lower energy state, emitting a photon with the same energy as the photon it just absorbed. If the atom reemits a photon just like the one it absorbed, why does the absorption matter? The photon that was taken out of the passing light was replaced, but all the absorbed photons were originally traveling in the same direction, whereas the emitted photons are traveling in random directions. In other words, some photons with energies equal to $E_2 - E_1$ are diverted from their original paths by their interaction with atoms. If you look at a white light through the cloud in Figure 5.12b, you will observe an absorption line at a wavelength of $\lambda_{2\to1}$, but if you look at the cloud from another direction, you will observe an emission line at that same wavelength.

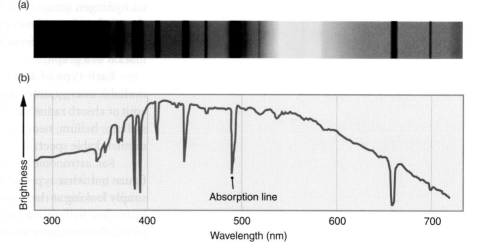

Figure 5.13 Absorption lines in the spectrum of a star as an image (a) and a graph (b).

AstroTour: Atomic Energy Levels and Light Emission and Absorption

Spectral Fingerprints of Atoms

In the mid-19th century, Gustav Kirchoff (1824–1887) first observed emission, continuous, and absorption spectra from the three types of sources shown in Figures 5.10 and 5.12. He summarized his findings as three empirical laws—but he did *not* know about energy levels in atoms. Kirchoff, together with Robert Bunsen (1811–1899), concluded that the dark-line (absorption) spectrum of the Sun was the "reverse" of the bright-line (emission) spectrum that would be produced by the Sun's atmosphere alone and identified some of the elements on the Sun. (A few years later the element helium was discovered in the solar spectrum.) Others observed the stars. Noteworthy among them were William (1824–1910) and Margaret Lindsay (1848–1915) Huggins, who published an atlas of stellar spectra in 1899 and who showed that the types of atoms seen in the stars are the same as those found on Earth.

The study of spectra of astronomical objects is fundamental to our understanding of the universe. Astronomers who study spectra will say that "a spectrum is worth a thousand pictures" because of the wealth of information that can come from it. We refer to spectra in every chapter in this book.

Real atoms can occupy many more than just two possible energy states; therefore, any given type of atom can emit and absorb photons at many wavelengths. An atom with three energy states, for example, might jump from state 3 to state 2, or from state 3 to state 1, or from state 2 to state 1. The three distinct emission lines in the spectrum from a gas made up of those atoms would have wavelengths of $hc/(E_3 - E_2)$, $hc/(E_3 - E_1)$, and $hc/(E_2 - E_1)$, respectively.

An atom's allowed energy states are determined by the interactions among the electrons and the nucleus. Every neutral hydrogen atom consists of a nucleus containing one proton, plus a single electron in a cloud surrounding the nucleus. Therefore, every hydrogen atom has the same energy states available to it, and

remain in their excited state for 1 minute before decaying and emitting a photon, then after 1 minute a 50-50 chance exists that any particular atom in the toy will have decayed and a 50-50 chance that the atom will remain in its excited state. Although it is impossible to say exactly which atoms will decay, about half of the trillions and trillions of atoms in the toy will decay within 1 minute, and the brightness of the glow from the toy will have dropped to half of what it was. After each minute, half of the remaining excited atoms decay, and the glow from the toy drops to half of what it was 1 minute earlier. Thus, the glow from the toy slowly fades away.

In deep space, where atoms can remain undisturbed for long periods, certain excited states of atoms last, on average, for tens of millions of years or even longer. An atom may have been in such an excited energy state for a few seconds, a few hours, or 50 million years when, in an instant, it decays to the lower energy state without anything having caused it to do so. Physicists can only calculate the *probabilities* that certain events will take place.

CHECK YOUR UNDERSTANDING 5.2

How can spectra tell us the chemical composition of a distant star?

. . . .

5.3 The Doppler Shift Indicates Motion Toward or Away from Us

Astronomy in Action: Doppler Shift

AstroTour: The Doppler Effect

You have already seen that light is a tightly packed bundle of information that can reveal a wealth of information about the physical state of material located tremendous distances away. Here in Section 5.3, we explain how light can be used to determine whether a distant astronomical object is moving away from us or toward us, and at what speed.

Have you ever listened to an ambulance speed by with sirens blaring? As the ambulance approaches, its siren has a certain high pitch, but as it passes by, the pitch of the siren drops noticeably. If you close your eyes and listen, you have no trouble knowing when the ambulance passed; the change in the pitch of its siren indicates that it has passed you by. You do not even need an ambulance to hear that effect. The sound of normal traffic behaves in the same way. As a car drives past, the pitch of the sound that it makes suddenly drops. (The same effect would happen if you were the one driving past a stationary siren).

The pitch of a sound is like the color of light: it is determined by the wavelength or, equivalently, the frequency of the sound wave. What you perceive as higher in pitch corresponds to sound waves with higher frequencies and shorter wavelengths. Sounds that you perceive as lower in pitch are waves with lower frequencies and longer wavelengths. When an object is moving toward you, the waves that it emits, whether light or sound or waves in the water, "crowd together" in front of the object. You can see how that process works in **Figure 5.16**, which shows the locations of successive wave crests emitted by a moving object. The waves that reach you have a shorter wavelength and therefore a higher frequency than the waves given off by the object when it is not moving. Conversely, if an object is moving away from you, the waves reaching you from the object are spread out (longer λ, lower f). That change in frequency as a result of motion is known as the **Doppler effect** and is named after physicist Christian Doppler (1803–1853).

The Doppler effect causes a shift in the light emitted from a moving object. If the object were at rest, it would emit light with the **rest wavelength** (λ_{rest}), as

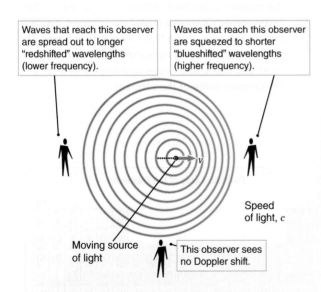

Waves that reach this observer are spread out to longer "redshifted" wavelengths (lower frequency).

Waves that reach this observer are squeezed to shorter "blueshifted" wavelengths (higher frequency).

Speed of light, *c*

Moving source of light

This observer sees no Doppler shift.

Figure 5.16 Motion of a light or sound source relative to an observer may cause waves to be spread out (*redshifted*, or lower in pitch) or squeezed together (*blueshifted*, or higher in pitch). A change in the wavelength of light or the frequency of sound is called a *Doppler shift*.

shown in **Figure 5.17a**. If an object such as a star is moving *toward* you, the light reaching you from the object has a shorter wavelength than its rest wavelength—the light is "bluer" than the rest wavelength, and the light is described as **blueshifted**, as shown by the blue waves in **Figure 5.17b**. In contrast, light from a source moving *away* from you is shifted to longer wavelengths. The light that you see is "redder" than if the source were not moving away from you and is described as **redshifted**, as shown by the red waves in **Figure 5.17c**. The faster the object is moving with respect to you, the larger the shift. The amount by which the wavelength of light is shifted by the Doppler effect is called the light's **Doppler shift**, which depends on the speed of the object emitting the light.

The Doppler shift provides information only about the **radial velocity** (v_r) of the object, which is the part of the motion that is toward you or away from you. The radial velocity is the rate at which the distance between you and the object is changing: if v_r is positive, the object is getting farther away from you; if v_r is negative, the object is getting closer. At the moment the ambulance is passing you, it is getting neither closer nor farther away, so the pitch you hear is the same as the pitch heard by the crew riding on the truck. Similarly, a distant object moving slowly across the sky does not move toward or away from you, and so its light will not be Doppler shifted from your point of view.

Doppler shifts become especially useful when you are looking at an object that has emission or absorption lines in its spectrum. Those spectral lines enable astronomers to determine how rapidly the object is moving toward or away from Earth. To determine that velocity, astronomers first identify the spectral line as being from a certain chemical element, which has a unique rest wavelength (λ_{rest}) measured in a lab on Earth. They then measure the observed wavelength (λ_{obs}) in the spectrum of the distant object. The difference between the rest wavelength and the observed wavelength indicates the object's radial velocity. See **Working It Out 5.2**.

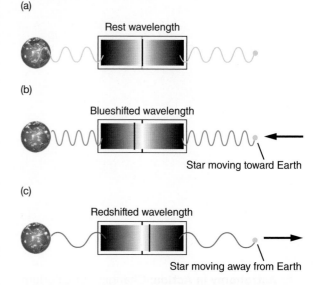

Figure 5.17 From their rest wavelength (a), spectral lines of astronomical objects are blueshifted if they are moving toward the observer (b) and redshifted if they are moving away from the observer (c).

5.2 Working It Out Making Use of the Doppler Effect

The Doppler formula for objects moving at a radial velocity (v_r) much less than the speed of light is given by

$$v_r = \frac{\lambda_{obs} - \lambda_{rest}}{\lambda_{rest}} \times c$$

A prominent spectral line of hydrogen atoms has a rest wavelength, λ_{rest}, of 656.3 nm (see Figure 5.14b). Suppose that you measure the wavelength of that line in the spectrum of a distant object and find that instead of seeing the line at 656.3 nm, you see the line at a wavelength, λ_{obs}, of 659.0 nm. What is its radial velocity? Using the above equation,

$$v_r = \frac{659.0 \, \text{nm} - 656.3 \, \text{nm}}{656.3 \, \text{nm}} \times (3 \times 10^5 \, \text{km/s})$$

$$v_r = 1,200 \, \text{km/s}$$

Thus, the object is moving away from you with a radial velocity of 1,200 km/s.

Suppose instead that you know the velocity and want to compute the wavelength at which you would observe the spectral line. Earth's nearest stellar neighbor, Proxima Centauri, is moving toward us at a radial velocity of −21.6 km/s. What is the observed wavelength, λ_{obs}, of a magnesium line in Proxima Centauri's spectrum that has a rest wavelength, λ_{rest}, of 517.27 nm? We can rearrange the Doppler formula to solve for λ_{obs}:

$$\lambda_{obs} = \left(1 + \frac{v_r}{c}\right) \times \lambda_{rest}$$

$$\lambda_{obs} = \left(1 + \frac{-21.6 \, \text{km/s}}{3 \times 10^5 \, \text{km/s}}\right) \times 517.27 \, \text{nm} = 517.23 \, \text{nm}$$

Although the observed Doppler blueshift ($\lambda_{obs} - \lambda_{rest} = 517.23 - 517.27$) is only −0.04 nm, it is easily measured with modern instrumentation.

5.4 Temperature Affects the Spectrum of Light That an Object Emits

The balance between heating and cooling in an object determines its temperature. If an object's temperature is constant, those two must be in balance. Here in Section 5.4, we examine that balance and see how we can use it to predict the temperatures of planets and stars.

Equilibrium and Balance

Your body is heated by the release of chemical energy inside it. Sometimes your body is also heated by energy from your surroundings. If you are standing in sunshine on a hot day, the hot air around you and the sunlight falling on you both heat you. In response to that heating, your body cools itself off by perspiring: water seeps from the pores in your skin and evaporates. The energy to evaporate the water comes from your body. As the perspiration evaporates, it cools you down because it removes heat energy from your body. For your body temperature to remain stable, the heating must be balanced by the cooling. If more heating than cooling occurs, your body temperature increases; if more cooling than heating, your body temperature decreases.

In a tug-of-war contest between two perfectly matched teams, each team pulls on the rope, but the force of one team's pull is only enough to match, not overcome, the force exerted by the other team. A picture taken now and another taken 5 minutes from now would not differ in any significant way. In that static equilibrium, opposing forces balance each other exactly. Static equilibrium can be stable, unstable, or neutral. A marble in a bowl is in a stable equilibrium: if it moves, it will return to its original position at the bottom of the bowl. A book standing on its edge, unsupported on either side, is in an unstable equilibrium: if you nudge it, it will fall over, not settle back into its original position. When an unstable equilibrium is disturbed, it moves further away from equilibrium rather than back toward it.

Equilibrium also can be dynamic, which means the system is constantly changing so that one source of change is exactly balanced by another source of change, and the configuration of the system remains the same. In **Figure 5.18**, placing a can with a hole cut in the bottom under an open water faucet provides a simple example of dynamic equilibrium. The depth of the water in the can determines how fast water pours out through the hole in the bottom of the can. When the water reaches just the right depth (Figure 5.18a), water pours out of the hole in the bottom of the can at the same rate it pours into the top of the can from the faucet. The water leaving the can balances the water entering, and equilibrium is established. If you took a picture now and another picture in a few minutes, little of the water in the can would be the same, but the pictures would be indistinguishable.

If a system is not in equilibrium, its configuration will change. If the level of the water in the can is too low (Figure 5.18b), water will not flow out of the bottom of the can fast enough to balance the water flowing in. The water level will begin to rise until the amount coming in equals the amount going out. A picture taken

Astronomy in Action: Changing Equilibrium

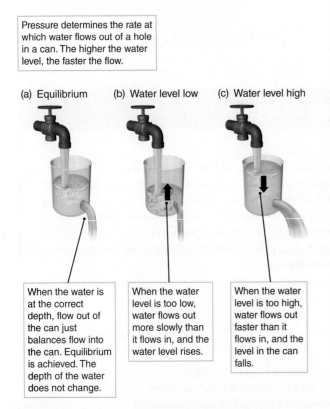

Pressure determines the rate at which water flows out of a hole in a can. The higher the water level, the faster the flow.

(a) Equilibrium   (b) Water level low   (c) Water level high

When the water is at the correct depth, flow out of the can just balances flow into the can. Equilibrium is achieved. The depth of the water does not change.

When the water level is too low, water flows out more slowly than it flows in, and the water level rises.

When the water level is too high, water flows out faster than it flows in, and the level in the can falls.

Figure 5.18 The relative rates at which water flows into and out of a can determine the water level in the can. This is an example of dynamic equilibrium.

now and another taken a short time later would not look the same. Conversely, if the water level in the can is too high (Figure 5.18c), water will flow out of the can faster than it flows in. The water level will begin to fall until the amount coming in equals the amount going out. Once again, if the system is not in equilibrium, its configuration will change.

When heating is balanced by cooling, we call it **thermal equilibrium**. Planets have a dynamic but stable thermal equilibrium, and electromagnetic radiation plays a crucial role in maintaining that balance. Energy from sunlight heats the surface of a planet, driving its temperature up, and the planet emits thermal radiation into space, cooling it down. For a planet to remain at the same average temperature over time, the energy it radiates into space must exactly balance the energy it absorbs from the Sun. **Figure 5.19** shows how the equilibrium temperature of a planet is analogous to the water level in Figure 5.18. We return to planetary equilibrium later in the chapter. Many kinds of equilibrium exist besides thermal equilibrium, some of which we encounter later in the book.

Temperature

In everyday life, we define hot and cold subjectively: something is hot or cold when it feels hot or cold. When we measure *temperature*, we specify degrees on a thermometer, but the way we define a degree is arbitrary. If you grew up in the United States, you probably think of temperatures in degrees Fahrenheit (°F), whereas if you grew up almost anywhere else in the world, you think of temperatures in degrees Celsius (°C).

Temperature is a measurement of how energetically the atoms that make up an object are moving about. The air around us is composed of vast numbers of atoms and molecules. Those molecules are moving about every which way. Some move slowly; some move more rapidly. All atoms and molecules are constantly in motion. The average kinetic energy (E_K) is given by $E_K = 1/2\,mv^2$, where m is the mass of an atom or molecule and v is its velocity. The more energetically the atoms or molecules are bouncing about, the higher the object's temperature. In fact, the random motions of atoms and molecules are often called their **thermal motions** to emphasize the connection between those motions and temperature. **Figure 5.20** shows that when the temperature of a gas is increased, the kinetic energy is increased, in which case the atoms move faster.

The atoms and molecules in a solid body (like you) cannot move about freely like a gas, but they still move back and forth around their average location, and temperature measures the amount of that movement. If an object is hotter than you are, thermal energy flows from that object into you. At the atomic level, that means the object's atoms are bouncing more energetically than the atoms in your body are, so if you touch the object, its atoms collide with your atoms, causing the atoms in your body to move faster. Your body gets hotter as thermal energy flows from the object to you. At the same time, those collisions rob the particles in the object of some of their energy. Their motions slow down, and the hotter object cools. Heating processes increase the average thermal energy of an object's particles, whereas cooling processes decrease the average thermal energy of those particles.

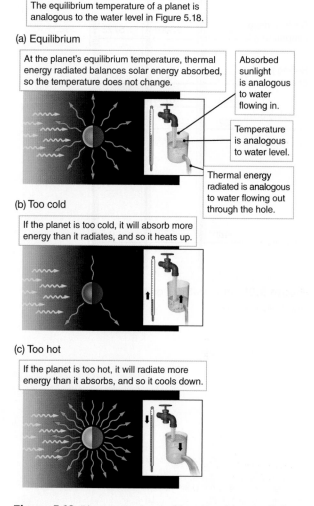

The equilibrium temperature of a planet is analogous to the water level in Figure 5.18.

(a) Equilibrium

At the planet's equilibrium temperature, thermal energy radiated balances solar energy absorbed, so the temperature does not change.

Absorbed sunlight is analogous to water flowing in.

Temperature is analogous to water level.

Thermal energy radiated is analogous to water flowing out through the hole.

(b) Too cold

If the planet is too cold, it will absorb more energy than it radiates, and so it heats up.

(c) Too hot

If the planet is too hot, it will radiate more energy than it absorbs, and so it cools down.

Figure 5.19 Planets are heated by absorbing sunlight (and sometimes by internal heat sources) and cooled by emitting thermal radiation into space. In the absence of other sources of heating or means of cooling, the equilibrium between those processes determines the temperature of the planet.

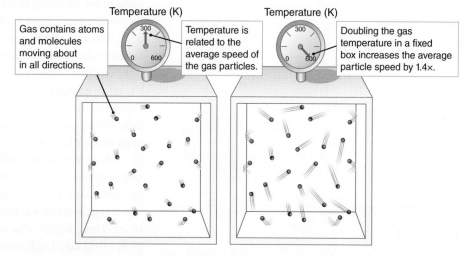

Gas contains atoms and molecules moving about in all directions.

Temperature (K)

Temperature is related to the average speed of the gas particles.

Temperature (K)

Doubling the gas temperature in a fixed box increases the average particle speed by 1.4×.

Figure 5.20 Higher gas temperatures correspond to faster-moving atoms.

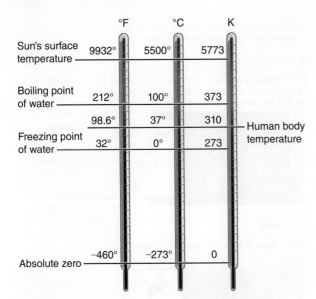

Figure 5.21 Temperatures on the Fahrenheit, Celsius, and Kelvin scales.

The three most common temperature scales are shown in **Figure 5.21**. On the Fahrenheit scale, water at sea level has a freezing point of 32°F and a boiling point of 212°F. On the Celsius scale, water freezes at 0°C and boils at 100°C. Because the number of degrees between freezing and boiling on those two scales is different (180°F vs. 100°C), a 1-degree change measured in °F is not the same as a 1-degree change measured in °C.

As the motions of the particles in an object slow down, the temperature decreases more and more. The lowest possible temperature, at which all thermal motions would stop, is called **absolute zero**. Absolute zero corresponds to −273.15°C and −459.57°F. It also is the starting point of the **Kelvin temperature scale**. The size of one unit on the Kelvin scale, called a **kelvin (K)**, is the same as the Celsius degree. To convert between °C and K, just add 273.15 to the temperature in °C. Thus, water freezes at 273.15 K and water boils at 373.15 K. The Kelvin scale has no negative temperatures.

When temperatures are measured in kelvins, the average thermal energy of particles is proportional to the measured temperature. Thus, the average thermal energy of the atoms in an object with a temperature of 200 K is twice the average thermal energy of the atoms in an object with a temperature of 100 K.

Temperature, Luminosity, and Color

We have seen the way discrete atoms emit and absorb radiation, which leads to a useful understanding of emission lines and absorption lines that tell us about the physical state and motion of distant objects. But not all objects have spectra dominated by discrete spectral lines. As you saw in Figure 5.12a, if you pass the light from a lightbulb through a prism, instead of discrete bright and dark bands you will see light spread out smoothly from the blue end of the spectrum to the red. Similarly, if you look closely at the spectrum of the Sun, you will see absorption lines, and you will see light smoothly spread out across all colors of the spectrum—the continuous spectrum noted earlier.

We can think of a dense material as being composed of a collection of charged particles being jostled as their thermal motions cause them to run into their neighbors. The hotter the material is, the more violently its particles are being jostled. A charged particle radiates anytime it accelerates, so the jostling of particles that results from their thermal motions causes them to give off a continuous spectrum of electromagnetic radiation. That is why any material dense enough for its atoms to be jostled by their neighbors emits light simply because of its temperature. That kind of radiation is called **thermal radiation**.

The radiation from an object changes as the object heats up or cools down. **Luminosity** is the amount of light *leaving* a source—that is, the total amount of light emitted (energy per second, measured in watts, W). The hotter the object, the more energetically the charged particles within it move, and the more energy they emit in the form of electromagnetic radiation. So, as an object gets hotter, the light that it emits becomes more intense. In other words, *an object is more luminous when it is hotter.*

Now, what color light does an object emit? As the object gets hotter, the thermal motions of its particles become more energetic, producing more energetic photons. As the average energy of the photons that it emits increases, the average wavelength of the emitted photons decreases, and the light from the object gets bluer. Thus, *hotter objects are bluer.* If you heat a piece of metal, the metal will glow—first a dull red, then orange, and then yellow. The hotter the metal becomes, the more the highly energetic blue photons become mixed with the less energetic red photons, and the color of the light shifts from red toward blue. The light becomes more intense and bluer as the metal becomes hotter.

Blackbodies are objects that emit electromagnetic radiation only because of their temperature, not their composition. Blackbodies emit just as much thermal radiation as they absorb from their surroundings. Physicist Max Planck (1858–1947) graphed the intensity of the emitted radiation across all wavelengths and obtained the characteristic curves that we now call **Planck spectra** or **blackbody spectra**. **Figure 5.22** shows blackbody spectra for objects at several temperatures.

Blackbody Laws

In the real world, the light from stars such as the Sun and the thermal radiation from a planet often come close to having blackbody spectra. Those objects follow the Stefan-Boltzmann law, which relates luminosity with temperature, and Wien's law, which relates temperature with color.

STEFAN-BOLTZMANN LAW As the temperature of an object increases, the object gives off more radiation at every wavelength, so the luminosity of the object should increase. Adding up all the energy in a blackbody spectrum shows that the increase in luminosity is proportional to the fourth power of the temperature. That relationship, luminosity $\propto T^4$, is known as the **Stefan-Boltzmann law**. It was discovered in the laboratory by physicist Josef Stefan (1835–1893) and derived mathematically by his student Ludwig Boltzmann (1844–1906).

Measuring all the photons emitted by Earth or the Sun in all possible directions is hard, but measuring the *flux* is easier. The **flux** ($\mathcal{F}$) is the amount of energy radiated by each square meter of the surface of an object each second. Flux is proportional to luminosity. You can find the luminosity by multiplying the flux by the total surface area. According to the Stefan-Boltzmann law, the flux is given by $\mathcal{F} = \sigma T^4$, where σ (the Greek letter sigma) is the **Stefan-Boltzmann constant**. It equals 5.67×10^{-8} W/(m² K⁴), where 1 watt = 1 joule per second (J/s).

The Stefan-Boltzmann law says that an object rapidly becomes more luminous as its temperature increases. If the temperature of an object doubles, the amount of energy being radiated each second increases by a factor of $2^4 = 16$. If the temperature of an object increases by a factor of 3, the energy being radiated by the object each second increases by a factor of $3^4 = 81$. A lightbulb with a filament temperature of 3000 K radiates 16 times as much light as it would if the filament temperature were 1500 K. Even modest changes in temperature can result in large changes in the luminosity of an object.

WIEN'S LAW Look again at Figure 5.22. The wavelength where the blackbody spectrum is at its peak, λ_{peak}, corresponds to where the electromagnetic radiation from an object is greatest. As the temperature, T, increases, the peak of the spectrum shifts toward shorter wavelengths. For example, $\lambda_{peak} = 970$ nm for a 3000 K object, but only 480 nm for a 6000 K object. Photon energy and wavelength are inversely related; thus, as the peak wavelength decreases, the average photon energy increases. The object becomes bluer. The physicist Wilhelm Wien (1864–1928) found that the peak wavelength in the spectrum is inversely proportional to the temperature of the object ($\lambda_{peak} \propto 1/T$). **Wien's law** states that if you double the temperature, the peak wavelength becomes half of what it was. If you increase the temperature by a factor of 3, the peak wavelength becomes a third

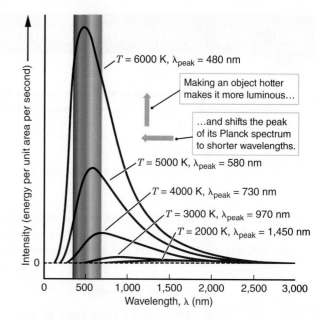

Figure 5.22 This graph shows blackbody spectra emitted by sources with temperatures of 2000, 3000, 4000, 5000, and 6000 K. At higher temperatures, the peak of the spectrum shifts toward shorter wavelengths, and the amount of energy radiated per second from each square meter of the source increases.

Astronomy in Action: Wien's Law

The Stefan-Boltzmann law can be used to estimate the flux and luminosity of Earth. Earth's average temperature is 288 K, so the flux from its surface is

$$\mathcal{F} = \sigma T^4$$

$$\mathcal{F} = (5.67 \times 10^{-8} \text{ W/m}^2 \text{ K}^4) \times (288 \text{ K})^4$$

$$\mathcal{F} = 390 \text{ W/m}^2$$

The luminosity is the flux multiplied by the surface area (A) of Earth. The surface area of a spherical object is given by $4\pi R^2$, and the radius of Earth is 6,378 km, or 6.378×10^6 meters. The luminosity, then, is

$$L = \mathcal{F} \times A = \mathcal{F} \times 4\pi R^2$$

$$L = (390 \text{ W/m}^2) \times [4\pi(6.378 \times 10^6 \text{ m})^2]$$

$$L \approx 2 \times 10^{17} \text{ W}$$

Earth emits the equivalent of the energy used by 2,000,000,000,000,000 (2 million billion) hundred-watt lightbulbs. That value is still not anywhere close to the amount emitted by the Sun.

If astronomers measure the spectrum of an object emitting thermal radiation and find where the peak in the spectrum is, Wien's law can be used to calculate the temperature of the object. Wien's law can be written as

$$T = \frac{2,900,000 \text{ nm K}}{\lambda_{\text{peak}}}$$

What, for example, is the surface temperature of the Sun? The spectrum of the light coming from the Sun peaks at a wavelength of $\lambda_{\text{peak}} = 500$ nm, so

$$T = \frac{2,900,000 \text{ nm K}}{500 \text{ nm}} = 5800 \text{ K}$$

What is the peak wavelength at which Earth radiates? Using Earth's average temperature of 288 K in Wien's law gives

$$\lambda_{\text{peak}} = \frac{2,900,000 \text{ nm K}}{288 \text{ K}} = 10,100 \text{ nm} = 10.1 \text{ μm}$$

Thus, Earth's radiation peaks in the infrared region of the spectrum.

of what it was. Stefan-Boltzmann's law and Wien's law are further explored in **Working It Out 5.3**. We use those laws later in the chapter to estimate the temperatures of the planets.

CHECK YOUR UNDERSTANDING 5.4

When you look at the sky on a dark night you see stars of different colors. Rank them from coolest to hottest. (a) orange; (b) red-orange; (c) yellow; (d) red; (e) blue

5.5 The Brightness of Light Depends on the Luminosity and Distance of the Source

Whereas luminosity is the amount of light *leaving* a source, the **brightness** of electromagnetic radiation is the amount of light *arriving* at a particular location. Therefore, the observed brightness depends on the luminosity and the distance of the light source. For example, replacing a 50-W lightbulb with a 100-W bulb makes a room twice as bright because it doubles the light reaching any point in the room. But brightness also depends on the distance from a source of electromagnetic radiation. For example, if you needed more light to read this book, you could replace the bulb in your lamp with a more luminous bulb or you could move the book closer to the light. Conversely, if a light were too bright, you could move away from it. Our everyday experience teaches us that as we move away from a light, its brightness decreases.

The particle description of light offers another way to think about the brightness of radiation and how brightness depends on distance. Suppose you had a piece of cardboard that measured 1 meter by 1 meter. To make the light falling on the cardboard twice as bright, you would need to double the number of photons that hit the cardboard each second. Tripling the brightness of the light would mean increasing the number of photons hitting the cardboard each second by a factor of 3, and so on. Brightness depends on the number of photons falling on each square meter of a surface each second.

Now imagine a lightbulb sitting at the center of a spherical shell (**Figure 5.23**). Photons from the bulb travel in all directions and land on the inside of the shell. To find the number of photons landing on each square meter of the shell during each second—that is, to determine the brightness of the light—take the *total* number of photons given off by the lightbulb each second and divide by the number of square meters over which those photons have to be spread. The surface area of a spherical shell is given by the formula $A = 4\pi d^2$, where d is the distance between the bulb and the surface of the sphere. The number of photons striking one square meter each second is equal to the total number of photons emitted each second divided by the surface area $4\pi d^2$.

Now increase the size of the spherical shell while keeping the total number of photons given off by the lightbulb each second the same. As the shell becomes larger, the photons from the lightbulb must spread out to cover a larger surface area. Each square meter of the shell receives fewer photons each second, so the brightness of the light decreases. If the shell's surface is moved twice as far from the light, the area over which the light must spread increases by a factor of $2^2 = 2 \times 2 = 4$. The photons from the bulb spread out over 4 times as much area, so the number of photons falling on each square meter each second becomes 1/4 of what it was. If the surface of the sphere is 3 times as far from the light, the area over which the light must spread increases by a factor of $3^2 = 3 \times 3 = 9$, and the number of photons per second falling on each square meter becomes 1/9 of what it was originally. That is the same kind of inverse square relationship you saw for gravity in Chapter 4. The brightness of the light from an object is inversely proportional to the square of the distance from the object. Twice as far means one-fourth as bright.

The idea of photons streaming and spreading onto a surface from a light explains why brightness follows an inverse square law. In practice, however, talking about the average *energy* falling on a surface each second, rather than the number of photons, is usually more more useful to an astronomer.

The luminosity of an object is the total number of photons given off by the object multiplied by the energy of each photon. So, instead of thinking about how the number of photons must spread out to cover the surface of a sphere, we can think about how the energy carried by the photons must spread out to cover the surface of a sphere. The brightness of the light is the amount of energy falling on a square meter in a second, and it equals the luminosity L divided by the area of the sphere, which depends on the radius squared. That tells us, for example, that the brightness of the Sun depends on the inverse square of the planet's distance from the Sun. That relationship factors in when we estimate the equilibrium temperatures of the planets in **Working It Out 5.4**.

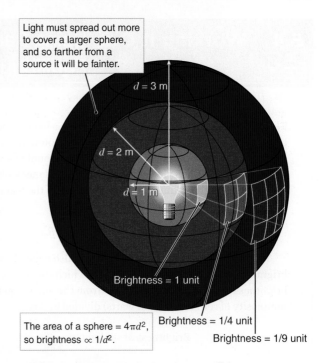

Light must spread out more to cover a larger sphere, and so farther from a source it will be fainter.

$d = 3$ m

$d = 2$ m

$d = 1$ m

Brightness = 1 unit

The area of a sphere = $4\pi d^2$, so brightness $\propto 1/d^2$.

Brightness = 1/4 unit

Brightness = 1/9 unit

Figure 5.23 Light obeys an inverse square law as it spreads away from a source. Twice as far means one-fourth as bright, three times as far means one-ninth as bright, and so on.

Astronomy in Action: Inverse Square Law

CHECK YOUR UNDERSTANDING 5.5

The average distance of Mars from the Sun is 1.4 AU. How bright is the Sun on Mars compared with its brightness on Earth? (a) 1.4 times brighter; (b) about 2 times brighter; (c) about 2 times fainter; (d) 1.4 times fainter

5.4 Working It Out Using Radiation Laws to Calculate Equilibrium Temperatures of Planets

The temperature of a planet is determined by a balance between the amount of sunlight being absorbed and the amount of energy being radiated back into space. We begin with the amount of sunlight being absorbed. When viewed from the Sun, a planet looks like a circular disk with a radius equal to the radius of the planet, R_{planet}. The area of the planet that is lit by the Sun is

$$\text{Absorbing area of planet} = \pi R_{planet}^2$$

The amount of energy striking a planet also depends on the brightness of sunlight at the distance at which the planet orbits. The brightness of sunlight at a distance d from the Sun is equal to the luminosity of the Sun (L_{Sun}, in watts) divided by $4\pi d^2$:

$$\text{Brightness of sunlight} = \frac{L_{Sun}}{4\pi d^2}$$

A planet does not absorb *all* the sunlight that falls on it. **Albedo**, a, is the fraction of the sunlight that reflects from a planet. The corresponding fraction of the sunlight absorbed by the planet is 1 minus the albedo ($1 - a$). A planet covered in snow would have a high albedo (close to 1), whereas a planet covered by black rocks would have a low albedo, close to 0:

$$\text{Fraction of sunlight absorbed} = 1 - a$$

We can now calculate the energy absorbed by the planet each second. Writing that relationship as an equation, we say that

$$\begin{pmatrix} \text{Energy absorbed} \\ \text{by the planet} \\ \text{each second} \end{pmatrix} = \begin{pmatrix} \text{Absorbing} \\ \text{area of} \\ \text{the planet} \end{pmatrix} \times \begin{pmatrix} \text{Brightness} \\ \text{of sunlight} \end{pmatrix} \times \begin{pmatrix} \text{Fraction} \\ \text{of sunlight} \\ \text{absorbed} \end{pmatrix}$$

$$= \pi R_{planet}^2 \times \frac{L_{Sun}}{4\pi d^2} \times (1 - a)$$

Now let's turn to the other piece of the equilibrium: the amount of energy that the planet radiates away into space each second. We can calculate that amount by multiplying the number of square meters of the planet's total surface area by the energy radiated by each square meter each second. The surface area for the planet is given by $4\pi R_{planet}^2$. According to the Stefan-Boltzmann law, the energy radiated by each square meter each second is given by σT^4. Thus,

$$\begin{pmatrix} \text{Energy radiated} \\ \text{by the planet} \\ \text{each second} \end{pmatrix} = \begin{pmatrix} \text{Surface} \\ \text{area of} \\ \text{the planet} \end{pmatrix} \times \begin{pmatrix} \text{Energy radiated} \\ \text{per square meter} \\ \text{per second} \end{pmatrix}$$

$$= 4\pi R_{planet}^2 \times \sigma T^4$$

If the planet's temperature is to remain stable—not heating up or cooling down—then each second the "energy radiated" must be equal to "energy absorbed":

$$\begin{pmatrix} \text{Energy radiated} \\ \text{by the planet} \\ \text{each second} \end{pmatrix} = \begin{pmatrix} \text{Energy absorbed} \\ \text{by the planet} \\ \text{each second} \end{pmatrix}$$

When we set those two quantities equal to each other, we arrive at the following expression:

$$4\pi R_{planet}^2 \sigma T^4 = \pi R_{planet}^2 \frac{L_{Sun}}{4\pi d^2}(1 - a)$$

Canceling out πR_{planet}^2 on both sides, and rearranging the equation to put T on one side and everything else on the other, gives

$$T^4 = \frac{L_{Sun}(1 - a)}{16\sigma\pi d^2}$$

If we take the fourth root of each side, we get

$$T = \left(\frac{L_{Sun}(1 - a)}{16\sigma\pi d^2}\right)^{1/4}$$

Putting in the appropriate numbers for the known luminosity of the Sun, L_{Sun}, and the constants π and σ yields a simpler equation:

$$T = 279\,\text{K} \times \left(\frac{1 - a}{d_{AU}^2}\right)^{1/4}$$

where d_{AU} is the distance of the planet from the Sun in astronomical units.

To use that equation, we need to know a planet's distance from the Sun and its average albedo. For a blackbody ($a = 0$) at 1 AU from the Sun, the temperature is 279 K. For Earth, with an albedo of 0.3 and a distance from the Sun of 1 AU, the temperature is

$$T = 279\,\text{K} \times \left(\frac{1 - 0.3}{1^2}\right)^{1/4} = 255\,\text{K}$$

(Calculator hint: To take a fourth root, you can take the square root twice, or use the x^y button with $y = 0.25$.)

Earth is cooler than a blackbody at 1 AU from the Sun because its average albedo is greater than zero. If Earth's albedo changed or the Sun's luminosity changed, that would affect the result. When we examine planets around other stars, we must use the luminosity of the particular star in the equation, instead of the Sun's luminosity, so the temperature at 1 AU will be different from what it is for Earth.

Origins Temperatures of Planets

In the previous chapters, we discussed how a planet's axial tilt and its orbital shape affect its temperature and thus its prospects for life. Now let's get more specific about the temperatures of planets, using what you learned in this chapter about thermal radiation. For a planet at an equilibrium temperature, the energy radiated by a planet exactly balances the energy absorbed by the planet. If the planet is hotter than that equilibrium temperature, it will radiate energy faster than it absorbs sunlight, and its temperature will decrease. If the planet is cooler than that temperature, it will radiate energy slower than it absorbs sunlight, and its temperature will increase.

Planets at different distances from the Sun will have different temperatures, and the temperature should be inversely proportional to the square root of the distance, as you saw in Working It Out 5.4. **Figure 5.24** plots the actual and predicted temperatures of nine solar system objects. Each vertical orange bar shows the range of temperatures found on the surface of the planet or, for the giant planets, at the top of the planet's clouds. The black dots show the predictions made using the equation in Working It Out 5.4. For most planets, the predictions are not too far off, indicating that our basic understanding of *why* planets have the temperatures they have is probably pretty good. The data for Mercury, Mars, and Pluto agree particularly well.

Sometimes, however, the predictions are wrong. For Earth, the actual measured temperature is a bit higher than the predicted temperature, and for Venus the actual surface temperature is much higher than the prediction.

The predicted values assume that the temperature of the planet is the same everywhere, but planets are likely to be hotter on the day side than on the night side. The predictions also assume that a planet's only source of energy is sunlight and that the fraction of sunlight reflected is constant over the surface of each planet. The values also incorporate the assumption that the planets absorb and radiate energy into space as blackbodies.

The discrepancies between the calculated and the measured temperatures of some of the planets indicate that for those planets, some or all of those assumptions are incorrect. For example, the planet may have its own source of energy besides sunlight, or it may have an atmosphere that traps more solar energy and increases the temperature of the planet. Understanding the temperatures of planets makes it possible to hypothesize which planets are more suitable for life.

Figure 5.24 Predicted temperatures for the planets and dwarf planet Pluto are based on the equilibrium between absorbed sunlight and thermal radiation into space. Those temperatures are compared with ranges of observed surface temperatures.

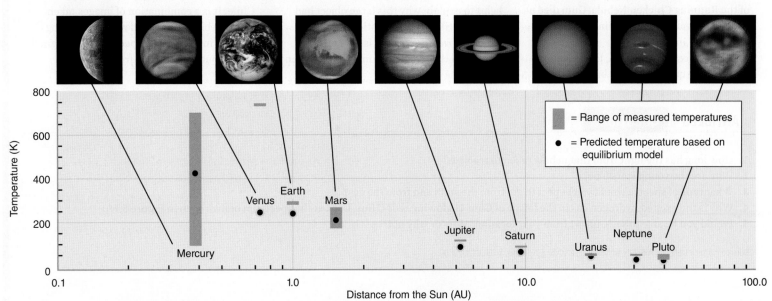

A Study in Scarlet

ESO

This new image from ESO's La Silla Observatory in Chile reveals a cloud of hydrogen called Gum 41 (**Figure 5.25**). In the middle of this little-known nebula, brilliant hot young stars are giving off energetic radiation that causes the surrounding hydrogen to glow with a characteristic red hue.

This area of the southern sky, in the constellation of Centaurus (The Centaur), is home to many bright nebulae, each associated with hot newborn stars that formed out of the clouds of hydrogen gas. The intense radiation from the stellar newborns excites the remaining hydrogen around them, making the gas glow in the distinctive shade of red typical of star-forming regions. Another famous example of this phenomenon is the Lagoon Nebula, a vast cloud that glows in similar bright shades of scarlet.

The nebula in this picture is located some 7,300 light-years from Earth. Australian astronomer Colin Gum discovered it on photographs taken at the Mount Stromlo Observatory near Canberra, and included it in his catalog of 84 emission nebulae, published in 1955. Gum 41 is actually one small part of a bigger structure called the Lambda Centauri Nebula, also known by the more exotic name of the Running Chicken Nebula. Gum died at a tragically early age in a skiing accident in Switzerland in 1960.

In this picture of Gum 41, the clouds appear to be quite thick and bright, but this is actually misleading. If a hypothetical human space traveler could pass through this nebula, it is likely that they would not notice it as—even at close quarters—it would be too faint for the human eye to see. This helps to explain why this large object had to wait until the mid-twentieth century to be discovered—its light is spread very thinly and the red glow cannot be well seen visually.

This new portrait of Gum 41—likely one of the best so far of this elusive object—has been created using data from the Wide Field Imager (WFI) on the MPG/ESO 2.2-meter telescope at the La Silla Observatory in Chile. It is a combination of images taken through blue, green, and red filters, along with an image using a special filter designed to pick out the red glow from hydrogen.

Figure 5.25 The Gum 41 Nebula.

1. How long has the light from that nebula taken to reach us?
2. Why are the young stars blue?
3. What type of spectra would you expect to get from the stars and from the gas?
4. Why is the excited hydrogen gas in the image of Gum 41 glowing red? (Hint: Look at the spectrum of hydrogen in Figure 5.14.)
5. Would you be able to see Gum 41 from your location? Why or why not?

Source: "Photo Release: A Study in Scarlet," ESO.org, April 16, 2014. CC by 4.0.

Summary

Light carries both information and energy throughout the universe. The speed of light in a vacuum is 300,000 km/s; nothing can travel faster. Visible light is only a tiny portion of the entire electromagnetic spectrum. Atoms absorb and emit radiation at unique wavelengths, giving them spectral fingerprints. A planet's temperature depends on its distance from its star, its albedo, and the luminosity of its star.

LG 1 **Describe the wave and particle properties of light, and describe the electromagnetic spectrum.** Light is simultaneously a stream of particles called photons and an electromagnetic wave. Different types of electromagnetic radiation, from gamma rays to visible light to radio waves, are electromagnetic waves that differ in frequency and wavelength.

LG 2 **Describe how to measure the chemical composition of distant objects by using the unique spectral lines of different types of atoms.** Nearly all matter is composed of atoms, and light can reveal the identity of the chemical elements present in matter. The electron energy levels of each element have different (unique) spacings, and the wavelengths of the photons each element emits correspond to the differences in those levels. As a result, we can identify different chemical elements and molecules in distant objects.

LG 3 **Describe the Doppler effect and how it can be used to measure the motion of distant objects.** Because of the Doppler effect, light from receding objects is redshifted to longer wavelengths, and light from approaching objects is blueshifted to shorter wavelengths. The wavelength shifts of the spectral lines indicate how fast an astronomical object is moving toward or away from Earth.

LG 4 **Explain how the spectrum of light that an object emits depends on its temperature.** Temperature is a measure of how energetically particles are moving in an object. A light source that emits electromagnetic radiation because of its temperature is called a blackbody. A blackbody emits a continuous spectrum. The total amount of energy emitted is proportional to the temperature to the fourth power, and the peak wavelength, which determines its color, is inversely proportional to the temperature.

LG 5 **Differentiate luminosity from brightness, and illustrate how distance affects each.** The luminosity of an object is the amount of light it emits. The brightness of an object is proportional to its luminosity divided by its distance squared. Thus, the brightness of the Sun is different when measured from each Solar System planet, but the luminosity is the same.

 Unanswered Questions

- Has the speed of light always been 300,000 km/s? Some theoretical physicists have questioned whether light traveled much faster earlier in the history of our universe. The observational evidence that may test that idea comes from studying the spectra of the most distant objects—whose light has been traveling for billions of years—and determining whether billions of years ago chemical elements absorbed light differently from how they do today. So far, no evidence exists that the speed of light changes.

- Will traveling faster than the speed of light ever be possible? Our current understanding of the science says no. A staple of science fiction films and stories is spaceships that go into "warp speed" or "hyperdrive"—moving faster than light—to traverse the huge distances of space (and visit a different planetary system every week). If that premise is simply fictional and the speed of light is a true universal limit, travel between the stars will take many years. Because all electromagnetic radiation travels at the speed of light, even an electromagnetic signal sent to another planetary system would take many years to get there. Interstellar visits (and interstellar conversations) will be quite prolonged.

Questions and Problems

Test Your Understanding

1. If the Sun instantaneously stopped giving off light, what would happen on Earth?
   a. Earth would immediately get dark.
   b. Earth would get dark 8 minutes later.
   c. Earth would get dark 27 minutes later.
   d. Earth would get dark 1 hour later.

2. Why is an iron atom a different element from a sodium atom?
   a. A sodium atom has fewer neutrons in its nucleus than an iron atom has.
   b. An iron atom has more protons in its nucleus than a sodium atom has.
   c. A sodium atom is bigger than an iron atom.
   d. A sodium atom has more electrons.

3. Suppose an atom has three energy levels, specified in arbitrary units as 10, 7, and 5. In those units, which of the following energies might an emitted photon have? (Select all that apply.)
   a. 3
   b. 2
   c. 5
   d. 4

4. When a boat moves through the water, the waves in front of the boat bunch up, whereas the waves behind the boat spread out. That is an example of
   a. the Bohr model.
   b. the wave nature of light.
   c. emission and absorption.
   d. the Doppler effect.

5. As a blackbody becomes hotter, it also becomes _____ and _____.
   a. more luminous; redder
   b. more luminous; bluer
   c. less luminous; redder
   d. less luminous; bluer

6. Which of the following factors does *not* directly influence the temperature of a planet?
   a. the luminosity of the Sun
   b. the distance from the planet to the Sun
   c. the albedo of the planet
   d. the size of the planet

7. Two stars are of equal luminosity. Star A is 3 times as far from you as star B. Star A appears _____ star B.
   a. 9 times brighter than
   b. 3 times brighter than
   c. the same brightness as
   d. 1/3 as bright as
   e. 1/9 as bright as

8. When less energy radiates from a planet, its _____ increases until a new _____ is achieved.
   a. temperature; equilibrium
   b. size; temperature
   c. equilibrium; size
   d. temperature; size

9. How does the speed of light in a medium compare with the speed in a vacuum?
   a. The speed is the same in both a medium and a vacuum because the speed of light is a constant.
   b. The speed in the medium is always faster than the speed in a vacuum.
   c. The speed in the medium is always slower than the speed in a vacuum.
   d. The speed in the medium may be faster or slower, depending on the medium.

10. When an electron moves from a higher energy level in an atom to a lower energy level,
    a. a continuous spectrum is emitted.
    b. a photon is emitted.
    c. a photon is absorbed.
    d. a redshifted spectrum is emitted.

11. In Figure 5.14, the red photons come from the transition from E_3 to E_2. Those photons will have the _____ wavelengths because they have the _____ energy compared with that of the other photons.
    a. shortest; least
    b. shortest; most
    c. longest; least
    d. longest; most

12. Star A and star B appear equally bright in the sky. Star A is twice as far away from Earth as star B. How do the luminosities of stars A and B compare?
    a. Star A is 4 times as luminous as star B.
    b. Star A is 2 times as luminous as star B.
    c. Star B is 2 times as luminous as star A.
    d. Star B is 4 times as luminous as star A.

13. What is the surface temperature of a star that has a peak wavelength of 290 nm?
    a. 1000 K
    b. 2000 K
    c. 5000 K
    d. 10,000 K
    e. 100,000 K

14. If a planet is in thermal equilibrium,
    a. no energy is leaving the planet.
    b. no energy is arriving on the planet.
    c. the amount of energy leaving equals the amount of energy arriving.
    d. the temperature is very low.

15. The temperature of an object has a specific meaning as it relates to the object's atoms. A high temperature means that the atoms
    a. are very large.
    b. are moving very fast.
    c. are all moving together.
    d. have a lot of energy.

Thinking about the Concepts

16. The speed of light in a vacuum is 3×10^5 km/s. Can light travel at a lower speed? Explain your answer.

17. Is light a wave or a particle or both? Explain your answer.

18. If any of the experiments mentioned in the Process of Science Figure had *not* agreed with the others, what would that mean for the conclusion that light has a finite, constant speed?

19. If photons of blue light have more energy than photons of red light, how can a beam of red light carry as much energy as a beam of blue light?

20. Patterns of emission or absorption lines in spectra can uniquely identify individual atomic elements. How can the positive identification of atomic elements be used to test the validity of the cosmological principle discussed in Chapter 1?

21. An atom in an excited state can drop to a lower energy state by emitting a photon. Can we predict exactly how long the atom will remain in the higher energy state? Explain your answer.

22. Spectra of astronomical objects show both bright and dark lines. Describe what those lines indicate about the atoms responsible for the spectral lines.

23. Astronomers describe certain celestial objects as being *redshifted* or *blueshifted*. What do those terms indicate about the objects?

24. An object somewhere near you is emitting a pure tone at middle C on the octave scale (262 Hz). You, having perfect pitch, hear the tone as A above middle C (440 Hz). Describe the motion of that object relative to where you are standing.

25. During a popular art exhibition, the museum staff finds it necessary to protect the artwork by limiting the total number of viewers in the museum at any particular time. New viewers are admitted at the same rate that others leave. Is that system an example of static equilibrium or of dynamic equilibrium? Explain.

26. A favorite object for amateur astronomers is the double star Albireo, with one of its components a golden yellow and the other a bright blue. What do those colors tell you about the relative temperatures of the two stars?

27. The stars you see in the night sky cover a large range of brightness. What does that range tell you about the distances of the various stars? Explain your answer.

28. Why is it not surprising that sunlight peaks in the "visible"?

29. In Figure 5.24, why is the range of temperatures so much greater for Mercury than for the other planets?

30. Suppose you want to find a planet with the same temperature as Earth. What could you say about the size of the orbit of such a planet if it is orbiting a cooler red star? A yellow star like the Sun? A hotter blue star?

Applying the Concepts

31. You are tuned to 790 on AM radio. That station is broadcasting at a frequency of 790 kHz (7.90×10^5 Hz). You switch to 98.3 on FM radio. That station is broadcasting at a frequency of 98.3 MHz (9.83×10^7 Hz).
    a. What are the wavelengths of the AM and FM radio signals?
    b. Which broadcasts at higher frequencies, AM or FM?
    c. What are the photon energies of the two broadcasts?

32. Your microwave oven cooks by vibrating water molecules at a frequency of 2.45 gigahertz (GHz), or 2.45×10^9 Hz. What is the wavelength, in centimeters, of the microwave's electromagnetic radiation?

33. Assume that an object emitting a pure tone of 440 Hz is on a vehicle approaching you at a speed of 25 m/s. If the speed of sound at this particular atmospheric temperature and pressure is 340 m/s, what will be the frequency of the sound that you hear? (Hint: Keep in mind that frequency is inversely proportional to wavelength.)

34. You observe a spectral line of hydrogen at a wavelength of 502.3 nm in a distant galaxy. The rest wavelength of that line is 486.1 nm. What is the radial velocity of that galaxy? Is it moving toward you or away from you?

35. If half of the phosphorescent atoms in a glow-in-the-dark toy give up a photon every 30 minutes, how bright (relative to its original brightness) will the toy be after 2 hours?

36. How bright would the Sun appear from Neptune, 30 AU from the Sun, compared with its brightness as seen from Earth? The spacecraft *Voyager 1* is now about 140 AU from the Sun and heading out of the Solar System. Compare the brightness of the Sun as seen from *Voyager 1* with that seen from Earth.

37. On a dark night you notice that a distant lightbulb happens to have the same brightness as a firefly 5 meters away from you. If the lightbulb is a million times more luminous than the firefly, how far away is the lightbulb?

38. Two stars appear to have the same brightness, but one star is 3 times more distant than the other. How much more luminous is the more distant star?

39. A panel with an area of 1 square meter (m^2) is heated to a temperature of 500 K. How many watts is the panel radiating into its surroundings?

40. The Sun has a radius of 6.96×10^5 km and a blackbody temperature of 5780 K. Calculate the Sun's luminosity.

41. Some of the hottest stars known have a blackbody temperature of 100,000 K. What is the peak wavelength of their radiation? What type of radiation is it?

42. Your body emits radiation at a temperature of about 98.6°F.
    a. What is that temperature in kelvins? What is the peak wavelength, in microns, of your emitted radiation? What region of the spectrum is this?
    b. If you assume an exposed body surface area of 0.25 m^2, how many watts of power do you radiate?

43. A planet with no atmosphere at 1 AU from the Sun would have an average blackbody surface temperature of 279 K if it absorbed all the Sun's electromagnetic energy falling on it (albedo = 0).
    a. What would the average temperature on that planet be if its albedo were 0.1, typical of a rock-covered surface?
    b. What would the average temperature be if its albedo were 0.9, typical of a snow-covered surface?

44. The orbit of Eris, a dwarf planet, carries it out to a maximum distance of 97.7 AU from the Sun. If you assume an albedo of 0.8, what is the average temperature of Eris when it is farthest from the Sun?

45. Suppose our Sun had 10 times its current luminosity. What would the average blackbody surface temperature of Earth be if Earth had the same albedo?

USING THE WEB

46. a. Go to the website for NASA's Astronomy Picture of the Day (https://apod.nasa.gov/apod/ap101027.html) and study the picture of the Andromeda Galaxy in visible light and in ultraviolet light (slide the mouse on and off to flip between the two pictures). Which light represents a hotter temperature? What differences do you see in the two images?
    b. Go to the APOD archive (https://apod.nasa.gov/cgi-bin/apod/apod_search) and enter "false color" in the search box. Examine a few images that come up in the search. What does *false color* mean in that context? What wavelength(s) were the pictures exposed in? What is the color coding; that is, what wavelength does each color in the image represent? You can read more about false color at http://chandra.harvard.edu/photo/false_color.html.

47. Crime scene investigators may use different types of light to examine a crime scene. Search for "forensic lighting" in your browser. What wavelengths of light are used to search for blood and saliva? For fingerprints? Why is having access to different kinds of light useful for an investigator? Search on "forensic spectroscopy" and select a recent report. How is spectroscopy being used in crime scene investigations?

48. Using Google Images or an equivalent website, search for "night vision imaging" and "thermal imaging." How do night vision goggles and thermal-imaging devices work differently from regular binoculars or cameras? When are night vision or thermal-imaging tools useful?

49. The Transportation Security Administration (TSA) uses several types of imaging devices to screen passengers in airports. Search for "TSA imaging" in your browser. What wavelengths of light are being used in those devices? What concerns do passengers have about some of those imaging devices?

50. Go to the NASA Earth Observations website (https://neo.sci.gsfc.nasa.gov) and look at a map of Earth's albedo (click on "albedo" in the menu for "energy" or "land" if it didn't come up). Compare that map with those of 2, 4, 6, 8, and 10 months earlier. Which parts of Earth have the lowest and highest albedos? In which parts do the albedos seem to change the most with the time of the year? Would you expect ice, snow, oceans, clouds, forests, and deserts to add or subtract in each case from the total Earth albedo? Which parts of Earth are not showing up on this map?

EXPLORATION

digital.wwnorton.com/astro6

Visit the Digital Resources Page and on the Student Site open the "Light as a Wave, Light as a Photon" AstroTour in Chapter 5. Watch the first section and then click through, using the "Play" button, until you reach "Section 2 of 3."

Here we explore the following questions: How many properties does a wave have? Are any of those properties related to each other?

Work your way to the experimental section, where you can adjust the properties of the wave. Watch the simulation for a moment to see how fast the frequency counter increases.

1 Increase the wavelength by pressing the arrow key. What happens to the rate of the frequency counter?

_____

_____

2 Reset the simulation and then decrease the wavelength. What happens to the rate of the frequency counter?

_____

_____

3 How are the wavelength and frequency related to each other?

_____

_____

4 Imagine that you increase the frequency instead of the wavelength. How should the wavelength change when you increase the frequency?

_____

_____

5 Reset the simulation, and increase the frequency. Did the wavelength change in the way you expected?

_____

_____

_____

6 Reset the simulation, and increase the amplitude. What happens to the wavelength and the frequency counter?

_____

_____

_____

7 Decrease the amplitude. What happens to the wavelength and the frequency counter?

_____

_____

_____

8 Is the amplitude related to the wavelength or frequency?

_____

_____

_____

9 Why can't you change the speed of this wave?

_____

_____

_____

6

The Tools of the Astronomer

I n Chapter 5, we explained that astronomers learn about the physical and chemical properties of distant planets, stars, and galaxies by studying the light they emit. That electromagnetic radiation, though, must first be collected and processed before it can be analyzed and converted to useful knowledge. Here in Chapter 6, we describe the tools that astronomers use to capture and scrutinize that information.

LEARNING GOALS

By the end of this chapter, you should be able to:

LG 1 Compare how the two main types of optical telescopes gather and focus light.

LG 2 Summarize the main types of detectors that are used on telescopes.

LG 3 Explain why some wavelengths of radiation must be observed from high, dry, and remote observatories on Earth, or from space.

LG 4 Explain the benefits of sending spacecraft to study the planets and moons of our Solar System.

LG 5 Describe other astronomical tools that contribute to the study of the universe.

The twin 10-meter Keck reflectors on Mauna Kea, Hawaii, have a multiple-mirror, compact design. ▶▶▶

Why are most telescopes on remote mountaintops?

(a)

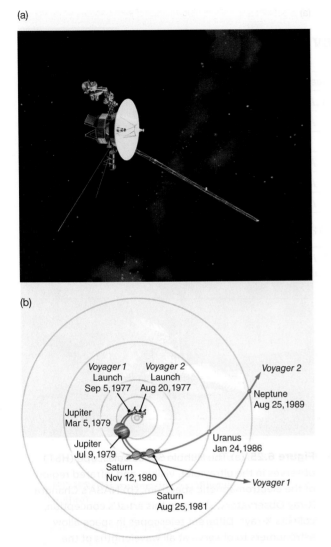

(b)

Figure 6.26 The two *Voyager* spacecraft (a) flew past the outer planets and (b) have recently passed the boundary of our Solar System.

Figure 6.27 The robotic rover *Curiosity* took this selfie on the surface of Mars.

such as *Voyager 1* and *Voyager 2,* can sometimes visit several worlds during their travels (**Figure 6.26**). The downside of flyby missions is that because of the physics of orbits, the spacecraft must move by very swiftly. They are limited to just a few hours or at most a few days in which to conduct close-up studies of their targets. Flyby spacecraft give astronomers their first close-up views of Solar System objects, and sometimes the data obtained are then used to plan follow-up studies.

More detailed reconnaissance work is done by spacecraft known as **orbiters** because they orbit around their target. Such missions are intrinsically more difficult than flyby missions because they have to make risky maneuvers and use their limited fuel to change their speed to enter an orbit. But orbiters can linger, looking in detail at more of the surfaces of the objects they are orbiting and studying things that change with time, such as planetary weather.

Orbiters use remote-sensing instrumentation like that used by Earth-orbiting satellites to study our own planet. Those instruments include tools such as cameras that take images at different wavelength ranges, radar that can map surfaces hidden beneath obscuring layers of clouds, and spectrographs that analyze the electromagnetic spectrum. Those instruments enable planetary scientists to map other worlds, measure the heights of mountains, identify geological features and rock types, watch weather patterns develop, measure the composition of atmospheres, and get a general sense of the place. Additional instruments make measurements of the extended atmospheres and space environment through which they travel.

Landers, Rovers, and Atmospheric Probes

Reconnaissance spacecraft provide a wealth of information about a planet, but no better way exists to explore a planet than doing so within a planet's atmosphere or on solid ground. Spacecraft have landed on the Moon, Mars, Venus, Saturn's large moon Titan, several asteroids, and a comet. Those spacecraft have taken pictures of planetary surfaces, measured surface chemistry, and conducted experiments to determine the physical properties of the surface rocks and soils.

Using **landers**—spacecraft that touch down and remain on the surface—has several disadvantages. Because of the expense, only a few landings in limited areas are practical. With that limitation, the results may apply only to the small area around the landing site. Imagine, for example, what a different picture of Earth you might get from a spacecraft that landed in Antarctica, as opposed to a spacecraft that landed in a volcano or on the floor of a dry riverbed. Sites to be explored with landed spacecraft must be carefully chosen on the basis of reconnaissance data. Some landers have wheels and can explore the vicinity of the landing site. Such remote-controlled vehicles, called **rovers**, were used first by the Soviet Union on the Moon four decades ago and more recently by the United States on Mars. **Figure 6.27** shows a self-portrait of the *Curiosity* rover on Mars.

Atmospheric probes descend into the atmospheres of planets and continually measure and send back data on temperature, pressure, and wind speed, along with other properties, such as chemical composition. Atmospheric probes have survived all the way to the solid surfaces of Venus and of Saturn's moon Titan (**Figure 6.28**), sending back streams of data during their descent. An atmospheric probe sent into Jupiter's atmosphere never reached that planet's surface because, as we discuss later in the book, Jupiter does not have a solid surface in the same sense that terrestrial planets and moons do. After sending back its data, the Jupiter probe eventually melted and vaporized as it descended into the hotter layers of the planet's atmosphere.

Sample Returns

If you pick up a rock from the side of a road, you might learn a lot from the rock by using tools that you could carry in your pocket or in your car. Much better, though, would be to pick up a few samples and carry them back to a laboratory equipped with a full range of state-of-the-art instruments that can measure chemical composition, mineral type, age, and other information needed to reconstruct the story of your rock sample's origin and evolution. The same is true of Solar System exploration. One of the most powerful methods for investigating remote objects is to collect samples of the objects and bring them back to Earth for study. So far, only samples of the Moon, a comet, and streams of charged particles from the Sun have been collected and returned to Earth. Scientists have found meteorites on Earth that are pieces of Mars that were blasted loose when objects crashed into that world. In the next 10 years, robotic "sample and return" missions may go to Mars.

The missions discussed so far in this section have all been conducted with robotic spacecraft. The only spacecraft that took people to another world were the *Apollo* missions to the Moon. That program ran from 1961 to 1972 and included several missions before the actual Moon landings. The *Apollo 8* astronauts brought back the famous picture of Earth viewed over the surface of the Moon (see the opening figure of Chapter 1). Each mission from *Apollo 11* through *Apollo 17* had three astronauts—two to land on the Moon and one to remain in orbit. *Apollo 13* did not reach the Moon but returned to Earth safely. Twelve American astronauts walked on the Moon between 1969 and 1972 and brought back a total of 382 kg of rocks and other material (**Figure 6.29**).

The return of extraterrestrial samples to Earth is governed by international treaties and standards to ensure that the samples do not contaminate our world. For example, before the lunar samples that the *Apollo* missions brought back could be studied, they (and the astronauts) had to be quarantined and tested for the presence of alien life-forms. The same international standards apply to spacecraft landing on other planets. The goal of those standards is to avoid transporting life-forms from Earth to another planet; we do not want to "discover" life that we, in fact, introduced. In addition, we do not want to potentially harm life that may exist on other planets.

With many missions under way and others on the horizon, robotic exploration of the Solar System is an ongoing, dynamic activity. Appendix 5 summarizes some recent and current missions. We will discuss some of those missions in the relevant chapter. Information on the latest discoveries can be found on mission websites and in science news sources.

Figure 6.28 Artist's illustration showing the Huygens space probe descending and landing on Saturn's moon Titan in 2005.

Figure 6.29 *Apollo 11* astronaut Buzz Aldrin on the moon, as photographed by mission commander Neil Armstrong, in 1969.

CHECK YOUR UNDERSTANDING 6.4

Spacecraft are the most effective way to study planets in our Solar System because: (a) planets move too fast across the sky for us to image them well from Earth; (b) planets cannot be imaged from Earth; (c) spacecraft can collect more information than is available just from images from Earth; (d) space missions are easier than long observing campaigns.

6.5 Other Tools Contribute to the Study of the Universe

High-profile space missions have sent back stunning images and data from across the electromagnetic spectrum. But astronomers use other tools as well, including particle accelerators and colliders, neutrino and gravitational-wave detectors, and super-computers.

Figure 6.30 The ATLAS particle detector at CERN's Large Hadron Collider near Geneva, Switzerland. The instrument's enormous size is evident from the person standing near the bottom center of the picture.

Particle Accelerators

Ever since the early years of the 20th century, physicists have been peering into the structure of the atom by observing what happens when small particles collide. By the 1930s, physicists had developed the technology to accelerate charged subatomic particles such as protons to very high speeds and then observe what happens when they slam into a target. From such experiments, physicists have discovered many kinds of subatomic particles and learned about their physical properties. High-energy particle colliders have proved to be an essential tool for physicists studying the basic building blocks of matter.

Astronomers have realized that to understand the very largest structures seen in the universe, it is important to understand the physics that took place during the earliest moments in the universe, when everything was extremely hot and dense. High-energy particle colliders that physicists use today are designed to approach the energies of the early universe. The effectiveness of particle accelerators is determined by the energy they can achieve and the number of particles they can accelerate. Modern particle colliders such as the Large Hadron Collider near Geneva, Switzerland (**Figure 6.30**), reach very high energies. Particles also can be studied from space. The Alpha Magnetic Spectrometer, installed on the International Space Station in 2011, searches for some of the most exotic forms of matter, such as dark matter, antimatter, and high-energy particles called cosmic rays.

Neutrinos and Gravitational Waves

The **neutrino** is an elusive elementary particle that plays a major role in the physics of the interiors of stars. Neutrinos are extremely difficult to detect. In less time than you take to read this sentence, a thousand trillion (10^{15}) solar neutrinos from the Sun are passing through your body, even during the night. Neutrinos are so nonreactive with matter that they can pass right through Earth (and you) as though it (or you) weren't there at all. To be observed, a neutrino has to interact with a detector. Neutrino detectors typically record only one of every 10^{22} (10 billion trillion) neutrinos passing through them, but that's enough to reveal processes deep within the Sun or the violent death of a star 160,000 light-years away.

Experiments designed to look for neutrinos originating outside Earth are buried deep underground in mines or caverns or under the ocean or ice to ensure that only neutrinos are detected. For example, the ANTARES experiment uses the Mediterranean Sea as a neutrino telescope. Detectors located 2.5 km under the sea, off the coast of France, observe neutrinos that originated in objects visible in southern skies and passed through Earth. In the IceCube neutrino observatory located at the South Pole in Antarctica, the neutrino detectors are 1.5–2.5 km under the ice, and they observe neutrinos that originated in objects visible in northern skies (**Figure 6.31**).

Another elusive phenomenon is the **gravitational wave**. Gravitational waves are disturbances in a gravitational field, similar to the waves that spread out from the disturbance you create when you toss a pebble onto the quiet surface of a pond (**Process of Science Figure**). Several facilities, including the Laser Interferometer Gravitational-Wave Observatory (LIGO) and the European VIRGO interferometer, have been constructed to detect gravitational waves. Recently, LIGO has detected several pairs of coalescing binary black holes and neutron stars, as we explore in Chapter 18.

Figure 6.31 The IceCube neutrino telescope at the South Pole, Antarctica.

Technology and Science Are Symbiotic

Scientists have been searching for waves that carry gravitational information for nearly 100 years, but the accuracy of their measurements has been limited by the available technology.

Einstein predicts that changing gravitational fields should produce gravitational waves. He calculates the size of the effect, and concludes that it will be impossible to detect them.

1916

c. 1970

Joseph Weber constructs precision-machined bars of metal that should "ring" as a gravitational wave passes by. His results were never confirmed.

1990s

Construction begins on LIGO, an observatory which uses lasers to precisely measure the change in the length of long tubes as gravitational waves pass by.

2015

LIGO detects gravitational waves for the first time, announcing the discovery in early 2016, 100 years after Einstein's prediction. This first detection was followed by several more in the next two years.

Future (2034?)

LISA, a gravitational wave observatory in space, will be more sensitive; it will be able to detect much weaker events than LIGO.

Technology and science develop together. New technologies enable humans to ask new scientific questions. Asking new scientific questions pushes the development of better instrumentation and new technologies.

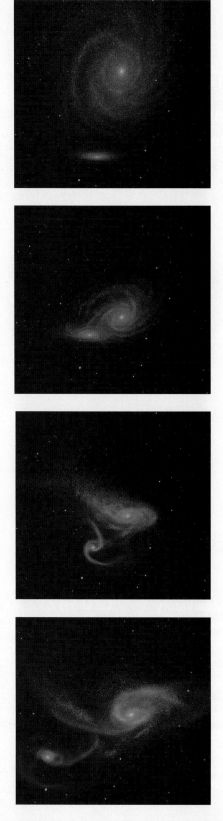

Computers

Astronomers use powerful computers to gather, analyze, and interpret data. A single CCD image may contain millions of pixels, with each pixel displaying roughly 30,000 levels of brightness. That adds up to several trillion pieces of information in each image. To analyze their data, astronomers typically do calculations for *every single pixel* of an image to remove unwanted contributions from Earth's atmosphere or to correct for instrumental effects. Astronomers conduct many types of sky surveys—in which one or more telescopes survey a specific part of the sky—yielding thousands of images that need to be analyzed.

High-performance computers also play an essential role in generating and testing theoretical models of astronomical objects. Even when we completely understand the underlying physical laws that govern the behavior of a particular object, often the object is so complex that calculating its properties and behavior would be impossible without the assistance of high-performance computers. As discussed in Chapter 4, for example, you can use Newton's laws to compute the orbits of two stars that are gravitationally bound to each other because their orbits take the form of simple ellipses. However, understanding the orbits of the several hundred billion stars that make up the Milky Way Galaxy is not so easy, even though the underlying physical laws are the same.

Computer modeling is used to determine the interior properties of stars and planets, including Earth. Although astronomers cannot see beneath the surfaces of those bodies, they have a surprisingly good understanding of their interiors, which we describe in later chapters. Astronomers start a model by assigning well-understood physical properties to tiny volumes within a planet or star. The computer assembles an enormous number of those individual elements into an overall representation. The result is a rather good picture of what the interior of the star or planet is like.

Astronomers also use high-performance computers to study how astronomical objects, systems of objects, and the universe as a whole evolve. For example, astronomers create models of galaxies and then run computer simulations to study how those galaxies might change over billions of years. **Figure 6.32** shows a simulation of the collision of two galaxies. The results of the computer simulations are then compared with telescopic observations. If the simulations do not match the observations, the model is adjusted and the simulations are run again until general agreement exists between them.

CHECK YOUR UNDERSTANDING 6.5

High-performance computers have become one of an astronomer's most important tools. Which of the following require the use of that type of computer? (Choose all that apply.) (a) analyzing images taken with very large CCDs; (b) generating and testing theoretical models; (c) pointing a telescope from object to object; (d) studying how astronomical objects or systems evolve.

Figure 6.32 These images show supercomputer simulations of the collision of two galaxies. Astronomers compare simulations such as those with telescopic observations.

Origins Microwave Telescopes Detect Radiation from the Big Bang

In this chapter, we explored the tools of the astronomer, from basic optical telescopes to instruments that observe in different wavelengths. Now let's examine in more detail one type of telescope that has aided the study of the origin of the universe. Recall from Chapter 1 that astronomers think the universe originated with a hot Big Bang. The multiple strands of evidence for that conclusion are discussed in Chapter 21. Here, we look at one piece: the observation of faint microwave radiation left over from the early hot universe. Two Bell Laboratories physicists, Arno Penzias (1933–) and Robert Wilson (1936–), were working on satellite communications when they first detected that radiation in 1964 with a microwave dish antenna in New Jersey. Today, we routinely use cell phones and handheld GPS devices that communicate directly with satellites, but at the time, that capability was at the limit of technology.

Penzias and Wilson needed a very sensitive microwave telescope for the work they were doing for Bell Labs because any spurious signals coming from the telescope itself might wash out the faint signals bounced off a satellite. To that end, they were working hard to eliminate all possible sources of interference originating from within their instrument, including keeping the telescope free of bird droppings. No matter how carefully they tried to eliminate sources of extraneous noise, they always still detected a faint signal at microwave wavelengths. That faint signal was the same in every direction and turned out to be from the Big Bang. Penzias and Wilson shared the 1978 Nobel Prize in Physics for discovering the **cosmic microwave background radiation** (**CMB**) left over from the Big Bang itself.

Since 1964, astronomers from around the world have designed increasingly precise instruments to measure that radiation from the ground, from high-altitude balloons, from rockets, and from satellites. The Russian experiment RELIKT-1, launched in 1983, found some limits on the variation of the CMB. The COBE (Cosmic Background Explorer) satellite, launched in 1989, showed that the spectrum of that radiation precisely matched that of a blackbody with a temperature of 2.73 K—exactly what was predicted for the radiation left over from the Big Bang. (Compare **Figure 6.33** with the curves in Figure 5.22.) The data also showed some slight differences in temperature—small fractions of a degree—over the map of the sky. Those slight variations tell us about how the universe evolved from one dominated by radiation to one that contains structures such as galaxies, stars, planets, and us. John Mather and George Smoot shared the 2006 Nobel Prize in Physics for that work.

In 1998 and 2003, a high-altitude balloon experiment called BOOMERANG (short for "balloon observations of millimetric extragalactic radiation and geophysics") flew over Antarctica at an altitude of 42 km to study CMB variations and estimate the overall geometry of the universe. The *WMAP* (Wilkinson Microwave Anisotropy Probe) satellite, launched in 2001, created an even more detailed map of the temperature variations in that radiation, yielding more precise values for the age and shape of the universe and the presence of dark matter and dark energy. The Planck space telescope, operated from 2009 to 2013 by the European Space Agency, was much more sensitive than *WMAP* and studied those CMB variations in even more precise detail. The Atacama Cosmology Telescope and Simons Array (Chile) and the South Pole Telescope (Antarctica) study that radiation to look for evidence of when galaxy clusters formed. Those experiments and observations have opened up the current era of precision cosmology, in which astronomers can make detailed models of how the universe was born, eventually leading to stars, planets, and us.

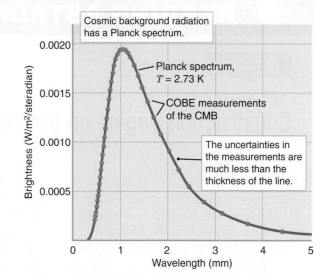

Figure 6.33 This graph shows the spectrum of the cosmic microwave background radiation (CMB) as measured by the COBE satellite (red dots). A steradian is a unit of solid angle. The uncertainty in the measurement at each wavelength is much less than the size of a dot. The line running through the data is a Planck blackbody spectrum with a temperature of 2.73 K.

The Extremely Large Telescope will be located high on a mountain in Chile.

Construction Begins on the World's First Super Telescope

University of Oxford

Scientists are a step closer to understanding the inner-workings of the universe following the laying of the first stone, and construction starting on the world's largest optical and infrared telescope.

With a main mirror 39 metres in diameter, the Extremely Large Telescope (ELT) is going to be, as its name suggests, enormous. Unlike any other before it, ELT is also designed to be an adaptive telescope and has the ability to correct atmospheric turbulence, taking telescope engineering to another level.

To mark the construction's milestone, a ceremony was held at ESO's Paranal residencia in northern Chile, close to the site of the future giant telescope which will be on top of Cerro Armazones, a 3,046-metre peak mountain.

Among many other representatives from industry, the significance of the project was highlighted by the attendance of the Director General of ESO, Tim de Zeeuw, and President of the Republic of Chile, Michelle Bachelet Jeria.

The ELT is being built by the European Southern Observatory (ESO), an international collaboration supported by the UK's Science and Technology Facilities Council (STFC). Oxford University scientists are playing a key role in the project, and are responsible for the design and construction of its spectrograph, 'HARMONI', an instrument designed to simultaneously take 4,000 images, each in a slightly different colour. The visible and near-infrared instrument will harness the telescope's adaptive optics to provide extremely sharp images.

'HARMONI' will enable scientists to form a more detailed picture of the formation and evolution of objects in the Universe. Supporting researchers to view everything from the planets in our own solar system and stars in our own and nearby galaxies with unprecedented depth and precision, to the formation and evolution of distant galaxies that have never been observed before.

Niranjan Thatte, Principal Investigator for 'HARMONI' and Professor of Astrophysics at Oxford's Department of Physics, said: 'For me, the ELT represents a big leap forward in capability, and that means that we will use it to find many interesting things about the Universe that we have no knowledge of today.

'It is the element of "exploring the unknown" that most excites me about the ELT. It will be an engineering feat, and its sheer size and light grasp will dwarf all other telescopes that we have built to date.'

A time capsule, created by members of the ESO team and sealed at the event, will serve as a lasting memory of the research and the scale of ambition and commitment behind it. Contents include a copy of a book describing the original scientific aims of the telescope, images of the staff that have and will play a role in its construction and a poster of an ELT visualisation. The cover of the time capsule is engraved with a hexagon made of Zerodur, a one-fifth scale model of one of the ELT's primary mirror segments.

Tim de Zeeuw, Director General of ESO, said: 'The ELT will produce discoveries that we simply cannot imagine today, and it will surely inspire numerous people around the world to think about science, technology and our place in the Universe. This will bring great benefit to the ESO member states, to Chile, and to the rest of the world.'

1. Why do astronomers want to build at this particular location?
2. What are the advantages of larger telescopes?
3. How can adaptive optics make images sharper?
4. Why will this telescope observe only in optical and infrared wavelengths?
5. Do a Web search to see the status of this project. What countries are partners? Does any local opposition to the project exist in the host country?

Source: "Construction begins on the world's first super telescope," by the University of Oxford, May 26, 2017. Reprinted with permission.

Summary

Earth's atmosphere blocks many spectral regions and distorts telescopic images. Telescopes are sited to be above as much of the atmosphere as possible. Telescopes are matched to the wavelengths of observation, with different technologies required for each region of the spectrum. The aperture of a telescope both determines its light-gathering power and limits its resolution; larger telescopes are better in both measures. Modern CCD cameras have better quantum efficiency and longer integration times, allowing astronomers to study fainter and more distant objects than were observable with earlier detectors. Telescopes observing at microwave wavelengths have detected radiation left over from the Big Bang.

LG 1 **Compare how the two main types of optical telescopes gather and focus light.** The telescope is the astronomer's most important tool. Ground-based telescopes that observe in visible wavelengths are either refractors (lenses) or reflectors (mirrors). All large astronomical telescopes are reflectors. Large telescopes collect more light and have greater resolution. The diffraction limit is the limiting resolution of a telescope.

LG 2 **Summarize the main types of detectors that are used on telescopes.** Photography improved the ability of astronomers to record details of faint objects seen in telescopes. CCDs are today's astronomical detector of choice because they are much more linear, have a broader spectral response, and can send electronic images directly to a computer. Spectrographs are specialized instruments that take the spectrum of an object to reveal what the object is made of and many other physical properties.

LG 3 **Explain why some wavelengths of radiation must be observed from high, dry, and remote observatories on Earth, or from space.** Radio, near-infrared, and optical telescopes can see through our atmosphere. Those types of telescopes can be arrayed to greatly increase angular resolution. Putting telescopes in space solves problems created by Earth's atmosphere.

LG 4 **Explain the benefits of sending spacecraft to study the planets and moons of our Solar System.** Most of what is known about the planets and their moons comes from observations by spacecraft. Flyby and orbiting missions obtain data from space, and landers and rovers collect data from the ground.

LG 5 **Describe other astronomical tools that contribute to the study of the universe.** Astronomers also use particle accelerators, neutrino detectors, and gravitational-wave detectors to study the universe. High-performance computers are essential to acquiring, analyzing, and interpreting astronomical data.

? Unanswered Questions

- Will telescopes be placed on the Moon? The Moon has no atmosphere to make stars twinkle, cause weather, or block certain wavelengths of light from reaching its surface. All parts of the Moon have days and nights that last for 2 Earth weeks each. China had a small (15 cm) ultraviolet telescope operated from Earth on their *Chang'e 3* lander. One proposal is for a small array of radio telescopes on the side of the Moon facing Earth that would study the Sun and the solar wind. Eventually, telescopes would go on the far side of the Moon, which faces away from the light and radio radiation of Earth. On the far side one might put an array of hundreds of radio telescopes that would be deployed to study the earliest formation of stars and galaxies. Another proposal is for the Lunar Liquid Mirror Telescope (LLMT), with a diameter of 20–100 meters, to be located at one of the Moon's poles. Gravity would settle the rotating liquid into the necessary parabolic shape, and those liquid mirror telescopes are much simpler than are arrays of telescopes with large glass mirrors. Astronomers debate whether telescopes on the Moon would be easier to service and repair than those in space and whether problems caused by lunar dust would outweigh any advantages.

- Will human exploration of the Solar System occur within your lifetime? Since the *Apollo* program, humans have not returned to the Moon or traveled to other planets or moons in the Solar System. Sending humans to the worlds of the Solar System is much more complicated, risky, and expensive than sending robotic spacecraft. Humans need life support such as air, water, and food. Radiation in space can be dangerous. Furthermore, human explorers would expect to return to Earth, whereas most spacecraft do not come back. Astronomers and space scientists have heated debates about human spaceflight versus robotic exploration. Some argue that true exploration requires that human eyes and brains actually go there; others argue that the costs and risks are too high for the potential additional scientific knowledge. Beyond basic exploration, we also do not know whether humans will ever permanently colonize space.

Questions and Problems

Test Your Understanding

1. If one telescope has an aperture of 20 cm, and another has an aperture of 30 cm, and if aperture size is the only difference, then which should you choose, and why?
   a. The 20 cm, because the light-gathering power will be better.
   b. The 20 cm, because the image size will be larger.
   c. The 30 cm, because the light-gathering power will be better.
   d. The 30 cm, because the image size will be larger.

2. Which of the following can be observed from Earth's surface? (Choose all that apply.)
   a. radio waves
   b. gamma radiation
   c. far UV light
   d. X-ray light
   e. visible light

3. Match the following properties of telescopes (lettered) with their corresponding definitions (numbered).
   a. aperture
   b. resolution
   c. focal length
   d. chromatic aberration
   e. diffraction
   f. interferometer
   g. adaptive optics

   (1) two or more telescopes connected to act as one
   (2) distance from lens to focal plane
   (3) diameter
   (4) ability to distinguish close objects
   (5) computer-controlled correction for atmospheric distortion
   (6) color-separating effect
   (7) smearing effect due to sharp edge

4. Two 10-meter telescopes, separated by 85 meters, can operate as an interferometer. What is its resolution when it observes in the infrared at a wavelength of 2 microns?
   a. 0.01 arcsec
   b. 0.005 arcsec
   c. 0.2 arcsec
   d. 0.05 arcsec

5. Arrays of radio telescopes can produce much better resolution than single-dish telescopes can because they work based on the principle of
   a. reflection.
   b. refraction.
   c. diffraction.
   d. interference.

6. Refraction is caused by
   a. light bouncing off a surface.
   b. light changing colors as it enters a new medium.
   c. light changing speed as it enters a new medium.
   d. two light beams interfering.

7. The light-gathering power of a 4-meter telescope is _____ than that of a 2-meter telescope.
   a. 4 times larger
   b. 8 times larger
   c. 16 times smaller
   d. 2 times smaller

8. Improved resolution is helpful to astronomers because
   a. they often want to look in detail at small features of an object.
   b. they often want to look at very distant objects.
   c. they often want to look at many objects close together.
   d. all of the above

9. The part of the human eye that acts as the detector is the
   a. retina.
   b. pupil.
   c. lens.
   d. iris.

10. Cameras that use adaptive optics provide higher-spatial-resolution images primarily because
   a. they operate above Earth's atmosphere.
   b. deformable mirrors are used to correct the blurring due to Earth's atmosphere.
   c. composite lenses correct for chromatic aberration.
   d. they simulate a much larger telescope.

11. The advantage of an interferometer is that
   a. the resolution is dramatically improved.
   b. the focal length is dramatically increased.
   c. the light-gathering power is dramatically increased.
   d. diffraction effects are dramatically decreased.
   e. chromatic aberration is dramatically decreased.

12. The angular resolution of a ground-based telescope is usually determined by
   a. diffraction.
   b. the focal length.
   c. refraction.
   d. atmospheric seeing.

13. A grating can spread white light out into a spectrum of colors because of the property of
   a. reflection.
   b. interference.
   c. dispersion.
   d. diffraction.

14. Why would astronomers put telescopes in airplanes?
    a. to get the telescopes closer to the stars
    b. to get the telescopes above most of the water vapor in Earth's atmosphere
    c. to be able to observe one object for more than 24 hours without stopping
    d. to allow the telescopes to observe the full spectrum of light

15. If we could increase the quantum efficiency of the human eye, doing so would
    a. allow humans to see a larger range of wavelengths.
    b. allow humans to see better at night or in other low-light conditions.
    c. increase the resolution of the human eye.
    d. decrease the resolution of the human eye.

Thinking about the Concepts

16. Galileo's telescope used simple lenses. What is the primary disadvantage of using a simple lens in a refracting telescope?

17. The largest astronomical refractor has an aperture of 1 meter. List several reasons why building a larger refractor with twice that aperture would be impractical.

18. Your camera may have a zoom lens, ranging between wide angle (short focal length) and telephoto (long focal length). How does the size of an object in the camera's focal plane differ between wide angle and telephoto?

19. Optical telescopes reveal much about the nature of astronomical objects. Why do astronomers also need information provided by gamma-ray, X-ray, infrared, and radio telescopes?

20. For light reflecting from a flat surface, the angles of incidence and reflection are the same. That is also true for light reflecting from the curved surface of a reflecting telescope's primary mirror. Sketch a curved mirror and several of those reflecting rays.

21. Consider two optically perfect telescopes having different diameters but the same focal length. Is the image of a star larger or smaller in the focal plane of the larger telescope? Explain your answer.

22. Study the Process of Science Figure. Make a flowchart for the symbiosis between technology and science that led to the detection of gravitational waves.

23. Explain adaptive optics and how they improve a telescope's image quality.

24. Explain integration time and quantum efficiency and how each contributes to the detection of faint astronomical objects.

25. Some people believe that we put astronomical telescopes on high mountaintops or in orbit because doing so gets them closer to the objects they are observing. Explain what is wrong with that popular misconception, and give the actual reason telescopes are located in those places.

26. Humans have sent various kinds of spacecraft—including flybys, orbiters, and landers—to all the planets in our Solar System. Explain the advantages and disadvantages of each of those types of spacecraft.

27. If Earth has meteorites that are pieces of Mars, why is going to Mars and bringing back samples of the martian surface so important?

28. Humans had a first look at the far side of the Moon as recently as 1959. Why had we not seen it earlier—when Galileo first observed the Moon with his telescope in 1610?

29. Where are neutrino detectors located? Why are neutrinos so difficult to detect?

30. Why do telescopes in space give a better picture of the left-over radiation from the Big Bang?

Applying the Concepts

31. Compare the light-gathering power of the Thirty Meter Telescope with that of the dark-adapted human eye (aperture, 8 mm) and with that of one of the 10-meter Keck telescopes.

32. Study the photograph of light entering and leaving a block of refractive material in Figure 6.2b. Use a protractor to measure the angles of the green light as it enters the block and as it leaves the block. How are those angles related?

33. Many amateur astronomers start out with a 4-inch (aperture) telescope and then graduate to a 16-inch telescope. By what factor does the light-gathering power of the telescope increase with that upgrade? How much fainter are the faintest stars that can be seen in the larger telescope?

34. The resolution of the human eye is about 1.5 arcmin. What would the aperture of a radio telescope (observing at 21 cm) have to be to have that resolution? Even though the atmosphere is transparent at radio wavelengths, humans do not see light in the radio range. Using your calculations and logic, explain why.

35. Assume that you have a telescope with an aperture of 1 meter. Compare the telescope's theoretical resolution when you are observing in the near-infrared region of the spectrum ($\lambda = 1,000$ nm) with that when you are observing in the violet region of the spectrum ($\lambda = 400$ nm).

36. Assume that the maximum aperture of the human eye, D, is approximately 8 mm and the average wavelength of visible light, λ, is 5.5×10^{-4} mm.
    a. Calculate the diffraction limit of the human eye in visible light.
    b. How does the diffraction limit compare with the actual resolution of 1–2 arcmin (60–120 arcsec)?
    c. To what do you attribute the difference?

37. The diameter of the full Moon in the focal plane of an average amateur's telescope (focal length, 1.5 meters) is 13.8 mm. How big would the Moon be in the focal plane of a very large astronomical telescope (focal length, 250 meters)?

38. One of the earliest astronomical CCDs had 160,000 pixels, each recording 8 bits (256 levels of brightness). A new generation of astronomical CCDs may contain a billion pixels, each recording 15 bits (32,768 levels of brightness). Compare the number of bits of data that each of those two CCD types produces in a single image.

39. Consider a CCD with a quantum efficiency of 80 percent and a photographic plate with a quantum efficiency of 1 percent. If an exposure time of 1 hour is required to photograph a celestial object with a given telescope, how much observing time would be saved by substituting a CCD for the photographic plate?

40. The VLBA uses an array of radio telescopes ranging across 8,000 km of Earth's surface from the Virgin Islands to Hawaii.
    a. Calculate the angular resolution of the array when radio astronomers are observing interstellar water molecules at a microwave wavelength of 1.35 cm.
    b. How does that resolution compare with the angular resolution of two large optical telescopes separated by 100 meters and operating as an interferometer at a visible wavelength of 550 nm?

41. When operational, the Space VLBI may have a baseline of 100,000 km. What will be the angular resolution when studying interstellar molecules emitting at a wavelength of 17 mm from a distant galaxy?

42. The *Mars Reconnaissance Orbiter* (*MRO*) flies at an average altitude of 280 km above the martian surface. If its cameras have an angular resolution of 0.2 arcsec, what is the size of the smallest objects that the *MRO* can detect on the martian surface?

43. At this writing, *Voyager 1* is about 140 astronomical units (AU) from Earth, continuing to record its environment as it departed the boundary of our Solar System.
    a. How far away is *Voyager 1*, in kilometers?
    b. How long do observational data take to come back to us from *Voyager 1*?
    c. How does *Voyager 1*'s distance from Earth compare with that of the nearest star (other than the Sun)?

44. Gravitational waves travel at the speed of light. Their speed, wavelength, and frequency are related as $c = \lambda \times f$. If we were to observe a gravitational wave from a distant cosmic event with a frequency of 10 hertz (Hz), what would be the wavelength of the gravitational wave?

45. Compute the peak of the blackbody spectrum with a temperature of 2.73 K. What region of the spectrum is this?

USING THE WEB

46. A webcast for the International Year of Astronomy 2009 called "Around the World in 80 Telescopes" can be accessed at http://eso.org/public/events/special-evt/100ha.html. The 80 telescopes are situated all over, including Antarctica and space. Pick two of the telescopes and watch the videos. Do you think those videos are effective for public outreach for the observatory in question? For astronomy in general? For each telescope you choose, answer the following questions: Does the telescope observe in the Northern Hemisphere or the Southern Hemisphere? What wavelengths does the telescope observe? What are some of the key science projects at the telescope?

47. Most major observatories have their own websites. Use the link in Question 46 to find a master list of telescopes, and click on a telescope name to link to an observatory website (or run a search on names from Tables 6.1 and 6.2). For the telescope you choose, answer the following questions: What is that telescope's "claim to fame"—is it the largest? At the highest altitude? At the driest location? With the darkest skies? The newest? Does the observatory website have news releases? What is a recent discovery from that telescope?

48. Go to the website for the International Dark Sky Association (http://darksky.org/). Watch the short "Losing the Dark" film (available under the "Resources" tab). Is a "dark sky place" located near you? What are the ecological arguments against too much light at night?

49. What is the current status of the James Webb Space Telescope (https://jwst.nasa.gov)? How is it different from the Hubble Space Telescope? What are some of the instruments for the JWST and its planned projects? What is the current estimated cost of the JWST?

50. Pick a mission from Appendix 5, go to its website, and see what's new. For the mission you choose, answer the following questions: Is the spacecraft still active? Is it sending images? What new science is coming from this mission?

EXPLORATION

digital.wwnorton.com/astro6

Visit the Digital Resources Page and on the Student Site open the "Geometric Optics and Lenses" AstroTour in Chapter 6. Read through the animation until you reach the optics simulation, pictured in **Figure 6.34**. The simulator shows a converging lens and a pencil. Rays come from the pencil on the left of the converging lens, pass through the lens, and make an image to the right of the lens. The view that would be seen by an observer at the position of the eye is shown in the circle at upper right. Initially, when the pencil is at position 2.3 and the eye is at position 2.0, the pencil is out of focus and blurry.

1 Is the eraser at the top or the bottom of the actual pencil? (That answer becomes important later.)

_____

Using the red slider in the upper left of the window, try moving the pencil to the right. Pause when the observer's eye sees a recognizable pencil (even if it's still blurry).

2 Does the eye see the pencil right side up or upside down?

_____

3 This is somewhat analogous to the view through a telescope. The objects are very far from the lenses, and the observer sees things upside down in the telescope. If an object in your field of view is at the top of the field and you want it in the center, should you move the telescope up or down?

_____

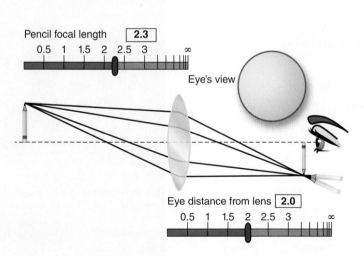

Figure 6.34 Use this simulation to change the position of the object of the eye to explore what will be seen for various configurations.

Now return the pencil to position 2.3. Use the red slider in the lower right of the window to move the eye closer to the lens (to the left).

4 At what distance does the image of the pencil first become crisp and clear?

_____

5 Is the pencil right side up or upside down?

_____

6 In practice at the telescope, we do not move the observer back (away from the eyepiece) to bring the image into focus. Why not?

_____

7 Instead of moving the observer, we use a focusing knob to move the lens in the eyepiece, which brings the image into focus. Imagine that you are looking through the eyepiece of a telescope and the image is blurry. You turn the focusing knob and things get blurrier! What should you try next?

_____

8 Now imagine that you get the image focused just right, so it is crisp and sharp. The next person to use the telescope wears glasses and insists that the image is blurry. When you look through the telescope again, though, the image is still crisp. Explain why your experiences differ.

_____

Step through the animation to the next picture. Carefully study the two telescopes shown and the path the light takes through them.

9 Which telescope has a longer focal length: the top one or the bottom one?

_____

10 Which telescope produces an image with the red and the blue stars more separated: the top one or the bottom one?

_____

11 A longer focal length is an advantage in one sense, but it's not the entire story. What are some disadvantages of a telescope with a very long focal length?

_____

7

The Formation of Planetary Systems

The planetary system containing Earth—our Solar System—is a by-product of the birth of the Sun, but the physical processes that shaped the formation of the Solar System are not unique to it. The same processes have formed many other multiplanet systems. Here in Chapter 7, we examine how planetary systems are born and evolve.

LEARNING GOALS

By the end of this chapter, you should be able to:

LG 1 Describe how our understanding of planetary system formation developed from the work of both planetary and stellar scientists.

LG 2 Discuss the role of gravity and angular momentum in explaining why planets orbit the Sun in a plane and why they revolve in the same direction that the Sun rotates.

LG 3 Explain how temperature at different locations in the protoplanetary disk affects the composition of planets, moons, and other bodies.

LG 4 Discuss the processes that resulted in the formation of planets and other objects in our Solar System.

LG 5 Describe how astronomers find planets around other stars and derive exoplanet properties.

From clouds of gas and dust, planetary systems are born. ▶▶▶

How did our
Solar System
form?

7.1 Planetary Systems Form around a Star

Earth is part of a collection of **planets**—large, round isolated bodies that orbit a star. Astronomers call a system of planets surrounding a star a **planetary system**. The Solar System, shown in **Figure 7.1**, is the planetary system that includes Earth, seven other planets, and the Sun. The system also includes moons that orbit planets and small bodies that occupy particular regions of the Solar System, such as the asteroid belt or the Kuiper Belt. Our Solar System is a tiny part of our galaxy, which is a tiny part of the universe. Review Figure 1.3 to remind yourself of the size scales involved. Light takes about 4 hours to travel to Earth from Neptune, the outermost known planet in the Solar System, but light from the most distant galaxies has taken more than 13 *billion* years to reach Earth.

Until the latter part of the 20th century, the origin of the Solar System remained speculative. Over the past century, with the aid of spectroscopy, astronomers have determined that the Sun is an ordinary star, one of hundreds of billions in its galaxy, the Milky Way, and that the Milky Way is an ordinary galaxy, one of hundreds of billions in the universe. In the past few decades, stellar astronomers studying the formation of stars and planetary scientists analyzing clues about the history of the Solar System have arrived at the same picture of the early Solar System—but from two very different directions. That unified understanding provides the foundation for the way astronomers now think about the Sun and the objects that orbit it. In this section, we look at how the work of stellar and planetary scientists converged to inform our understanding of planetary system formation.

AstroTour: Solar System Formation

The Nebular Hypothesis

The first plausible theory for the formation of the Solar System, the **nebular hypothesis**, was proposed in 1755 by the German philosopher Immanuel Kant (1724–1804) and conceived independently in 1796 by the French astronomer Pierre-Simon Laplace (1749–1827). Kant and Laplace argued that a rotating cloud of interstellar gas, or **nebula** (Latin for "cloud"), gradually collapsed and flattened to form a disk with the Sun at its center. Surrounding the Sun were rings of material from which the planets formed. That configuration would explain why the planets orbit the Sun in the same direction in the same plane. The nebular hypothesis remained popular throughout the 19th century, and those basic principles of the hypothesis are still retained today.

Our modern theory of planetary system formation calculates the conditions required for a cloud of interstellar gas to collapse under the force of their own self-gravity to form stars. Recall from Chapter 4 that self-gravity is the gravitational attraction between the parts of an object, such as a planet or star, that pulls all the parts toward the object's center. That inward force is opposed by either structural strength (for rocks that make up terrestrial planets) or the outward force resulting from gas pressure and radiation pressure within a star. If the outward force is less than self-gravity, the object contracts; if it is greater,

Figure 7.1 Our Solar System includes planets, moons, and other small bodies. Sizes and distances shown are not to scale.

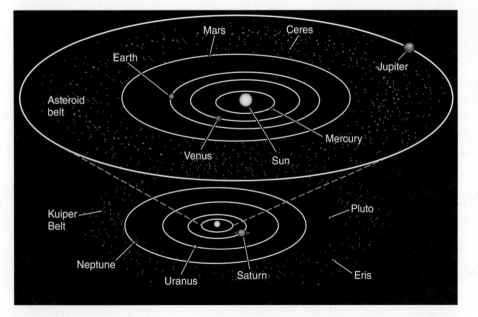

the object expands. In a stable object, the inward and outward forces are balanced.

In support of the nebular hypothesis, disks of gas and dust have been observed surrounding young stellar objects (**Figure 7.2**). From that observational evidence, stellar astronomers have shown that, much like a spinning ball of pizza dough spreads out to form a flat crust, the cloud that produces a star—the Sun, for example—collapses first into a rotating disk. Material in the disk eventually suffers one of three fates: it travels inward onto the forming star at its center, it remains in the disk itself to form planets and other objects, or it is ejected back into interstellar space.

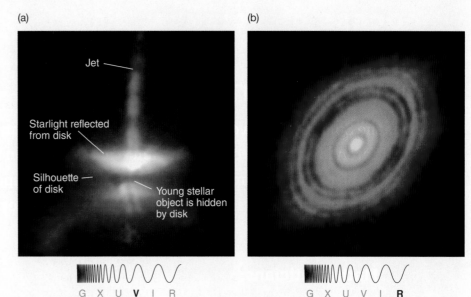

(a)

Jet

Starlight reflected from disk

Silhouette of disk

Young stellar object is hidden by disk

G X U **V** I R

(b)

G X U V I **R**

Figure 7.2 (a) Hubble Space Telescope image of a disk around newly formed stars. The dark band is the silhouette of the disk seen edge on. Bright regions are dust illuminated by the star's light. The jets are shown in green. (b) The Atacama Large Millimeter/submillimeter Array (ALMA) obtained this image of a protoplanetary disk around the star HL Tau. This image shows substructures and possibly planets in the system's dark patches.

Planetary Scientists and the Convergence of Evidence

While astronomers were working to understand star formation, other groups of scientists with very different backgrounds were piecing together the history of the Solar System. Planetary scientists, geochemists, and geologists looking at the current structure of the Solar System inferred what some of its early characteristics must have been. The orbits of all the planets lie very close to a single plane, so the early Solar System must have been flat. In addition, all the planets orbit the Sun in the same direction, so the material from which the planets formed must have been orbiting the Sun in the same direction as well.

To find out more, scientists study samples of the very early Solar System. Rocks that fall to Earth from space, known as **meteorites**, include pieces of material left over from the Solar System's youth. Many meteorites, such as the one in **Figure 7.3**, resemble a piece of concrete in which pebbles and sand are mixed with a much finer filler, suggesting that the larger bodies in the Solar System must have grown from the aggregation of smaller bodies. That finding suggests an early Solar System in which the young Sun was surrounded by a flattened disk of both gaseous and solid material. Our Solar System formed from that swirling disk of gas and dust.

As astronomers and planetary scientists compared notes, they realized they had arrived at the same picture of the early Solar System from two completely different directions. The rotating disk from which the planets formed was the remains of the disk that had accompanied the formation of the Sun. Earth, along with all the other orbiting bodies that make up the Solar System, formed from the remnants of an *interstellar cloud* that collapsed to form the local star, the Sun. The connection between the formation of stars and the origin and later evolution of the Solar System is one of the cornerstones of both astronomy and planetary science—a central theme of our understanding of our Solar System (see the **Process of Science Figure**).

CHECK YOUR UNDERSTANDING 7.1

Which of the following pieces of evidence support the nebular hypothesis? (Choose all that apply.) (a) Planets orbit the Sun in the same direction. (b) The Solar System is relatively flat. (c) Earth has a large Moon. (d) We observe disks of gas and dust around other stars.

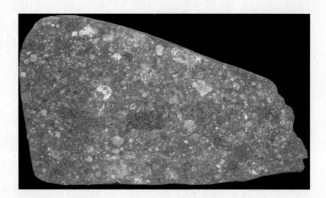

Figure 7.3 Meteorites are the surviving pieces of Solar System fragments that land on planets. This meteorite formed from many smaller components that stuck together.

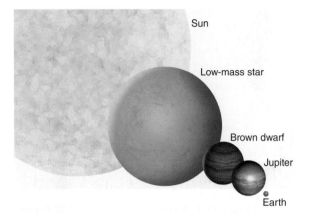

Figure 7.15 A comparison of the diameters of the Sun, a low-mass star, a brown dwarf, Jupiter, and Earth.

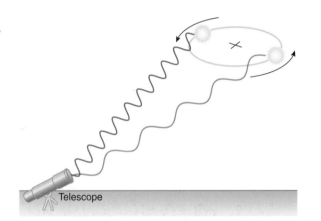

Figure 7.16 Doppler shifts observed in the spectrum of a star as it moves around its common center of gravity with its planet. The spectral lines are blueshifted as the star moves toward us and redshifted as it moves away from us.

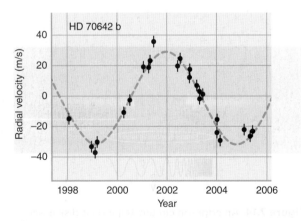

Figure 7.17 Radial velocity data for a star with a planet. A positive velocity is motion away from the observer, whereas a negative velocity is motion toward the observer. The plot repeats as the planet completes another orbit around the star.

physical processes that led to the formation of the Solar System should be commonplace wherever new stars are being born. Compared with stars, however, planets are small and dim objects. They shine primarily by reflection and therefore are millions to billions of times fainter than their host stars. Thus, they were difficult to find until advances in telescope detector technology in the 1990s enabled astronomers to discover them through indirect methods. In 1995, astronomers announced the first confirmed **extrasolar planet**, usually shortened to **exoplanet**—a planet orbiting around a star other than the Sun. Today, the number of known exoplanets has grown to the thousands, and new discoveries occur almost daily.

The IAU defines an exoplanet as an object that orbits a star other than the Sun and has a mass less than 10–13 Jupiter masses (10–13 M_{Jup}). Objects more massive than 10–13 M_{Jup} but less massive than 0.08 solar masses (0.8 M_{Sun}; about 80 M_{Jup}) are **brown dwarfs**. Objects more massive than 0.08 M_{Sun} are defined as stars. **Figure 7.15** compares the diameters of typical objects of each class.

The Search for Exoplanets

The first planets were discovered indirectly, by observing their gravitational tug on the central star. As technology has improved, other methods have become more productive. Astronomers now have direct images of planets orbiting stars and have taken spectra of some exoplanets to observe the composition of their atmospheres. Almost certainly, between the time we write this and the time you read it, new discoveries will have been made. The field is advancing extremely quickly: more than 100 projects are searching for exoplanets from the ground and from space. We will now look at each discovery method.

THE RADIAL VELOCITY METHOD   As a planet orbits a star, the planet's gravity tugs the star around ever so slightly. If the star has a radial velocity (Chapter 5), an observable Doppler shift may appear in the spectrum of the star. **Figure 7.16** illustrates that motion. When the star is moving toward us (negative radial velocity), the light is blueshifted; when the star is moving away from us (positive radial velocity), the light is redshifted. That pattern of radial velocity repeats. After detecting those changes in the radial velocity (**Figure 7.17**), astronomers can infer the existence of the planet and find the planet's mass and distance from the star.

The smallest planet that can be found depends on the precision of the observations. For example, in our Solar System, Jupiter's mass is greater than the mass of all the other planets, asteroids, and comets combined, so Jupiter is the planet with the largest effect on the Sun. Both the Sun and Jupiter orbit a common center of gravity (sometimes called center of mass—the location where the effect of one mass balances the other) that lies just outside the surface of the Sun, as shown in **Figure 7.18**. Jupiter tugs the Sun around in a circle with a speed of 12 m/s. Alien astronomers would find that the Sun's radial velocity varies by ±12 m/s, with a period equal to Jupiter's orbital period of 11.86 years. From that information, the astronomers would rightly conclude that the Sun has at least one planet with a mass comparable to Jupiter's. Without greater precision, the observers would be unaware of the other, less massive planets. To detect Saturn, they would need to improve the precision of their measurements to 2.7 m/s. Earth would not be detectable unless the aliens could detect motions as small as 0.09 m/s.

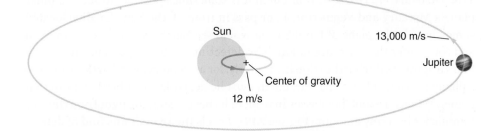

Figure 7.18 The Sun and Jupiter orbit around a common center of gravity (+), which lies just outside the Sun's surface. Spectroscopic measurements made by an extrasolar astronomer would reveal the Sun's radial velocity varying by ±12 m/s over 11.86 years, which is Jupiter's orbital period. Jupiter travels around its orbit at a speed of 13,000 m/s.

The precision of radial velocity instruments has been about 0.3 m/s. The technique enabled astronomers to detect giant planets around solar-type stars, but not yet to find planets with masses similar to Earth's. Finding the signal of the Doppler shift in the noise of the observation requires the star to be quite bright in our sky. A new instrument at the Very Large Telescope is expected to improve precision and be able to detect smaller planets. **Working It Out 7.2** explains more about the spectroscopic radial velocity method.

Astronomy in Action: Doppler Shift

Simulation: Radial Velocity

7.2 Working It Out Estimating the Size of the Orbit of a Planet

In the spectroscopic radial velocity method, the star is moving about its center of mass, and its spectral lines are Doppler-shifted accordingly. Recall from Figure 7.18 that an alien astronomer looking toward the Solar System would observe a shift in the wavelengths of the Sun's spectral lines—caused by the presence of Jupiter—of ~12 m/s.

Figure 7.17 showed the radial velocity data for a star with a planet discovered with that method. How do astronomers use that method to estimate the distance (A) of the planet from the star? Recall from Chapter 4 that Newton generalized Kepler's law relating the period of an object's orbit to the orbital semimajor axis:

$$P^2 = \frac{4\pi^2}{G} \times \frac{A^3}{M}$$

where A is the semimajor axis of the orbit, P is its period, and M is the combined mass of the two objects. To find A, we rearrange the equation as follows:

$$A^3 = \frac{G}{4\pi^2} \times M \times P^2$$

According to the graph of radial velocity observations in Figure 7.17, the period of the orbit is 5.7 years. A year has 3.16×10^7 seconds, so

$$P = 5.7 \text{ yr} \times (3.16 \times 10^7 \text{ s/yr}) = 1.8 \times 10^8 \text{ s}$$

The mass of the star is much greater than the mass of the planet, so the combined masses of the star and the planet can be approximated as the mass of the star, which here is about equal to the mass of the Sun, 2×10^{30} kg. (Stellar masses can be estimated from their spectra.) The gravitational constant G is 6.67×10^{-20} km³/(kg s²). Plugging in the numbers gives

$$A^3 = \frac{6.67 \times 10^{-20} \dfrac{\text{km}^3}{\text{kg s}^2}}{4\pi^2} \times (2 \times 10^{30} \text{ kg}) \times (1.8 \times 10^8 \text{ s})^2$$

$$A^3 = 1.1 \times 10^{26} \text{ km}^3$$

taking the cube root,

$$A = 4.8 \times 10^8 \text{ km}$$

If we convert that number into astronomical units (where 1 AU = 1.5×10^8 km), the semimajor axis of the orbit of this planet is

$$A = \frac{4.8 \times 10^8 \text{ km}}{1.5 \times 10^8 \text{ km/AU}} = 3.2 \text{ AU}$$

The planet is 3.2 times farther from its star than Earth is from the Sun.

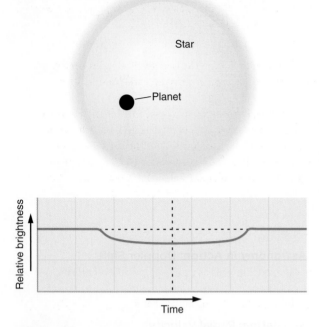

Figure 7.19 A planet that passes in front of a star blocks some of the light coming from the star's surface, causing the brightness of the star to decrease slightly. (The decrease in brightness is exaggerated here.)

Figure 7.20 Multiple planets can be detected through multiple transits with different changes in brightness. The arrows point to the changes in the total light as the three planets transit the star.

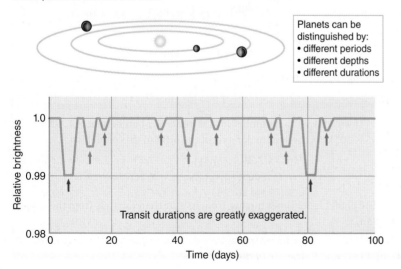

THE TRANSIT METHOD   From Earth it is sometimes possible to see the inner planets Mercury and Venus transit, or pass in front of, the Sun. An alien located somewhere in the plane of Earth's orbit would see Earth pass in front of the Sun and could infer the existence of Earth by detecting the 0.009 percent drop in the Sun's brightness during the transit. Similarly, for astronomers on Earth to observe a planet passing in front of a star, Earth must lie nearly in the orbital plane of that planet. When an exoplanet passes in front of its parent star, the light from the star diminishes by a tiny amount (**Figure 7.19**). This is the **transit method** of detecting exoplanets, in which we observe the dimming of starlight as a planet passes in front of its parent star. Whereas the radial velocity method gives us the mass of the planet and its orbital distance from a star, the transit method provides the radius of a planet. **Working It Out 7.3** shows how the radii are estimated.

Current ground-based technology limits the sensitivity of the transit method to about 0.1 percent of a star's brightness. Telescopes in space improve the sensitivity because smaller dips in brightness can be measured. The small Fench COROT telescope (27 cm) discovered 32 planets during its 6 years of operation (2007–2013). NASA's 0.95-meter Kepler telescope has discovered many planets and has found thousands more candidates that are being investigated further. **Figure 7.20** illustrates how systems of multiple planets are identified with that method: if one planet is found, then multiple sets of transits can indicate that other planets are orbiting the same star. Several thousand exoplanets have been detected from ground-based and space telescopes by using the transit method.

OTHER METHODS   The gravitational field of an unseen planet can act like a lens, bending the light from a distant star in such a way that it causes the star to brighten temporarily while the planet is passing in front of it. Because the effect is small, it is usually called *microlensing*. Like the radial velocity method, microlensing provides an estimate of the mass of the planet. To date, about 60 exoplanets have been found with that technique.

Planets also may be detected by *astrometry*—precisely measuring the position of a star in the sky. If the system is viewed from "above," the star moves in a mini-orbit as the planet pulls it around. That motion is generally tiny and therefore very difficult to measure. For systems viewed from above the plane of the planet's orbit, however, none of the prior methods will work because the planet neither passes in front of the star nor causes a shift in its speed along the line of sight. Space missions such as the Gaia observatory, launched in 2013 by the European Space Agency, conduct such observations.

Direct imaging involves taking a picture of the planet directly. The technique is conceptually straightforward but is technically difficult because it involves searching for a relatively faint planet in the overpowering glare of a bright star—a challenge far more difficult than looking for a star in a clear, bright daytime sky. Even when an object is detected by direct imaging, an astronomer must still determine whether the observed object is actually a planet. Suppose we detect a faint object near a bright star. Could it be a more distant star that just happens to be in the line of sight? Future observations could tell whether the object shares the bright star's motion through space, but it also could be a brown dwarf rather than a true planet. An astronomer would need to make further observations to determine the object's mass.

The masses of exoplanets can often be estimated using Kepler's laws and the conservation of angular momentum. When planets are detected with the transit method, astronomers can estimate the radius of an exoplanet. In that method, astronomers look for planets that eclipse their stars and then observe how much the star's light decreases during that eclipse (see Figure 7.19). When Venus or Mercury transits the Sun, a black circular disk is visible on the face of the circular Sun. During the transit, the amount of light from the transited star is reduced by the area of the circular disk of the planet divided by the area of the circular disk of the star:

$$\% \text{ reduction in light} = \frac{\text{Area of disk of planet}}{\text{Area of disk of star}} = \frac{\pi R_{\text{planet}}^2}{\pi R_{\text{star}}^2} = \frac{R_{\text{planet}}^2}{R_{\text{star}}^2}$$

Then, to solve for the radius of the planet, astronomers need an estimate of the radius of the star and a measurement of the percentage of reduction in light during the transit. The radius of a star is estimated from the surface temperature and the luminosity of the star.

Kepler-11, for example, is a system of at least six planets that transit a star. The radius of the star, R_{star}, is estimated to be 1.1 times the radius of the Sun, or $1.1 \times (7.0 \times 10^5 \text{ km}) = 7.7 \times 10^5 \text{ km}$. The light from the Kepler-11 star is observed to decrease by 0.077 percent, or 0.00077 (see Figure 7.19), from planet Kepler-11c. What is the radius of Kepler-11c?

$$0.00077 = \frac{R_{\text{Kepler-11c}}^2}{R_{\text{star}}^2} = \frac{R_{\text{Kepler-11c}}^2}{(7.7 \times 10^5 \text{ km})^2}$$

$$R_{\text{Kepler-11c}}^2 = 4.5 \times 10^8 \text{ km}^2 \quad \text{thus} \quad R_{\text{Kepler-11c}} = 2.1 \times 10^4 \text{ km}$$

Dividing $R_{\text{Kepler-11c}}$ by R_{Earth} (6,400 km) shows that $R_{\text{Kepler-11c}} = 3.3 \, R_{\text{Earth}}$.

Some planets have been discovered through that method with large ground-based telescopes operating in the infrared region of the spectrum, using adaptive optics. **Figure 7.21** is an infrared image of Beta Pictoris b. A related form of direct observation involves separating the spectrum of a planet from the spectrum of its star to obtain information about the planet directly. Large ground-based telescopes have obtained spectra of the atmospheres of some exoplanets and have found, for example, carbon monoxide and water in those atmospheres.

Types of Exoplanets

Searches for exoplanets have been remarkably successful. Between the discovery of the first (in 1995) and this writing, nearly 4,000 more have been confirmed, and thousands more candidates are under investigation. As the number of observed systems with single and multiple planets increases, astronomers can compare those worlds with Solar System planets, finding more variation than they expected. The field is changing so fast that the most up-to-date information can be found only online.

The first discoveries included many **hot Jupiters**—Jupiter-sized planets orbiting solar-type stars in circular or highly eccentric orbits that bring them closer to their parent stars than Mercury is to our own Sun. Those planets were among the first to be detected because they are relatively easy targets for the spectroscopic radial velocity method. The large mass of a nearby hot Jupiter tugs the star very hard, creating large radial velocity variations in the star. In addition, those large planets orbiting close to their parent stars are more likely to pass in front of the star periodically and reveal themselves via the transit method.

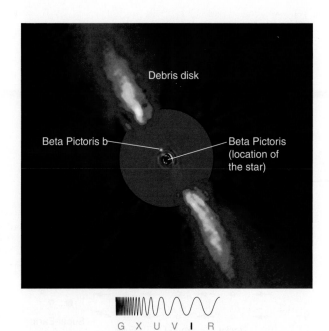

G X U V I R

Figure 7.21 The planet Beta Pictoris b is seen orbiting within a dusty debris disk that surrounds the bright naked-eye star Beta Pictoris. The planet's estimated mass is 8 times that of Jupiter. The star is hidden behind an opaque mask, and the planet appears through a semitransparent mask used to subdue the brightness of the dusty disk.

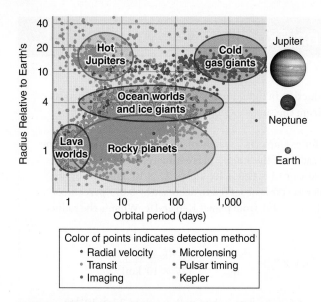

Figure 7.22 Different populations of exoplanets. Color of points indicates detection method. Jupiter-sized planets can be close to their star and hot, or they can be far from their star and cold. Neptune-sized planets are probably composed of water and ice. Earth-sized planets have higher density, indicating that they are made of rock.

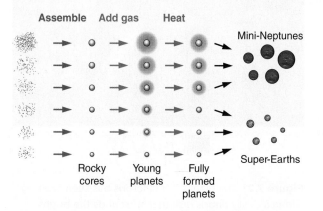

Figure 7.23 Many exoplanets are between the sizes of Earth and Neptune ($\sim$4 R_E). "Super-Earths" began with smaller rocky cores and less material, whereas "mini-Neptunes" had larger cores and collected more gas while forming.

Therefore, those hot Jupiter systems are easier to find than smaller, more distant planets. Astronomers realized that those hot Jupiter systems are not representative of most planetary systems; they were just easier to find. Scientists call that type of bias a *selection effect*. **Figure 7.22** illustrates the types of exoplanet populations.

Astronomers were surprised by the hot Jupiters because, according to the theory of planet formation described earlier in the chapter, those giant, volatile-rich planets should not have been able to form so close to their parent stars. From theories based on the Solar System, astronomers expected that Jupiter-type planets should form in the more distant, cooler regions of the protoplanetary disk, where the volatiles that make up much of their composition can survive. Hot Jupiters may form much farther from their parent stars and later migrate inward to a closer orbit. That migration may be caused by an interaction with gas or planetesimals in which orbital angular momentum is transferred from the planet to its surroundings, allowing it to spiral inward.

The radial velocity method yields an estimate of the mass of a planet (see Working It Out 7.2), and the transit method yields an estimate of the size of a planet (see Working It Out 7.3). With both pieces of information, the density (mass divided by the volume) of the planet can be computed to determine whether the planet is mostly gaseous (low density) or mostly rock (high density). Many of the new planets being discovered by Kepler are mini-Neptunes (gaseous planets with masses of 2–10 M_{Earth}) or super-Earths (rocky planets more massive than Earth). **Figure 7.23** illustrates the formation of those two types.

Planets with longer orbital periods, and therefore larger orbits, can be discovered only when the observations have gone on long enough to observe more than one complete orbit. Some of the exoplanets have highly elliptical orbits compared with those in the Solar System. Planets have been found with orbits that are highly tilted with respect to the plane of the rotation of their star, and some planets move in orbits whose direction is opposite that of their star's rotation. Multiple-planet systems have been observed in which the larger mini-Neptunes alternate with smaller super-Earths. The multiple-planet systems that have been found with the transit method reside in flat systems like our own, offering further evidence that the planets formed in a flat protoplanetary disk around a young star. But the current hypothesis to explain the Solar System's inner, small rocky planets and outer, large gaseous planets may not apply to all other planetary systems.

In addition, some planets detected through microlensing seem to be wandering freely through the Milky Way. Those planets may have been ejected from their solar systems after they formed and are no longer in gravitationally bound orbits around their stars.

We return to exoplanets at several points in this book—namely, when we review the planets in our own Solar System, when we consider the types of stars, and when we discuss the search for Earth-like planets in Chapter 24. The frequent new discoveries requiring revisions of existing theories make exoplanets one of the most exciting topics in astronomy today.

CHECK YOUR UNDERSTANDING 7.5

What is the most common method for discovering an Earth-mass planet around a star?

(a) Doppler spectroscopy; (b) direct imaging; (c) transit; (d) astrometric

Origins The Search for Earth-Sized Planets

The discovery of planetary systems, many different from the Solar System, shows us that the formation of planets often, and perhaps always, accompanies the formation of stars. The implications of that conclusion are profound. Planets are a common by-product of star formation. In a galaxy of 200 billion stars and a universe of hundreds of billions of galaxies, how many planets (and moons) might exist? And with all those planets in the universe, how many might have conditions suitable for the particular category of chemical reactions that we refer to as "life"? (We return to that point in Chapter 24.)

The Kepler Mission was developed by NASA to find Earth-sized and larger planets in orbit about a variety of stars. Kepler is a 0.95-meter telescope with 42 CCD detectors that observes approximately 150,000 stars in 100 square degrees of sky to look for planetary transits. To confirm a planetary detection, transits need to be observed three times with repeatable changes in brightness, duration of transit times, and computed orbital period. Kepler could detect a dip of 0.01 percent in the brightness of a star—sensitive enough to detect an Earth-sized planet. Kepler identified the first Earth-sized planets in 2011. Stars with transiting planets detected by Kepler also are observed spectroscopically to obtain radial velocity measurements that can lead to an estimate of a planet's mass. If a planet's radius and mass are known, the planet's density (mass per volume) can be estimated, too. From the density, astronomers can get a sense of whether the planet is composed primarily of gas, rock, ice, water, or some mixture of those.

On Earth, liquid water was essential for life to form and evolve. Because life on Earth is the only example of life for which we have evidence, we do not know whether liquid water is a cosmic requirement, but it is a place to start. The Kepler Mission searched for rocky planets at the right distance from their stars to permit the existence of liquid water, a distance known as the **habitable zone**. The location and width of the zone will vary depending on the temperature of the star. If a planet is too close to its star, water will exist only as a vapor; if too far, water will be frozen as ice. In the Solar System, Earth is the only planet currently in the habitable zone. Although announcements of new planets often state whether the planet is in the habitable zone, just being in the zone doesn't guarantee that the planet actually has liquid water— or that the planet is inhabited! An example of an Earth-sized planet in a habitable zone is shown in **Figure 7.24**.

Kepler identified thousands of planet candidates, some in the habitable zones of their respective stars. The candidates must be confirmed by follow-up observations of more transits or of radial velocities before they are officially announced as planet detections. Citizen scientists contribute to that search through the Exoplanet Explorers project; recently, the project announced the discovery of a system with five planets.

Astronomers are looking forward to results from new missions such as the Transiting Exoplanet Survey Satellite (TESS), launched in 2018, and the James Webb Space Telescope (JWST), scheduled for launch in 2020. TESS will monitor 200,000 nearby stars for dimming caused by planetary transits, whereas JWST will observe in the infrared and should be able to detect gases in the atmospheres of exoplanets.

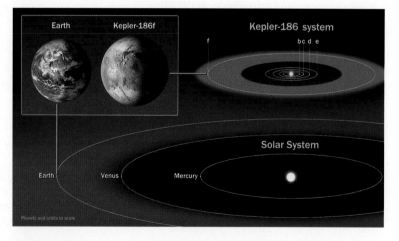

Figure 7.24 Artist's conception of the Kepler-186 system. Located about 500 light-years from Earth, that system has a planet in its habitable zone. The Solar System is shown for comparison. The Kepler-186 star is cooler than the Sun; thus, its habitable zone is closer to the star.

Four Earth-Sized Planets Detected Orbiting the Nearest Sun-like Star

August 9, 2017, by TIM STEPHENS

A new study by an international team of astronomers reveals that four Earth-sized planets orbit the nearest Sun-like star, Tau Ceti, which is about 12 light years away and visible to the naked eye. These planets have masses as low as 1.7 Earth mass, making them among the smallest planets ever detected around nearby Sun-like stars. Two of them are super-Earths located in the habitable zone of the star, meaning they could support liquid surface water.

The planets were detected by observing the wobbles in the movement of Tau Ceti. This required techniques sensitive enough to detect variations in the movement of the star as small as 30 centimeters per second.

"We are now finally crossing a threshold where, through very sophisticated modeling of large combined data sets from multiple independent observers, we can disentangle the noise due to stellar surface activity from the very tiny signals generated by the gravitational tugs from Earth-sized orbiting planets," said coauthor Steven Vogt, professor of astronomy and astrophysics at UC Santa Cruz.

According to lead author Fabo Feng of the University of Hertfordshire, UK, the researchers are getting tantalizingly close to the 10-centimeter-per-second limit required for detecting Earth analogs. "Our detection of such weak wobbles is a milestone in the search for Earth analogs and the understanding of the Earth's habitability through comparison with these analogs," Feng said. "We have introduced new methods to remove the noise in the data in order to reveal the weak planetary signals."

The outer two planets around Tau Ceti are likely to be candidate habitable worlds, although a massive debris disk around the star probably reduces their habitability due to intensive bombardment by asteroids and comets.

The same team also investigated Tau Ceti four years ago in 2013, when coauthor Mikko Tuomi of the University of Hertfordshire led an effort in developing data analysis techniques and using the star as a benchmark case. "We came up with an ingenious way of telling the difference between signals caused by planets and those caused by star's activity. We realized that we could

see how star's activity differed at different wavelengths and use that information to separate this activity from signals of planets," Tuomi said.

The researchers painstakingly improved the sensitivity of their techniques and were able to rule out two of the signals the team had identified in 2013 as planets. "But no matter how we look at the star, there seem to be at least four rocky planets orbiting it," Tuomi said. "We are slowly learning to tell the difference between wobbles caused by planets and those caused by stellar active surface. This enabled us to essentially verify the existence of the two outer, potentially habitable planets in the system."

Sun-like stars are thought to be the best targets in the search for habitable Earth-like planets due to their similarity to the Sun. Unlike more common smaller stars, such as the red dwarf stars Proxima Centauri and Trappist-1, they are not so faint that planets would be tidally locked, showing the same side to the star at all times. Tau Ceti is very similar to the sun in its size and brightness, and both stars host multi-planet systems.

1. How do planets cause their stars to "wobble"?
2. With the spectroscopic method, why is finding Earth-sized planets harder than finding giant planets?
3. How did the astronomers improve their ability to find small planets?
4. What is a debris disk?
5. What would be the disadvantage to a planet's being tidally locked to its star? (Hint: Refer to Chapter 2.)

Summary

Stars and their planetary systems form from collapsing interstellar clouds of gas and dust, following the laws of gravity and conservation of angular momentum. Conservation of angular momentum produces an accretion disk around a protostar that often fragments to form multiple planets, as well as smaller objects such as asteroids and dwarf planets, through the gradual accumulation of material into larger and larger objects. Multiple methods exist for finding planets around other stars, and those planets are now thought to be very common. This field of study is evolving very quickly as technology advances.

LG 1 **Describe how our understanding of planetary system formation developed from the work of both planetary and stellar scientists.** Planets are a common by-product of star formation, and many stars are surrounded by planetary systems. Gravity pulls clumps of gas and dust together, causing them to shrink and heat up. Angular momentum must be conserved, leading to both a spinning central star and an accretion disk that rotates and revolves in the same direction as the central star. Solar System meteorites show that larger objects build up from smaller objects.

LG 2 **Discuss the role of gravity and angular momentum in explaining why planets orbit the Sun in a plane and why they revolve in the same direction that the Sun rotates.** As particles orbit the forming star, the cloud of dust and gas flattens into a plane. Conservation of angular momentum determines both the speed and the direction of the revolution of the objects in the forming system. Dust grains in the protoplanetary disk first stick together because of collisions and static electricity. As those objects grow, they eventually have enough mass to attract other objects gravitationally. Once that occurs, they begin emptying the space around them. Collisions of planetesimals lead to the formation of planets.

LG 3 **Explain how temperature at different locations in the protoplanetary disk affects the composition of planets, moons, and other bodies.** The temperature is higher near the central protostar, forcing volatile elements, such as water, to evaporate and leave the inner part of the disk. Planets in the inner part of the disk will have fewer volatiles than those in the outer part of the disk. The gas that a planet captures when it forms is the planet's primary atmosphere. Less massive planets lose their primary atmospheres and then form secondary atmospheres.

LG 4 **Discuss the processes that resulted in the formation of planets and other objects in our Solar System.** In the current model of the formation of the Solar System, solid terrestrial planets formed in the inner disk, where temperatures were high, whereas giant gaseous planets formed in the outer disk, where temperatures were low. Dwarf planets such as Pluto formed in the asteroid belt and in the region beyond the orbit of Neptune. Asteroids and comet nuclei remain today as leftover debris.

LG 5 **Describe how astronomers find planets around other stars and derive exoplanet properties.** Astronomers find planets around other stars by using the radial velocity method, the transit method, microlensing, astrometry, and direct imaging. As technology has improved, the number and variety of known exoplanets has increased dramatically, with thousands of planets and planet candidates discovered orbiting other stars near the Sun within the Milky Way Galaxy in just the past few years.

? Unanswered Questions

- How typical is the Solar System? Only within the past decade have astronomers found other systems containing four or more planets, and so far the observed distributions of large and small planets in those multiplanet systems have looked different from those of the Solar System. Computer simulations of planetary system formation suggest that a system with stable orbits and a planetary distribution like those of the Solar System may develop only rarely. Improved supercomputers can run more complex simulations, which can be compared with the observations to better understand how solar systems are configured.

- How Earth-like must a planet be before scientists declare it to be "another Earth"? An editorial in the science journal *Nature* cautioned that scientists should define "Earth-like" in advance—before multiple discoveries of planets "similar" to Earth are announced and a media frenzy ensues. Must a planet be of similar size and mass, be located in the habitable zone, and have spectroscopic evidence of liquid water before we call it "Earth 2.0"?

Questions and Problems

Test Your Understanding

1. Place the following events in the order that corresponds to the formation of a planetary system.
   a. Gravity collapses a cloud of interstellar gas.
   b. A rotating disk forms.
   c. Small bodies collide to form larger bodies.
   d. A stellar wind "turns on" and sweeps away gas and dust.
   e. Primary atmospheres form.
   f. Primary atmospheres are lost.
   g. Secondary atmospheres form.
   h. Dust grains stick together by static electricity.

2. If the radius of an object's orbit is halved, and angular momentum is conserved, what must happen to the object's speed?
   a. It must be halved.
   b. It must stay the same.
   c. It must be doubled.
   d. It must be squared.

3. Unlike the giant planets, the terrestrial planets formed when
   a. the inner Solar System was richer in heavy elements than the outer Solar System.
   b. the inner Solar System was hotter than the outer Solar System.
   c. the outer Solar System took up more volume than the inner Solar System, so more material was available to form planets.
   d. the inner Solar System was moving faster than the outer Solar System.

4. The terrestrial planets and the giant planets have different compositions because
   a. the giant planets are much larger.
   b. the terrestrial planets formed closer to the Sun.
   c. the giant planets are made mostly of solids.
   d. the terrestrial planets have few moons.

5. The spectroscopic radial velocity method preferentially detects
   a. large planets close to the central star.
   b. small planets close to the central star.
   c. large planets far from the central star.
   d. small planets far from the central star.
   e. The method detects all those planets equally well.

6. The concept of disk instability was developed to solve the problem that
   a. Jupiter-like planets migrate after formation.
   b. not enough gas was in the Solar System to form Jupiter.
   c. the early solar nebula probably dispersed too soon to form Jupiter.
   d. Jupiter consists mostly of volatiles.

7. Because angular momentum is conserved, an ice-skater who throws her arms out will
   a. rotate more slowly.
   b. rotate more quickly.
   c. rotate at the same rate.
   d. stop rotating.

8. Clumps grow into planetesimals by
   a. gravitationally pulling in other clumps.
   b. colliding with other clumps.
   c. attracting other clumps with opposite charge.
   d. conserving angular momentum.

9. The transit method preferentially detects
   a. large planets close to the central star.
   b. small planets close to the central star.
   c. large planets far from the central star.
   d. small planets far from the central star.
   e. The method detects all those planets equally well.

10. If the radius of a spherical object is halved, what must happen to the period so that the spin angular momentum is conserved?
    a. It must be divided by 4.
    b. It must be halved.
    c. It must stay the same.
    d. It must double.
    e. It must be multiplied by 4.

11. The amount of angular momentum in a spherical object does *not* depend on
    a. its radius.
    b. its mass.
    c. its rotation speed.
    d. its temperature.

12. The planets in the inner part of the Solar System are made primarily of refractory materials; the planets in the outer Solar System are made primarily of volatiles. That difference occurs because
    a. refractory materials are heavier than volatiles, so they sank farther into the nebula.
    b. no volatiles were in the inner part of the accretion disk.
    c. the volatiles on the inner planets were lost soon after the planets formed.
    d. the outer Solar System has gained more volatiles from space since formation.

13. If scientists want to find out about the composition of the early Solar System, the best objects to study are
    a. the terrestrial planets.
    b. the giant planets.
    c. the Sun.
    d. asteroids and comets.

14. The direction of revolution in the plane of the Solar System was determined by
    a. the plane of the galaxy in which the Solar System sits.
    b. the direction of the gravitational force within the original cloud.
    c. the direction of rotation of the original cloud.
    d. the amount of material in the original cloud.

15. A planet in the "habitable zone"
    a. is close to the central star.
    b. is far from the central star.
    c. is the same distance from its star as Earth is from the Sun.
    d. is at a distance where liquid water can exist on the surface.

Thinking about the Concepts

16. What is the source of the material that now makes up the Sun and the rest of the Solar System?

17. Describe the different ways by which stellar astronomers and planetary scientists each came to the same conclusion about how planetary systems form.

18. What is a protoplanetary disk? What are two reasons that the inner part of the disk is hotter than the outer part?

19. Physicists describe certain properties, such as angular momentum and energy, as being *conserved*. What does that mean? Do conservation laws imply that an individual object can never lose or gain angular momentum or energy? Explain your reasoning.

20. The Process of Science Figure in this chapter makes the point that different areas of science must agree with one another. Suppose that a few new exoplanets are discovered that appear not to have formed from the collapse of a stellar nebula (for example, the planetary orbits might be in random orientations). What will scientists do with that new information?

21. How does the law of conservation of angular momentum control a figure-skater's rate of spin?

22. What is an accretion disk?

23. Describe how tiny grains of dust grow to become massive planets.

24. Look under your bed, the refrigerator, or any similar place for dust bunnies. Once you find them, blow one toward another. Watch carefully and describe what happens as they meet. What happens if you repeat that action with additional dust bunnies? Will those dust bunnies ever have enough gravity to begin pulling themselves together? If they were in space instead of on the floor, might that happen? What force prevents their mutual gravity from drawing them together into a "bunny-tesimal" under your bed?

25. Why do we find rocky material everywhere in the Solar System but find large amounts of volatile material only in the outer regions?

26. Why could the four giant planets collect massive gaseous atmospheres, whereas the terrestrial planets could not? Explain the source of the secondary atmospheres surrounding the terrestrial planets.

27. Describe four methods that astronomers use to search for exoplanets. What are the limitations of each method; that is, what circumstances are necessary to detect a planet by each method?

28. Why is it difficult to obtain an image of an exoplanet?

29. Many of the first exoplanets that astronomers found orbiting other stars were giant planets with Jupiter-like masses and with orbits located very close to their parent stars. Explain why those characteristics are a selection effect of the discovery method.

30. How has the Kepler telescope found Earth-like planets, and what do astronomers mean by "Earth-like"?

Applying the Concepts

31. In Figure 7.17, what is the maximum radial velocity of HD 70642 b in meters per second? Convert that number to miles per hour (mph). How does that value compare with the speed at which Earth orbits the Sun (67,000 mph)?

32. Use Appendix 4 to answer the following:
    a. What is the total mass of all the planets in the Solar System, expressed in Earth masses (M_{Earth})?
    b. What fraction of that total planetary mass is Jupiter?
    c. What fraction does Earth represent?

33. Compare Earth's orbital angular momentum with its spin angular momentum by using the following values: $m = 5.97 \times 10^{24}$ kg, $v = 29.8$ kilometers per second (km/s), $r = 1$ AU, $R = 6,378$ km, and $P = 1$ day. Assume that Earth is a uniform body. What fraction does each component (orbital and spin) contribute to Earth's total angular momentum? Refer to Working It Out 7.1.

34. Venus has a radius 0.949 times that of Earth and a mass 0.815 times that of Earth. Venus's rotation period is 243 days. What is the ratio of Venus's spin angular momentum to that of Earth? Assume that Venus and Earth are uniform spheres.

35. Jupiter has a mass equal to 318 times Earth's mass, an orbital radius of 5.2 AU, and an orbital velocity of 13.1 km/s. Earth's orbital velocity is 29.8 km/s. What is the ratio of Jupiter's orbital angular momentum to that of Earth?

36. The asteroid Vesta has a diameter of 530 km and a mass of 2.7×10^{20} kg.
    a. Calculate the density (mass/volume) of Vesta.
    b. The density of water is 1,000 kg/m³, whereas that of rock is about 2,500 kg/m³. What does that difference tell you about the composition of Vesta?

37. Using data in Appendix 4, compute the densities of Venus, Jupiter, and Neptune. Compare with the densities of rock, water, and gas.

38. Recalling Kepler's laws, put the three planets in Figure 7.20 in order from fastest to slowest. Compare the duration of the transits in Figure 7.20. Why does the outermost planet have the longest duration?

39. Suppose you can measure radial velocities of about 0.3 m/s. Suppose you are observing a spectral line with a wavelength of 575 nanometers (nm). How large a shift in wavelength would a radial velocity of 0.3 m/s produce?

40. Earth tugs the Sun around as it orbits, but that effect (only 0.09 m/s) is much smaller than that of any known exoplanet. How large a shift in wavelength does that effect cause in the Sun's spectrum at 500 nm?

41. If an alien astronomer observed a plot of the light curve as Jupiter passed in front of the Sun, by how much would the Sun's brightness drop during the transit?

42. A planet has been found to orbit a 1-M_{Sun} star in 200 days.
    a. What is the orbital radius of that exoplanet?
    b. Compare its orbit with that of the planets around our own Sun. What temperatures must that planet experience?

43. One of the planets orbiting the star Kepler-11 has an orbital radius of 1.1 R_{Sun}. If the planet has a radius of 4.5 R_{Earth}, by how much does the brightness of Kepler-11 decrease when that planet transits the star?

44. Kepler detected a planet with a diameter of 1.7 Earth (D_{Earth}).
    a. How much larger is the volume of that planet than Earth's?
    b. Assume that the density of the planet is the same as Earth's. How much more massive is it than Earth?

45. The planet COROT-11b was discovered using the transit method, and astronomers have followed up with radial velocity measurements, so both its radius (1.43 R_{Jup}) and its mass (2.33 M_{Jup}) are known. The density provides a clue about whether the object is gaseous or rocky.
    a. What is the mass of the planet in kilograms?
    b. What is the planet's radius in meters?
    c. What is the planet's volume?
    d. What is the planet's density? How does that density compare with the density of water (1,000 kg/m³)? Is the planet likely to be rocky or gaseous?

USING THE WEB

46. Using the exoplanet catalogs:
    a. Go to the "Catalog" Web page (http://exoplanet.eu/catalog) of the Extrasolar Planets Encyclopedia. Look for a star that has multiple planets. Make a graph showing the distances of the planets from that star, and note the masses and sizes of the planets. Put the Solar System planets on the same axis. How does that extrasolar planet system compare with the Solar System?
    b. Go to the "Exoplanets Data Explorer" website (http://exoplanets.org) and click on "Table." That website lists planets that have detailed orbital data published in scientific journals, and it may have a smaller total count than the website in part (a). Pick a planet discovered recently, as specified in the "First Reference" column. What is the planet's minimum mass? What is its semimajor axis and the period of its orbit? What is the eccentricity of its orbit? Click on the star name in the first column to get more information. Does that planet have a radial velocity curve? Was it observed in transit, and if so, what is the planet's radius and density? Is it more like Jupiter or more like Earth?

47. Space missions:
    a. Go to the website for the Kepler Mission (https://nasa.gov/mission_pages/kepler/main/). How many confirmed planets has Kepler discovered? How many planet candidates are there? When did the mission end? Search for the latest version of the "Kepler Orrery," an animation that shows multiplanet systems discovered by Kepler. Do most of those systems look like our own?
    b. Go to the website for the TESS mission (https://tess.gsfc.nasa.gov/). Are there any results?

48. Citizen science projects I:
    a. Go to the "Exoplanet Explorers" website (https://zooniverse.org/projects/ianc2/exoplanet-explorers) Exoplanet Explorers is part of the Zooniverse, a citizen science project that invites individuals to participate in a major science project by using their own computers. To participate in that or any of the other Zooniverse projects mentioned in later chapters, you will need to sign up for an account. Read through the sections under "About," including the FAQ. What are some of the advantages to crowd-sourcing Kepler data analysis? Back on the Exoplanet Explorers home page, click on "Classify" and then "Show the project tutorial" under the example on the right. Also click on the "Field Guide" to see more examples. Then look at the data presented and classify.
    b. Go to the "Planet Hunters" website (http://planethunters.org), an older Zooniverse project. Read through the sections under "Science," including the FAQ. Back on the Planet Hunters home page, click on "Start Classifying" and go through the tutorial. Classify a few images.

49. Citizen science projects II: Go to the "Disk Detective" website at http://diskdetective.org/, another Zooniverse project for which you will need to open an account as in problem 48. In that project, you will look at observations of young stars to see whether evidence exists for a planetary disk. Under "Menu," read "Science," "About," "FAQ," and then "Classify." Work through an example and then classify a few images.

50. Go to the "Super Planet Crash" Web page (http://stefanom.org/spc/) or the App. Read "Help" to see the rules. First build a system like ours with four Earth-sized planets in the inner 2 AU. Is that system stable? What happens if you add in super-Earths or "ice giants"? Build up a few completely different planetary systems and see what happens. What types of situations cause instability in the inner 2 AU of those systems?

digital.wwnorton.com/astro6

Visit the Digital Resources Page and on the Student Site open the "Radial Velocity" Interactive Simulation in Chapter 7. That applet has several panels that allow you to experiment with the variables important for measuring radial velocities. First, in the window labeled "Visualization Controls," check the box to show multiple views. Compare the views shown in panels 1–3 with the colored arrows in the last panel to see where an observer would stand to see the view shown. Start the animation (in the "Animation Controls" panel), and allow it to run while you watch the planet orbit its star from each of the views shown. Stop the animation, and in the "Presets" panel, select "Option A" and then click "set."

1 Is Earth's view of that system most nearly like the "side view" or most nearly like the "orbit view"?

_____

2 Is the orbit of that planet circular or elongated?

_____

3 Study the radial velocity graph in the upper right panel. The blue curve shows the radial velocity of the star over a full period. What is the maximum radial velocity of the star?

_____

4 The horizontal axis of the graph shows the "phase," or fraction of the period. A phase of 0.5 is halfway through a period. The vertical red line indicates the phase shown in views in the upper left panel. Start the animation to see how the red line sweeps across the graph as the planet orbits the star. The period of the planet is 365 days. How many days pass between the minimum radial velocity and the maximum radial velocity?

_____

5 When the planet moves away from Earth, the star moves toward Earth. The sign of the radial velocity tells the direction of the motion (toward or away). Is the radial velocity of the star positive or negative at that time in the orbit? If you could graph the radial velocity of the planet at that point in the orbit, would it be positive or negative?

_____

In the "Presets" window, select "Option B" and then click "set."

6 What has changed about the orbit of the planet as shown in the views in the upper left panel?

_____

7 When is the planet moving fastest: when it is close to the star or when it is far from the star?

_____

8 When is the star moving fastest: when the planet is close to it or when the planet is far away?

_____

9 Explain how an astronomer would determine, from a radial velocity graph of the star's motion, whether the orbit of the planet was circular or elongated.

_____

10 Study the "Earth view" panel at the top of the window. Would that planet be a good candidate for a transit observation? Why or why not?

_____

In the "System Orientation" panel, change the inclination to 0.0.

11 Now is Earth's view of the system most nearly like the "side view" or most nearly like the "orbit view"?

_____

12 How does the radial velocity of the star change as the planet orbits?

_____

13 Click the box that says "show simulated measurements," and change the "noise" to 1.0 m/s. The gray dots are simulated data, and the blue line is the theoretical curve. Use the slider bar to change the inclination. What happens to the radial velocity as the inclination increases? (Hint: Pay attention to the vertical axis as you move the slider, not just the blue line.)

_____

14 What is the smallest inclination for which you would find the data convincing? That is, what is the smallest inclination for which the theoretical curve is in good agreement with the data?

_____

8

The Terrestrial Planets and Earth's Moon

The objects that formed in the inner part of the protoplanetary disk around the Sun are relatively small, rocky worlds, one of which is Earth. A comparison of those worlds reveals the forces that shape a planet. The past six decades have been an exciting time for the exploration of and discovery about Earth and the other planets in the Solar System. Robotic probes have visited every planet, and astronauts walked on the Moon. In addition to discoveries from new space missions and telescopes, improved analytical techniques applied to the rocks and soil brought back from the Moon 50 years ago have led to surprising new results. The information from those missions has revolutionized the understanding of the Solar System, offering insights into the current state of each of the neighboring planets and clues about their histories.

LEARNING GOALS

By the end of this chapter, you should be able to look at an image of a planet and identify which geological features occurred early in the history of that world and which occurred late. You also should be able to:

LG 1 Describe how impacts have affected the evolution of the terrestrial planets.

LG 2 Explain how radiometric dating is used to measure the ages of rocks and terrestrial planetary surfaces.

LG 3 Explain how scientists use both theory and observation to determine the structure of terrestrial planetary interiors.

LG 4 Describe tectonism and volcanism and the forms they take on different terrestrial planets.

LG 5 Summarize what is known about the presence of water on the terrestrial planets.

Mars Reconnaissance Orbiter image of Newton Crater on Mars. The dark streaks may indicate flowing water. ▶▶▶

Does water exist
on other planets?

Certainty Is Sometimes Out of Reach

There are several hypotheses for how the Moon formed. One of these fits the data better than the others, but none has been absolutely ruled out.

Did the Moon split off from Earth?

Was the Moon captured?

Did the Moon and Earth form together?

Did the Moon form from an impact of another object with Earth?

In some cases, hypotheses cannot be definitively falsified (at least not yet). The working hypothesis, then, is the one that best fits the data, but other ideas are kept in mind.

8.2 Radioactive Dating Tells Us the Age of the Moon and the Solar System

The number of visible craters on a planet is determined by the rate at which those craters are destroyed. Geological activity on Earth, Mars, and Venus erased most evidence of early impacts. By contrast, the Moon's surface still preserves the scars of craters dating from about 4 billion years ago. The lunar surface has remained essentially unchanged for more than a billion years because the Moon has no atmosphere or surface water and a cold, geologically dead interior. Mercury also has well-preserved craters, although recent evidence from the *Messenger* mission shows tilted crater floors that are higher on one side than the other—evidence that internal forces lifted the floors unevenly after the craters formed.

Planetary scientists use that cratering record to estimate the ages and geological histories of planetary surfaces: extensive cratering indicates an older planetary surface that remains relatively unchanged because of minimal geological activity. The amount of cratering can serve as a clock to measure the relative ages of surfaces, but to determine the exact age of a surface from the number of craters, we need to know how fast the clock runs. In other words, we need to "calibrate the cratering clock."

To assign real dates to the layers in rock, scientists use a technique called radioactive (or radiometric) dating. A geologist can find the age of a rock by measuring the relative amounts of a radioactive element, known as a **radioisotope**, and the decay products it turns into. Recall that the nuclei of atoms contain neutrons and protons. An isotope is an atom that has the same number of protons as other atoms of the same chemical element but has a different number of neutrons. The radioactive element is known as the **parent element**, and the decay products are called **daughter products**. Chemical analysis of a rock containing radioactive elements immediately after its formation would reveal the presence of the radioactive parents, but the daughter products of the radioactive decay would be absent because they would not have formed yet. As radioactive atoms decay, however, the amount of parent elements decreases and the amount of daughter products builds up. Chemical analysis reveals both the remaining radioactive parent atoms and the daughter products trapped within the structure of the mineral.

The time interval over which a radioactive isotope decays to half its original amount is called that isotope's **half-life**. With every half-life that passes, the remaining amount of the radioisotope decreases by a factor of 2. For example, after 3 half-lives, the remaining amount of a parent radioisotope will be $1/2 \times 1/2 \times 1/2 = 1/8$ of its original amount (**Figure 8.7**). At formation, 100 percent of the material is radioactive parent isotope (in red), with no daughter isotope (in blue) yet. After 1 half-life has passed, half of the parent isotope has decayed, and parent and daughter isotopes are present in equal numbers. After another half-life has passed, the sample is now only 1/4 parent and 3/4 daughter isotopes, and so on. By comparing the percentages of parent and daughter isotopes in a mineral, scientists can determine the number of half-lives that have passed and thus the age of the mineral. Some numerical examples are discussed in **Working It Out 8.1**.

The age of the Solar System is estimated from radioactive dating of meteorites found on Earth that are 4.5 billion to 4.6 billion years old. Earth may be as young as 4.4 billion years. The age of the Moon is determined from radioactive dating of lunar rocks. Between 1969 and 1976, *Apollo* astronauts and Soviet unmanned probes brought back samples from nine locations on the lunar surface.

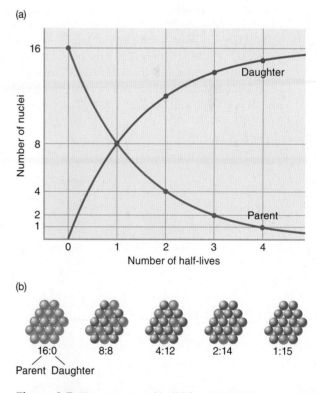

(a)

(b)

16:0   8:8   4:12   2:14   1:15

Parent Daughter

Figure 8.7 The concept of half-life. A parent population of 16 radioactive nuclei decays over several half-lives. That information can be presented (a) graphically or (b) as collections of particles.

8.1 Working It Out Computing the Ages of Rocks

With every half-life that passes, the remaining amount of radioactive isotopes will decrease by a factor of 2. If we express the number of half-lives more generally as n, we can translate that relationship into math:

$$\frac{P_F}{P_O} = \left(\frac{1}{2}\right)^n$$

where P_O and P_F are the original and final amounts, respectively, of a parent radioisotope, and n is the number of half-lives that have gone by, which equals the time interval of decay (its age) divided by the half-life of the isotope.

For example, the most abundant isotope of the element uranium (uranium-238, or ^{238}U—the parent) decays through a series of intermediate daughters to an isotope of the element lead (lead-206, or ^{206}Pb—its final, stable daughter). The half-life of ^{238}U is 4.5 billion years. Therefore, after 4.5 billion years, a sample that originally contained the uranium isotope (the parent) but no lead (its final daughter) would be found instead to contain equal numbers of uranium and lead atoms. If we were to find a mineral with such composition, we would know that half the uranium atoms had turned to lead and that the mineral formed 4.5 billion years ago.

Another isotope of uranium (^{235}U) decays to a different lead isotope (^{207}Pb) with a half-life of 700 million years. Suppose that a lunar mineral brought back by astronauts has 15 times as much ^{207}Pb (the daughter product) as ^{235}U (the parent radioisotope). Therefore, 15/16 of the parent radioisotope (^{235}U) has decayed to the daughter product (^{207}Pb), leaving only 1/16 of the parent remaining in the mineral sample. By noting that 1/16 is $(1/2)^4$, we see that 4 half-lives have elapsed since the mineral was formed, in which case the lunar sample is 4 × 700 million years = 2.8 billion years old.

Because the measured quantity of the isotope is not always a neat power of 2, looking at how we would solve the equation mathematically is worthwhile. We do that by taking the logarithm on both sides:

$$\log_{10}\frac{P_F}{P_O} = \log_{10}\left(\frac{1}{2}\right)^n$$

$$\log_{10}\frac{P_F}{P_O} = n\log_{10}\left(\frac{1}{2}\right)$$

$$\log_{10}\frac{P_F}{P_O} = -0.3n$$

Putting that back into words, we can write the relationship as

$$\log_{10}\left(\frac{\begin{array}{c}\text{Actual measured}\\\text{quantity of isotope}\end{array}}{\begin{array}{c}\text{Original quantity}\\\text{of isotope}\end{array}}\right) = -0.3 \times \left(\frac{\begin{array}{c}\text{Time it has been}\\\text{decaying (age)}\end{array}}{\text{Half-life}}\right)$$

Solving for age,

$$\text{Age} = -3.3 \times \text{Half-life} \times \log_{10}\left(\frac{\begin{array}{c}\text{Actual measured}\\\text{quantity of isotope}\end{array}}{\begin{array}{c}\text{Original quantity}\\\text{of isotope}\end{array}}\right)$$

(Most calculators have a button called "log" or "$\log_{10}$" for calculating such numbers.)

For the ^{235}U that decays to ^{207}Pb with a half-life of 700 million years, the lunar mineral is measured to have 15 times as much lead as uranium, so the mineral now contains only 1/16 of the original quantity of uranium:

$$\text{Age of mineral} = -3.3 \times (700 \times 10^6 \text{ yr}) \times \log_{10}\left(\frac{1}{16}\right) = 2.8 \times 10^9 \text{ yr}$$

The mineral is 2.8 billion years old.

By measuring relative amounts of various radioactive elements and the elements into which they decay, scientists assigned ages to those lunar regions. The oldest, most heavily cratered regions on the Moon date back to about 4.4 billion years ago, whereas most of the smoother parts of the lunar surface are typically 3.1 billion to 3.9 billion years old. That finding suggests that the Moon formed after Earth, and the heavy cratering suggests heavy bombardment at that time. As the graph in **Figure 8.8** shows, almost all the major cratering in the Solar System took place within its first billion years.

CHECK YOUR UNDERSTANDING 8.2

If radioactive element A decays into radioactive element B with a half-life of 20 seconds, then after 40 seconds: (a) none of element A will remain; (b) none of element B will remain; (c) half of element A will remain; (d) one-quarter of element A will remain.

8.3 The Surface of a Terrestrial Planet Is Affected by Processes in the Interior

Whereas impact cratering is driven by forces outside a planet, two other important processes, tectonism and volcanism, are determined by conditions in the interior of the planet. To understand those processes, we must understand the structure and composition of the interiors of planets. But how do we know what the interiors of planets are like? On Earth, the deepest holes drilled are about 12 km deep; that is tiny in comparison with Earth's radius of 6,378 km. It is impossible to drill down into Earth's core to observe Earth's interior structure directly. Scientists have determined a lot about the interior of Earth but less about the interiors of the other terrestrial planets.

Probing the Interior of Earth

The composition of Earth's interior can be determined in two ways. In one approach, Kepler's or Newton's laws are used to find the mass of Earth—for example, by applying Kepler's third law to a satellite orbiting Earth. Dividing the mass by the volume of Earth gives an average density of 5,500 kilograms per cubic meter (kg/m^3), or 5.5 times the density of water. But rocky surface material averages only 2,900 kg/m^3. Because the density of the whole planet is greater than the density of the surface, the interior must contain material denser than surface rocks. Another approach to determine the composition of Earth's interior comes from studies of meteorites. Because meteorites are left over from a time when the Solar System was young and Earth was forming from similar materials, the overall composition of Earth should resemble the composition of meteorite material. That material includes minerals with large amounts of iron, which has a density of nearly 8,000 kg/m^3. From those considerations, planetary scientists can determine the composition of Earth's interior.

The most important source of information about the structure of Earth's interior comes from monitoring the vibrations from earthquakes. When an earthquake occurs, vibrations spread out through and across the planet as **seismic waves**.

Surface waves travel across the surface of a planet, much like waves on the ocean. If conditions are right, surface waves from earthquakes can be seen rolling across the countryside like ripples on water. Those waves are responsible for much of the heaving of Earth's surface during an earthquake, causing damage such as the buckling of roadways.

The other types of seismic waves travel *through* Earth, probing the interior of the planet, faster than surface waves travel. **Primary waves** (P waves) are a type of **longitudinal wave** resulting from alternating compression and decompression of a material. Imagine a stretched-out spring (**Figure 8.9a**). A quick push along its length will make a longitudinal wave. P waves distort the material they travel through, much as compression waves do when they move along the length of a spring. **Secondary waves** (S waves) are a type of **transverse wave** that result from the sideways motion of material (**Figure 8.9b**).

The progress of seismic waves through Earth's interior depends on the characteristics of the material

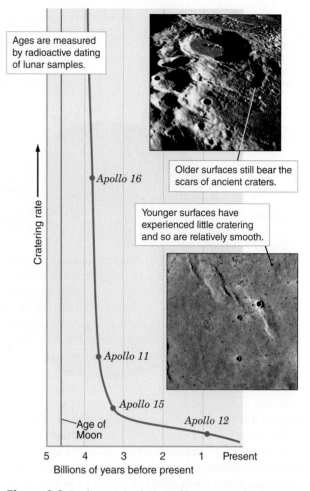

Ages are measured by radioactive dating of lunar samples.

Older surfaces still bear the scars of ancient craters.

Younger surfaces have experienced little cratering and so are relatively smooth.

Apollo 16

Apollo 11

Apollo 15

Apollo 12

Age of Moon

Cratering rate

5   4   3   2   1   Present
Billions of years before present

Figure 8.8 Radiometric dating of lunar samples returned from *Apollo* missions that landed at different sites was used to determine how the cratering rate has changed. Cratering records can then be used to establish the age of other parts of the lunar surface.

Figure 8.9 (a) A longitudinal wave involves oscillations along the direction of travel of the wave. (b) A transverse wave involves oscillations perpendicular to the direction in which the wave travels. Primary seismic waves are longitudinal; secondary seismic waves are transverse.

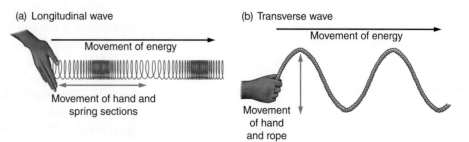

(a) Longitudinal wave

Movement of energy

Movement of hand and spring sections

(b) Transverse wave

Movement of energy

Movement of hand and rope

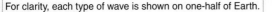

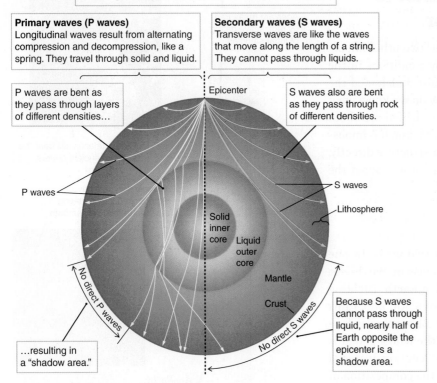

For clarity, each type of wave is shown on one-half of Earth.

Primary waves (P waves)
Longitudinal waves result from alternating compression and decompression, like a spring. They travel through solid and liquid.

Secondary waves (S waves)
Transverse waves are like the waves that move along the length of a string. They cannot pass through liquids.

P waves are bent as they pass through layers of different densities…

S waves also are bent as they pass through rock of different densities.

Epicenter

P waves

S waves

Lithosphere

Solid inner core

Liquid outer core

Mantle

Crust

No direct P waves

…resulting in a "shadow area."

No direct S waves

Because S waves cannot pass through liquid, nearly half of Earth opposite the epicenter is a shadow area.

Figure 8.10 Primary and secondary seismic waves move through the interior of Earth in distinctive ways. Measurements of when and where different types of seismic waves arrive after an earthquake enable scientists to test predictions from detailed models of Earth's interior. Note the "shadow areas" caused by the refraction of primary waves (yellow) at the outer boundary of the liquid outer core and the inability of secondary waves (blue) to pass through the liquid outer core.

through which they are moving. Primary waves (white; **Figure 8.10**) can travel through either solids or liquids, but secondary waves (blue) cannot travel through liquids. Seismic waves travel at different speeds, depending on the density and composition of the rocks they encounter. As a result, seismic waves moving through rocks of various densities or composition are bent in much the same way that waves of light are bent when they enter or leave glass. The speed of seismic waves provides additional information about Earth's interior. The refraction of primary waves at the outer edge of Earth's liquid outer core and the inability of secondary waves to penetrate the liquid outer core create "shadows" of the liquid core on the side of Earth opposite an earthquake's epicenter (Figure 8.10). Much of scientists' knowledge of Earth's liquid outer core comes from studies of those waves.

Scientists use instruments called seismometers to measure the distinctive patterns of seismic waves. For more than 100 years, thousands of seismometers scattered around the globe have measured the vibrations from countless earthquakes and other seismic events, such as volcanic eruptions and nuclear explosions. A single seismometer can record ground motion at only one place on Earth, but when combined with the recordings of many other seismometers placed all over Earth, scientists can use the data to get a comprehensive picture of the planet's interior.

Building a Model of Earth's Interior

To model the structure of Earth's interior, geologists use the laws of physics and the properties of materials and how they behave at different temperatures and pressures. The pressure at any point in Earth's interior must be just high enough that the outward forces balance the inward force of the weight of all the material above that point. If the outward pressure at some point within a planet were *less* than the weight per unit area of the overlying material, that material would fall inward, crushing what was underneath it. If the pressure at some point within a planet were *greater* than the weight per unit area of the overlying material, the material would expand and push outward, lifting the overlying material. The situation is stable only when the weight of matter above is just balanced by the pressure within the whole interior of the planet. The balance between pressure and weight is known as **hydrostatic equilibrium**, which is important to the structure of planetary interiors, planetary atmospheres, and the structure and evolution of stars.

From considerations of hydrostatic equilibrium and from seismic wave measurements, scientists construct a layered model of Earth's interior. They then test their model by comparing its predictions of how seismic waves would propagate through Earth with actual observations of seismic waves from real earthquakes. The extent to which the predictions agree with observations identifies both strengths and weaknesses of the model. Geologists adjust the model—always remaining consistent with the known physical properties of materials—until a good match is found between prediction and observation.

That is the method geologists used to arrive at the current picture of the interior of Earth (Figure 8.10). The innermost region of Earth's interior consists of a **core**. Earth's solid inner core is at a temperature of about 6000 K and is composed primarily of iron, nickel, and other dense metals. The liquid outer core is cooler, about 4000 K, and is composed of liquid metals. Outside the outer core is Earth's **mantle**, a rocky shell made of solid, medium-density materials such as silicates. That hot, rocky shell is under pressure, and over long timescales—hundreds of years—it flows like a viscous fluid. Covering the mantle is the **crust**, a thin, hard layer of lower-density materials that is chemically distinct from the interior.

The cross sections in **Figure 8.11** show the interior structures of each terrestrial planet and Earth's Moon. Earth's interior is not uniform. The materials have been separated according to density, a process known as **differentiation**. When rocks of different types are mixed, they tend to stay mixed. Once that rock melts, however, the denser materials sink to the center and the less dense materials float toward the surface. Today, little of Earth's interior is molten, but the differentiated structure shows that Earth was once much hotter, and its interior was liquid throughout. The cores of all the terrestrial planets and the core of the Moon were once molten. Using new, improved methods to reanalyze 8 years of data from seismometers left on the Moon by the *Apollo* astronauts, planetary scientists found that the Moon has a solid inner core, possibly a liquid outer core, and a partially melted layer between the core and the mantle.

The Evolution of Planetary Interiors

The balance between energy received and energy produced and emitted governs the temperature within a planet. The interiors of planets evolve as their temperatures change. Factors that influence how the temperature changes include the size of the planet, the composition of the material, and heating from various sources. Here, we are concerned with thermal energy—the kinetic energy of particles within a substance that determines the temperature. In general, the interior of a planet cools as heat is emitted from the surface. Because heat takes time to travel through rock, the temperature increases the deeper we go within a planet. That trend is similar to the effect of taking a hot pie out of the oven. The pie radiates heat from the surface and eventually cools, but the filling takes much longer to cool than the crust.

Planets lose thermal energy from their surfaces primarily through radiation. Recall from Chapter 5 that the hotter an object is, the more energy it radiates. The relative proportions of the type of energy radiated (infrared, optical, ultraviolet, and so forth) depend on the temperature of the object. The rate at which a planet cools depends on its size. A larger planet has a larger volume of matter and more thermal energy trapped inside. Thermal energy has to escape through the planet's surface, so the planet's surface area determines the rate at which energy is lost. Smaller planets have more surface area in comparison with their small volumes, so they cool faster, whereas larger planets have a smaller surface-area-to-volume ratio and cool more slowly (**Working It Out 8.2**). Because geological activity is powered by heat, smaller objects become geologically inactive sooner. Major geological activity ended on Mercury and the Moon first, but it continued on the larger terrestrial planets—Venus, Earth, and Mars.

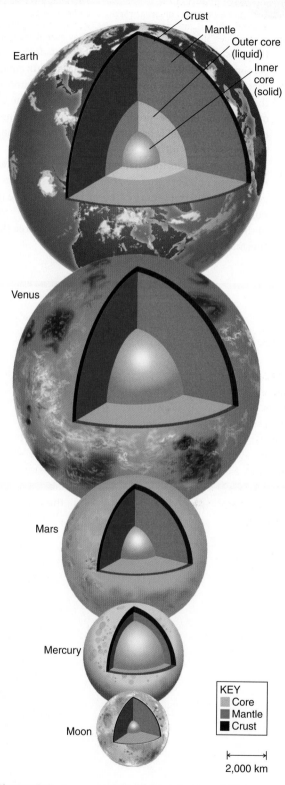

Figure 8.11 A comparison of the interiors of the terrestrial planets and Earth's Moon. Some fractions of the cores of Mercury, Venus, the Moon, and Mars are probably liquid.

8.2 Working It Out How Planets Cool

If we assumed that all the terrestrial planets formed with the same percentage of radioactive materials in their bulk composition and that those radioactive materials are their sole source of internal thermal energy, a planet's volume would determine the total amount of the thermal energy–producing material it contains. The volume of a spherical planet is proportional to the *cube* of the planet's radius (volume = $4/3 \times \pi R^3$).

A planet loses its internal energy by radiating it away at its surface, so a planet's surface area determines the rate at which it can get rid of its thermal energy. The surface area of the planet is proportional to the *square* of the radius (surface area = $4\pi R^2$). The ratio of the two—the volume divided by the surface area through which thermal energy can escape—is given by

$$\frac{\text{Volume}}{\text{Surface area}} = \frac{\frac{4}{3} \times \pi R^3}{4\pi R^2} = \frac{R}{3}$$

so the time it takes to lose internal energy is proportional to the planet's radius.

A planet's ability to transfer internal energy from its hot core to its cooling surface also depends on some of its own internal properties. Nevertheless, all things being equal, planets with larger radii retain their internal energy longer than smaller planets do. For example, Mars has a radius about half that of Earth, so it has been losing its internal thermal energy to space about twice as fast as Earth has. That is one reason that Mars is less geologically active than Earth.

Some of the thermal energy in the interior of Earth is left over from when Earth formed. The tremendous energy of collisions and the energy from short-lived radioactive elements melted the planet, leading to the differentiated structure. As the surface of Earth radiated energy into space, it cooled rapidly. A solid crust formed above a molten interior. Because a solid crust does not conduct thermal energy well, it helped to retain the remaining heat. Over a long time, energy from the interior of the planet continued to leak through the crust and radiate into space. As a result, the interior of the planet cooled slowly, and the mantle and the inner core solidified.

If the thermal energy from Earth's formation were the only source of heating in Earth's interior, Earth would have long ago solidified completely. Most of the rest of the thermal energy in Earth's interior comes from long-lived radioactive elements trapped in the mantle. As those radioactive elements decay, they release energy, which heats the planet's interior. Today, the temperature of Earth's interior is determined by a dynamic equilibrium between the radioactive heating of the interior and the loss of energy to space. As the radioactive elements decay, the amount of thermal energy generated declines, and Earth's interior cools as it ages. A small amount of additional heating of Earth's interior is due to friction generated by tidal effects of the Moon and Sun.

Although temperature plays an important role in a planet's interior structure, whether a material is solid or liquid also depends on pressure. Higher pressure forces atoms and molecules closer together and makes the material more likely to become a solid. Toward the center of Earth, the effects of temperature and pressure oppose each other: the higher temperatures make it more likely that material will melt, but the higher pressure favors a solid form. In the outer core of Earth, the high temperature wins, allowing the material to exist in a molten state. At the center of Earth, even though the temperature is higher, the pressure is so great that Earth's inner core is solid.

CHECK YOUR UNDERSTANDING 8.3A

Differentiation refers to materials that are separated on the basis of their: (a) weight; (b) mass; (c) volume; (d) density.

Magnetic Fields

A magnetic field is created by moving charges and exerts a force on magnetically reactive objects, such as iron and other charged particles. In a navigation compass, for example, the compass needle lines up with Earth's magnetic field and points "north" and "south," as shown in **Figure 8.12a**. In the north, a compass needle points to a location in the Arctic Ocean off the coast of northern Canada, near but not at the geographic North Pole (about which Earth spins). In the south, a compass needle points to a location off the coast of Antarctica, 2,800 km from the geographic South Pole. Earth behaves as though it contained a giant bar magnet that were slightly tilted with respect to the planet's rotation axis and had its two endpoints near the two magnetic poles (**Figure 8.12b**). Earth also has a **magnetosphere**, the region surrounding a planet that is filled with relatively intense magnetic fields and charged particles.

Earth's magnetic field is not actually the result of a bar magnet buried within the planet. A magnetic field is the result of moving electric charges. Earth's magnetic field is created by the combination of Earth's rotation about its axis and a liquid, electrically conducting, circulating outer core. From that combination, Earth converts mechanical energy into magnetic energy. The magnetic field of a planet is an important probe into its internal structure.

Earth's magnetic field is constantly changing. At the moment, the north magnetic pole is traveling several tens of kilometers per year toward the northwest. If that rate and direction continue, the north magnetic pole could be in Siberia before the end of the century. The magnetic pole tends to wander, constantly changing direction as a result of changes in the core.

The geological record shows that much more dramatic changes in the magnetic field have occurred over the history of our planet. When a magnet made of material such as iron gets hot enough, it loses its magnetization. As the material cools, it again becomes magnetized by any magnetic field surrounding it. Thus, iron-bearing minerals record the direction of Earth's magnetic field at the time that they cooled. A memory of that magnetic field thus becomes "frozen" into the material. For example, lava extruded from a volcano carries a record of Earth's magnetic field at the moment the lava cooled. By using radiometric techniques to date those materials, geologists obtain a record of how Earth's magnetic field has changed. Although Earth's magnetic field has probably existed for at least 3.5 billion years, the north and south magnetic poles switch from time to time—on average, about every half-million years.

The general idea of how Earth's magnetic field (and those of other planets) originates is called the *dynamo theory*. In general, magnetic fields result from electric currents, which are moving electric charges. Earth's magnetic field is thought to be a side effect of three factors: Earth's rotation about its axis; an electrically conducting, liquid outer core; and fluid motions within the outer core. That model has been tested with computer simulations, which also produce the pole reversals. The theory suggests that any rotating planet with an internal heat source will have a magnetic field.

During the *Apollo* program, astronauts measured the Moon's local magnetic fields, and small satellites have searched for global magnetism. The Moon has a

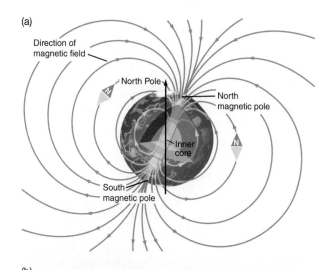

(a)

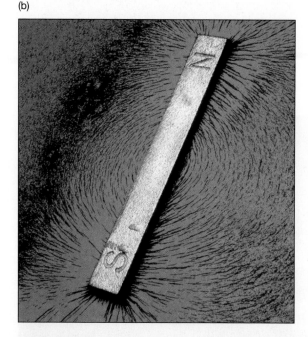

(b)

Figure 8.12 (a) Earth's magnetic field can be visualized as though it were a giant bar magnet tilted relative to Earth's axis of rotation. Compass needles line up along magnetic-field lines and point toward Earth's north magnetic pole. (b) Iron filings sprinkled around a bar magnet help us visualize such a magnetic field.

very weak field, possibly none at all, because the Moon is very small and therefore has a solid (not liquid and rotating) inner core. The Moon also has a very small core. However, remnant magnetism is preserved in lunar rocks from an earlier time before the lunar surface and rocks solidified. Recent analysis of the oldest lunar rocks brought back on *Apollo 17* suggests that 4.2 billion years ago, the Moon had a liquid core with a generated magnetic field that lasted a billion years or more. The Moon no longer generates a global magnetic field, so what is detected is the "fossil" remains of its early field. Data from India's *Chandrayaan-1* spacecraft suggest that the Moon has a very weak and localized mini-magnetosphere on its far side.

Other than Earth, Mercury is the only terrestrial planet with a significant global magnetic field today—although its field is only about 1 percent as strong as Earth's. Slow rotation and a large iron core, parts of which are molten and circulating, cause Mercury's magnetic field. Because the field is so weak, Mercury's magnetosphere is small, meeting the solar wind about 1,700 km above its surface. At that boundary, twisted bundles of magnetic fields transfer magnetic energy from the planet to space.

Planetary scientists expected that Venus would have a magnetic field because the planet's mass and distance from the Sun imply an iron-rich core and partly molten interior like Earth's. Venus's lack of a magnetic field might be attributed to its extremely slow rotation (see Table 8.1) or to its not having a giant impact early in its history as Earth did. Or perhaps Venus's magnetic field is temporarily dormant—a condition that Earth is believed to have experienced at times of magnetic field reversals.

Mars has a weak magnetic field, presumably frozen in place early in its history. The magnetic signature occurs only in the ancient crustal rocks, showing that early in the history of Mars, some sort of an internally generated magnetic field must have existed. Geologically younger rocks lack that residual magnetism, so the planet's original magnetic field has long since disappeared. The lack of a strong magnetic field today on Mars might be the result of its small core. Or Mars might have lost its ability to generate a magnetic field after a series of giant impacts early in the planet's history, which could have heated the mantle of Mars enough to reduce the flow of heat out of the core to the mantle. The oldest large impact basins on Mars appear to be magnetized; newer ones are not.

CHECK YOUR UNDERSTANDING 8.3B

According to the dynamo theory, a planet will have a strong magnetic field if it has: (a) fast rotation and a solid core; (b) slow rotation and a liquid core; (c) fast rotation and a liquid core; (d) slow rotation and a solid core; (e) fast rotation and a gaseous core.

8.4 Planetary Surfaces Evolve through Tectonism

Now that we have looked at planetary interiors, we can connect the interior conditions to the processes that shape the surface. The crust and part of the upper mantle form the **lithosphere** of a planet. **Tectonism**, the deformation of a planet's lithosphere, warps, twists, and shifts the lithosphere to form visible surface features. If you have driven through mountainous or hilly terrain, you may have seen places like the one shown in **Figure 8.13**, where the roadway has

Figure 8.13 Tectonic processes fold and warp Earth's crust, as seen in these rocks along a roadside north of Denver, Colorado.

been cut through rock. The exposed layers tell the story of Earth through the vast expanse of geological time. In this section, we look at tectonic processes that create those layers and play an important part in shaping the surface of a planet.

The Theory of Plate Tectonics

Early in the 20th century, some scientists recognized that Earth's continents could be fit together like pieces of a giant jigsaw puzzle. In addition, the layers in the rock and the fossil records they hold on the east coast of South America match those on the west coast of Africa. On the basis of that evidence, Alfred Wegener (1880–1930) proposed a hypothesis that the continents were originally joined in one large landmass that broke apart as the continents began to "drift" away from one another over millions of years. That hypothesis was further developed into the theory known today as **plate tectonics**. Geologists now recognize that Earth's outer shell is composed of several relatively brittle segments, or **lithospheric plates**. About seven major plates and about a half dozen smaller plates are floating on the mantle. The motion of those plates is constantly changing the surface of Earth.

Originally, the idea of plate tectonics was met with great skepticism among geologists because they could not imagine a mechanism that could move such huge landmasses. In the late 1950s and early 1960s, however, studies of the ocean floor provided compelling evidence for plate tectonics. Those surveys showed surprising characteristics in bands of basalt—a type of rock formed from cooled lava—that were found on both sides of the ocean rifts. Ocean floor rifts such as the Mid-Atlantic Ridge are **spreading centers**. As **Figure 8.14** shows, hot material in those rifts rises toward Earth's surface, becoming new ocean floor. When that hot material cools, it becomes magnetized along the direction of Earth's magnetic field, thus recording the direction of Earth's magnetic field at that time. Greater distance from the rift indicates the ocean floor is older and formed earlier. Combined with radiometric dates for the rocks, that magnetic record proved that the spreading of the seafloor and the motions of the plates have continued over long geological time spans.

Precise surveying techniques and global positioning systems (GPS) can now determine locations on Earth to within a few centimeters. Those measurements confirm that Earth's lithosphere is moving. Some areas are being pulled apart by more than 15 centimeters (cm) each year. Over millions of years, such motions add up. Over 10 million years—a short time by geological standards—15 cm/yr becomes 1,500 km, and maps definitely need to be redrawn.

The theory of plate tectonics is perhaps the greatest advance in 20th century geology. Plate tectonics is responsible for a variety of geological features on our planet, including the continental drift that Wegener hypothesized.

The Role of Convection

Moving lithospheric plates requires immense forces. Those forces are the result of thermal energy escaping from the interior of Earth. The transport of thermal energy by the movement of packets of gas or liquid is known as **convection**.

AstroTour: Continental Drift

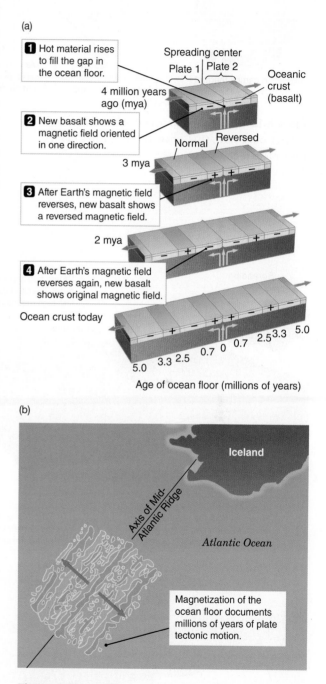

Figure 8.14 (a) New seafloor is formed at a spreading center. The cooling rock becomes magnetized and is then carried away by tectonic motions. (b) Maps such as this one of banded magnetic structure in the seafloor near Iceland support the theory of plate tectonics.

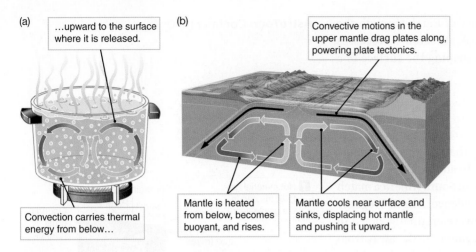

(a) ...upward to the surface where it is released.

Convection carries thermal energy from below...

(b) Convective motions in the upper mantle drag plates along, powering plate tectonics.

Mantle is heated from below, becomes buoyant, and rises.

Mantle cools near surface and sinks, displacing hot mantle and pushing it upward.

Figure 8.15 (a) Convection occurs when a fluid is heated from below. (b) Convection in Earth's mantle drives plate tectonics.

Figure 8.15a illustrates the process. If you have ever watched water in a heated pot on a stovetop, you have observed convection. Thermal energy from the stove warms water at the bottom of the pot. The warm water expands slightly, becoming less dense than the cooler water above it, and the cooler water with higher density sinks, displacing the warmer water upward. When the lower-density water reaches the surface, it gives up part of its energy to the air and cools; as the water cools, it then becomes denser and sinks back toward the bottom of the pot. Water rises in some locations and sinks in others, forming convection cells. As discussed in later chapters, convection also plays an important role in planetary atmospheres and in the structure of the Sun and stars.

Figure 8.15b shows how convection works in Earth's mantle, where radioactive decay supplies the heat to drive convection. We know the mantle is not molten—if it were, secondary seismic waves could not travel through it—but the mantle is viscous, having the consistency of hot molten glass. That consistency allows convection to take place very slowly. Convection cells in Earth's mantle drive the plates, carrying both continents and ocean crust along with them. Convection also creates crust along rift zones in the ocean basins, where mantle material rises up, cools, and slowly spreads out.

Figure 8.16 illustrates plate tectonics and some of its consequences. If material rises and spreads in one location, it must converge and sink in another. Locations where plates converge and convection currents turn downward are called **subduction zones**. In a subduction zone, one plate slides beneath the other, and convection drags the submerged lithospheric material back down into the mantle. The

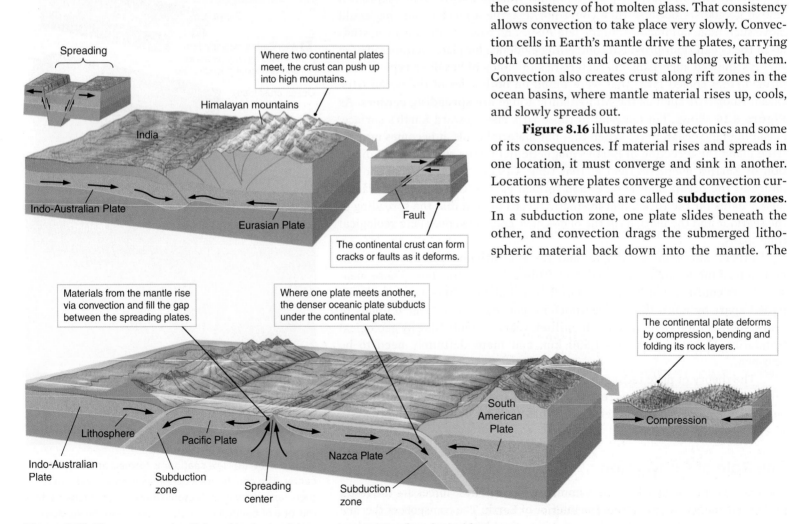

Spreading

Where two continental plates meet, the crust can push up into high mountains.

Himalayan mountains

India

Indo-Australian Plate

Eurasian Plate

Fault

The continental crust can form cracks or faults as it deforms.

Materials from the mantle rise via convection and fill the gap between the spreading plates.

Where one plate meets another, the denser oceanic plate subducts under the continental plate.

The continental plate deforms by compression, bending and folding its rock layers.

Lithosphere

Indo-Australian Plate

Pacific Plate

Subduction zone

Spreading center

Nazca Plate

Subduction zone

South American Plate

Compression

Figure 8.16 Divergence and collision of tectonic plates create a variety of geological features.

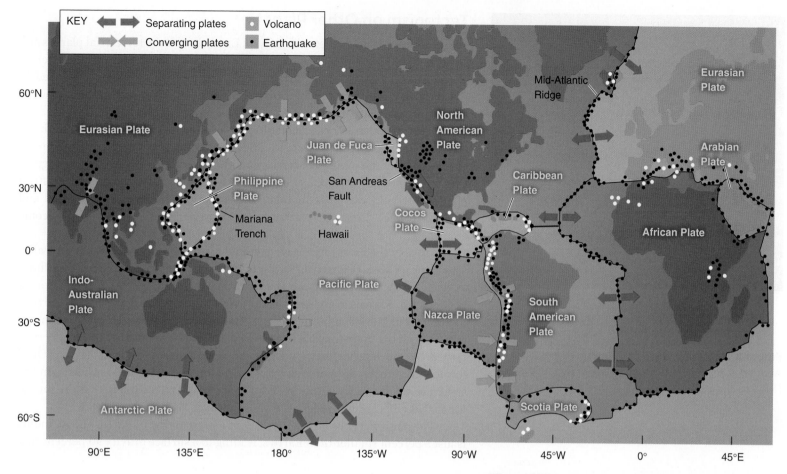

Figure 8.17 Major earthquakes and volcanic activity are often concentrated along the boundaries of Earth's principal tectonic plates.

Mariana Trench—the deepest part (11 km) of Earth's ocean floor—is a subduction zone. Much of the ocean floor lies between spreading centers and subduction zones, so the ocean floor is the youngest portion of Earth's crust. In fact, the *oldest* seafloor rocks are less than 200 million years old.

In some places, the plates are not sinking but instead are colliding and, consequently, shoved upward. The highest mountains on Earth, the Himalayas, grow a half-meter per century as the Indo-Australian subcontinental plate collides with the Eurasian Plate. In other places, plates meet at oblique angles and slide along past each other. One such place is the San Andreas Fault in California, where the Pacific Plate slides past the North American Plate. A **fault** is a fracture in a planet's crust along which material can slide.

Locations where plates meet tend to be very active geologically. One of the best ways to see the outline of Earth's plates is to look at a map of where earthquakes and volcanism occur (**Figure 8.17**). Where plates meet, enormous stresses build up. Earthquakes occur when a portion of the boundary between two plates suddenly slips, relieving the stress. Volcanoes are created when friction between plates melts rock, which is then pushed up through cracks to the surface. Earth also has many **hot spots**, where hot deep-mantle material rises, releasing thermal energy. As plates shift, some parts move more rapidly than others, causing the plates to stretch, buckle, or fracture. Those effects are seen on the surface as folded and faulted rocks. Mountain chains are common near converging plate boundaries, where plates buckle and break.

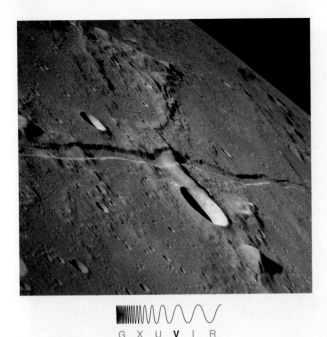

G X U V I R

Figure 8.18 This *Apollo 10* photograph shows Rima Ariadaeus, a 2-km-wide valley between two tectonic faults on the Moon.

Tectonism on Other Planets

We have observed plate tectonics only on Earth, but all the terrestrial planets and some moons show evidence of tectonic disruptions. Fractures have cut the crust of the Moon in many areas, leaving fault valleys such as the one pictured in **Figure 8.18**. Many of those features are the result of large impacts that cracked and distorted the lunar crust.

Mercury has fractures and faults similar to those on the Moon. In addition, Mercury has many cliffs that are hundreds of kilometers long. They appear to be the result of the shrinking of Mercury; recent observations by *Messenger* suggest that the planet has shrunk by about 10–14 km across. Like the other terrestrial planets, Mercury was once molten. As it shrank, Mercury's crust cracked and buckled in much the same way that a grape skin wrinkles as it shrinks to become a raisin.

Possibly the most impressive tectonic feature in the Solar System is Valles Marineris on Mars (**Figure 8.19**). Stretching nearly 4,000 km, and nearly 4 times as deep as the Grand Canyon, that canyon system is as long as the distance between San Francisco and New York. Valles Marineris includes massive cracks in the crust of Mars that formed as local forces, perhaps related to mantle convection, pushed the crust upward from below. The surface could not be equally supported by the interior everywhere, and unsupported segments fell in. Once formed, the cracks were eroded by wind, water, and landslides, resulting in the structure we see today. Other parts of Mars have faults similar to those on the Moon, but cliffs as high and long as those seen on Mercury are absent.

The mass of Venus is only 20 percent less than that of Earth, and its radius is just 5 percent smaller than Earth's, leading to a surface gravity 90 percent that

(a) (b)

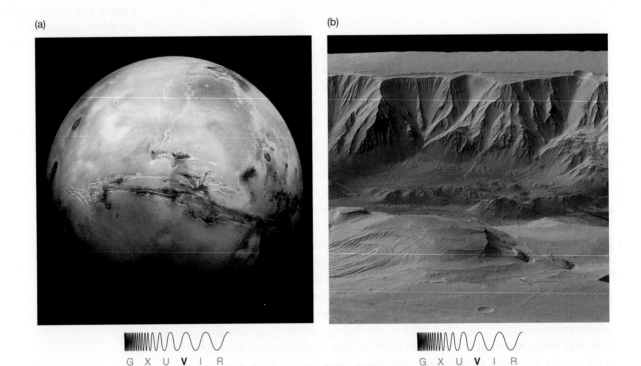

G X U V I R   G X U V I R

Figure 8.19 (a) A mosaic of *Viking Orbiter* images shows Valles Marineris, the major tectonic feature on Mars, stretching across the center of the image from left to right. This canyon system is more than 4,000 km long. The dark spots on the left are huge shield volcanoes. (b) This close-up perspective view of the canyon wall was photographed by the European Space Agency's *Mars Express* orbiting spacecraft.

of Earth. Because of the similarities between the planets, many scientists predicted that Venus might also show evidence of plate tectonics. The NASA *Magellan* mission orbited Venus from 1990 to 1994, and *Magellan* mapped nearly the entire surface of Venus, providing the first high-resolution radar views of the planet's surface (**Figure 8.20**).

Venus Express, launched by the European Space Agency (ESA), orbited Venus from 2006 to 2014. The probe mapped Venus in the infrared, which can penetrate the clouds to enable a view of the surface. The impact craters on Venus seem to be evenly distributed, suggesting that the surface is all about the same age—about a billion years old. Venus is mostly covered with smooth lava, but the planet has two highland regions: Ishtar Terra in the north and Aphrodite Terra in the south. The highland rocks are less smooth and older than those on the rest of Venus and may be similar to granite rocks on Earth. Because granite results from plate tectonics and water, the data hint at the possibility that those highlands on Venus are ancient continents, created by volcanic activity, on a planet that once had liquid water oceans.

Because of the similarities between Venus and Earth, the interior of Venus should be very much like the interior of Earth, and convection should be occurring in its mantle. On Earth, mantle convection and plate tectonism release the most thermal energy from the interior. On Venus, however, hot spots may be the principal way that thermal energy escapes from the planet's interior. Circular fractures called coronae on the surface of Venus, ranging from a few hundred kilometers to more than 2,500 km across, may be the result of upwelling plumes of hot mantle that have fractured Venus's lithosphere. Alternatively, energy may build up in the interior until large chunks of the lithosphere melt and overturn, releasing an enormous amount of energy. Then, the surface cools and solidifies. It is uncertain why Venus and Earth are so different with regard to plate tectonics.

Figure 8.20 The atmosphere of Venus blocks our view of the surface in visible light. This false-color view of Venus is a radar image made by the *Magellan* spacecraft. Bright yellow and white areas are mostly fractures and ridges in the crust. Some circular features seen in the image may be regions of mantle upwelling, or *hot spots*. Most of the surface is formed by lava flows, shown in orange.

CHECK YOUR UNDERSTANDING 8.4

On which of the following does plate tectonics occur now? (Select all that apply.)
(a) Mercury; (b) Venus; (c) Earth; (d) the Moon; (e) Mars

8.5 Volcanism Signifies a Geologically Active Planet

The molten rock that a volcano spews onto the surface of Earth is known as **magma**. It originates deep in the crust and in the upper mantle, where sources of thermal energy combine. Those sources include rising convection cells in the mantle, heating by friction from movement in the crust, and concentrations of radioactive elements. In this section, we look at the occurrence of volcanic activity on a planet or moon, called **volcanism**. Volcanism not only shapes planetary surfaces but also is a key indicator of a geologically active planet.

Terrestrial Volcanism Is Related to Tectonism

Volcanoes are usually located along plate boundaries and over hot spots. Maps of geological activity such as the one in Figure 8.17 leave little doubt that most terrestrial volcanism is linked to the same forces responsible for plate motions. A tremendous amount of friction is generated as plates slide under each other. That friction raises the temperature of rock toward its melting point.

Material at the base of a lithospheric plate is under a great deal of pressure because of the weight of the plate pushing down on it. That pressure increases the melting temperature of the material, forcing it to remain solid even at high temperature. As the material is forced up through the crust, its pressure drops, in which case the material's melting temperature drops, too. Material that was solid at the base of a plate becomes molten as it nears the surface. Places where convection carries hot mantle material toward the surface are frequent sites of eruptions. Iceland, one of the most volcanically active regions in the world, sits astride one such spreading center—the Mid-Atlantic Ridge (see Figure 8.17). In recent years, volcano eruptions in Iceland have disrupted travel plans for thousands of people as the airborne volcanic ash made it unsafe for airplanes to fly near the eruption.

Once lava reaches the surface of Earth, it can form many types of structures. Flows often form vast sheets, especially if the eruptions come from long fractures called fissures. If very fluid lava flows from a single *point source*, it spreads out over the surrounding terrain or ocean floor, forming a **shield volcano** (**Figure 8.21a**). A **composite volcano** (**Figure 8.21b**) forms when thick lava flows alternating with explosively generated rock deposits build a steep-sided structure.

Terrestrial volcanism also occurs where convective plumes rise toward the surface in the interiors of lithospheric plates, creating local hot spots. Volcanism over hot spots works much like volcanism elsewhere at a spreading center, except that the convective upwelling occurs at a single spot rather than along the edge of a plate. Those hot spots force mantle and lithospheric material toward the surface, where the material emerges as liquid lava.

Earth has many hot spots, including the regions around Yellowstone Park and the Hawaiian Islands (**Figure 8.21c**). The Hawaiian Islands are a chain of shield volcanoes that formed as the lithospheric plate moved across a hot spot. Volcanoes erupt over a hot spot, building an island. The island ceases to grow as the plate motion carries the island away from the hot spot. Meanwhile, a new island grows over the hot spot. Today, the Hawaiian hot spot is located off the southeast coast of the Big Island of Hawaii, where it continues to power the active volcanoes. On top of the hot spot, the newest Hawaiian island, Loihi, is forming. Loihi is already a massive shield volcano, rising more than 3 km above the ocean floor. Loihi will eventually break the surface of the ocean and merge with the Big Island of Hawaii—but not for another 100,000 years.

AstroTour: Hot Spot Creating a Chain of Islands

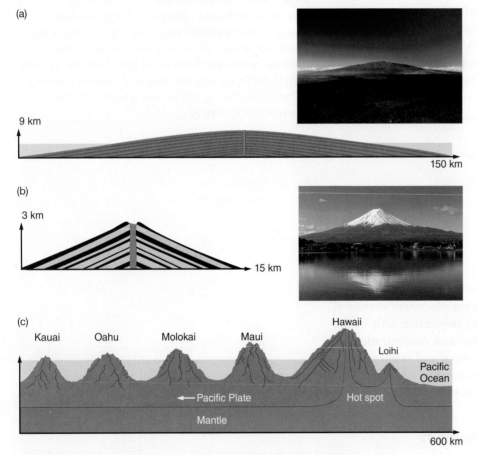

(a)

(b)

(c)

Figure 8.21 Magma reaching Earth's surface commonly forms (a) shield volcanoes, such as Mauna Loa, which have gently sloped sides built up by fluid lava flows; and (b) composite volcanoes, such as Mt. Fuji, which have steeply symmetric sides built up by viscous lava flows. (c) Hot spots are convective plumes of lava that can form a successive series of volcanoes as the plate above them slides by.

Volcanism in the Solar System

THE MOON Although Earth is the only planet on which plate tectonics is an important process, evidence of volcanism appears throughout the Solar System, including several moons of the outer planets. Some of the first observers to use telescopes to view the Moon noted dark areas that looked like bodies of water—thus they were named **maria** (singular: *mare*), Latin for "seas." Early photographs showed flowlike features in the dark regions of the Moon. We now know that the maria are actually vast, hardened lava flows, similar to volcanic rocks known as basalts on Earth. Because the maria contain relatively few craters, those volcanic flows must have occurred after the period of heavy bombardment ceased.

Many of the rock samples that the *Apollo* astronauts brought back from the lunar maria contained gas bubbles typical of volcanic materials (**Figure 8.22**). The lava that flowed across the lunar surface must have been relatively fluid. That fluidity, due partly to the lava's chemical composition, explains why lunar basalts form vast sheets that fill low-lying areas such as impact basins (**Figure 8.23**). The fluidity also partly explains the Moon's lack of classic volcanoes: the lava was too fluid to pile up—the way motor oil spreads out when poured from a container.

The lunar rock samples also showed that most of the lunar lava flows are older than 3 billion years. Samples from the heavily cratered terrain of the Moon also originated from magma, indicating that the young Moon went through a molten stage. Those rocks cooled from a "magma ocean" and are more than 4 billion years old, thus preserving the early history of the Solar System. Most of the sources of heating and volcanic activity on the Moon must have shut down some 3 billion years ago—unlike on Earth, where volcanism continues. That conclusion is consistent with the idea that smaller objects and planets cool more efficiently (see Working It Out 8.2) and are thus less active than larger planets.

Only in a few limited areas of the Moon are younger lavas thought to exist, but most of those have not been sampled directly. The *Lunar Reconnaissance Orbiter* observed volcanic cones that were probably built up from volcanic rocks erupting from the surface. Those volcanic rocks are far different from the mare basalt rocks and contain silica and thorium. Those domes could have been formed as recently as 800 million years ago, which would make them the result of the most recent volcanic activity found on the Moon.

MERCURY Mercury also shows evidence of past volcanism. The *Mariner 10* and *Messenger* missions revealed smooth plains on Mercury that looked similar to the lunar maria. Those sparsely cratered plains are the youngest areas on Mercury and, like those on the Moon, are almost certainly volcanic in origin, created when fluid lavas flowed into and filled huge impact basins. Many of the volcanic plains on Mercury also are associated with impact scars. The volcanic activity that created the plains probably ceased 3.8 billion years ago, possibly from the shrinking of the planet as it cooled. The ending of the late heavy bombardment might also have been a factor. High-resolution imaging by *Messenger* has also identified several volcanoes. Vents that could be from explosive volcanism have been found around the large, old impact basin Caloris (**Figure 8.24**) and may be as young as 1 billion to 2 billion years.

MARS Mars also has been volcanically active. More than half the surface of Mars is covered with volcanic rocks. Lavas covered huge regions of Mars, flooding the older, cratered terrain. Most of the vents or long cracks that created those flows are buried under the lava that poured forth from them. Among the most impressive features on Mars are its enormous shield volcanoes—the largest mountains in the Solar System. Olympus Mons, standing 27 km high at its peak

Figure 8.22 This rock sample from the Moon, collected by the *Apollo 15* astronauts from a lunar lava flow, shows gas bubbles typical of gas-rich volcanic materials. The rock is about 6 × 12 cm.

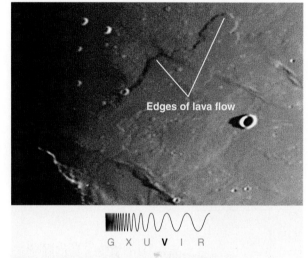

G X U V I R

Figure 8.23 The lava flowing across the surface of Mare Imbrium on the Moon must have been extremely fluid to spread out for hundreds of kilometers in sheets only tens of meters thick.

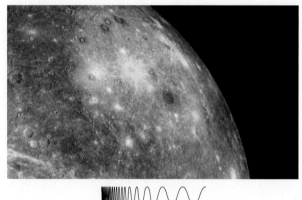

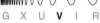

G X U V I R

Figure 8.24 The Caloris Basin on Mercury (yellow) is one of the largest impact basins in the Solar System, with a span of about 1,500 km. The orange regions may be volcanic vents. The false color is enhanced to show more detail.

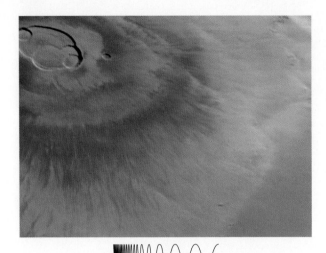

G X U V I R

Figure 8.25 The largest known volcano in the Solar System, Olympus Mons is a 27-km-high shield-type volcano on Mars, similar to but much larger than Hawaii's Mauna Loa. This partial view of Olympus Mons was taken by the *Mars Global Surveyor*.

(a)

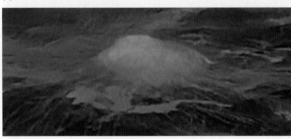

G X U V I R

(b)

G X U V I R

Figure 8.26 The volcanic peak Idunn Mons on Venus is 200 km in diameter and 2.5 km high. (a) Radar data on top of the topographic data from NASA's *Magellan* mission. (b) Visible and infrared data from ESA's *Venus Express* orbiter. The red-orange area at the summit is the warmest, suggesting the presence of some volcanic outflow activity.

and 550 km wide at its base (**Figure 8.25**), would tower over Earth's largest mountains. Olympus Mons and its neighbors grew as the result of hundreds of thousands of individual eruptions. In the absence of plate tectonics on Mars, its volcanoes have remained over their hot spots for billions of years, growing ever taller and broader in the lower surface gravity with each successive eruption.

Lava flows and other volcanic landforms span nearly the entire history of Mars, estimated to extend from the formation of the crust some 4.4 billion years ago to geologically recent times and to cover more than half of the red planet's surface. Although some "fresh appearing" lava flows have been identified on Mars, we will not know the age of those latest eruptions until rock samples are radiometrically dated. Mars could, in principle, experience eruptions today.

VENUS   Of the terrestrial planets, Venus has the most volcanoes. Radar images reveal a variety of volcanic landforms. They include highly fluid flood lavas covering thousands of square kilometers, shield volcanoes approaching the size and complexity of those on Mars, and lava channels thousands of kilometers long. Those lavas must have been extremely hot and fluid to flow for such long distances. Some of the volcanic eruptions on Venus are thought to have been associated with deformation of Venus's lithosphere above hot spots such as the circular features mentioned earlier.

Lavas on Venus are basalts, much like the lavas on Earth, the Moon, Mars, and possibly Mercury. The *Venus Express* spacecraft imaged three of the nine Hawaii-like hot spots in infrared wavelengths (**Figure 8.26**). Each hot spot has several volcanoes, with altitudes of 500–1,200 meters above the nearby plains. Some of the volcanic regions were radiating heat more efficiently than the regions nearby. That finding suggests that those volcanic regions have younger material and that volcanic activity had taken place within the past 2.5 million years—perhaps as recently as a few thousand years ago. As we describe in Chapter 9, Venus has some of the volcanic gas sulfur dioxide in its atmosphere, so Venus may still be cooling its interior through volcanic activity.

A geological timescale for Venus has not yet been devised. But from its relative lack of impact craters, most of the surface is considered less than 1 billion years old, and some of it may be much more recent. When volcanism began on Venus and how much active volcanism exists today remain unanswered questions.

CHECK YOUR UNDERSTANDING 8.5

Which is *not* a reason for the large size of volcanoes on Mars in comparison with the smaller ones on Earth? (a) absence of plate tectonics; (b) distance from the Sun; (c) lower surface gravity than Earth's; (d) many repeated eruptions

· ·

8.6 The Geological Evidence for Water

Today, Earth is the only planet in the Solar System where the temperature and atmospheric conditions allow extensive liquid surface water to exist. The other inner planets do not have extensive liquid surface water, but evidence exists for water ice in deep craters or in the polar regions and for water below the surface in permafrost, subsurface glaciers, or possibly as liquid in the core.

Life, as we know it on Earth, requires water as a solvent and as a delivery mechanism for essential chemistry. Therefore, the search for water is central to the search for life in the Solar System. In addition, if humans are ever going to live on another terrestrial planet, they will need a source of water. In this section, we look at how water modifies the surface of a planet and then discuss the search for water in the Solar System.

Water and Erosion

Tectonism, volcanism, and impact cratering affect Earth's surface by creating variations in the height of the surface. **Erosion** is the wearing away of a planet's surface by mechanical action. The term *erosion* covers a variety of processes. Erosion, not only by running water but also by wind and by the actions of living organisms, wears down hills, mountains, and craters; the resulting debris fills in valleys, lakes, and canyons. If erosion were the only geological process operating, it would eventually smooth out the surface of the planet completely. Because Earth is a geologically and biologically active world, however, its surface is an ever-changing battleground between processes that build up topography and those that tear it down.

Weathering is the first step in erosion. During weathering, rocks are broken into smaller pieces and may be chemically altered. For example, rocks on Earth are physically weathered along shorelines, where the pounding waves break them into beach sand. Other weathering processes include chemical reactions, such as when oxygen in the air combines with iron in rocks to form a type of rust. One of the most efficient forms of weathering is caused by water: liquid water runs into crevices and then freezes. As the water freezes, it expands and shatters the rock.

After weathering, the resulting debris can be carried away by flowing water, glacial ice, or blowing wind and deposited in other areas as sediment. Where material is eroded, we can see features such as river valleys, wind-sculpted hills, or mountains carved by glaciers. Where eroded material is deposited, we see features such as river deltas, sand dunes, or piles of rock at the bases of mountains and cliffs. Erosion is most efficient on planets with water and wind. On Earth, where water and wind are prevalent, most impact craters on land are worn down and filled in.

Even though the Moon and Mercury have almost no atmosphere and no running water, a type of slow erosion is still at work. Radiation from the Sun and from deep space slowly decomposes some types of minerals, effectively weathering the rock. Such effects are only a few millimeters deep at most. Impacts of micrometeoroids also chip away at rocks. In addition, landslides can occur wherever gravity and differences in elevation are present. Although water enhances landslide activity, landslides also are seen on dry bodies such as Mercury and the Moon.

As we discuss in Chapter 9, Earth, Mars, and Venus have atmospheres, and all three planets show the effects of windstorms. Images of Mars and Venus returned by spacecraft landers show surfaces that have been subjected to the forces of wind. Sand dunes are common on Earth and Mars (**Figure 8.27**), and some have been identified on Venus. Orbiting spacecraft also have found wind-eroded hills and surface patterns called wind streaks. Those surface patterns appear, disappear, and change in response to winds blowing sediments around hills, craters, and cliffs. The surface patterns serve as local weather vanes, telling planetary scientists about the direction of local prevailing surface winds. Planet-encompassing windy dust storms have been seen on Mars.

The Search for Water in the Solar System

A priority of recent planetary exploration missions is the search for water on the terrestrial planets and the Moon. Some of the evidence for past water comes from the geological processes discussed earlier in the chapter. Water was brought in by impacts in the early Solar System and is affected by geological and atmospheric activity on a planet. The search for water includes examination of images of the

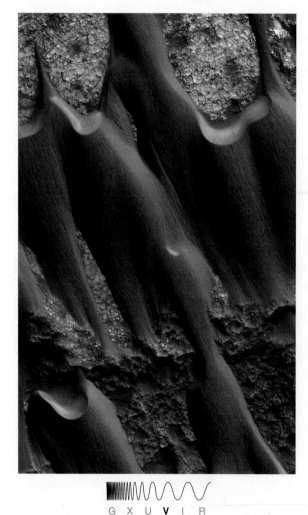

G X U V I R

Figure 8.27 A *Mars Reconnaissance Orbiter* (*MRO*) image of the Nili Patera dune field on Mars. The dunes change over months because of winds on Mars.

G X U V I R

Figure 8.28 A photograph of gully channels in a crater on Mars taken by the *MRO*. The gullies coming from the rocky cliffs near the crater's rim (out of the image, to the upper left) show meandering and braided patterns similar to those of water-carved channels on Earth.

G X U V I R

Figure 8.29 The water-worn gravel embedded in sand shown in this Mars image from the *Curiosity* rover is evidence of an ancient streambed.

terrain obtained by flybys, orbiters, and landers. For the Moon, the search has included reanalyzing lunar rocks and soil brought back to Earth during the *Apollo* missions (1969–1972) and crashing spacecraft into the surface to analyze the resulting debris.

MARS The search for water on Mars goes back almost a century and a half. Even small telescopes show polar ice caps, which change with the martian seasons. In 1877, Italian astronomer Giovanni Schiaparelli (1835–1910) observed what appeared to be linear features on Mars and dubbed them *canali* (Italian for "channels"). Unfortunately, some other observers, including the U.S. astronomer Percival Lowell (1855–1916), incorrectly translated Schiaparelli's *canali* as "canals," implying that they were artificially constructed by intelligent life rather than naturally formed as a result of geological processes. Lowell strongly advocated the theory that those "canals" were built to move water around a drying planet Mars. Other observers of the time disputed that idea, arguing that the features were optical illusions seen in telescopes. Larger telescopes and astrophotographs did not show canals, astronomical spectroscopy did not find water vapor, and the idea of artificial canals went out of favor after Lowell died.

Scientists debate about how recently significant liquid water flow has occurred on the surface of Mars. The geological evidence suggests that water once flowed across the surface of Mars in vast quantities. Canyons and huge, dry riverbeds attest to tremendous floods that poured across the martian surface. In addition, many regions on Mars show small networks of valleys that probably were carved by flowing water (**Figure 8.28**). Large deposits of ice have been detected under the surface. Mars may have contained oceans at one time, including one ocean that might have covered a third of the planet's surface.

In 2004, NASA sent two instrument-equipped roving vehicles, *Opportunity* and *Spirit*, to search for evidence of water on Mars. *Opportunity* landed inside a crater. For the first time, martian rocks were available for study in the original order in which they had been laid down. Previously, the only rocks that landers and rovers had come across were those that had been dislodged from their original settings by either impacts or river floods. The layered rocks at the *Opportunity* site revealed that they had once been soaked in or transported by water. The form of the layers was typical of layered sandy deposits laid down by gentle currents of water. Magnified images of the rocks showed "blueberries"—small spheres a few millimeters across that probably formed in place among the layered rocks. Analysis of the spheres revealed abundant hematite, an iron-rich mineral that forms in the presence of water.

Observations by ESA's *Mars Express* and NASA's *Mars Odyssey* and *Mars Reconnaissance Orbiter* (*MRO*) have shown the hematite signature and the presence of sulfur-rich compounds in a vast area surrounding *Opportunity*'s landing site. Those observations suggest the existence of an ancient martian sea larger than the combined area of the Great Lakes and as deep as 500 meters.

In 2012, NASA's Mars rover *Curiosity* landed in Gale Crater, a large (150 km), old (3.8 billion years) crater just south of the equator of Mars. *Curiosity* found evidence of a stream that flowed at a rate of about 1 meter per second and was as much as 2/3 meter deep. The streambed is identified by water-worn gravel (**Figure 8.29**). The rover, about the size of a car, includes cameras, a drill, and an instrument to measure chemical composition. When the rover drilled into a rock, it found sulfur, nitrogen, hydrogen, oxygen, phosphorus, and carbon, together with clay minerals that formed in a water-rich environment that was not very salty. That crater may well have once been a large lake that held water in the ground long after the lake evaporated.

Where did the water go? Some escaped into the thin atmosphere of Mars, and some is locked up as ice in the polar regions, just as the ice caps on Earth hold much of its water. Unlike Earth's polar caps of frozen water, those on Mars are a mixture of frozen carbon dioxide and frozen water. Water must be hiding elsewhere on Mars. Small amounts of water can be found on the surface, and in 2008 NASA's *Phoenix* lander found water ice just a centimeter or so beneath surface soils at high northern latitudes (**Figure 8.30**). However, most of the water on Mars appears to be trapped well below the surface. Radar imaging by *Mars Express* and the *MRO* indicates huge quantities of subsurface water ice, not only in the polar areas as expected but also at lower latitudes under craters, even as far south as the equator. In addition, *MRO* images suggest that seasonal salt water might flow on the surface far from the poles. Salt water freezes at a lower temperature, so some sites could be warm enough to have temporary liquid salt water. Other *MRO* images show underground ice exposed on a steep slope (**Figure 8.31**). Thus, Mars may have had conditions suitable to support Earth-like microbial life in the distant past.

VENUS Evidence for liquid water on Venus comes primarily from water vapor in its atmosphere, but some geological indications of past water are apparent, such as color differences between the highland and lowland regions. As noted earlier, on Earth such a difference indicates the presence of granite, which requires water for its formation. We return to the subject of what happened to the water on Venus and Mars in Chapter 9, when we discuss their atmospheres.

THE MOON Infrared measurements of the Moon by the U.S. *Clementine* mission in 1994 supported the possibility of ice at the lunar poles. In 1998, NASA's

G X U V I R

Figure 8.30 Water ice appears a few centimeters below the surface in this trench dug by a robotic arm on the *Phoenix* lander. The trench measures about 20 × 30 cm.

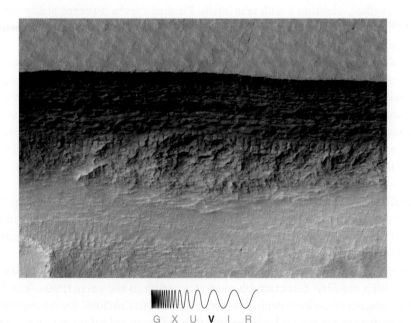

G X U V I R

Figure 8.31 Color-enhanced *MRO* image of a steep slope on Mars in which underground water ice is exposed (and appears bright blue). *MRO* observed eight of these slopes, suggesting that Mars has places where water is accessible.

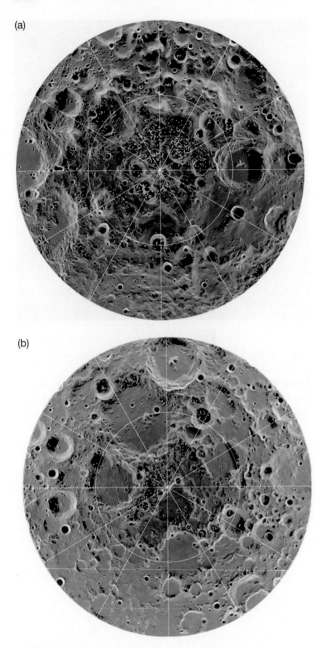

(a)

(b)

Figure 8.32 NASA's Moon Minerology Mapping instrument found surface ice at the poles of the moon. In this image, the blue areas show the location of the ice, in the coldest and darkest locations at the South Pole (top) and North Pole.

Lunar Prospector observations suggested subsurface water ice in the polar regions. When the lunar orbiter's primary mission was completed, NASA crashed *Lunar Prospector* into a crater near the Moon's south pole while ground-based telescopes searched for evidence of water vapor above the impact site, but none was seen. Since 2007, several new missions have been sent to the Moon, including Japan's *Kaguya*, India's *Chandrayaan*, and NASA's *Lunar Reconnaissance Orbiter* (*LRO*) and its companion *Lunar Crater Observation and Sensing Satellite* (*LCROSS*).

In 2009, NASA crashed the *LCROSS* launch vehicle into the Cabeus Crater at the lunar south pole, which sent up dust and vapor that the *LCROSS* and *LRO* spacecraft analyzed. More than 5 percent of the resulting plume was water, which makes that part of the Moon wetter than many Earth deserts. Other volatiles also were detected. Measurements of hydrogen by the *LRO* suggest that a fair amount of buried water ice exists in the cold southern polar region. In mid-2018, data from an instrument on *Chandrayaan* definitively confirmed that there was water ice in the shadows of craters in the polar regions (**Figure 8.32**).

Those space observations of lunar ice sent planetary scientists back to the collections of lunar rocks and soil returned to Earth decades ago by the *Apollo* mission astronauts. New analysis of volcanic glass beads in lunar soil suggests that the interior of the Moon may have a much larger amount of volatiles than previously believed. Reanalysis of the *Apollo* lunar rocks also found evidence of water. One way to distinguish whether water on the Moon originated from its interior or from impacts is to look at the ratio of water molecules composed of regular hydrogen (with one proton) and oxygen to water molecules composed of oxygen, hydrogen, and an isotope of hydrogen called deuterium (hydrogen with one proton and one neutron). The ratio is higher in the water in lunar rocks than in the water on Earth. Water that originated in comets or in meteorites rich in water has a different ratio. Alternatively, protons from the solar wind or from high-energy cosmic rays in space could have combined with oxygen on the lunar surface, yielding a different ratio. Findings from a recent study suggest that the Moon's water came from the interactions between solar wind protons and oxygen.

Those results on lunar water are preliminary. Considerable debate continues among scientists about exactly how much water ice exists on the lunar surface and how much liquid water is in the interior, how the Moon acquired that water, and how the presence and origin of water affect the current theories of lunar formation. Several countries—and some private companies—are considering proposals to send robotic spacecraft to the Moon to collect additional lunar material and bring it back to Earth for analysis.

MERCURY Water ice also has been detected in the polar regions of Mercury. Some deep craters in the polar regions of Mercury have floors that are in perpetual shadow and thus receive no sunlight. Temperatures in those permanently shadowed areas remain very cold—below 180 K. For many years, planetary scientists had speculated that ice could be found in those polar craters, with a possible detection by radar occurring in the early 1990s. The *Messenger* spacecraft, which orbited Mercury from 2011 to 2015, found deposits of ice at craters at the planet's north pole. Other frozen volatiles also were seen in the polar craters. The icy areas have sharp boundaries, which indicates they are relatively recent, either from comet impacts or from some ongoing process on the planet.

Which of the following worlds show evidence of the current presence of liquid or frozen water? (Choose all that apply.) (a) Mercury; (b) Venus; (c) Earth; (d) the Moon; (e) Mars

Origins   The Death of the Dinosaurs

When large impacts happen on Earth, they can have far-reaching consequences for Earth's climate and for terrestrial life. One of the biggest and most significant impacts happened at the end of the Cretaceous Period, which lasted from 146 million years ago to 65 million years ago. At the end of the Cretaceous Period, more than half of all living species, including the dinosaurs, became extinct. That mass extinction is marked in Earth's fossil record by the Cretaceous-Paleogene boundary, or *K-Pg boundary* (the *K* comes from *Kreide*, German for "Cretaceous"). Fossils of dinosaurs and other now-extinct life-forms are found in older layers below the K-Pg boundary. Fossils in the newer rocks above the K-Pg boundary lack more than half of all previous species but contain a record of many other newly evolving species. Big winners in the new order were the mammals—distant ancestors of humans—that moved into the ecological niches that extinct species vacated.

How do scientists know that an impact was involved? The K-Pg boundary is marked in the fossil record in many areas by a layer of clay. Studies at more than 100 locations around the world have found large amounts of the element iridium in that layer, as well as traces of soot. Iridium is very rare in Earth's crust but is common in meteorites. The soot at the K-Pg boundary possibly indicates that widespread fires burned the world over. The thickness of the layer of clay at the K-Pg boundary and the concentration of iridium increases toward what is today the Yucatán Peninsula in Mexico. Although erosion has largely erased the original crater, geophysical surveys and rocks from drill holes in the area show a deeply deformed subsurface rock structure, similar to that seen at known impact sites. Those results provide compelling evidence that 65 million years ago an asteroid about 10 km in diameter struck the area, throwing great clouds of red-hot dust and other debris into the atmosphere (**Figure 8.33**) and possibly igniting a worldwide conflagration. The energy of the impact is estimated to have been more than that released by 5 billion nuclear bombs.

An impact of that energy would have devastated terrestrial life. In addition to the possible firestorm ignited by the impact, computer models suggest that earthquakes and tsunamis would have occurred. Dust from the collision and soot from the firestorms thrown into Earth's upper atmosphere would have remained there for years, blocking sunlight and plunging Earth into decades of a cold and dark "impact winter." Recent measurements of ancient microbes in ocean sediments suggest that Earth may have cooled by 7°C. The firestorms, temperature changes, and decreased food supplies could have led to a mass starvation that would have been especially hard on large animals such as the dinosaurs.

However, not all paleontologists believe that mass extinction event was the result of an impact; some think volcanic activity was important as well. However, the evidence is compelling that a great impact did occur at the end of the Cretaceous Period. Life on our planet has had its course altered by sudden and cataclysmic events when asteroids and comets have slammed into Earth. We quite possibly owe our existence to the luck of our remote ancestors—small rodent-like mammals—that could live amid the destruction after such an impact 65 million years ago.

Figure 8.33 This artist's rendition depicts an asteroid or comet, perhaps 10 km across, striking Earth 65 million years ago in what is now the Yucatán Peninsula in Mexico. The lasting effects of the impact might have killed off most forms of terrestrial life, including the dinosaurs.

The Moon had a magnetic field for at least 2 billion years, or maybe longer.

Moon Had a Magnetic Field for at Least a Billion Years Longer than Thought

By ASHLEY YEAGER, *ScienceNews*

Analysis of a relatively young rock collected by *Apollo* astronauts reveals the moon had a weak magnetic field until 1 billion to 2.5 billion years ago, at least a billion years later than previous data showed. Extending this lifetime offers insights into how small bodies generate magnetic fields, researchers report August 9 in *Science Advances*. The result may also suggest how life could survive on tiny planets or moons.

"A magnetic field protects the atmosphere of a planet or moon, and the atmosphere protects the surface," says study coauthor Sonia Tikoo, a planetary scientist at Rutgers University in New Brunswick, N.J. Together, the two protect the potential habitability of the planet or moon, possibly those far beyond our solar system.

The moon does not currently have a global magnetic field. Whether one ever existed was a question debated for decades. On Earth, molten rock sloshes around the outer core of the planet over time, causing electrically conductive fluid moving inside to form a magnetic field. This setup is called a dynamo. At 1 percent of Earth's mass, the moon would have cooled too quickly to generate a long-lived roiling interior.

Magnetized rocks brought back by *Apollo* astronauts, however, revealed that the moon must have had some magnetizing force. The rocks suggested that the magnetic field was strong at least 4.25 billion years ago, early on in the moon's history, but then dwindled and maybe even got cut off about 3.1 billion years ago.

Tikoo and colleagues analyzed fragments of a lunar rock collected along the southern rim of the moon's Dune Crater during the *Apollo 15* mission in 1971. The team determined the rock was 1 billion to 2.5 billion years old and found it was magnetized. The finding suggests the moon had a magnetic field, albeit a weak one, when the rock formed, the researchers conclude.

A drop in the magnetic field strength suggests the dynamo driving it was generated in two distinct ways, Tikoo says. Early on, Earth and the moon would have sat much closer together, allowing Earth's gravity to tug on and spin the rocky exterior of the moon. That outer layer would have dragged against the liquid interior, generating friction and a very strong magnetic field.

Then slowly, starting about 3.5 billion years ago, the moon moved away from Earth, weakening the dynamo. But by that point, the moon would have started to cool, causing less dense, hotter material in the core to rise and denser, cooler material to sink, as in Earth's core. This roiling of material would have sustained a weak field that lasted for at least a billion years, until the moon's interior cooled, causing the dynamo to die completely, the team suggests.

The two-pronged explanation for the moon's dynamo is "an entirely plausible idea," says planetary scientist Ian Garrick-Bethell of the University of California, Santa Cruz. But researchers are just starting to create computer simulations of the strength of magnetic fields to understand how such weaker fields might arise. So it is hard to say exactly what generated the lunar dynamo, he says.

If the idea is correct, it may mean other small planets and moons could have similarly weak, long-lived magnetic fields. Having such an enduring shield could protect those bodies from harmful radiation, boosting the chances for life to survive.

1. Refer to Working It Out 8.2. Is the smaller *mass* of the Moon the primary reason it cooled quicker? Explain.
2. What is the other cause of the dynamo's weakening?
3. How might the presence of a magnetic field affect the possibility of habitability?
4. Why did the Moon move away from the Earth billions of years ago? (Hint: see Chapter 4.)

Source: Ashley Yeager, "The Moon had a magnetic field for at least billion years longer than thought," Science News, August 9, 2017. © Society for Science & the Public. Reprinted with permission.

Summary

The terrestrial planets in the Solar System are Mercury, Venus, Earth, and Mars, all of which have evidence of past or present water. The Moon is usually included in discussions of the terrestrial worlds because it is similar to them in many ways. Comparative planetology is the key to understanding the planets. Four geological processes—impacts, volcanism, tectonism, and erosion—are responsible for the topography on the terrestrial planets. Active volcanism and tectonics are the results of a "living" planetary interior: one that is still hot inside. Over time, the interiors cool, and tectonics and volcanism weaken. On Earth, radioactive decay and tidal effects from the Moon help heat the interior. Erosion is a surface phenomenon that results from weathering by wind or water. Surface features on the terrestrial planets, such as tectonic plates, volcanoes, mountain ranges, or canyons, are the result of the interplay among those four processes. Evolution on Earth may have been affected by impacts, such as the impact of an asteroid 65 million years ago that might have led to the death of the dinosaurs.

LG 1 **Describe how impacts have affected the evolution of the terrestrial planets.** Impact cratering is the result of a direct interaction of an astronomical object with the surface of a planet. The layering of craters gives their relative ages, with more recent craters found superimposed on older ones. Crater densities can be used to find the relative ages of regions on a surface; more heavily cratered regions are older than less cratered ones. Planets protected by atmospheres, such as Earth and Venus, have fewer small impact craters. The Moon was probably created when a Mars-sized protoplanet collided with Earth.

LG 2 **Explain how radiometric dating is used to measure the ages of rocks and terrestrial planetary surfaces.** Radioactive isotopes found in rocks can be used to measure their age. The oldest rocks measured, from the Moon and from meteorites, give the Solar System an age of 4.5 billion to 4.6 billion years.

LG 3 **Explain how scientists use both theory and observation to determine the structure of terrestrial planetary interiors.** Models of Earth's interior are used to predict how seismic waves should propagate through the interior, and those predictions are compared with observations of seismic waves. The interiors of other planets are modeled using physical principles, along with observational data of their magnetic fields. Earth has a strong magnetic field, but Venus and Mars do not. The cause for that difference between the terrestrial planets is still uncertain.

LG 4 **Describe tectonism and volcanism and the forms they take on different terrestrial planets.** Tectonism folds, twists, and cracks the outer surface of a planet. Plate tectonics is unique to Earth currently, although other types of tectonic disruptions are observed on the other terrestrial planets, such as cracking and buckling on the surface. Smooth areas on the Moon and Mercury are ancient lava flows. While Venus has the most volcanoes, the largest mountains in the Solar System are volcanoes on Mars. Earth's surface is still changing as volcanic hot-spot activity forms new members of island chains.

LG 5 **Summarize what is known about the presence of water on the terrestrial planets.** Geological features observed in orbiting and surface missions to Mars suggest that liquid water was once on the surface. Mars today has large amounts of subsurface water ice. Venus might have had liquid oceans early in its history. Space mission data indicate that water ice exists near the poles of the Moon and Mercury. The search for water is important to both the search for extraterrestrial life and the possibilities of human colonization of space.

? Unanswered Questions

- Why is Venus so different from Earth? The two planets have similar size, mass, and composition, but they are very different geologically with respect to magnetic fields, plate tectonics, and recent activity, and it is not yet known why. In addition, how did Venus end up rotating in the direction opposite that of its revolution around the Sun? Did it form with a different orbit or rotation? Was it the result of an impact early in its history? Did it change very slowly because of tidal effects from other planets?

- Will humans someday "live off the land" on the Moon? Recent space missions have provided evidence of some water ice on the Moon. If the Moon has water, that would certainly make living there more practical than if water had to be brought from Earth or synthesized from hydrogen and oxygen extracted from the lunar soil. Scientists and engineers have been studying several methods to see whether oxygen can be extracted from lunar rocks to make breathable air for people. Other researchers have looked at using lunar rock as a building material (to make concrete, for example). But the most valuable material on the Moon might turn out to be an isotope of helium, helium-3 (^{3}He, which is helium with two protons and one neutron). On Earth, that isotope exists only as a by-product of nuclear weapons, but up to a million tons of it may be on the Moon. Some scientists and engineers think that helium-3 could be used in a "clean" type of nuclear energy. That helium-3 could be brought back from the Moon for use on Earth or possibly even used in a power plant on the Moon to create energy for a lunar colony.

Questions and Problems

Test Your Understanding

1. _____, _____, and _____ build up structures on the terrestrial planets, whereas _____ tears them down.
   a. impacts, erosion, volcanism; tectonism
   b. impacts, tectonism, volcanism; erosion
   c. tectonism, volcanism, erosion; impacts
   d. tectonism, impacts, erosion; volcanism

2. Geologists can determine the relative age of features on a planet because
   a. the ones on top must be older.
   b. the ones on top must be younger.
   c. the larger ones must be older.
   d. the larger ones must be younger.

3. Scientists can learn about the interiors of the terrestrial planets from
   a. seismic waves.
   b. satellite observations of gravitational fields.
   c. physical arguments about cooling.
   d. satellite observations of magnetic fields.
   e. all of the above

4. Earth's interior is heated by
   a. angular momentum and gravity.
   b. radioactive decay and gravity.
   c. radioactive decay and tidal effects.
   d. angular momentum and tidal effects.
   e. gravity and tidal effects.

5. If a radioactive element has a half-life of 10,000 years, what fraction of it is left in a rock after 40,000 years?
   a. 1/2   c. 1/8   e. 1/32
   b. 1/4   d. 1/16

6. Lava flows on the Moon and Mercury created large, smooth plains. We don't see similar features on Earth because
   a. Earth has less lava.
   b. Earth had fewer large impacts in the past.
   c. Earth has plate tectonics that recycle the surface.
   d. Earth is large with respect to the size of those plains, so they are not as noticeable.
   e. Earth rotates much faster than either of those other worlds.

7. Scientists know the history of Earth's magnetic field because
   a. the magnetic field hasn't changed since Earth formed.
   b. they see today's changes and project backward in time.
   c. the magnetic field becomes frozen into rocks, and plate tectonics spreads those rocks apart.
   d. they compare the magnetic fields on other planets to Earth's.

8. Suppose an earthquake occurs on an imaginary planet. Scientists on the other side of the planet detect primary waves but not secondary waves after the quake. That suggests that
   a. part of the planet's interior is liquid.
   b. the planet's entire interior is solid.
   c. the planet has an iron core.
   d. the planet's interior consists entirely of rocky materials.
   e. the planet's mantle is liquid.

9. Geologists can determine the actual age of features on a planet by
   a. radiometric dating of rocks retrieved from the planet.
   b. comparing cratering rates on one planet with those on another.
   c. assuming that all features on a planetary surface are the same age.
   d. both a and b
   e. both b and c

10. Impacts on the terrestrial planets and the Moon
    a. are more common than they used to be.
    b. have occurred at approximately the same rate since the Solar System formed.
    c. are less common than they used to be.
    d. periodically become more common and then less common.
    e. never occur anymore.

11. Earth has fewer craters than Venus because
    a. Earth's atmosphere protects better than Venus's.
    b. Earth is a smaller target than Venus.
    c. Earth is closer to the asteroid belt.
    d. Earth's surface experiences more erosion.

12. Scientists propose an early period of heavy bombardment in the Solar System because
    a. the Moon is heavily cratered.
    b. all the craters on the Moon are old.
    c. the smooth part of the Moon is nearly as old as the heavily cratered part.
    d. all the craters on the Moon are young.

13. Scientists know that Earth was once completely molten because
    a. the surface is smooth.
    b. the interior layers are denser.
    c. the chemical composition indicates that.
    d. volcanoes exist today.

14. What is the main reason that Earth's inner core is liquid today?
    a. tidal force of the Moon on Earth
    b. seismic waves that travel through Earth's interior
    c. decay of radioactive elements
    d. convective motions in the mantle
    e. pressure on the core from Earth's outer layers

15. Mars has about half of Earth's diameter. If the interiors of those planets are heated by radioactive decays, how does the heating rate of the interior of Mars compare with that of Earth?
    a. The heating rate of Mars is 0.125 times that of Earth.
    b. The heating rate of Mars is 8 times that of Earth.
    c. The heating rate of Mars is 0.5 times that of Earth.
    d. The heating rate of Mars is 4 times that of Earth.
    e. The heating rates are about the same.

Thinking about the Concepts

16. When discussing the terrestrial planets, why do we include Earth's Moon?

17. Can all rocks be dated with radiometric methods? Explain.

18. Explain how scientists know that rock layers at the bottom of the Grand Canyon are older than those found on the rim.

19. Describe the sources of heating responsible for generating Earth's magma.

20. Explain why the Moon's core is cooler than Earth's.

21. Explain the difference between longitudinal waves and transverse waves.

22. How do we know that Earth's core includes a liquid zone?

23. Study the Process of Science Figure. What evidence makes the impactor theory the currently preferred explanation for the origin of the Moon? What evidence remains to be found to rule out the competing theories?

24. Compare tectonism on Venus, Earth, and Mercury.

25. Explain plate tectonics and identify the only planet on which that process has been observed.

26. Volcanoes have been found on all the terrestrial planets. Where are the largest volcanoes in the inner Solar System?

27. Explain the criteria you would apply to images (assume adequate resolution) to distinguish between a crater formed by an impact and one formed by a volcanic eruption.

28. What are the primary reasons scientists have decided that the surfaces of Venus, Earth, and Mars are younger than those of Mercury and the Moon?

29. Explain some of the geological evidence suggesting that Mars once had liquid water on its surface.

30. What evidence supports the theory suggesting that a mass extinction occurred as a consequence of an enormous impact on Earth 65 million years ago?

Applying the Concepts

31. Use Figure 8.8 to answer the following questions.
    a. How has the cratering rate changed? Has it fallen off gradually or abruptly?
    b. At present, what is the cratering rate compared with that about 4 billion years ago?
    c. Why does that falloff in cratering rate fit nicely into the theory of planet formation?

32. Are the vertical and horizontal axes in Figure 8.7 linear or logarithmic? After how many half-lives does the number of parent isotopes equal the number of daughter isotopes? Is the result unique to that example? Why or why not?

33. The destruction of the parent isotope is an example of exponential decay (Figure 8.7). Is the growth of the daughter isotope an example of exponential growth? How can you tell?

34. Compare Figures 8.18 and 8.23. Which region is older? How do you know?

35. Assume that Earth and Mars are perfect spheres with radii of 6,371 km and 3,390 km, respectively.
    a. Calculate the surface area of Earth.
    b. Calculate the surface area of Mars.
    c. If 0.72 (72 percent) of Earth's surface is covered with water, compare the amount of Earth's land area with the total surface area of Mars.

36. Compare the kinetic energy ($= 1/2\,mv^2$) of a 1-gram piece of ice (about half the mass of a dime) entering Earth's atmosphere at a speed of 50 km/s with that of a 2-metric-ton SUV (mass $= 2 \times 10^3$ kg) speeding down the highway at 90 km/h.

37. The object that created Arizona's Meteor Crater was estimated to have a radius of 25 meters and a mass of 300 million kg. Calculate the density of the impacting object, and explain what that may tell you about its composition.

38. Using the information in Table 8.1 and Working It Out 8.2, determine the relative rates of internal energy loss experienced by Earth and the Moon.

39. Earth's mean radius is 6,371 km, and its mass is 6.0×10^{24} kg. The Moon's mean radius is 1,738 km, and its mass is 7.2×10^{22} kg.
    a. Calculate Earth's average density. Show your work; do not look the value up.
    b. The average density of Earth's crust is 2,600 kg/m³. What does that tell you about the composition of Earth's interior?
    c. Compute the Moon's average density. Show your work.
    d. Compare the average densities of the Moon, Earth, and Earth's crust. What do those values tell you about the Moon's composition in comparison with that of Earth and of Earth's crust?

40. The glaze on a piece of ancient pottery contains radium, a radioactive element that decays to radon and has a half-life of 1,620 years. Radon could not have been in the glaze when the pottery was being fired, but now it contains three atoms of radon for each atom of radium. How old is the pottery?

41. Archaeological samples are often dated by using radiocarbon dating. The half-life of carbon-14 is 5,700 years.
    a. After how many half-lives will the sample have only 1/64 as much carbon-14 as it originally contained?
    b. How much time will have passed?
    c. If the daughter product of carbon-14 is present in the sample when it forms (even before any radioactive decay happens), you cannot assume that every daughter product you see is the result of carbon-14 decay. If you did make that assumption, would you overestimate or underestimate the age of a sample?

42. Different radioisotopes have different half-lives. For example, the half-life of carbon-14 is 5,700 years, the half-life of uranium-235 is 704 million years, the half-life of potassium-40 is 1.3 billion years, and the half-life of rubidium-87 is 49 billion years.
    a. Why would an isotope with a half-life like that of carbon-14 be a poor choice to get the age of the Solar System?
    b. The age of the universe is approximately 14 billion years. Does that mean that no rubidium-87 has decayed yet?

43. Assume that the east coast of South America and the west coast of Africa are separated by an average distance of 4,500 km. Assume also that GPS measurements indicate that those continents are now moving apart at 3.75 cm/yr. If that rate has been constant over geological time, how long ago were the continents joined as part of a supercontinent?

44. Shield volcanoes are shaped like flattened cones. The volume of a cone is equal to the area of its base multiplied by one-third of its height. The largest volcano on Mars, Olympus Mons, is 27 km high and has a base diameter of 550 km. Compare its volume with that of Earth's largest volcano, Mauna Loa, which is 9 km high and has a base diameter of 120 km.

45. Using the data in Table 8.1, compare the surface gravity on Mars with that on Earth. How does that difference help explain why volcanoes on Mars can grow so high?

USING THE WEB

46. Go to the U.S. Geological Survey's "Earthquake" website (https://earthquake.usgs.gov/earthquakes/map). Set "Zoom to . . ." to "World," set the "Settings" icon in the upper right to "7 Days, Magnitude 2.5+," and look at the earthquakes for the past week. Did any really large ones occur? Compare the map of recent earthquakes with Figure 8.17 in the text. Are any of the quakes in surprising locations? Where was the most recent one? Now change the "Zoom to . . ." to the United States and change the settings to "30 Days, Magnitude 2.5+." Has seismic activity occurred, and if so, where?

47. Citizen science:
    a. Go to http://planetfour.org, a Zooniverse Citizen Science Project in which people examine images of the surface of Mars. Choose a project that has data to analyze. Log in or create a Zooniverse account if you don't have one. Read through what is available of "About," "Guide," "FAQs," "Learn More," and "Classify." Where did those data come from? What are the goals of this project? Why is having many people look at the data useful? Now classify some images.

    b. Go to the website for Cosmoquest (http://cosmoquest.org) and click on "Map Mercury." You will need to create an account for the Cosmoquest projects. Click on the circled question mark under the blue check box, and read the FAQ and watch the tutorial. What is the goal of this project? Where did the data come from? Classify some images.
    c. Go to the website for Cosmoquest (http://cosmoquest.org) and click on "Map the Moon." As in part (b), you will need an account. Click on the circled question mark under the blue check box and read the FAQ and watch the four tutorials. What are some of the basic features? How does the angle of the sunlight and the direction of illumination affect what you see? Now classify a few craters.

48. Space missions: Mars
    a. Go to the website for the *Mars Science Laboratory Curiosity* (https://mars.nasa.gov/msl/), which landed in 2012. What are the latest science results?
    b. Go to the website for the Mars InSight mission (https://mars.nasa.gov/insight/). Did it successfully land on Mars? Are any science results available?
    c. Go to the website for the Mars 2020 mission (https://mars.nasa.gov/mars2020/). What are the goals for that mission? What is its status? If it is on Mars, are any science results available yet?

49. Space Missions: The Moon
    a. India's *Chandrayaan-2* mission is planned to go to a southern region of the Moon. Are any results available from that mission?
    b. NASA, SpaceX, and China are planning new missions to send people to the Moon. Search the Internet to see the status of those missions. Are any happening soon?

50. Video:
    a. Watch one of the available documentaries about the *Apollo* missions to the Moon (such as *In the Shadow of the Moon*, 2008). Why did the United States decide to send astronauts to the Moon? Why did the *Apollo* program end?
    b. The first science fiction film was the short *A Trip to the Moon* (Georges Méliès, 1902). A version with an English narration can be viewed at https://archive.org/details/Levoyagedanslalune. A restored digitized and colorized version was released in 2011 and can be found at http://vimeo.com/39275260. Where do the "Selenians" live on the Moon? In that first cinematic depiction of contact with life from outside Earth, what do the human astronomers do to the Selenians? Compare what the astronomers in the film find on the Moon with what the *Apollo* astronauts actually saw.

EXPLORATION

digital.wwnorton.com/astro6

Knowledge of the exponential function is critical to understanding life in modern times. That function shows up in contexts as diverse as economics, population studies, and climate change.

Radioactive decay exhibits exponential decay, where the amount of a radioactive material that remains after an elapsed time is proportional to the amount present at the beginning of the period. In this Exploration, you will use a small (1.69-ounce) bag of plain M&M's to investigate that behavior.

That size bag of M&M's contains approximately 56 pieces of candy. Each piece has an M stamped on one side. Thus, each piece can fall in two ways when dropped: M side up or M side down. That is much like what happens to a radioactive nucleus: for any given time, either it decays or it doesn't.

Instead of acquiring 10 bags of M&M's, with all the mess that would make, you will use the same bag 10 times and then add your results together. That approach makes the sample large enough that the exponential behavior becomes apparent.

Step 1 For the first trial, shake all the M&M's into your hands and then pour them onto the table.

Step 2 Count how many land M side up, and record that number in the provided table under Trial 1, Time Step 1.

Step 3 Set the M-side-up candies to one side.

Step 4 Repeat steps 1–3 for time steps 2, 3, 4, and so on, until the last candy lands M side up.

Step 5 Repeat steps 1–4 for nine more trials (for a total of 10 trials).

Step 6 Add the results of the trials for each time step, and record them in the Sum column of the table.

Step 7 Making sense of numbers in tabular form like this is hard, so plot the results on a graph with the sum on the y-axis and the time step on the x-axis.

Time Step	Trial 1	Trial 2	Trial 3	Trial 4	Trial 5	Trial 6	Trial 7	Trial 8	Trial 9	Trial 10	Sum
1											
2											
3											
4											
5											
6											
7											
8											
9											
10											
11											
12											

1 Study your graph. At first, does the number of M&M's landing M side up decrease slowly or quickly? In later time steps (such as 8 or 9), does the number decrease slowly or quickly?

_____

2 Generalize your answer to question 1 to the behavior of radioactive sources. Does the radioactivity fall off slowly or quickly at the beginning of an observation? How about at later times?

_____

3 As time goes by, what happens to the number of M&M's that remain? What happens to the number of M&M's in the pile set aside?

_____

4 Generalize your answer to question 3 to the behavior of radioactive sources. What happens to the number of radioactive isotopes over time? What happens to the number of daughter products?

_____

5 Imagine that you walk into a room and observe another student performing this experiment. She pours the M&M's onto the table and counts 10 candies that landed M side up. About how many time steps have passed since she started the experiment?

_____

_____

6 Apply your answer to question 5 to the behavior of radioactive sources. When scientists study radioactive sources, they generally study the ratio of the number of radioactive isotopes to the number of daughter products. Explain how that method, though different from what you did in step 5, contains the same information about the amount of time that has passed.

_____

_____

9

Atmospheres of the Terrestrial Planets

Earth's atmosphere surrounds its inhabitants like an ocean of air. It is evident in the blueness of the sky and in the breezes in the air. Without Earth's atmosphere, neither clouds nor oceans would exist. Without an atmosphere, Earth would look something like the Moon, and life would not exist on our planet. Among the five terrestrial bodies that we discussed in Chapter 8, only Venus and Earth have dense atmospheres. Mars has a very low-density atmosphere, and the atmospheres of Mercury and the Moon are so sparse that they can hardly be detected. To understand the origins of the atmospheres of Earth, Venus, and Mars—how they have changed, how they compare, and how they are likely to evolve—requires us to look back nearly 5 billion years to a time when the planets were just completing their growth.

LEARNING GOALS

Here in Chapter 9, we compare the atmospheres of the terrestrial planets. By the end of this chapter, you should be able to:

LG 1 Identify the processes that cause primary and secondary atmospheres to be formed, retained, and lost.

LG 2 Compare the strength of the greenhouse effect on Earth, Venus, and Mars and how it contributes to differences among the atmospheres of those planets.

LG 3 Describe the layers of the atmospheres on Earth, Venus, and Mars.

LG 4 Explain how Earth's atmosphere has been reshaped by the presence of life.

LG 5 Describe the evidence that shows Earth's climate is changing and how comparative planetology contributes to a better understanding of those changes.

Earth from space. ▶ ▶ ▶

Why can you
breathe only
on Earth?

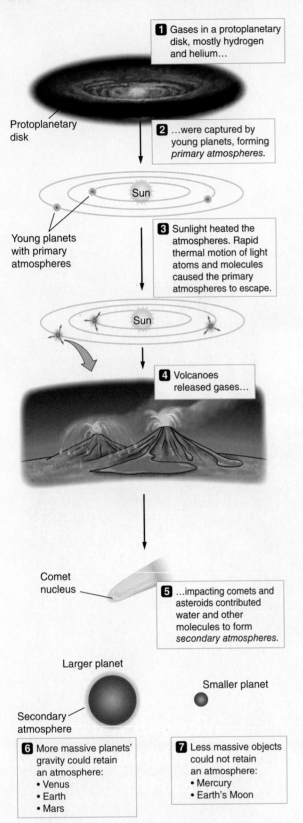

1 Gases in a protoplanetary disk, mostly hydrogen and helium…

Protoplanetary disk

2 …were captured by young planets, forming *primary atmospheres.*

Sun

Young planets with primary atmospheres

3 Sunlight heated the atmospheres. Rapid thermal motion of light atoms and molecules caused the primary atmospheres to escape.

Sun

4 Volcanoes released gases…

Comet nucleus

5 …impacting comets and asteroids contributed water and other molecules to form *secondary atmospheres.*

Larger planet

Smaller planet

Secondary atmosphere

6 More massive planets' gravity could retain an atmosphere:
• Venus
• Earth
• Mars

7 Less massive objects could not retain an atmosphere:
• Mercury
• Earth's Moon

Figure 9.1 Planetary atmospheres form and evolve in phases.

9.1 Atmospheres Change over Time

An **atmosphere** is a layer of gas that sits above the surface of a body such as a terrestrial planet. A blanket of atmosphere warms and sustains Earth's temperate climate. On Venus, a thick atmosphere of carbon dioxide pushes the planet's surface temperature very high. A thin atmosphere leaves the surface of Mars unprotected and frozen. Mercury and the Moon have essentially no atmosphere. Why do some terrestrial planets have dense atmospheres, whereas others have little or none? In this section, we look at the formation of planetary atmospheres.

Formation and Loss of Primary Atmospheres

Planetary atmospheres formed in phases (**Figure 9.1**). When they formed, the young planets were initially enveloped by the remaining hydrogen and helium that filled the protoplanetary disk surrounding the Sun, and they captured some of that surrounding gas. Gas capture continued until soon after the planets formed, when the gaseous disk dissipated and the supply of gas ran out. The gaseous atmosphere collected by a newly formed planet is called its primary atmosphere. The terrestrial planets lost that primary atmosphere as the lightweight atoms and molecules escaped from the planet's gravity. To understand that, we must consider how particles move within a planetary atmosphere.

How can gas molecules escape from a planet? Any object—from a molecule to a spacecraft—can escape a planet if the object exceeds the escape velocity (Chapter 4) and is pointed in the right direction. The escape velocity depends on the mass of the planet. In Chapter 7 you saw that the terrestrial planets are less massive than the giant planets. Therefore, the terrestrial planets have weaker gravitational attraction and could not hold light gases such as hydrogen and helium. When the supply of gas in the protoplanetary disk ran out, the primary atmospheres of the terrestrial planets began leaking back into space. At the same time, intense radiation from the Sun—the primary source of thermal (kinetic) energy in the atmospheres of the terrestrial planets—raised the temperature of the atmosphere and increased the speed of atmospheric atoms and molecules enough for some to escape. In addition, giant impacts by large planetesimals early in the history of the Solar System may have blasted away some of their primary atmospheres.

How do molecules move within a planetary atmosphere? Imagine a large box that contains air. In thermal equilibrium, each type of molecule or atom in the box, from the lightest to the most massive, will have the same average kinetic energy. That average kinetic energy of the gas atoms or molecules is directly proportional to the temperature of the gas. The kinetic energy of a molecule (or any object) depends on its mass and its speed. If each type has the same average energy, the lightest molecules must be moving faster than the more massive ones. So for a gas at a given temperature, the less massive molecules must be moving faster than the more massive ones if each type of molecule has the same average kinetic energy. For example, in a mixture of hydrogen and oxygen at room temperature, hydrogen molecules will be rushing around the box at about 2,000 meters per second (m/s) on average, whereas the much more massive oxygen molecules will be moving at only 500 m/s. Remember, though, that those are average speeds. A few molecules will always be moving much faster or much slower than average.

Deep within a planet's atmosphere, fast molecules near the ground will almost certainly collide with other molecules before the fast molecules can escape. The upper levels of the atmosphere, though, contain fewer molecules.

Therefore, fast molecules there are less likely to collide with other molecules and have a better chance of escaping (as long as they are heading more or less upward). Because the giant planets were farther from the Sun, they were much more massive and were cooler: stronger gravity and lower temperatures enabled them to retain nearly all their massive primary atmospheres.

AstroTour: Atmospheres: Formation and Escape

The Formation of Secondary Atmospheres

Although their primary atmospheres were lost, some of the terrestrial planets do have an atmosphere today, known as a secondary atmosphere. Where did that secondary atmosphere come from? Accretion, volcanism, and impacts are responsible for the atmospheres of Earth, Venus, and Mars today. During the planetary accretion process, minerals containing water, carbon dioxide, and other volatile matter collected in the planetary interiors. Later, as the interior heated up, those gases were released from the minerals that had held them. Volcanism then brought the gases to the surface, where they accumulated and created a secondary atmosphere (see step 4 of Figure 9.1).

Impacts by comets and asteroids were another important source of gases. Huge numbers of comets formed in the outer Solar System and were therefore rich in volatiles. As the giant planets there grew to maturity or migrated their orbits, their gravitational perturbations stirred up the comets and asteroids that orbited relatively nearby. Many of those icy bodies were flung outward by the giant planets to join other planetesimals in the Kuiper Belt, 30–50 astronomical units (AU) from the Sun. Other bodies joined the Oort Cloud, a spherical cloud of icy planetesimals that surrounds the Sun at a distance ranging from the Kuiper Belt to about 50,000 AU from the Sun—nearly one-quarter of the way to the nearest star. Other comets were scattered into the inner Solar System. Upon impact with the terrestrial planets, those objects brought ices such as water, carbon monoxide, methane, and ammonia. On the terrestrial planets, cometary water mixed with the water that had been released into the atmosphere by volcanism (see step 5 of Figure 9.1). On Earth, and perhaps Mars and Venus, most of the water vapor then condensed as rain and flowed into the lower areas to form the earliest oceans.

Sunlight also influenced the composition of secondary atmospheres. Ultraviolet (UV) light from the Sun easily breaks down molecules such as ammonia (NH_3) and methane (CH_4). Ammonia, for example, breaks down into hydrogen and nitrogen. When that happens, the lighter hydrogen atoms quickly escape to space, leaving behind the much heavier nitrogen atoms. Nitrogen atoms then combine in pairs to form more massive nitrogen molecules (N_2), and those molecules are even less likely to escape into space. Decomposition of ammonia by sunlight became the primary source of molecular nitrogen in the atmospheres of the terrestrial planets. Molecular nitrogen makes up ~78% of Earth's atmosphere.

Among the terrestrial planets, today only Venus, Earth, and Mars have significant secondary atmospheres. What happened to Mercury and the Moon? Even if those two bodies experienced less volcanism than the other terrestrial planets (see Chapter 8), they would have had the same early bombardment of comet nuclei from the outer Solar System. Some carbon dioxide and water must have accumulated during volcanic eruptions and comet impacts. A body can lose a secondary atmosphere through the same processes that cause the loss of a primary atmosphere. Large impacts were less frequent as the Solar System aged, but atmospheric escape continued over the past 4 billion years. In addition, decreases in the magnetic field as Mercury and the Moon cooled might have contributed to atmospheric escape. With a weaker magnetic field, the planet became less protected from the **solar wind**—a constant stream of charged particles from the Sun. The solar wind can accelerate atmospheric particles to escape velocity.

Both the Moon and Mercury have virtually no atmosphere today. Mercury lost nearly its entire secondary atmosphere to space, just as it had previously lost its primary atmosphere. Even molecules as massive as carbon dioxide can escape from a small planet if the temperature is high enough, as it is on Mercury's sunlit side. Furthermore, intense UV radiation from the Sun can break molecules into less massive fragments, which are lost to space even more quickly. Because the distance from the Sun to the Moon is much farther than the distance from the Sun to Mercury, the Moon is much cooler than Mercury, but its mass is so small that molecules easily escaped even at relatively low temperatures. The ability of planets to hold on to their atmospheres is explored further in **Working It Out 9.1**.

CHECK YOUR UNDERSTANDING 9.1

Which are reasons Mercury has so little gas in its atmosphere? (Choose all that apply.) (a) Its mass is small. (b) It has a high temperature. (c) It is close to the Sun. (d) Its escape velocity is low. (e) It has no moons.

9.2 Secondary Atmospheres Evolve

Although Venus, Earth, and Mars most likely started out with atmospheres of similar composition, they ended up being very different from one another. Earth is volcanically active, Venus might still be volcanically active, and Mars has been volcanically active in the recent past. All three planets must have shared the intense cometary showers of the early Solar System. Their similar geological histories suggest that their early secondary atmospheres might also have been similar. The development of life, however, increased the amount of oxygen in Earth's secondary atmosphere. Earth's atmosphere is now made up primarily of nitrogen and oxygen, with only a trace of carbon dioxide; the compositions of the atmospheres of Venus and Mars today are nearly identical—mostly carbon dioxide, with much smaller amounts of nitrogen. The atmospheres of those planets differ for two main reasons—planetary mass and the greenhouse effect.

Planetary Mass Affects a Planet's Atmosphere

Table 9.1 shows that the *compositions* of the atmospheres of Venus and Mars today are nearly identical. They are both composed mostly of carbon dioxide, with much smaller amounts of nitrogen. Carbon dioxide and water vapor came from volcanic gases, and nitrogen came from decomposed cometary ammonia. The total *amount* of atmosphere, however, differs widely among the three planets. The atmospheric pressure on Venus is nearly 100 times greater than that on Earth. By contrast, the average surface pressure on Mars is less than a hundredth of that on Earth. Venus is nearly 8 times as massive as Mars, so Venus probably had about 8 times as much carbon within its interior to produce carbon dioxide, the principal secondary-atmosphere component of both planets.

The large difference in atmospheric mass comes from the relative strengths of each planet's surface gravity, which involves both the mass and the radius of a planet. Venus has the gravitational pull necessary to hang on to its atmosphere; Mars has less gravitational attraction to keep its atmosphere (see Working It Out 9.1). Furthermore, when a planet such as Mars began to lose its atmosphere to space, the process began to take on a *runaway* behavior. With a thinner

| TABLE 9.1 | **Atmospheres of the Terrestrial Planets** Physical Properties and Composition |

| | PLANET | | |
Property	Venus	Earth	Mars
Surface Pressure (bars)	92	1.0	0.006
Atmospheric Mass (kg)	4.8×10^{20}	5.1×10^{18}	2.5×10^{16}
Surface Temperature (K)	740	288	210
Carbon Dioxide (%)	96.5	0.041	95.3
Nitrogen (%)	3.5	78.1	2.7
Oxygen (%)	0.00	20.9	0.13
Water (%)	0.002	0.1 to 3	0.02
Argon (%)	0.007	0.93	1.6
Sulfur Dioxide (%)	0.015	0.02	0.00

9.1 Working It Out Atmosphere Retention

To estimate a planet's ability to retain its atmosphere, we compare the escape velocity from the planet (which depends on the planet's gravity, determined from its mass and radius) with the average speed of the molecules in a gas (which depends on the temperature of the gas and the mass of the molecules that make up the gas). The escape velocity is

$$v_{esc} = \sqrt{\frac{2GM}{R}}$$

and the values of v_{esc}, in kilometers per second (km/s), are given for the inner planets in Table 8.1.

The temperature, T, of a gas is proportional to the kinetic energy of the particles, $1/2 \, mv^2$ (Chapter 4). We can rearrange that relationship to solve for v and insert the constants of proportionality to get the average speed of a molecule in a gas:

$$v_{molecule} = \sqrt{\frac{3kT}{m}}$$

where T is the temperature of the gas in kelvins, m is the mass of the molecule in kilograms (kg), and k is the Boltzmann constant. The atomic mass of a molecule is found by adding the atomic masses of its composite atoms as specified in the periodic table. (Atomic masses of atoms come from the total number of neutrons and protons; the electron weighs little in comparison.) An oxygen molecule (O_2), for example, has an atomic weight of 32 (16 for each O atom), whereas a hydrogen molecule (H_2) has an atomic weight of 2 (1 for each H atom). Thus, O_2 has 16 times the mass of H_2.

If we put in the value of the Boltzmann constant ($k = 1.38 \times 10^{-23}$ joule per kelvin, or J/K) and the mass of the hydrogen atom ($m = 1.67 \times 10^{-27}$ kg), then $v_{molecule}$, in kilometers per second, is given by

$$v_{molecule} = 0.157 \text{ km/s} \times \sqrt{\frac{\text{Temperature of gas}}{\text{Atomic weight of molecule}}}$$

The higher the temperature, the higher the average kinetic energy of the individual molecules and the faster the average speed of the particles. That difference explains why Earth can hold on to the oxygen in its atmosphere but loses hydrogen to space.

In our example of hydrogen and oxygen molecules, H_2 has an atomic weight of 2 (1 for each hydrogen atom), and O_2 has an atomic weight of 32 (16 for each oxygen atom, from 8 protons and 8 neutrons). Earth has an average temperature of 288 K. The average speeds of H_2 and O_2 are thus

$$\text{For } H_2: v_{molecule} = 0.157 \times \sqrt{\frac{288}{2}} = 1.88 \text{ km/s}$$

$$\text{For } O_2: v_{molecule} = 0.157 \times \sqrt{\frac{288}{32}} = 0.47 \text{ km/s}$$

Thus, in a gas containing both H_2 and O_2, the hydrogen molecule will, on average, be moving 4 times faster than the average oxygen molecule. At any given temperature, the lighter molecules will be moving faster. Not all molecules in a gas are moving at the average speed: some move faster and some move slower (**Figure 9.2**). The general rule is that over the age of the Solar System, a planet can keep its atmosphere if, for that type of gas molecule,

$$v_{molecule} \leq 1/6 \, v_{esc}$$

The escape velocity from Earth is 11.2 km/s, and one-sixth of that is 1.87 km/s (that is, slightly less than the average speed of H_2). Those values explain why Earth has kept its O_2 but not its H_2. A similar analysis shows that on the Moon, with its lower v_{esc} value, both H_2 and O_2 escape. Jupiter, with its much colder temperatures and higher v_{esc} value, keeps both H_2 and O_2.

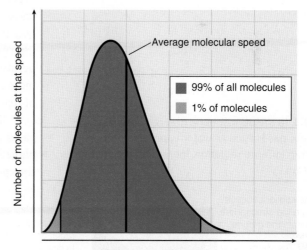

Figure 9.2 The distribution of the speeds of molecules in a gas. The shape of the curve and the exact numbers depend on the temperature of the gas and the masses of the molecules. In all cases, some of the speedier molecules may escape if they exceed the escape velocity.

atmosphere, fewer slow molecules were present to keep fast molecules from escaping, and the rate of escape increased. That process led, in turn, to even less atmosphere and still greater escape rates.

Another possibility is that Mars lost atmosphere in a giant impact. In addition, scientists debate how much atmospheric loss arises from the solar wind's interaction with planetary atmospheres, especially in the absence of a planetary magnetic field. Even though Earth has a magnetic field and Venus and Mars do not, all three planets are losing some atmosphere to space. The extent to which the lack of a magnetic field on Mars played a role in its atmospheric loss is being studied by NASA's *Mars Atmosphere and Volatile EvolutioN* (*MAVEN*) mission, which arrived at Mars in September 2014.

The Atmospheric Greenhouse Effect

Astronomy in Action: Changing Equilibrium

Differences in the present-day masses of the atmospheres of Venus, Earth, and Mars have a large effect on their surface temperatures. In Chapter 5, we calculated the temperature of a planet by finding the equilibrium between the amount of energy it absorbs from sunlight and the amount of energy it radiates back into space. Recall that we found that this calculation gives a good result for planets without atmospheres. But Earth is warmer than expected, and Venus is very much hotter than that simple model predicted. When the predictions of a model fail, the implication is that something was left out of the model. Here, that something was the *atmospheric greenhouse effect*, which traps solar radiation.

The atmospheric greenhouse effect in planetary atmospheres and the conventional **greenhouse effect** operate differently, although the results are much the same. Planetary atmospheres and the interiors of greenhouses are both heated by trapping the Sun's energy, but there the similarities end. The conventional greenhouse effect is the rise in temperature in a car on a sunny day when you leave the windows closed, or what allows plants to grow in the winter in a greenhouse. Sunlight pours through the glass, heating the interior and raising the internal air temperature. With the windows closed, hot air is trapped, and temperatures can climb as high as 80°C (about 180°F). Heating by solar radiation is most efficient when an enclosure is transparent, which is why the walls and roofs of real greenhouses are made mostly of glass.

The atmospheric greenhouse effect is illustrated in **Figure 9.3**. Atmospheric gases freely transmit visible light (Figure 6.20), allowing the Sun to warm the planet's surface. The warmed surface radiates the energy in the infrared region of the spectrum. Some of the atmospheric gases strongly absorb that infrared radiation and convert it to thermal energy, which is released in random directions. Some of the thermal energy continues into space, but much of it returns to the ground, raising the planet's surface temperature. As a result of that radiation, the planet's surface receives thermal energy from both the Sun and the atmosphere. Gases that transmit visible radiation but absorb infrared radiation are known as **greenhouse gases**. Examples include water vapor, carbon dioxide, methane, and nitrous oxide, as well as industrial chemicals such as halogens. The presence of greenhouse gases in a planet's atmosphere will cause its surface temperature to rise.

That rise in temperature continues until the surface becomes hot enough—and therefore radiates enough energy—that the fraction of infrared radiation leaking out through the atmosphere balances the absorbed sunlight, and equilibrium is reached. Convection

Figure 9.3 In the atmospheric greenhouse effect, greenhouse gases such as water vapor and carbon dioxide trap infrared radiation, increasing a planet's temperature.

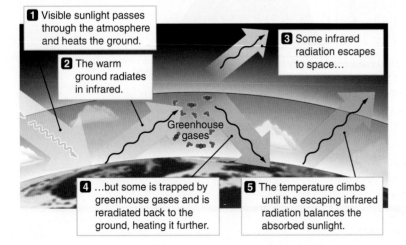

1 Visible sunlight passes through the atmosphere and heats the ground.

2 The warm ground radiates in infrared.

3 Some infrared radiation escapes to space...

Greenhouse gases

4 ...but some is trapped by greenhouse gases and is reradiated back to the ground, heating it further.

5 The temperature climbs until the escaping infrared radiation balances the absorbed sunlight.

also helps maintain equilibrium by transporting thermal energy to the top of the atmosphere, where it can be more easily radiated to space. If the amount of greenhouse gases increases in the atmosphere, the trapping effect increases, and the temperature at which energy input and output balances also increases. Even though the mechanisms are different, the conventional greenhouse effect and the atmospheric greenhouse effect produce the same result: the local environment is heated by trapped solar radiation.

AstroTour: Greenhouse Effect

Similarities and Differences among the Terrestrial Planets

How does the atmospheric greenhouse effect operate on Mars, Earth, and Venus? What really matters is the actual number of greenhouse molecules in a planet's atmosphere, not the fraction they represent. For example, even though the atmosphere of Mars is composed almost entirely of carbon dioxide (see Table 9.1)—an effective greenhouse molecule—the atmosphere is very thin and contains fewer greenhouse molecules than the atmosphere of Venus or Earth. As a result, the atmospheric greenhouse effect is relatively weak on Mars and raises the average surface temperature by only about 5 K (9°F). At the other extreme, Venus's massive atmosphere of carbon dioxide and sulfur compounds raises its average surface temperature by more than 400 K, to about 740 K (467°C, 870°F). At such high temperatures, any remaining water and most carbon dioxide locked up in surface rocks are driven into the atmosphere, further enhancing the atmospheric greenhouse effect.

The atmospheric greenhouse effect on Earth is not as severe as it is on Venus—the average global temperature near Earth's surface is about 288 K (15°C, 59°F). Temperatures on Earth's surface are about 35 K (35°C, 63°F) warmer than they would be without an atmospheric greenhouse effect, mainly because of water vapor and carbon dioxide. Without that comparatively small greenhouse effect, though, Earth's average global temperature would be 255 K (−18°C, 0°F)— well below the freezing point of water, leaving us with a world of frozen oceans and ice-covered continents.

How has the atmospheric greenhouse effect made the composition of Earth's atmosphere so different from the high–carbon dioxide atmospheres of Venus and Mars? The answer lies in Earth's location in the Solar System. Earth and Venus have about the same mass, but Venus orbits the Sun closer than Earth, at 0.7 AU. Volcanism and cometary impacts produced large amounts of carbon dioxide and water vapor to form early secondary atmospheres on both planets. Most of Earth's water quickly rained out of the atmosphere to fill vast ocean basins. But because Venus was closer to the Sun, its surface temperatures were higher than those of Earth. As the Sun itself aged and brightened, Venus got warmer, and most of the rainwater on Venus immediately reevaporated, much as water does in Earth's desert regions. Venus was left with a surface that contained very little liquid water and an atmosphere filled with water vapor. The water vapor caused even higher temperatures, which led to the release of more carbon dioxide from the rocks to the atmosphere. The continuing buildup of both water vapor and carbon dioxide in the atmosphere of Venus led to a runaway atmospheric greenhouse effect that drove up the surface temperature of the planet even more. Ultimately, the surface of Venus became so hot that no liquid water could exist on it.

That early difference between a watery Earth and an arid Venus forever changed how their atmospheres and surfaces evolved. On Earth, water erosion caused by rain and rivers continually exposed fresh minerals, which then reacted chemically with atmospheric carbon dioxide to form solid carbonates. That reaction removed some of the atmospheric carbon dioxide, burying it within

Earth's crust as a component of a rock called limestone. Later, the development of life in Earth's oceans accelerated the removal of atmospheric carbon dioxide. Tiny sea creatures built their protective shells of carbonates, and as they died they built up massive beds of limestone on the ocean floors. Water erosion and the chemistry of life tied up all but a trace of Earth's total inventory of carbon dioxide in limestone beds. Earth's particular location in the Solar System seems to have spared it from the runaway atmospheric greenhouse effect. If those reactions had not locked up all the carbon dioxide now in limestone beds, Earth's atmosphere would be composed of about 98 percent carbon dioxide, similar to that of Venus or Mars. The atmospheric greenhouse effect would be much stronger, and Earth's temperature would be much higher.

Scientists study the details of the differences in the amount of water on Venus, Earth, and Mars. Geological evidence indicates that liquid water was once plentiful on the surface of Mars. Several of the spacecraft orbiting Mars have found evidence that significant amounts of water still exist on Mars as subsurface ice—far more than the atmospheric abundance indicated in Table 9.1. Earth's liquid and solid water supply is even greater, about 0.02 percent of its total mass. More than 97 percent of Earth's water is in the oceans, which have an average depth of about 4 km. Earth today has 100,000 times more water than Venus.

Some scientists think that Venus once had as much water as Earth—as liquid oceans or as more water vapor than is measured today. As the Sun aged and became brighter, and the planets received more solar energy, water molecules high in the atmosphere of Venus were broken apart into hydrogen and oxygen by solar UV radiation. The low-mass hydrogen atoms were quickly lost to space. Oxygen escaped more slowly, so some eventually migrated downward to the planet's surface, where it was removed from the atmosphere by bonding with minerals on the surface. In support of that theory, the *Venus Express* spacecraft has measured hydrogen and some oxygen escaping from the upper levels of Venus's atmosphere.

CHECK YOUR UNDERSTANDING 9.2

The main greenhouse gases in the atmospheres of the terrestrial planets are: (a) oxygen and nitrogen; (b) methane and ammonia; (c) carbon dioxide and water vapor; (d) hydrogen and helium.

9.3 Earth's Atmosphere Has Detailed Structure

Now that we have considered some of the overall processes that have influenced how the terrestrial planet atmospheres evolved, we look in depth at each of them. We begin with the composition and structure of Earth's atmosphere, not only because we know it best but also because it helps us better understand the atmospheres of other worlds.

Life and the Composition of Earth's Atmosphere

Earth's atmosphere is about four-fifths nitrogen (N_2) and one-fifth oxygen (O_2) (see Table 9.1). Many important minor constituents such as water vapor and carbon dioxide (CO_2) are also present, the amounts of which vary depending on global location and season. The composition of Earth's atmosphere is relatively uniform on a global scale, but temperatures can vary widely. Atmospheric temperatures near Earth's surface can range from as high as 60°C (140°F) in the deserts to as low as −90°C (−130°F) in the polar regions. The mean global temperature is about 15°C (59°F).

OXYGEN Table 9.1 shows that Earth's atmosphere contains abundant amounts of oxygen (O_2), whereas the atmospheres of other planets do not. Oxygen is a highly reactive gas: it chemically combines with, or oxidizes, almost any material it touches. The rust (iron oxide) that forms on steel is an example. The reddish surface of Mars is coated with oxidized iron-bearing minerals—one reason the martian atmosphere is almost completely free of oxygen. A planet with significant amounts of oxygen in its atmosphere requires a way to replace oxygen lost through oxidation. On Earth, plants perform that role.

The oxygen concentration in Earth's atmosphere has changed over the history of the planet (**Figure 9.4**). When Earth's secondary atmosphere first appeared about 4 billion years ago, it had very little oxygen because O_2 is not found in volcanic gases or comets. Studies of ancient sediments show that about 2.8 billion years ago, an ancestral form of cyanobacteria—single-celled organisms that contain chlorophyll, which enables them to obtain energy from sunlight—began releasing oxygen into Earth's atmosphere as a waste product of their metabolism. At first, that biologically generated oxygen combined with exposed metals and minerals in surface rocks and soils, so it was removed from the atmosphere as quickly as it formed. Ultimately, the explosive growth of cyanobacteria and then plant life accelerated the production of oxygen, building up atmospheric concentrations that approached today's levels only about 250 million years ago.

All plants, from tiny green algae to giant redwoods, use the energy of sunlight to build carbon compounds out of carbon dioxide and produce oxygen as a metabolic waste product in a process called photosynthesis. In that way, emerging life dramatically changed the very composition and appearance of Earth's surface—the first of many such widespread modifications imposed on Earth by living organisms. Earth's atmospheric oxygen content is held in a delicate balance primarily by plants. If plant life on the planet were to disappear, so, too, would nearly all Earth's atmospheric oxygen, and therefore all animal life—including us.

OZONE Ozone (O_3) is another constituent in Earth's atmosphere. Ozone is formed when UV light from the Sun breaks molecular oxygen (O_2) into its individual atoms. Those oxygen atoms can then recombine with other oxygen molecules to form ozone (the net reaction is $O_2 + O \rightarrow O_3$). Most of Earth's natural ozone is concentrated in the upper atmosphere at altitudes between 10 and 50 km (the **stratosphere**). There it absorbs UV sunlight very strongly. Without the ozone layer, that radiation would reach Earth's surface, where it would be lethal to nearly all forms of life. Ozone exists in the lower atmosphere, too, where it is primarily a by-product of power plants, factories, and automobiles. That ozone is a human-made pollutant and a health hazard that raises the risk of respiratory and heart problems.

In the mid-1980s, scientists began noticing that the measured amount of ozone in Earth's upper atmosphere had been decreasing seasonally since the 1970s, primarily over the polar latitudes during springtime in both the Northern and Southern hemispheres. Scientists called those depleted regions "ozone holes": reduced concentrations of ozone and not an actual hole in the ozone layer. Ozone depletion is caused by a seasonal buildup of atmospheric halogens—mostly chlorine, fluorine, and bromine—such as those found in industrial refrigerants, especially chlorofluorocarbons (CFCs). Halogens diffuse upward into the stratosphere, where they destroy ozone without themselves being consumed. Such agents are called catalysts—materials that participate in and accelerate chemical reactions but are not themselves modified in the process. Because they are not modified or used up, halogens may remain in Earth's upper atmosphere for decades or even centuries. Even though more of the chemicals originated in the north, the

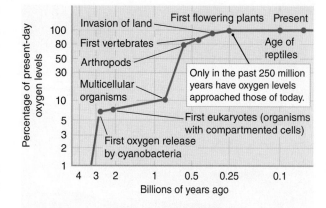

Figure 9.4 The amount of oxygen in Earth's atmosphere has built up as a result of photosynthesizing plant life.

Figure 9.5 Polar stratospheric clouds form in the polar springtime and serve as the surface upon which ozone destruction takes place.

depletions are greater in the Southern Hemisphere because the colder temperatures in the southern polar regions produce a type of cloud that provides a surface on which the ozone-destroying chemical reactions can take place (**Figure 9.5**).

Scientists predicted that the continuing removal of ozone from the high atmosphere could cause trouble for terrestrial life as more and more UV radiation reached the ground. Measured increases in the levels of UV radiation appear to be related to increases in skin cancer in humans, and the mutating effects it may have on other life-forms are not completely understood. By the late 1980s, international agreements on phasing out the production of CFCs and other ozone-depleting chemicals were signed (the Montreal Protocol), and world consumption has steadily declined. The largest Southern Hemisphere ozone hole occurred over Antarctica in 2006. The southern polar ozone holes have mostly stabilized; the hole was large in 2015 but smaller in 2016 and 2017. The ozone hole in the north has been smaller than in the south but would affect more populated areas when it grows, as it did in 2011. Ozone seems to be still decreasing slowly in the lower stratosphere in midlatitudes over more populated regions, perhaps from the release of other chemicals or from changes in the climate. Full recovery to 1980 levels is not expected before the late 21st century.

CARBON DIOXIDE Carbon dioxide (CO_2) is another variable component of Earth's atmosphere. How much carbon dioxide is in the atmosphere at any one time depends on a complex pattern of carbon dioxide *sources* (places where it originates) and *sinks* (places where it goes). Plants consume carbon dioxide in great quantities as part of their metabolic process. Coral reefs are colonies of tiny ocean organisms that build their protective shells with carbonates produced from dissolved carbon dioxide. Fires, decaying vegetation, and the human burning of fossil fuels all release carbon dioxide back into the atmosphere. That balance between carbon dioxide sources and sinks changes. As we describe later, the amount of carbon dioxide in the atmosphere has varied historically but has been increasing more rapidly since the industrial revolution (since 1840 or so). That recent increase in carbon dioxide, in turn, has directly affected global temperature because carbon dioxide is a powerful greenhouse gas.

WATER VAPOR Water vapor (H_2O) in Earth's atmosphere also affects daily life and is a powerful greenhouse gas. Over the range of temperatures on Earth, the amount of water in the atmosphere varies from time to time and from place to place. In warm, moist climates, water vapor may account for as much as 3 percent of the total atmospheric composition. In cold, arid climates, it may be less than 0.1 percent. The continual process of condensation and evaporation of water involves the exchange of thermal and other forms of energy, making water vapor a major contributor to Earth's weather.

The Layers of Earth's Atmosphere

Earth's atmosphere is a blanket of gas several hundred kilometers thick. It has a total mass of approximately 5×10^{18} kg, less than one-millionth of Earth's total mass. The weight of Earth's atmosphere creates a force of approximately 100,000 newtons (N) acting on each square meter of the planet's surface, equivalent to about 14.7 pounds pressing on every square inch. That amount of pressure is called a **bar** (from the Greek *baros*, meaning "weight" or "heavy"). Earth's average atmospheric pressure at sea level is approximately 1 bar. A millibar (mb) is one-thousandth of 1 bar and is more commonly used in meteorology and in weather reports. Underwater, each depth of 10 meters adds 1 bar of pressure. We are largely unaware of Earth's atmospheric pressure because the same pressure exists both inside and outside our bodies, so the force pushing out precisely balances the force pushing in.

Recall from Chapter 8 that the pressure at any point within a planet's interior must be great enough to balance the weight of the overlying layers. The same principle holds true in a planetary atmosphere. The atmospheric pressure on a planet's surface must be great enough to support the weight of the overlying atmosphere. Different forms of matter provide the pressure within a planet's interior and in its atmosphere. In the interior of a solid planet, solid materials exert pressure as they resist being compressed. In a planetary atmosphere, the motions of gas molecules exert enough pressure to support the atmosphere.

Earth's atmosphere consists of several distinct layers (**Figure 9.6**). Those layers are distinguished by the changes in temperature and pressure through the atmosphere. The lowest layer, the one in which humans live and breathe, is called the **troposphere**. It contains 90 percent of Earth's atmospheric mass and is the source of all our weather. At Earth's surface, usually called sea level, the troposphere has an average temperature of 15°C (288 K). Within the troposphere, atmospheric pressure, density, and temperature all decrease as altitude increases. For example, at an altitude of 5.5 km (18,000 feet, a few thousand feet below the

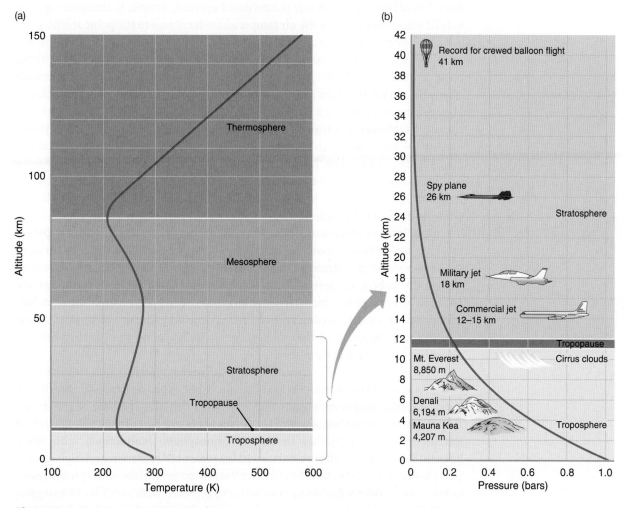

Figure 9.6 These graphs show (a) temperature and (b) pressure plotted for Earth's atmospheric layers as a function of altitude. Most human activities are confined to the bottom layers of Earth's atmosphere.

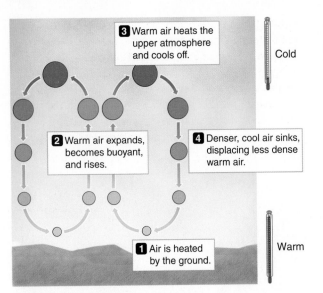

3 Warm air heats the upper atmosphere and cools off.

Cold

2 Warm air expands, becomes buoyant, and rises.

4 Denser, cool air sinks, displacing less dense warm air.

1 Air is heated by the ground.

Warm

Figure 9.7 Atmospheric convection carries thermal energy from the Sun-heated surface of Earth upward through the atmosphere.

summit of Denali in Alaska), the atmospheric pressure and density are only 50 percent of their sea-level values, and the average temperature has dropped to −20°C (253 K). In the lower stratosphere, at an altitude of 12–15 km, where commercial jets cruise, the temperature is −60°C (213 K), and the density and pressure are less than one-fifth what they are at sea level.

The atmosphere is warmer near Earth's surface because the air is closer to the sunlight-heated ground, which warms the air by infrared radiation. The atmosphere is cooler at very high altitudes because there the atmosphere freely radiates its thermal energy into space. In fact, it would get colder with increasing altitude even faster if not for convection. **Figure 9.7** illustrates how convection carries thermal energy upward through Earth's atmosphere. At a given pressure, cold air is denser than warm air. So, when cold air encounters warm air, the denser cold air slips under the less dense warm air, pushing the warm air upward. That convection sets up air circulation between the lower and upper levels of the atmosphere and tends to diminish the temperature extremes caused by heating at the bottom and cooling at the top.

Convection also affects the vertical distribution of atmospheric water vapor. Air's ability to hold water in the form of vapor depends very strongly on the air temperature: the warmer the air, the more water vapor it can hold. The amount of water vapor in the air relative to what the air could hold at a particular temperature is called the relative humidity. Air saturated with water vapor has a relative humidity of 100 percent. As air is convected upward, it cools, limiting its capacity to hold water vapor. When the air temperature decreases to the point at which the air can no longer hold all its water vapor, water begins to condense to tiny droplets or ice crystals. In large numbers those become visible as clouds. When those droplets combine to form large drops, convective updrafts can no longer support them, and they fall as rain or snow. Therefore, most of the water vapor in Earth's atmosphere stays within 2 km of the surface. At an altitude of 4 km, the Mauna Kea Observatories (see the Chapter 6 opening figure) are higher than approximately one-third of Earth's atmosphere, but they lie above nine-tenths of the atmospheric water vapor. That altitude is important for astronomers who observe in the infrared region of the spectrum because water vapor strongly absorbs infrared light. The water in the atmosphere is more often visible as condensed water in the form of clouds and ice.

The layer of the atmosphere above the troposphere and extending upward to an altitude of 50 km above sea level is the stratosphere. The boundary between the troposphere and stratosphere is called the **tropopause**. It varies between 10 and 15 km above sea level, depending on latitude, and is highest at the equator. Little convection takes place in the stratosphere because the temperature no longer decreases with increasing altitude. In fact, the temperature begins to *increase* with altitude because of the ozone layer, which warms the stratosphere by absorbing UV radiation from the Sun.

The region above the stratosphere is the **mesosphere**, which extends from an altitude of 50 km to about 90 km. The mesosphere has no ozone to absorb sunlight, so temperatures once again decrease with altitude. The base of the stratosphere and the upper boundary of the mesosphere are two of the coldest levels in Earth's atmosphere. Higher in Earth's atmosphere, interactions with space become important. At altitudes above 90 km, solar UV radiation and high-energy particles from the solar wind strip electrons from, or **ionize**, atmospheric molecules, causing the temperature once again to increase with altitude. That region, called the **thermosphere**, is the hottest part of the atmosphere. The temperature can reach 1000 K near the top of the thermosphere, at an altitude of 600 km.

The atoms and molecules in the gases within and beyond the thermosphere are ionized by UV photons and high-energy particles from the Sun. That region of ionized atmosphere is called the **ionosphere**, and it not only overlaps the thermosphere but also extends farther into space. The ionosphere reflects certain frequencies of radio waves back to the ground. For example, the frequencies used by AM radio bounce back and forth between the ionosphere and the surface, enabling radio receivers to pick up stations at great distances from the transmitters. Amateur radio operators can communicate with one another around the world by bouncing their signals off the ionosphere.

CHECK YOUR UNDERSTANDING 9.3A

List the layers of the atmosphere in order from nearest to the surface to farthest from the surface: (a) stratosphere; (b) thermosphere; (c) troposphere; (d) ionosphere; (e) mesosphere.

Earth's Magnetosphere

Even farther out than the ionosphere is Earth's magnetosphere, which surrounds Earth and its atmosphere. The magnetosphere is a large region filled with electrons, protons, and other charged particles from the Sun that have been captured by the planet's magnetic field. That region has a radius approximately 10 times that of Earth and fills a volume more than 1,000 times as large as the volume of the planet itself. Magnetic fields affect only moving charges. Charged particles move freely along the direction of the magnetic field but cannot cross magnetic field lines. If they try to move across the direction of the field, they experience a force that causes them to loop around the direction of the magnetic field, as illustrated in **Figure 9.8a**. That force is perpendicular both to the motion of the particle and to the direction of the magnetic field.

If the magnetic field is pinched together at some point, particles moving into the pinch will experience a magnetic force that reflects them back along the direction they came from, creating a sort of "magnetic bottle" that contains the charged particles. If charged particles are located in a region where the field is pinched on both ends, as shown in **Figure 9.8b**, they may bounce back and forth many times. Earth's magnetic field is pinched together at the two magnetic poles and spreads out around the planet.

Earth and its magnetic field are immersed in the solar wind. When the charged particles of the solar wind first encounter Earth's magnetic field, the smooth flow is interrupted and their speed suddenly drops—they are diverted by Earth's magnetic field like a river is diverted around a boulder. As they flow past, some of those charged particles become trapped by Earth's magnetic field, where they bounce back and forth between Earth's magnetic poles, as illustrated in **Figure 9.8c**.

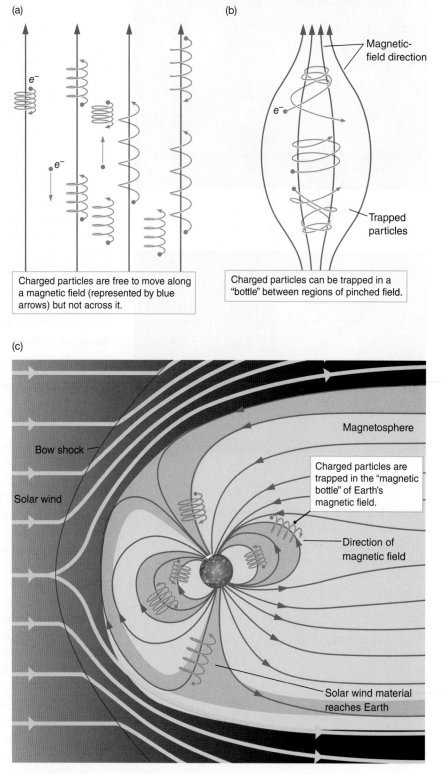

Figure 9.8 (a) Charged particles, here electrons, spiral in a uniform magnetic field. (b) When the field is pinched, charged particles can be trapped in a "magnetic bottle." (c) Earth's magnetic field acts like a bundle of magnetic bottles, trapping particles in Earth's magnetosphere. The bow shock is where the solar wind slows as it meets the magnetosphere. In all these images, the radius of the helix that the charged particle follows is greatly exaggerated.

(a)

e^-

e^-

Charged particles are free to move along a magnetic field (represented by blue arrows) but not across it.

(b)

Magnetic-field direction

e^-

Trapped particles

Charged particles can be trapped in a "bottle" between regions of pinched field.

(c)

Magnetosphere

Bow shock

Solar wind

Charged particles are trapped in the "magnetic bottle" of Earth's magnetic field.

Direction of magnetic field

Solar wind material reaches Earth

(a)

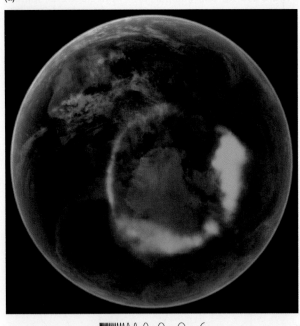

G X U V I R

(b)

G X U V I R

Figure 9.9 Auroras result when particles trapped in Earth's magnetosphere collide with molecules in the upper atmosphere. (a) An auroral ring around Earth's south magnetic pole, as seen from space. (b) Aurora borealis—the "northern lights"—viewed from the ground in Alaska.

Astronomy in Action: Charged Particles and Magnetic Fields

Regions in the magnetosphere that contain especially strong concentrations of energetic charged particles, called **radiation belts**, can be very damaging to both electronic equipment and astronauts. However, leaving the surface of the planet is not necessary to witness the dramatic effects of the magnetosphere. Disturbances in Earth's magnetosphere caused by changes in the solar wind can lead to changes in Earth's magnetic field that are large enough to trip power grids, cause blackouts, and disrupt communications.

Earth's magnetic field also funnels energetic charged particles down into the ionosphere in two rings located around the magnetic poles. Those charged particles (mostly electrons) collide with atoms and molecules such as oxygen, nitrogen, and hydrogen in the upper atmosphere, causing them to glow like the gas in a neon sign. Interactions with different atoms cause different colors. Those glowing rings, called **auroras**, can be seen from space (**Figure 9.9a**). When viewed from the ground (**Figure 9.9b**), auroras appear as eerie, shifting curtains of multicolored light. People living far from the equator are often treated to spectacular displays of the aurora borealis (the "northern lights") in the Northern Hemisphere or the aurora australis in the Southern Hemisphere. When the solar wind is particularly strong, auroras can be seen at lower latitudes, far from their usual zone. Auroras also have been seen on Venus, Mars, all the giant planets, and some moons.

The general structure we have described here is not limited to Earth's atmosphere. The major vertical structural components—troposphere, tropopause, stratosphere, and ionosphere—also exist in the atmospheres of Venus and Mars, as well as in the atmospheres of Titan and the giant planets. The magnetospheres of the giant planets are among the largest structures in the Solar System.

Wind and Weather

Weather is the local day-to-day state of the atmosphere. Local weather is caused by winds and convection. Recall from Chapter 5 that heating a gas increases its pressure, which causes it to push into its surroundings. Those pressure differences cause winds. Winds are the natural movement of air, both locally and on a global scale, in response to variations in temperature from place to place. The air is usually warmer in the daytime than at night, warmer in the summer than in winter, and warmer at the equator than in the polar regions. Large bodies of water, such as oceans, also affect atmospheric temperatures. The strength of the winds is governed by the size of the temperature difference from place to place.

Recall from Chapter 2 that the effect of Earth's rotation on winds—and on the motion of any object—is called the Coriolis effect (see Figure 2.12). As air in Earth's equatorial regions is heated by the warm surface, convection causes that air to rise. The warmed surface air displaces the air above it, which then has nowhere to go but toward the poles. That air becomes cooler and denser as it moves toward the poles, so it sinks back down through the atmosphere. In the process, it displaces the surface polar air, which is forced back toward the equator, completing the circulation. As a result, the equatorial regions remain cooler and the polar regions remain warmer than they otherwise would be. Air moves between the equator and poles of a planet in a pattern known as **Hadley circulation** (**Figure 9.10a**).

On Earth, other factors break up the planetwide flow into smaller Hadley cells. Most Solar System planets and their atmospheres rotate rapidly enough that the Coriolis effect strongly interferes with Hadley circulation by redirecting the horizontal flow (**Figure 9.10b**). The Coriolis effect creates winds that blow predominantly in an east–west direction and are often confined to relatively narrow bands of latitude. Meteorologists call those **zonal winds**. Planets that rotate

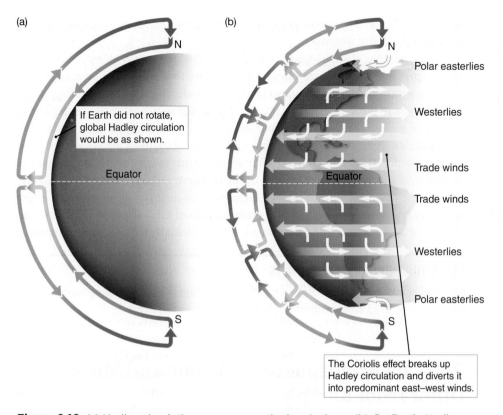

Figure 9.10 (a) Hadley circulation covers an entire hemisphere. (b) On Earth, Hadley circulation breaks up into smaller circulation cells because of the Coriolis effect, which diverts the north–south flow into east–west zonal flow.

faster have a stronger Coriolis effect and stronger zonal winds. Between the equator and the poles in most planetary atmospheres, the zonal winds alternate between winds blowing from the east toward the west (easterlies) and winds blowing from the west toward the east (westerlies).

In Earth's atmosphere, several bands of alternating zonal winds lie between the equator and each hemisphere's pole. That zonal pattern is called Earth's **global circulation** because its extent is planetwide. The best-known zonal currents are the subtropical trade winds—more or less easterly winds that once carried sailing ships from Europe westward to the Americas—and the midlatitude prevailing westerlies that carried them home again.

Embedded within Earth's global circulation pattern are systems of winds associated with large high-pressure and low-pressure regions. A combination of a low-pressure region and the Coriolis effect produces a circulating pattern called **cyclonic motion (Figure 9.11).** Cyclonic motion is associated with stormy weather, including hurricanes. Similarly, high-pressure systems are localized regions where the air pressure is higher than average. Owing to the Coriolis effect, high-pressure regions rotate in a direction opposite to that of low-pressure regions. Those high-pressure circulating systems experience **anticyclonic motion** and are generally associated with fair weather.

Earth has a water cycle in which water from the surface enters the air and later returns to the oceans. When liquid water in Earth's oceans, lakes, and rivers absorbs enough thermal energy from sunlight, it turns to water vapor. The water vapor carries that thermal energy as it circulates throughout the atmosphere, releasing the energy to its surroundings when the water vapor condenses back

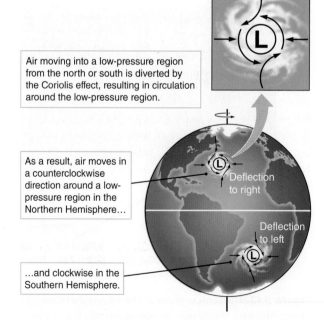

Figure 9.11 As a result of the Coriolis effect, air circulates around regions of low pressure on the rotating Earth.

into rain or snow. That process powers rainstorms, thunderstorms, hurricanes, and other dramatic weather.

For example, Coriolis forces acting on air rushing into regions of low atmospheric pressure create huge circulating systems that result in hurricanes. The conditions must be just right: warm tropical seawater, light winds, and a region of low pressure in which air spirals inward. Sustained winds near the center of the storm can exceed 300 kilometers per hour (km/h), causing widespread damage and fatalities. Tornadoes are small but violent circulations of air associated with storm systems. Dust devils are similar in structure to tornadoes, but generally smaller and less intense and usually occur in fair weather in places such as the deserts of the American Southwest. Diameters of dust devils range from a few meters to a few dozen meters, with average heights of several hundred meters. The lifetime of a typical tornado or dust devil is brief—usually a dozen or so minutes.

CHECK YOUR UNDERSTANDING 9.3B

All weather and wind on Earth are a result of convection in the: (a) troposphere; (b) stratosphere; (c) mesosphere; (d) ionosphere; (e) thermosphere.

....

9.4 The Atmospheres of Venus and Mars Differ from Earth's

The atmospheres of Venus, Earth, and Mars are very different (see Table 9.1). The lower atmosphere of Venus is very hot and dense compared with Earth's, whereas the atmosphere of Mars is very cold and thin. The greenhouse effect has turned Venus hellish, with extremely high temperatures and choking amounts of sulfurous gases. Compared with Venus, the surface of Mars is almost hospitable. Understanding why and how those atmospheres are so different helps us understand how Earth's atmosphere may evolve.

Venus

Venus and Earth are similar enough in size and mass that they were once thought of as sister planets. Indeed, when we used the laws of radiation in Chapter 5 to predict temperatures for the two planets, we concluded that they should be very similar. However, spacecraft visits to Venus in the 1960s revealed that the temperature, density, and pressure of Venus's atmosphere were all much higher than for Earth's atmosphere. Ninety-six percent of Venus's massive atmosphere is carbon dioxide, with only 3.5 percent nitrogen and lesser amounts of other gases. Those atmospheric properties are due to the greenhouse effect and the role of carbon dioxide in retaining the infrared radiation typically emitted by a planetary surface. That thick blanket of carbon dioxide traps the infrared radiation, raising the temperature at the surface of the planet to a sizzling 740 K (872°F) (**Figure 9.12**), hot enough to melt lead. The atmospheric pressure at the surface of Venus is 92 times greater than that at Earth's surface: that is equal to the water pressure at an ocean depth of 900 meters—more than enough to crush the hull of a submarine.

The graph in Figure 9.12 shows that the atmospheric temperature of Venus decreases continuously with altitude throughout the planet's troposphere—similar to Earth, dropping to a low of about 160 K at the tropopause. At an altitude of approximately 50 km, Venus's atmosphere has an average temperature and

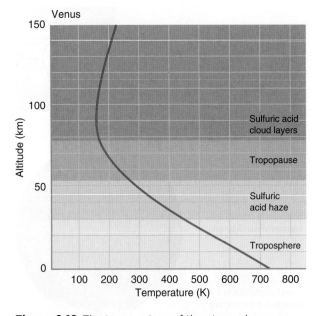

Figure 9.12 The temperature of the atmosphere on Venus primarily decreases as altitude increases, unlike temperatures in Earth's atmosphere, which fall and rise and fall again through the troposphere, stratosphere, and mesosphere (compare Figure 9.6a).

pressure similar to those of Earth's atmosphere at sea level. At altitudes between 50 and 80 km, the atmosphere is cool enough for sulfurous oxide vapors to react with water vapor to form dense clouds of concentrated droplets of sulfuric acid (H_2SO_4). Those dense clouds block the view of the surface of Venus (**Figure 9.13**). Large variations in the observed amounts of sulfurous compounds in the high atmosphere of Venus suggest that the sulfur arises from sporadic episodes of volcanic activity. That finding, along with some hot spots seen near a large shield volcano, strengthens the possibility that Venus is volcanically active.

In the 1960s, radio telescopes and spacecraft with cloud-penetrating radar provided low-resolution views of the surface of Venus. Not until 1975, when the Soviet Union landed cameras there, did scientists get a clear picture of the surface. Those images showed fields of rocks 30–40 centimeters (cm) across and basalt-like slabs surrounded by weathered material. Soviet landers in the 1980s revealed similar landscapes (**Figure 9.14**). Radar images taken by the *Magellan* spacecraft in the early 1990s (see Figure 8.20) produced a global map of the surface of Venus. The high atmospheric temperatures on Venus also mean that neither liquid water nor liquid sulfurous compounds can exist on its surface, leaving an extremely dry lower atmosphere with only 0.01 percent water and sulfur dioxide vapor.

Imagine standing on the surface of Venus. Because sunlight cannot easily penetrate the dense clouds above you, noontime is no brighter than a very cloudy day on Earth. High temperatures and very light winds keep the lower atmosphere free of clouds and hazes. The local horizon can be seen clearly, but strong scattering of light by molecules in the dense atmosphere would greatly soften any view you might have of distant mountains.

Unlike the other Solar System planets, Venus rotates on its axis in a direction opposite to its motion around the Sun. Astronomers call that *retrograde rotation*. Relative to the stars, Venus rotates on its axis once every 243 Earth days. However, a solar day on Venus—the time it takes for the Sun to return to the same place in the sky—is only 117 Earth days. The slow rotation means that Coriolis effects on the atmosphere are small. Global Hadley circulation seldom occurs in planetary atmospheres because other factors, such as planet rotation, break up the planetwide flow into smaller Hadley cells. Because of its slow rotation, however, Venus is an exception, making it the only planet with global circulation close to a classic Hadley pattern (see Figure 9.10a).

The massive atmosphere on Venus transports thermal energy around the planet very efficiently, so the polar regions are only a few degrees cooler than the equatorial regions, with almost no temperature difference between day and night. Because Venus's equator is nearly in the plane of the planet's orbit, seasonal effects are small, producing only negligible changes in surface temperature. Such small temperature variations also mean that wind speeds near the surface of Venus are low, typically about a meter per second, so wind erosion is weaker than that on Earth and Mars. High in the atmosphere, 70 km up, temperature differences are larger, contributing to superhurricane-force winds that reach speeds of 110 m/s (400 km/h), circling the planet in only 4 days. The variation of that high-altitude wind speed with latitude can be seen in the chevron, or V-shaped, cloud patterns.

When the *Pioneer Venus* spacecraft was orbiting Venus during the 1980s, its radio receiver picked up many bursts of lightning static—so many that Venus appears to have a rate of lightning activity comparable to that of Earth. On Venus, as on Earth, lightning is created in the clouds; but Venus's clouds are so high—typically 55 km above the surface of the planet—that the lightning bolts never hit

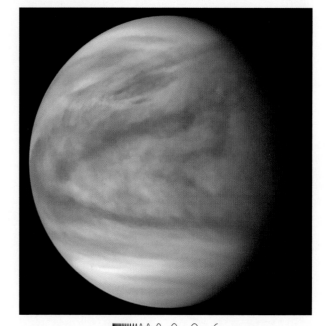

G X **U** V I R

Figure 9.13 This image from *Akatsuki* shows the thick clouds obscuring the view of the surface of Venus.

G X U **V** I R

Figure 9.14 This image of Venus is from the 1982 Soviet *Venera 14* mission. The spacecraft is in the foreground. Note the rocky ground and the orange sky.

Figure 9.15 This true-color image of the surface of Mars was taken by the rover *Spirit*. Without dust, the sky's thin atmosphere would appear deep blue. In this image, windblown dust turns the sky pinkish.

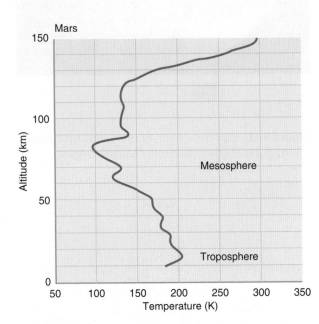

Figure 9.16 The temperature profile of the atmosphere of Mars. Note the differences in temperature and structure between this profile and the profile of the atmosphere of Venus in Figure 9.12.

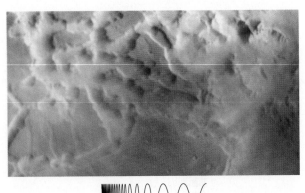

Figure 9.17 Patches of early-morning water vapor fog forming in canyons on Mars.

the ground. The Japanese Space Agency mission *Akatsuki* has been orbiting Venus since late 2015 to study its weather, lightning, and signs of volcanic activity.

Mars

Mars has a stark landscape, colored reddish by the oxidation of iron-bearing surface minerals. The sky is sometimes dark blue but more often has a pinkish color due to windblown dust (**Figure 9.15**). The lower density of the martian atmosphere makes it more responsive than Earth's atmosphere to heating and cooling, so Mars has greater temperature extremes. The surface near the equator at noontime is a comfortable 20°C—a cool room temperature (68°F) on Earth. However, nighttime temperatures typically drop to a frigid −100°C, and during the polar night the air temperature can reach −150°C—cold enough to freeze carbon dioxide out of the air in the form of a dry-ice frost. The temperature profile of the atmosphere of Mars (**Figure 9.16**) has a range of only 100 degrees up to about 125 km. Above that the temperature rises because sunlight is absorbed in the upper atmosphere. The temperature profile of Mars is more similar to that of Earth than that of Venus.

The average atmospheric surface pressure of Mars is equivalent to the pressure at 35 km above sea level on Earth (well into the stratosphere; see Figure 9.6). Mars has no "sea level" because it has no oceans. Surface pressure varies from 11.5 mb in the lowest impact basins of Mars to 0.3 mb at the summit of Olympus Mons. Earth's pressure at sea level is about 1 bar, so the highest pressure on Mars is only 1.1 percent of that. Like Earth, Mars has some water vapor in its atmosphere, but its low temperatures condense much of the water vapor out as clouds of ice crystals. Mars can have early-morning ice fog in the lowlands (**Figure 9.17**) and clouds hanging over the mountains.

In the absence of plants, Mars has only a tiny trace of oxygen, which is crucial to life on Earth. Like Venus, the atmosphere of Mars is composed almost entirely of carbon dioxide (95 percent) and a lesser amount of nitrogen (2.7 percent). The near absence of oxygen means that Mars has very little ozone. Without ozone, solar UV radiation reaches the surface. Those UV rays could be lethal to any surface life-forms, so any life would either need to develop protective layers or be located away from direct exposure on the surface of the planet, such as in caves or below the surface.

The tilt of the rotation axis of Mars is similar to Earth's at present, so both planets have similar seasons. Seasonal effects on Mars, though, are larger for two reasons. First, the elliptical orbit of Mars has a higher eccentricity, so the annual orbital distance of Mars from the Sun varies more than Earth's does. Second, the low density of the martian atmosphere makes it more responsive to seasonal change. The large daily, seasonal, and latitudinal surface temperature differences on Mars often create locally strong winds—some estimated to be greater than 100 m/s (360 km/h). High winds can stir up huge quantities of dust and distribute it around the planet's surface. For more than a century, astronomers have watched the seasonal development of springtime dust storms on Mars. The stronger storms spread quickly and can envelop the planet in a shroud of dust within a few weeks (**Figure 9.18**). Such large amounts of windblown dust can take many months to settle out of the atmosphere. Seasonal movement of dust from one area to another alternately exposes and covers large areas of dark, rocky surface. That phenomenon led some astronomers of the late 19th and early 20th centuries to believe erroneously that they were witnessing the seasonal growth and decay of vegetation on Mars.

The *Viking* landers first noticed dust devils on Mars in 1976. More recently, *Mars Reconnaissance Orbiter* spotted many dust devils that were visible because of the shadows they cast on the martian surface. **Figure 9.19** shows a 20-km-high, 70-meter-wide dust devil. Most martian dust devils leave dark meandering trails behind them where they have lifted bright surface dust, revealing the dark surface rock that lies beneath. Dust devils on Mars—typically higher, wider, and stronger than those on Earth—reach heights of up to 20 km and have diameters ranging from a few dozen to a few hundred meters.

Mars probably had a more massive secondary atmosphere in the distant past. Geological evidence strongly suggests that liquid water once flowed across its surface (Chapter 8), but the low incidence (or possibly cessation) of volcanism and the planet's low gravity, and perhaps the decrease of its magnetic field, was responsible for the loss of much of that earlier atmosphere. Scientists have not yet reached a consensus on how massive the martian atmosphere was in the past.

Recent analysis of five years of data from instruments on the Mars *Curiosity* Rover have indicated the presence of organic (carbon-based) compounds. These could have formed on the surface with or without life, or come by space on comets or meteors. One of these compounds, methane gas, has been detected in Mars' atmosphere, and varies with the seasons. The methane is thought to have originated on or below the surface of Mars.

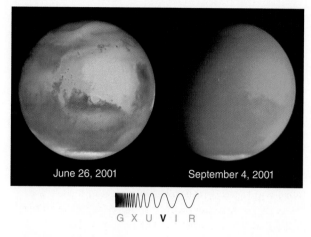

June 26, 2001   September 4, 2001

G X U **V** I R

Figure 9.18 Hubble Space Telescope images show the development of a global dust storm that enshrouded Mars in September 2001. The same region of the planet is shown in both images; surface features are obscured by the thick layer of dust.

Mercury and the Moon

The ultrathin atmospheres of Mercury and the Moon are known as **exospheres**, and they are less than a million-billionth (10^{-15}) as dense as Earth's atmosphere. The recent NASA *Lunar Atmosphere and Dust Environment Explorer* (*LADEE*) mission found helium, argon, and dust in the Moon's exosphere. Other atoms, such as sodium, calcium, and even water-related ions, were seen in Mercury's exosphere by the *Messenger* spacecraft, and they may have been blasted loose from Mercury's surface by the solar wind or micrometeoroids. The exospheres of Mercury and the Moon probably vary with the strength of the solar wind and the atoms of hydrogen and helium they capture from it. Exospheres do not affect local surface temperatures, but astronomers study how exospheres interact with the solar wind.

CHECK YOUR UNDERSTANDING 9.4

Rank, from greatest to smallest, the seasonal variations on (a) Mercury, (b) Venus, (c) Earth, and (d) Mars.

9.5 Greenhouse Gases Affect Global Climates

Climate is the *average* state of an atmosphere, including its temperature, humidity, winds, and so on. Climate describes the planet as a whole, over timescales of years or decades. That is an important distinction from weather, which is the state of an atmosphere at any given time and place. The study of climate change on Earth and Mars is not new to the 21st century. Nineteenth-century scientists found evidence of past ice ages and knew that Earth's climate had been very different earlier in its history. Observations of changes in the martian ice caps led to speculation about whether Mars also had ice ages. In this section, we look at the natural factors that can cause climates to change on planets, and we then examine the additional factors that affect Earth.

G X U **V** I R

Figure 9.19 This dust devil on Mars was imaged by the *Mars Reconnaissance Orbiter*.

Factors That Can Cause Climate Change on a Planet

Scientists study the astronomical, geological, and (on Earth) biological mechanisms controlling climate on the planets. Astronomical mechanisms that influence changes in planetary temperature include changes in the Sun's energy output, which has increased slowly as the Sun ages, and possibly changes in the galactic environment as the Sun travels in its orbit around the center of the Milky Way. Scientists have suggested that sporadic bursts of gamma rays or cosmic rays (fast-moving protons) from distant exploding stars could interact with planetary atmospheres. Those mechanisms would affect *all* planets in the Solar System at the same time.

Other astronomical mechanisms relevant to climate change and specific to each planet are the Milankovitch cycles, named for geophysicist Milutin Milanković (1879–1958). Milankovitch cycles may arise because a planet's energy balance can be related to a planet's motion and affected by periodic changes in its orbital eccentricity, the tilt of its rotational axis, and its precession. Recall from Working It Out 5.4 that many factors affect a planet's energy balance and therefore its temperature. If a planet's orbit becomes more eccentric, the amount of energy it receives from the Sun will vary more over its year. If the tilt of a planet increases, its seasons will become more extreme, and its temperature variation during the year will increase. The precession cycle affects which hemisphere is pointed toward the Sun at different times of the elliptical orbit, so that one hemisphere may have longer winters and the other, longer summers.

The tilt (obliquity) of Earth's axis varies from 22.1° to 24.5°, and Earth's relatively large Moon keeps that tilt from changing more than that. In contrast, the moons of Mars are small, and the gravitational influence of Jupiter is a greater factor on Mars. The tilt of Mars is thought to vary from 13° to 40° or possibly more. Given the precession of Mars and its more eccentric orbit, which creates differences in season length between its northern and southern hemispheres, Mars may have had very large swings in its climate throughout its history as the obliquity changed.

A second set of factors that can affect climate is the geological activity of a planet. Volcanic eruptions can produce dust, aerosol particles, clouds, or hazes that block sunlight over the entire globe and lower the temperature. Impacts by large objects can kick up sunlight-blocking particles. Tectonic activity also can affect climate. On Earth, for example, the shifting of the plates has led to different configurations of the oceans and the shifting continents, thus affecting global atmospheric and oceanic circulatory patterns. The albedo of a planet can increase if it has more clouds and ice or can decrease if ice melts or is covered by volcanic ash. Changes also may arise from variations in carbon cycles. On Earth, long-term interactions of the oceans, land, and atmosphere affect the levels of greenhouse gases such as carbon dioxide and water vapor.

A third set of mechanisms that trigger climate changes is biological. Over billions of years on Earth, photosynthesis by bacteria and later by plants removed carbon dioxide from the atmosphere and replaced it with oxygen (Section 9.3). Biological (and geological) activity on Earth can produce methane, a strong greenhouse gas. Certain microorganisms produce methane as a metabolic by-product. For example, bubbles rising to the surface of a stagnant pond—swamp gas—contain biologically produced methane. Methane also is emitted from the guts of grain-fed livestock (and, in the past, from some large dinosaurs) as well as from termites. Another biological effect on climate could come from phytoplankton. If the oceans get more solar energy and warm up because of one of the astronomical mechanisms, the phytoplankton in the ocean may grow faster, leading to the release of more aerosols from sea spray and the formation of more clouds, which increases albedo. Finally, human activities are triggering some major changes, as discussed in the next subsection.

In short, many factors affect the temperature of a planet. Earth's climate is the most complicated of those of the terrestrial planets because Earth is the most geologically and biologically active. How do scientists sort out all those factors? They use the scientific method. Scientists create mathematical models to simulate the general circulation and energy balance of a planet, incorporating all the appropriate factors. The goal is to create a global climate model that reproduces the empirical data from observations of a planet. Once the model correctly predicts past and present climate, it can be used to predict future climate. The first simple climate models for Earth were run on the earliest computers in the 1950s and 1960s. (Fifty years later, scientists are impressed with how well a climate model from 1967 quantified the factors affecting Earth's climate.) One set of models developed at NASA's Goddard Institute for Space Studies in the 1970s was a spinoff of a program originally designed to study Venus. The insights from comparative planetology are important for producing better models that will aid in scientific predictions of climate change on Earth.

Climate Change on Earth

Paleoclimatology is the study of changes in Earth's climate throughout history. Scientists use evidence from geology and paleontology, such as sediments, ice sheets, rocks, tree rings, coral, shells, and fossils, to get data on Earth's past climate. Researchers have found that Earth's climate has lengthy temperature cycles, some lasting hundreds of thousands of years and some tens of thousands of years. As you can see in the middle plot in **Figure 9.20**, Earth has had periods

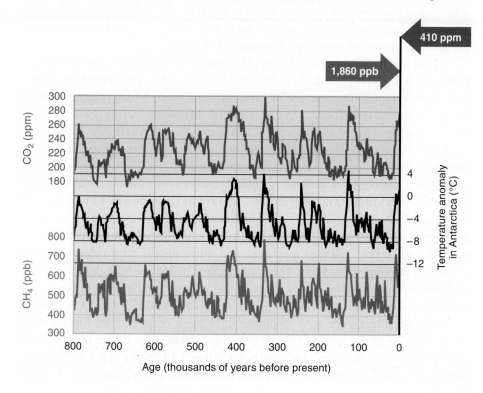

Figure 9.20 Global variations in carbon dioxide (CO_2), temperature, and methane (CH_4) concentrations over the past 800,000 years of Earth's history. Notice the multiple *y*-axes on this graph. The axis on the right relates to the temperature data (black); the axes on the left relate to the CO_2 (blue) and CH_4 (red) data. These data sets have been plotted on the same graph to make the similarities and differences easier to see. The low points correspond to ice ages. ppb = parts per billion; ppm = parts per million. 2017 values for CH_4 (red) and CO_2 (blue), indicated by the arrows, are larger than at any time in the past 800,000 years.

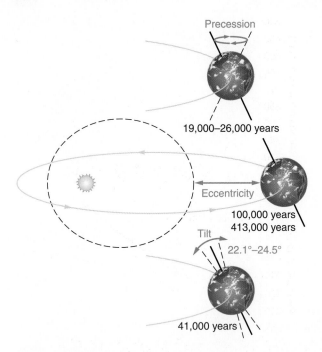

Figure 9.21 The Milankovitch cycles for Earth. The precession of Earth and the rotation of its elliptical orbit combine to yield a cyclic variation of about 21,000 years, its eccentricity varies with two cycles of about 100,000 and 413,000 years, and its tilt varies in cycles of 41,000 years. (Not to scale.)

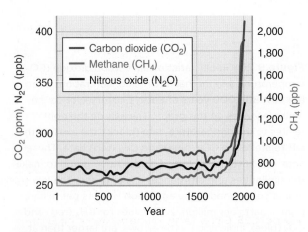

Figure 9.22 Concentrations of greenhouse gases from the year 1 to 2005, showing the increases beginning at the time of industrialization.

of colder temperatures, known as ice ages. Earth's atmosphere is so sensitive to even small temperature changes that a drop of only a few degrees in the mean global temperature can plunge the planet's climate into an ice age. Those oscillations in the mean global temperature are far smaller than typical geographic or seasonal temperature variations.

Periodic Milankovitch cycle changes in Earth's orbit correspond to some temperature cycles. As shown in **Figure 9.21**, Earth's axial tilt varies in cycles of 41,000 years, the eccentricity of Earth's orbit varies with two cycles of about 100,000 and 413,000 years, and the time of the year when Earth is closest to the Sun varies in cycles of about 21,000 years. Global climate models using those Milankovitch cycles have replicated much of the observed paleoclimatology data. Temperature changes that are not periodic may have been triggered internally by volcanic eruptions or long-term interactions between Earth's oceans and its atmosphere or by other factors already mentioned.

Some scientists think that at least four of the five major extinction events in Earth's history arose from changes in the level of greenhouse gases. Those mass extinctions occurred 439 million, 364 million, 250 million, and 200 million years ago. The increase in carbon dioxide emitted by active volcanoes, along with methane released from the ground, led to ocean acidification and acid rain and perhaps reduced oxygen in the ocean. Most species died off. An increase in CO_2 from volcanic activity may also have been a factor, along with the impacting asteroid discussed in Chapter 8, for the fifth major extinction, which led to the death of the dinosaurs 65 million years ago.

If Earth's climate has been changing naturally for most of its history, why are scientists especially concerned about the current trend in global climate? Figure 9.20 shows the carbon dioxide levels (top), methane levels (bottom), and temperature (middle) of Earth's atmosphere over the past 800,000 years, obtained from measuring deep ice cores in Antarctica. Notice that those three factors are correlated: when one rises, so do the others. Those data show the naturally occurring ranges since before the first humans existed. The temperature difference between ice ages and interglacial periods is only 10°C–15°C, and those changes are gradual, occurring over tens of thousands of years.

Two major changes have taken place on Earth during the past 150 years. First, the industrial revolution led to an increase in the production of greenhouse gases, especially from the burning of fossil fuels, which releases carbon dioxide into the atmosphere. In 1896, Svante Arrhenius (1859–1927), a Nobel Prize–winning chemist from Sweden, calculated that CO_2 released from burning fossil fuels could increase the greenhouse effect and raise Earth's surface temperature. The second change has been the rapid growth in human population. When populations increase, people burn more forests to clear land for agriculture and industry, reducing the amount of CO_2 absorbed by photosynthesizing plants, increasing CO_2 emissions, and locally changing Earth's albedo. A larger population means more agricultural soil that releases nitrous oxide, more livestock that releases methane, and more people who release carbon dioxide by burning fossil fuels. The data graphed in **Figure 9.22** show that concentrations of those greenhouse gases in the atmosphere have been increasing since the industrial revolution. The CO_2 level has risen and is higher than any of the levels seen in Figure 9.20. Zooming in to more recent times, **Figure 9.23** shows the rise in the level of CO_2 plotted with the average global temperature on Earth to show that they rise together. **Figure 9.24** shows that levels of carbon dioxide, nitrous oxide, and methane have steadily increased over the past decades.

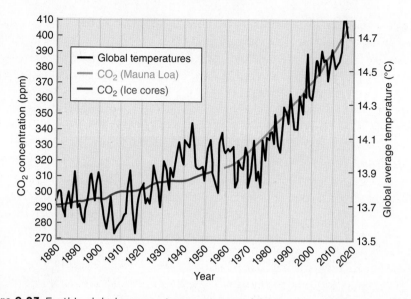

Figure 9.23 Earth's global average temperature and CO_2 concentrations since 1880. This graph shows that global temperatures are climbing along with concentrations of carbon dioxide. Annual variations in atmospheric CO_2 arises from seasonal variations in plant life and fossil fuel use, whereas the overall steady climb is due to human activities.

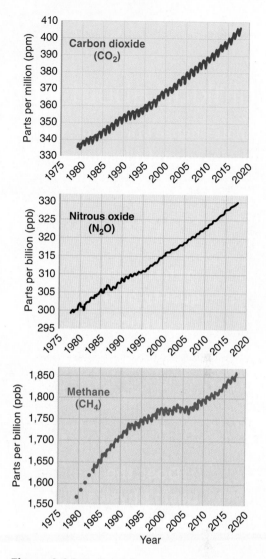

Figure 9.24 National Oceanic and Atmospheric Administration (NOAA) plots of global average amounts of the major greenhouse gases.

Nearly all climatologists accept the computer models indicating that this trend represents the beginning of a long-term change in temperature caused by the buildup of human-produced greenhouse gases. That anthropogenic (human-caused) change is happening much faster than the changes seen in Figure 9.20. Even an average increase of a few degrees can change the climate from an ice age to an interglacial period. Earth's atmosphere is a delicately balanced mechanism, and its climate is a complex system within which tiny changes can produce enormous and often unexpected results. To add to the complexity, Earth's climate is intimately tied to ocean temperatures and currents. Ocean currents are critical in transporting energy from one part of Earth to another, and how increased temperatures may affect those systems is uncertain. Warmer oceans evaporate more, leading to wetter air, which can mean more intense summer and winter storms (including more snow). We see examples of that connection in the periodic El Niño and La Niña conditions, in which small shifts in ocean temperature cause much larger global changes in air temperature and rainfall. **Figure 9.25** shows that in the past several decades, however, natural factors have been overwhelmed by the contribution from **anthropogenic climate change**: the release of greenhouse gases into the atmosphere from the burning of fossil fuels and other human activities. Models including only natural causes fail to fit the data as well as models that include both human activities and natural factors. The **Process of Science Figure** discusses how scientists think about such complex issues.

Changes in climate affect where plants and animals can live, and scientists have observed that the dates and locations of breeding, migration, hibernation, and so on, have changed. Agricultural growing seasons and pollination also are affected, as is the availability of freshwater. The melting of mountain glaciers and polar sea ice from the increase in temperature is already being observed. The levels of the oceans are predicted to rise not only from melted ice in Greenland and Antarctica but also from the thermal expansion of the water as the ocean

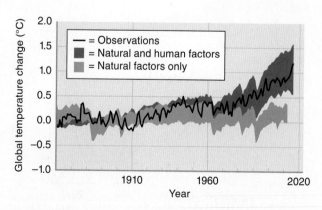

Figure 9.25 Global temperature variations on Earth since 1900, overlaid with model results that include only natural causes of temperature change (pink) and those that include both natural and human factors (purple).

Thinking about Complexity

Climate change is a complex scientific issue. When confronted with complex science, there are several questions you should ask.

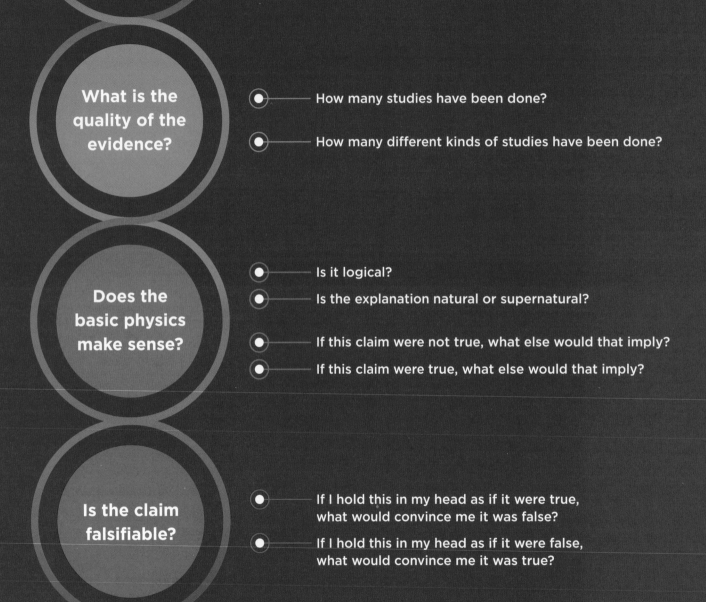

What is the quality of the evidence?

- How many studies have been done?
- How many different kinds of studies have been done?

Does the basic physics make sense?

- Is it logical?
- Is the explanation natural or supernatural?
- If this claim were not true, what else would that imply?
- If this claim were true, what else would that imply?

Is the claim falsifiable?

- If I hold this in my head as if it were true, what would convince me it was false?
- If I hold this in my head as if it were false, what would convince me it was true?

Scientific issues become remarkably complex when they relate to policy decisions with a broad reach. Keeping an open mind in such cases means thinking carefully about the quality of the evidence, as well as what data would make you change your mind. If your mind can*not* be changed, then you are not making decisions based on evidence.

temperature increases. That is a serious problem for the many people who live in coastal or low-lying regions, and it is already an issue in Miami. A warmer Arctic can lead to a greater release of methane from the permafrost. Less ice and snow can decrease Earth's albedo, allowing more sunlight to reach the surface, although albedo may rise because of an increase in cloud cover, caused by more water in the atmosphere.

The processes are so complex that scientists still cannot predict accurately all the long-term outcomes of small changes that humans are now making to the composition of Earth's atmosphere and to Earth's albedo. In a real sense, we are *experimenting* with Earth. We are asking the question, What happens to Earth's climate if we steadily increase the number of greenhouse molecules in its atmosphere? We do not yet know the full answer, but we are already seeing some of the consequences.

CHECK YOUR UNDERSTANDING 9.5

Over the past 800,000 years, ice-core data indicate which of the following are correlated? (Choose all that apply.) (a) temperature; (b) CO_2 levels; (c) methane levels; (d) the size of the ozone hole

. . . .

Origins Our Special Planet

Why are we here on Earth instead of elsewhere in the Solar System? The habitable zone of a planetary system (Chapter 7) is the range of distances from its star within which a planet could have a surface temperature such that large amounts of water could exist in liquid form. The temperature of a planet, though, can depend on more than just distance from the star. Temperature can change because of many factors, such as orbital variations, the atmospheric greenhouse effect, and the planet's ability to hold on to its atmosphere. Mercury and the Moon were too small to hold on to either their primary or their secondary atmospheres, leaving them as essentially airless rocks. Venus, Earth, and Mars all had some liquid water at one time, so early in the history of the Solar System all three might have been classified as habitable. Their atmospheres were probably similar then, too, before the atmosphere of Mars escaped, the atmosphere of Venus was heated by the greenhouse effect, and the atmosphere of Earth was oxygenated. Primitive life may even have developed on Venus and Mars at the same time it developed on Earth, about a billion years after the Solar System formed.

Ultimately, the three planets evolved differently, and only Earth now has the liquid water vital for our biology. Venus receives more energy from the Sun and was slightly warmer than the young Earth. The young Venus might have been too hot for liquid water to form oceans, or any liquid water might have quickly evaporated. Because Venus had a lot of water vapor in its atmosphere and lacked a liquid ocean to store carbon dioxide, the greenhouse effect created a thicker atmosphere. That thicker atmosphere fed back to a stronger greenhouse effect, which created an even thicker atmosphere. With that resulting runaway greenhouse effect, Venus became just too hot: the evaporated water molecules broke apart, the hydrogen escaped to space, and any water cycle was destroyed. Primitive bacteria may have lingered in water vapor in the clouds on Venus, but life did not evolve into anything more complex.

Mars is smaller and less massive than Venus or Earth, so it has a weaker gravity (**Figure 9.26**). The red planet has a larger orbit than Venus and Earth and so receives less energy from the Sun. Over time, much of its atmosphere escaped and was not replaced by volcanic emissions, and the atmospheric pressure became too low to maintain liquid water. Whereas Venus was too hot, the young Mars was too cold. Back then it had a thicker atmosphere and liquid rain. Images of the surface of Mars show flood basins, indicating huge rivers in the past. The process that prevented Earth from becoming a Venus-like hothouse continued further on Mars, and the temperature fell until the water froze. Hints of subsurface water exist on Mars, and perhaps some form of martian bacteria will be found beneath the ground.

Thus, our Solar System contains astronomical evidence of how the greenhouse effect can influence planetary atmospheres, including Earth's. Planetary scientists view those other planets as a cautionary tale, showing the results of varied "doses" of greenhouse gases. Only Earth stayed "just right" and could retain the liquid oceans in which more complex life evolved. We owe our lives to the blanket of atmosphere that covers the planet. The study of the Solar System reveals that Earth is maintained by the most delicate of balances. Over billions of years, life has shaped Earth's atmosphere, and today, through the activities of humans, life is reshaping the planet's atmosphere once again. Human civilization is much younger than Earth and has been brief in comparison with the cycles of climate on the planet. The past 10,000 years has been a relatively stable period of Earth's climate: many people argue that this stability enabled agriculture and civilization to develop. Scientists are uncertain what will happen to agriculture, civilization, or the planet itself if the climate undergoes fast, intense changes. The other planets in the Solar System are not places where billions of people from our planet can live. Earth is the only planet suitable for human life.

Figure 9.26 Physical and orbital differences among Venus, Earth, and Mars cause the planets to have very different atmospheres.

NASA's MAVEN Spacecraft Unlocks the Mystery of Mars's Lost Atmosphere

By JAY BENNETT

https://popularmechanics.com/space/moon-mars/a25898/nasa-maven-mars-lost-atmosphere/

Mars is about 4.5 billion years old, and after it formed out of the dust and debris of the early Solar System, it likely had a nice thick atmosphere with clouds that looked a bit like Earth. Temperatures would have been much higher than the current average of about negative 80 degrees Fahrenheit, and liquid water could have flowed over the surface of our planetary neighbor. It is very possible that basic life existed on the surface of Mars sometime in the planet's early history.

But that's all gone now. All that remains is a desolate freezing wasteland with an atmosphere too thin to insulate the planet, and any remaining liquid or life has been driven deep underground. For the last 4 billion years, Mars is thought to have had an ambient surface temperature below freezing, as determined by studying Martian meteorites that make it to Earth. The question is, what happened?

"We've determined that most of the gas ever present in the Mars atmosphere has been lost to space," said Bruce Jakosky in a press release, principal investigator for the Mars Atmosphere and Volatile Evolution Mission (MAVEN) spacecraft run by the Laboratory for Atmospheric and Space Physics

(LASP) at the University of Colorado Boulder. A paper detailing the team's latest findings was published in *Science* today, and the research shows that a whopping 65 percent of the argon that was ever in Mars's atmosphere has been blown out into space.

Argon was chosen for the study because it is a noble gas, meaning it virtually never reacts chemically. This is important because solar wind is only one way that a planet's atmosphere can dissipate. Rocks and minerals on the surface can chemically react with the gases of a planet's atmosphere and siphon them away, trapping the gas inside the surface of the planet, for example. Because argon won't react, it can only be stripped away from the planet by high-energy particles blowing from the sun, a process called "sputtering."

After the CU team determined the amount of argon that had been lost due to

sputtering, they were able to measure the amount of other gases that were lost to solar winds as well, including CO_2 which makes up 95 percent of the Martian atmosphere. In the early ages of the Solar System, the sun emitted much more volatile ultraviolet radiation and solar winds, and much of the atmosphere was likely blasted away during this time.

There are some wild ideas out there to artificially allow Mars to re-form a thick atmosphere, such as launching a magnetic shield in front of the Red Planet that would block and divert solar particles, an idea recently put forth by NASA Planetary Science Director Jim Green. It might just be possible in the centuries to come, but the first astronauts to walk on Mars are going to have to make sure they keep their spacesuits on tight.

Figure 9.27

1. Why did the scientists specifically study argon on Mars?
2. What factors led to Mars losing its atmosphere?
3. What is meant by *volatile ultraviolet radiation*—is that the same as volatile gases?
4. Why would astronauts on Mars need to "keep their spacesuits on tight"?
5. Do a news search for "Mars atmosphere." What are the most recent discoveries?

Source: Jay Bennett, "NASA's MAVEN Spacecraft Unlocks the Mystery of Mars's Lost Atmosphere," PopularMechanics.com, March 31, 2017. Reprinted with permission.

Summary

Earth, Venus, and Mars are warmer than they would be from solar illumination alone. Earth's atmosphere is thick enough to warm the surface to life-sustaining temperatures, but not so thick that Earth becomes overheated. Earth, Venus, and Mars all have significant atmospheres that are different from the original atmospheres they captured when they formed. Their atmospheres are complex, both in chemical composition and in physical characteristics such as temperature and pressure. The climates of Earth, Venus, and Mars are all determined by their individual atmospheres. The atmospheres of Earth, Venus, and Mars have different chemical compositions. They, in turn, led to dramatic differences in temperature and pressure. Life has altered Earth's atmosphere several times, most notably in the distant past from an increase in the amount of oxygen in the atmosphere and in modern times from an increase in greenhouse gases. Mars and Venus might have been habitable early in the history of the Solar System, but now only Earth has substantial liquid water on its surface.

LG 1   **Identify the processes that cause primary and secondary atmospheres to be formed, retained, and lost.** Planetary atmospheres evolve. The primary atmospheres consisted mainly of hydrogen and helium captured from the protoplanetary disk. The terrestrial planets lost their primary atmospheres soon after the planets formed. Secondary atmospheres were created by volcanic gases and from volatiles brought in by impacting comets and asteroids. Planetary bodies must have enough mass to hold on to their atmospheres.

LG 2   **Compare the strength of the greenhouse effect on Earth, Venus, and Mars and how it contributes to differences among the atmospheres of those planets.** Earth, Venus, and Mars have naturally occurring greenhouse gases, which increase the average surface temperature of each planet. The amount by which those greenhouse gases raise the temperature of a planet depends on the number of greenhouse gas molecules in the atmosphere. The differences in global temperatures among those planets can be explained in part by their distances from the Sun. However, their having different compositions and atmospheric densi-

ties is a highly significant factor in determining their global temperatures. The atmospheric greenhouse effect keeps Earth from freezing, but it turns Venus into an inferno.

LG 3   **Describe the layers of the atmospheres on Earth, Venus, and Mars.** The atmospheres of Earth, Venus, and Mars have different temperatures, pressures, and compositions. Earth's atmosphere, in particular, has many layers. The layers are determined by the vertical variations in temperature and absorption of solar radiation throughout the atmosphere. Temperature and pressure decrease with altitude in the tropospheres of Earth, Venus, and Mars. Earth's magnetosphere shields the planet from the solar wind. Venus has a massive, hot atmosphere of carbon dioxide and sulfur compounds. Venus has surprisingly fast winds for a slowly rotating planet. Mars has a thin, cold, carbon dioxide atmosphere that may have been much thicker in the past. Earth's atmosphere is thicker than that of Mars but thinner than that of Venus and is composed primarily of nitrogen.

LG 4   **Explain how Earth's atmosphere has been reshaped by the presence of life.** The oxygen levels in Earth's atmosphere have been enhanced through photosynthesis by bacteria and then by plants. Increased oxygen levels led to the formation of the ozone layer and to the development of more advanced forms of life.

LG 5   **Describe the evidence that shows Earth's climate is changing and how comparative planetology contributes to a better understanding of those changes.** Astronomical, geological, and biological processes can lead to large changes in the climate of planets. The study of climate on the terrestrial planets expands scientists' knowledge of Earth's past, present, and future conditions. Large variations in global temperature over the past 800,000 years correlate strongly with the number of greenhouse molecules in the atmosphere. The current level of greenhouse gases in Earth's atmosphere is higher than any seen during that period and correlates with a later increase in temperature.

• Was Venus once hospitable to life? Earlier in the history of the Solar System, the young Sun was fainter and Venus was cooler. One group is using three-dimensional climate simulations to model conditions on a young Venus and concluded Venus could have had moderate temperatures if it was rotating slowly but in a prograde direction. Another group modeled the conditions on early Venus and concluded that Venus may have had an ocean of liquid water early in its history, if it had the right balance of cloud cover and the same amount of carbon dioxide as today. Understanding the differences in the evolution of Venus and Earth is of great interest to astronomers as they detect exoplanets of that size near the edge of their stars' habitable zone.

• Will humans find a way to slow or stop the rise in greenhouse gases on Earth? Climate scientists worry that current changes are abrupt compared with the natural cycles of changes in climate that take place gradually over thousands of years. Will nations reach an agreement to reduce the production of those gases, as they did to reduce the use of chemicals that created the ozone hole?

Questions and Problems

Test Your Understanding

1. Place in chronological order the following steps in the formation and evolution of Earth's atmosphere.
   a. Plant life converts carbon dioxide (CO_2) to oxygen.
   b. Hydrogen and helium are lost from the atmosphere.
   c. Volcanoes, comets, and asteroids increase the inventory of volatile matter.
   d. Hydrogen and helium are captured from the protoplanetary disk.
   e. Oxygen enables the growth of new life-forms.
   f. Life releases CO_2 from the subsurface into the atmosphere.

2. On which of the following planets is the atmospheric greenhouse effect strongest?
   a. Venus
   b. Earth
   c. Mars
   d. Mercury

3. The oxygen molecules in Earth's atmosphere
   a. were part of the primary atmosphere.
   b. arose when the secondary atmosphere formed.
   c. are the result of life.
   d. are being rapidly depleted by the burning of fossil fuels.

4. The differences in the climates of Venus, Earth, and Mars are caused primarily by
   a. the composition of their atmospheres.
   b. their relative distances from the Sun.
   c. the thickness of their atmospheres.
   d. the time at which their atmospheres formed.

5. The words *weather* and *climate*
   a. mean essentially the same thing.
   b. refer to very different timescales.
   c. refer to very different-sized scales.
   d. both b and c

6. Less massive molecules tend to escape from an atmosphere more often than more massive molecules because
   a. the gravitational force on them is less.
   b. they are moving faster.
   c. they are more buoyant.
   d. they are smaller, so they experience fewer collisions on their way out.

7. Venus is hot and Mars is cold primarily because
   a. Venus is closer to the Sun.
   b. Venus has a much thicker atmosphere.
   c. the atmosphere of Venus is dominated by CO_2, but the atmosphere of Mars is not.
   d. Venus has stronger winds.

8. Studying climate on other planets is important to understanding climate on Earth because (select all that apply)
   a. the underlying physical processes are the same on every planet.
   b. other planets offer a variety of extremes to which Earth can be compared.
   c. comparing climates on other planets helps scientists understand which factors are important.
   d. other planets can be used to test atmospheric models.

9. The atmosphere of Mars is often pink-orange because
   a. it is dominated by carbon dioxide.
   b. the Sun is at a low angle in the sky.
   c. Mars has no oceans to reflect blue light to the sky.
   d. winds lift dust into the atmosphere.

10. Auroras are the result of
    a. the interaction of particles from the Sun and Earth's atmosphere and magnetic field.
    b. upper-atmosphere lightning strikes.
    c. the destruction of stratospheric ozone, which leaves a hole.
    d. the interaction of Earth's magnetic field with Earth's atmosphere.

11. The stratospheric ozone layer protects life on Earth from
    a. high-energy particles from the solar wind.
    b. micrometeorites.
    c. ultraviolet radiation.
    d. charged particles trapped in Earth's magnetic field.

12. Hadley circulation is broken into zonal winds by
    a. convection from solar heating.
    b. hurricanes and other storms.
    c. interactions with the solar wind.
    d. the planet's rapid rotation.

13. The _____ of greenhouse gas molecules affects the temperature of an atmosphere.
    a. percentage
    b. fraction
    c. number
    d. mass

14. Over the past 800,000 years, Earth's temperature has closely tracked with
    a. solar luminosity.
    b. oxygen levels in the atmosphere.
    c. the size of the ozone hole.
    d. carbon dioxide levels in the atmosphere.

15. Convection in the _____ causes weather on Earth.
    a. stratosphere
    b. mesosphere
    c. troposphere
    d. ionosphere

Thinking about the Concepts

16. Primary atmospheres of the terrestrial planets were composed almost entirely of hydrogen and helium. Explain why they contained only those gases and not others.

17. How were the secondary atmospheres of the terrestrial planets created?

18. Nitrogen, the principal gas in Earth's atmosphere, was not a significant component of the protostellar disk from which the Sun and planets formed. Where did Earth's nitrogen come from?

19. What are the likely sources of Earth's water?

20. In what way is the atmospheric greenhouse effect beneficial to terrestrial life?

21. In what ways does plant life affect the composition of Earth's atmosphere?

22. What is the difference between ozone in the stratosphere and ozone in the troposphere? Which is a pollutant, and which protects terrestrial life?

23. What is the principal cause of winds in the atmospheres of the terrestrial planets?

24. Global warming appears to be responsible for increased melting of the ice in Earth's polar regions.
    a. Why does the melting of Arctic ice, which floats on the Arctic Ocean, *not* affect the level of the oceans?
    b. How is the melting of glaciers in Greenland and Antarctica affecting the level of the oceans?

25. Why can we not get a clear view of the surface of Venus, as we have so successfully done with the surface of Mars?

26. What is the evidence that the greenhouse effect exists on Earth, Venus, and Mars?

27. Explain why surface temperatures on Venus hardly vary between day and night and between the equator and the poles.

28. Why do scientists think that Mars and Venus were once more habitable but no longer are?

29. The last step in the Process of Science Figure is one that anyone can carry out about any complex issue. Write down your current take on the issue of anthropogenic climate change: do you accept the evidence? Then write down a piece of scientific evidence that would persuade you to change your mind. This exercise may help you to think critically about any issue.

30. Given the current conditions on Venus and Mars, which planet might be easier to engineer to make it habitable to humans? Explain.

Applying the Concepts

31. Study Figure 9.20.
    a. Are the axes linear or logarithmic?
    b. Compare the CO_2 levels (top) with the temperature relative to present (middle). How would you describe the relationship between the graphs?
    c. How much higher is the current CO_2 level than the previous highest value?

32. Study Figure 9.22.
    a. According to the figure, when (approximately) did greenhouse gases begin rising exponentially?
    b. These graphs show that several greenhouse gases have behaved similarly in recent times. Was that true in the past (in general)?
    c. Speculate on possible causes for the common behavior of greenhouse gases in modern times.

33. Why do commercial jet planes fly at the altitudes shown in Figure 9.6b?

34. Commercial jets are pressurized. If you take an unopened bag of chips on a commercial jet airplane, the bag puffs up as you travel to the cruising altitude of 14 km.
    a. Is the pressure in the cabin higher or lower than the pressure on the ground?
    b. If a second bag of chips were attached to the outside of the plane, which bag would puff up more? About how much more (assume an unbreakable bag)? (See Figure 9.6b.)

35. Atmospheric pressure is caused by the weight of a column of air above you pushing down. At sea level on Earth, that pressure is equal to 10^5 newtons per square meter (N/m^2).
    a. Estimate the total force on the top of your head from that pressure. (Note: Force = pressure × area.)
    b. Recall that the acceleration due to gravity is 9.8 m/s^2. If the force in part (a) were caused by a kangaroo sitting on your head, what would the mass of the kangaroo be?
    c. Assume that a typical kangaroo has a mass of 60 kg. How many kangaroos would have to be sitting on your head to equal the mass of the extremely massive kangaroo in part (b)?
    d. Why are you *not* crushed by that astonishing force on your head?

36. Repeat the calculations in question 35 for Venus.

37. Repeat the calculations in question 35 for Mars.

38. Increasing the temperature of a gas inside a closed, rigid container increases the pressure. (That is why you should not put an unopened can of soup directly on the stove!) Explain how Figure 9.6b shows that phenomenon at work in Earth's atmosphere.

39. The total mass of Earth's atmosphere is 5×10^{18} kg. Carbon dioxide (CO_2) makes up about 0.06 percent of Earth's atmospheric mass.
    a. What is the mass of CO_2 (in kilograms) in Earth's atmosphere?
    b. The annual global production of CO_2 is now estimated to be 3×10^{13} kg. What annual fractional increase does that value represent?
    c. The mass of a molecule of CO_2 is 7.31×10^{-26} kg. How many molecules of CO_2 are added to the atmosphere each year?
    d. Why does an increase in CO_2 have such a big effect, even though it represents a small fraction of the atmosphere?

40. Wind's ability to erode the surface of a planet is related in part to the wind's kinetic energy.
    a. Compare the kinetic energy of a cubic meter of air at sea level on Earth (mass, 1.23 kg) moving at a speed of 10 m/s with a cubic meter of air at the surface of Venus (mass, 64.8 kg) moving at 1 m/s.
    b. Compare the kinetic-energy value you determined for Earth in part (a) with that of a cubic meter of air at the surface of Mars (mass, 0.015 kg) moving at a speed of 50 m/s.
    c. Why do you think more evidence of wind erosion on Earth is not apparent?

41. Suppose you seal a rigid container that has been open to air at sea level when the temperature is 0°C (273 K). The pressure inside the sealed container is now equal to the outside air pressure: 10^5 N/m^2.
    a. What would the pressure inside the container be if it were left sitting in the desert shade, where the surrounding air temperature was 50°C (323 K)?
    b. What would the pressure inside the container be if it were left sitting out in an Antarctic night, where the surrounding air temperature was −70°C (203 K)?
    c. What would you observe in each case if the walls of the container were not rigid?

42. Oxygen molecules (O_2) are 16 times as massive as hydrogen molecules (H_2). Carbon dioxide molecules (CO_2) are 22 times as massive as H_2.
    a. Compare the average speed of O_2 and CO_2 molecules in a volume of air.
    b. Does the ratio of the speeds in part (a) depend on air temperature?

43. Calculate the average speed of a carbon dioxide molecule in the atmospheres of Earth and Mars. Compare those speeds with their respective escape velocities. What does that tell you about each planet's hold on its atmosphere?

44. The average surface pressure on Mars is 6.4 mb. Using Figure 9.6, estimate how high you would have to go in Earth's atmosphere to experience the same atmospheric pressure that you would experience if you were standing on Mars.

45. Water pressure in Earth's oceans increases by 1 bar for every 10 meters of depth. Compute how deep you would have to go to experience pressure equal to the atmospheric surface pressure on Venus.

USING THE WEB

46. Look up the data on this year's ozone hole. NASA's "Ozone Watch" website (https://ozonewatch.gsfc.nasa.gov) shows a daily image of southern ozone, as well as animations for current and previous years and some comparative plots. Other comparative plots are available on NOAA's "Meteorological Conditions & Ozone in the Polar Stratosphere" Web page; click on "ozone hole size" (http://cpc.ncep.noaa.gov/products/stratosphere/polar/polar.shtml). At what time of year is the hole the largest, and why? How do the most recent ozone holes compare with previous ones in size and minima? Do they seem to be getting smaller?

47. Mars:
    a. Go to https://planetfour.org, a Zooniverse Citizen Science Project in which people examine images of the surface of Mars. Log in or create a Zooniverse account if you don't have one. Read through "About": Where did these data come from? What are the goals of the project? Why is having many people look at the data useful? Read through the sections in "About," and then return to the top page to "Start Exploring." Classify some images.
    b. Go to the website for the *MAVEN* mission, which entered the orbit of Mars in 2014 (http://lasp.colorado.edu/home/maven). What are the scientific goals of the mission? Is that mission a lander, an orbiter, or a flyby? What instruments are on the mission? How will the mission contribute to understanding climate change on Mars? Go to the NASA Web page for *MAVEN* (https://nasa.gov/mission_pages/maven/main/). What are some results of that mission?

48. Earth:
    a. Go to the National Snow & Ice Data Center (NSIDC) websites (https://nsidc.org/data/seaice_index/ and http://nsidc.org/arcticseaicenews/). What are the current status and the trend of the Arctic sea ice? How do those compare with previous years? Is anything new reported about Antarctic ice? Qualitatively, how might a change in the amount of ice at Earth's poles affect the albedo of Earth, and how does the albedo affect Earth's temperature?
    b. Go to the website for NASA's Goddard Institute for Space Studies (https://www.giss.nasa.gov), click on "Datasets & Images," and select "Surface Temperature." Click on "Graphs" in the upper right and scroll to "Global Annual Mean Surface Air Temperature Change." The graphs are updated every year. Note that the temperature is compared to a baseline of the average temperature in the period 1951–1980. What has happened with the temperature in the past few years? If the annual mean decreased, does that change the trend? What does the 5-year running mean show? How much warmer is it on average now than in 1880? Click on "News and Feature" (usually from January) to see a video map and report of the previous year's temperatures—how does the map compare with previous years'?
    c. Go to NOAA's "Trend in Atmospheric Carbon Dioxide" Web page on carbon dioxide levels at the observatory on Mauna Loa (https://esrl.noaa.gov/gmd/ccgg/trends/mlo.html). What is the current level of CO_2? How does that value compare with the level from 1 year ago? Since 1960? Why is this a good site for measuring CO_2? What exactly is measured?

49. Climate change:
    a. Go to the timeline on the "Discovery of Global Warming" Web page of the American Institute of Physics (https://history.aip.org/history/climate/timeline.htm). When did scientists first suspect that CO_2 produced by humans might affect Earth's temperature? When were other anthropogenic greenhouse gases identified? When did scientific opinions about global warming start to converge? Click on "Venus & Mars" (under 1971): How did observations of those planets add to an understanding of global climate change? Click on "Aerosols" (under 1970): How do those contribute to "global dimming"?
    b. The Fifth Assessment report from the Intergovernmental Panel on Climate Change (IPCC) was released in 2014. Go to the IPCC website section on the 2014 Synthesis report (http://ipcc.ch/report/ar5/syr/) and click through the "Presentation." What are some causes of the increase in warming? What are some effects of warming seen in the polar regions? How are measurements from the past and present used to predict the climate in the future? A report on the effects of 1.5-degree warming was published in 2018 (http://www.ipcc.ch/report/sr15/). What are some of the conclusions of the 2018 report?
    c. Advanced: Go to the website for "Educational Global Climate Modeling," or EdGCM (http://edgcm.columbia.edu). That is a version of the NASA GISS modeling software that will enable you to run a functional three-dimensional global climate model on your computers. Download the trial version and install it on your computer. What can you study with the program? What factors that contribute to global warming or to global cooling on Earth can you adjust in the model? Your instructor may give you an assignment to use this program and the Earth Exploration Toolbook (http://serc.carleton.edu/eet/envisioningclimatechange).

50. Mars movies:
    a. Watch a science fiction film about people going to Mars. How does the film handle the science? Can people breathe the atmosphere? Are the low surface gravity and atmospheric pressure correctly portrayed? Do the astronauts have access to water?
    b. At the end of the film *Total Recall* (1990), Arnold Schwarzenegger's character presses an alien button. The martian volcanoes start spewing, and within a few minutes the martian sky is blue, the atmospheric pressure is Earth-like, and the atmosphere is totally breathable. (You may be able to find the scene online.) What, scientifically, is wrong with that scene? That is, why would volcanic gases *not* quickly create a breathable atmosphere on Mars?

digital.wwnorton.com/astro6

One prediction about climate change is that as the planet warms, ice in the polar caps and in glaciers will melt. Such melting certainly seems to be occurring in almost all glaciers and ice sheets around the planet. Does that actually matter, and if so, why? In this Exploration, we explore several consequences of the melting ice on Earth.

Experiment 1: Floating Ice

For this experiment, you will need a permanent marker, a translucent plastic cup, water, and ice cubes. Place a few ice cubes in the cup and add water until the cubes float (so they don't touch the bottom). Mark the water level on the outside of the cup with the marker, and label that mark "initial water level."

1   As the ice melts, what do you expect to happen to the water level in the cup?

_____

_____

Wait for the ice to melt completely, and then mark the cup again.

2   What happened to the water level in the cup when the ice melted?

_____

_____

3   From the results of your experiment, what do you predict will happen to global sea levels when the Arctic ice sheet, which floats on the ocean, melts?

_____

_____

Experiment 2: Ice on Land

For this experiment, you will need the same materials as in experiment 1, plus a paper or plastic bowl. Fill the cup about halfway with water and then mark the water level, labeling the mark "initial water level." Poke a hole in the bottom of the bowl and set the bowl over the cup. Add some ice cubes to the bowl.

4   As the ice melts, what do you expect will happen to the water level in the cup?

_____

_____

Wait for the ice to melt completely, and then mark the cup again.

5   What happened to the water level in the cup when the ice melted?

_____

_____

6   In this experiment, the water in the cup is analogous to the ocean, and the ice in the bowl is analogous to ice on land. From the results of your experiment, what do you predict will happen to global sea levels when the Antarctic ice sheet, which sits on land, melts?

_____

_____

Experiment 3: Why Does It Matter?

Search online for the phrase "Earth at night" to find a satellite picture of Earth taken at night. The bright spots on the image trace out population centers. In general, the brighter a spot, the more populous the area (although a confounding factor relates to technological advancement).

7   Where do humans tend to live—near coasts or inland? Coastal regions are, by definition, near sea level. If both the Greenland and Antarctic ice sheets melted completely, sea levels would rise by as much as 80 meters. How would a sea-level rise of a few meters (in the range of reasonable predictions) over the next few decades affect the global population? (Note that each story of a building is about 3 meters.)

_____

_____

_____

10

Worlds of Gas and Liquid— The Giant Planets

Unlike the solid planets of the inner Solar System, the four worlds in the outer Solar System captured and retained gases and volatile materials from the Sun's protoplanetary disk and grew to enormous size and mass. Those planets have dense cores, have very large atmospheres, and rotate faster than Earth.

LEARNING GOALS

By the end of this chapter, you should be able to:

LG 1 Differentiate the giant planets from one another and from the terrestrial planets.

LG 2 Describe the atmosphere of each giant planet.

LG 3 Explain the extreme conditions deep inside the giant planets.

LG 4 Describe the magnetosphere of each giant planet.

LG 5 Compare the planets of our Solar System with those in exoplanetary systems.

This *Juno* image of Jupiter was taken in 2017. The color has been enhanced. The white ovals are part of eight large massive storms in the southern hemisphere. ▶▶▶

What causes the atmospheric features on Jupiter?

Scientific Laws Make Testable Predictions

How did astronomers predict that Neptune existed, before it was discovered? Newton's laws of motion and gravity do more than describe what we see. They also enable us to predict the existence of things yet unseen.

● **Uranus is discovered in 1781.**

● *That's odd: Uranus's orbit does not match predictions.*

● *Could another planet's gravitational pull be acting on Uranus?*

● **Mathematicians used observations of Uranus and Newton's laws to predict the location of an unknown planet.**

Neptune is discovered in 1846. ●

Newton's laws pass another test!

Laws make predictions that can be tested in order to verify or falsify them. Each test that does not falsify a scientific law strengthens scientists' confidence in its predictions.

The giant planets contain 99.5 percent of all the nonsolar mass in the Solar System. All other Solar System objects—terrestrial planets, dwarf planets, moons, asteroids, and comets—make up the remaining 0.5 percent. Even though Jupiter is only about a thousandth as massive as the Sun, it contains more than twice the mass of all the other planets combined. Jupiter is 318 times as massive as Earth, 3.3 times as massive as Saturn, and about 20 times as massive as either Uranus or Neptune.

The mass of a planet can be calculated by observing the motions and orbital size of a planet's moon. Newton's law of gravitation and Kepler's third law (Chapter 4) together show a relation between the motion of an orbiting object and the mass of the body it is orbiting. Planetary spacecraft now make it possible to measure the masses of planets even more accurately. As a spacecraft flies by a planet, the planet's gravity deflects it. By using several antennae on Earth to track and compare the spacecraft's radio signals, astronomers can detect tiny changes in the spacecraft's path and accurately measure the planet's mass.

Composition of the Giant Planets

The giant planets are made up primarily of gases and liquids. Jupiter and Saturn are composed of hydrogen and helium and are therefore known as **gas giants**. Uranus and Neptune are known as **ice giants** because they both contain much larger amounts of water and other ices than Jupiter and Saturn. On a giant planet, a relatively shallow atmosphere merges seamlessly into a deep liquid ocean, which merges smoothly, in turn, into a denser liquid or solid core. No abrupt transition exists from atmosphere to solid ground, as is found on the terrestrial planets. Although shallower than the liquid layers below, the atmospheres of giant planets are still much thicker than the atmospheres of the terrestrial planets—thousands of kilometers rather than hundreds. As with Venus, only the very highest levels of the atmospheres of the gas giants are visible to us. For Jupiter and Saturn, we see the top of a layer of thick clouds that obscures deeper layers (see the chapter-opening figure and **Figure 10.3a**). Only a few thin clouds are visible on Uranus, although atmospheric models suggest that thick cloud layers must lie below. Neptune displays a few high clouds with a deep, clear atmosphere showing between them (**Figure 10.3b**).

The terrestrial planets (Chapter 8) are composed mostly of rocky minerals, such as silicates, along with various amounts of iron and other metals. Although the atmospheres of the terrestrial planets contain lighter materials, the masses of those atmospheres—and even of Earth's oceans—are insignificant compared with the total planetary masses. The terrestrial planets are the densest objects in the Solar System, with densities ranging from 3.9 times that of water for Mars and 5.5 times for Earth, respectively. In contrast, the giant planets have lower densities because they are composed almost entirely of lighter materials, such as hydrogen, helium, and water. Among the giant planets, Neptune has the highest density, about 1.6 times that of water. Saturn has the lowest density, only 0.7 times that of water. Therefore, Saturn would float—if you had an immobile and deep enough body of water—with 70 percent of its volume submerged. The densities of Jupiter and Uranus are between those of Neptune and Saturn.

Jupiter's chemical composition is similar to that of the Sun—mostly hydrogen and helium. (Recall Figure 5.15, the astronomer's periodic table.) Only a few percent of Jupiter's mass is made up of **heavy elements**, which astronomers

(a) Saturn

G X U **V** I R

(b) Neptune

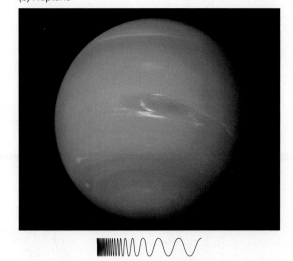

G X U **V** I R

Figure 10.3 (a) Saturn, imaged in visible light by *Cassini*. (b) Neptune, imaged in visible light by *Voyager*.

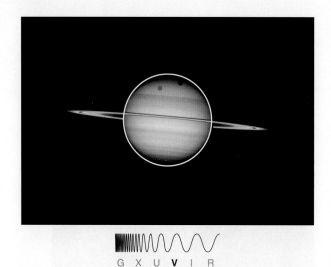

Figure 10.4 This Hubble Space Telescope image of Saturn was taken in 1999. The oblateness of the planet is apparent when compared with the white circle. The large orange moon Titan appears near the top of the disk of Saturn, along with its black shadow.

define as all elements more massive than helium. Many of those heavy elements combine chemically with hydrogen (H). For example, atoms of oxygen (O), carbon (C), nitrogen (N), and sulfur (S) combine with hydrogen to form molecules of water (H_2O), methane (CH_4), ammonia (NH_3), and hydrogen sulfide (H_2S). More complex combinations produce materials such as ammonium hydrosulfide (NH_4HS). Jupiter's core, which contains most of the planet's iron and silicates, is left over from the original rocky planetesimals around which Jupiter grew. Computer models of the density are required to understand the compositions deep in the cores of the giant planets.

The principal compositional differences among the four giant planets lie in the amounts of hydrogen and helium that each one contains. Because of its larger mass, Jupiter accumulated more hydrogen and helium when it formed than the other planets did. Saturn contains more heavy elements and less hydrogen and helium than Jupiter. Heavy elements are significant components of Uranus and Neptune. Methane is a particularly important molecule in the atmospheres of those two planets, giving them their characteristic blue-green color. Those differences in composition are important clues to understanding how the giant planets formed.

Days and Seasons on the Giant Planets

The giant planets rotate rapidly (Table 10.1), so their days are short (ranging from 10 to 17 hours) and their shapes are distorted from perfectly spherical to oblate—they bulge at their equators and have an overall flattened appearance. Saturn's appearance is noticeably oblate (**Figure 10.4**): its equatorial diameter is almost 10 percent greater than its polar diameter. By contrast, Earth's polar and equatorial diameters differ by only 0.3 percent.

The intensity of a planet's seasons is determined by the tilt of its axis (Chapter 2). With a tilt of only 3°, for example, Jupiter has almost no seasons at all. The tilts of Saturn (27°) and Neptune (28°), which are slightly larger than those of Earth (23.5°) and Mars (25°), cause moderate but well-defined seasons. Uranus spins on an axis, which lies nearly in the plane of its orbit—its tilt is about 98°. Uranus's high tilt causes its seasons to be extreme, with each polar region alternately experiencing 42 years of continuous sunshine followed by 42 years of total darkness. Averaged over an entire orbit, the poles receive more sunlight than the equator—a situation markedly different from that of any other Solar System planet.

Viewed from Earth, Uranus appears to be either spinning face-on or rolling along on its side (or something between), depending on where Uranus happens to be in its orbit. A tilt greater than 90° indicates that the planet rotates clockwise when seen from above its orbital plane. Why is Uranus tilted so differently from most other planets? One possible explanation is that Uranus was "knocked over" by the impact of one huge or several large planetesimals near the end of its accretion phase. Venus, Pluto, Pluto's moon Charon, and Neptune's moon Triton also have tilts greater than 90°.

CHECK YOUR UNDERSTANDING 10.1

How are Uranus and Neptune different from Jupiter and Saturn? (a) Uranus and Neptune have a higher percentage of ices in their interiors; (b) Uranus and Neptune have more hydrogen; (c) Uranus and Neptune have no storms; (d) Uranus and Neptune are closer to the Sun.

10.2 The Giant Planets Have Clouds and Weather

When we observe the giant planets through a telescope or in visible images from a spacecraft, we are seeing only the top layers of the atmosphere. Sometimes we can see a bit deeper into the clouds, but in essence, we are seeing a two-dimensional view of the cloud tops. The existence of deeper cloud layers on those giant planets is inferred from physical models of temperature as a function of depth. In this section, we explore the atmospheres of the giant planets.

Viewing the Cloud Tops

Even when viewed through small telescopes, Jupiter is very colorful. Parallel bands—ranging in hue from bluish gray to various shades of orange, reddish brown, and pink—stretch out across its large, pale yellow disk. Astronomers call the darker bands *belts* and the lighter bands *zones*. Many clouds—some dark and some bright, some circular and others more oval—appear along the edges of, or within, the belts. The most prominent cloud structure is a large, red, oval feature in Jupiter's southern hemisphere known as the **Great Red Spot** (**Figure 10.5**).

The Great Red Spot was first observed more than three centuries ago, shortly after the telescope was invented. Since then, it has varied unpredictably in size, shape, color, and motion as it drifts among Jupiter's clouds. In the 1800s, the Great Red Spot was greater than twice the diameter of Earth, but now it has shrunk to 1.3 Earth diameters. Observations of small clouds circling the perimeter of the Great Red Spot show that it is an enormous atmospheric whirlpool, swirling counterclockwise with a period of about a week. Its cloud pattern looks a lot like that of a terrestrial hurricane, but it rotates in the opposite direction—exhibiting *anticyclonic* rather than cyclonic flow, indicating a high-pressure system. Data from Juno suggest that the roots of the Great Red Spot go at least 300 kilometers (km) into the atmosphere, about 20 times more than the largest storms on Earth.

Comparable whirlpool-like behavior is observed in many of the smaller oval-shaped clouds found elsewhere in Jupiter's atmosphere and in similar clouds observed in the atmospheres of Saturn and Neptune. Whirlpool-like, swirling features are known as vortices (the singular is "vortex"). Those vortices are familiar to us on Earth as high- and low-pressure systems, hurricanes, and super-cell thunderstorms.

Jupiter's vortices are so complex that scientists still do not fully understand how they interact with one another. *Voyager 2* observed several Alaska-sized clouds being swept into the Great Red Spot. Some of those clouds were carried around the vortex a few times and then ejected, whereas others were swallowed up and never seen again. Other smaller clouds with structure and behavior similar to that of the Great Red Spot are seen in Jupiter's middle latitudes.

The *Juno* mission entered an elongated orbit around Jupiter in 2016. The spacecraft uses infrared and microwave instruments to further analyze the atmosphere. When closest to Jupiter, *Juno* is only a few thousand kilometers above the cloud tops. *Juno* has shown that the clouds are present even in Jupiter's polar regions and that the depths of the clouds vary greatly. For example, *Juno* observed a 4,000-km-wide cyclone near the north pole of Jupiter, with a ring of eight smaller cyclones surrounding it (**Figure 10.6**). In the southern polar region, five smaller ones circle the large cyclone.

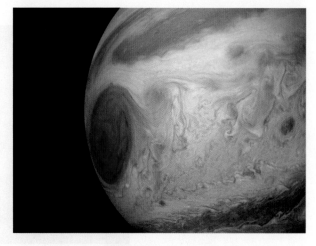

G X U **V** I R

Figure 10.5 A digitally enhanced image of Jupiter taken by *Juno*. The Great Red Spot is a hurricane about 1.3 times the diameter of Earth.

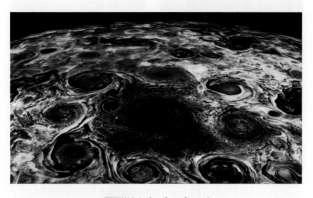

G X U V I R

Figure 10.6 *Juno* image of cyclones at the north pole of Jupiter. The colors represent heat: the thinner yellow clouds are about −13°C, and the dark red thicker clouds are about −118°C.

Figure 10.7 Saturn images from *Cassini*. (a) A northern hemisphere 2010 storm in Saturn's atmosphere is shown in true color. (b) The eye of this storm in Saturn's north polar region is about 2,000 km across and has wind speeds up to 150 meters per second. The false colors represent three near-infrared wavelengths; red shows low clouds and green indicates higher clouds.

(a)

G X U **V** I R

(b)

G X U **V** I R

(a)

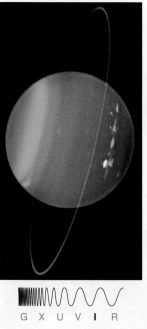

G X U **V** I R

(b)

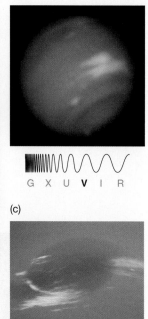

G X U **V** I R

(c)

G X U **V** I R

Figure 10.8 The ground-based Keck telescope image of Uranus (a) and a 2016 Hubble Space Telescope (HST) image of Neptune (b) were taken at wavelengths of light that are strongly absorbed by methane. The visible clouds are high in the atmosphere. (c) The Great Dark Spot on Neptune disappeared between the time *Voyager 2* flew by Neptune in 1989 and the time earlier HST images were obtained in 1994, but a new spot has now appeared.

Because Saturn is farther away than Jupiter and smaller in radius, from Earth Saturn appears less than half as large as Jupiter (see Figure 10.1b). Saturn also displays atmospheric bands, but they tend to be wider than those on Jupiter, with much more subdued colors and contrasts. The largest atmospheric features on Saturn are about the size of the continental United States, but many are smaller than terrestrial hurricanes. Close-up views from the *Cassini* spacecraft show immense lightning-producing storms in a region of Saturn's southern hemisphere known as "storm alley." In December 2010 a large storm appeared (**Figure 10.7a**) that was visible in even small amateur telescopes. Individual clouds are not seen often on Saturn, so that was an unusual event. That large storm eventually wrapped itself around the planet. *Cassini* also discovered a spinning vortex at Saturn's northern polar region (**Figure 10.7b**), but it is not known how long it has been active.

Through most telescopes on Earth, Uranus and Neptune look like tiny, featureless, pale bluish green disks. With the largest ground-based telescopes or the telescopes in space, however, optical and infrared imaging reveals several individual clouds and belts. Images (**Figure 10.8a**) show atmospheric bands and small clouds suggestive of those seen on Jupiter and Saturn, but more subdued. Methane's strong absorption of reflected sunlight causes the atmospheres of Uranus and Neptune to appear dark in the near infrared, allowing the highest clouds and bands to stand out against the dark background.

Several bright cloud bands appear in the Hubble Space Telescope (HST) image of Neptune's atmosphere (**Figure 10.8b**). Located near the planet's tropopause, those cloud bands cast their shadows downward through the clear upper atmosphere onto a dense cloud layer 50 km below. A large, dark, oval feature seen in the southern hemisphere first observed in images taken by *Voyager 2* in 1989 reminded astronomers of Jupiter's Great Red Spot, so they called it the Great Dark Spot (**Figure 10.8c**). However, the Neptune feature was gray rather than red, and it changed in length and shape more rapidly than the Great Red Spot. When HST observed Neptune in 1994, the Great Dark Spot had disappeared. A different spot of comparable size first seen in 2016 has continued to grow (Figure 10.8b), indicating that such storms come and go.

The Structure below the Cloud Tops

As we saw in discussing the terrestrial planets, atmospheric temperature, density, pressure, and even chemical composition vary with height and over horizontal distances. As a rule, atmospheric temperature, density, and pressure all decrease with increasing altitude, although temperature is sometimes higher at very high altitudes, as in Earth's thermosphere. The stratospheres above the cloud tops of the giant planets appear relatively clear, but closer inspection shows that they contain layers of thin haze that show up best when seen in profile above the edges of the planets. The composition of the haze particles remains unknown, but they may be smoglike products created when ultraviolet sunlight acts on hydrocarbon gases such as methane.

Water is the only substance in Earth's lower atmosphere that can condense into clouds, but the atmospheres of the giant planets have a much larger range of temperatures and pressures, so more kinds of volatiles (materials that become gases at moderate temperatures) can condense and form clouds. **Figure 10.9** shows how the ice layers are stacked in the tropospheres of the giant planets. Because each kind of volatile, such as water or ammonia, condenses at a particular temperature and pressure, each forms clouds at a different altitude. Convection carries volatile materials upward along with other atmospheric gases, and when each particular volatile reaches an altitude with its condensation temperature, most of that volatile condenses and separates from the other gases, so very little of it is carried higher aloft. Those volatiles form dense layers of cloud separated by regions of relatively clear atmosphere.

The farther the planet is from the Sun, the colder its troposphere will be. Therefore, the distance from the Sun determines the altitude at which a particular volatile, such as ammonia or water, will condense to form a cloud layer on each planet. If temperatures are too high, some volatiles may not condense at all. The highest clouds in the frigid atmospheres of Uranus and Neptune are crystals of methane ice. The highest clouds on Jupiter and Saturn are made up of ammonia ice. Methane never freezes to ice in the warmer atmospheres of Jupiter and Saturn.

In 1995, an atmospheric probe on the *Galileo* spacecraft descended slowly via parachute into the atmosphere of Jupiter. Near the top of Jupiter's troposphere at a temperature of about 130 K (about −140°C), the probe found that ammonia had condensed. Next it found a layer of ammonium hydrosulfide clouds at a temperature of about 190 K (about −80°C). Soon after descending to an atmospheric pressure of 22 bars and a temperature of about 373 K (100°C), the *Galileo* probe failed, presumably because its transmitter got too hot. In 2017, *Juno* detected a deep plume of ammonia arising from the depths near Jupiter's equator.

Why are some clouds so colorful, especially Jupiter's? In their purest form, the ices that make up the clouds of the giant planets are all white, similar to snow on Earth. The colorful tints and hues must come from impurities in the ice crystals. Those impurities are elemental sulfur and phosphorus, as well as various organic materials produced when ultraviolet sunlight breaks up hydrocarbons such as methane, acetylene, and ethane. The molecular fragments can then recombine to form complex organic compounds that condense into solid particles, many of which are colorful. Such reactions also occur in Earth's atmosphere. Some of the photochemical products produced close to the ground on Earth are called *smog*.

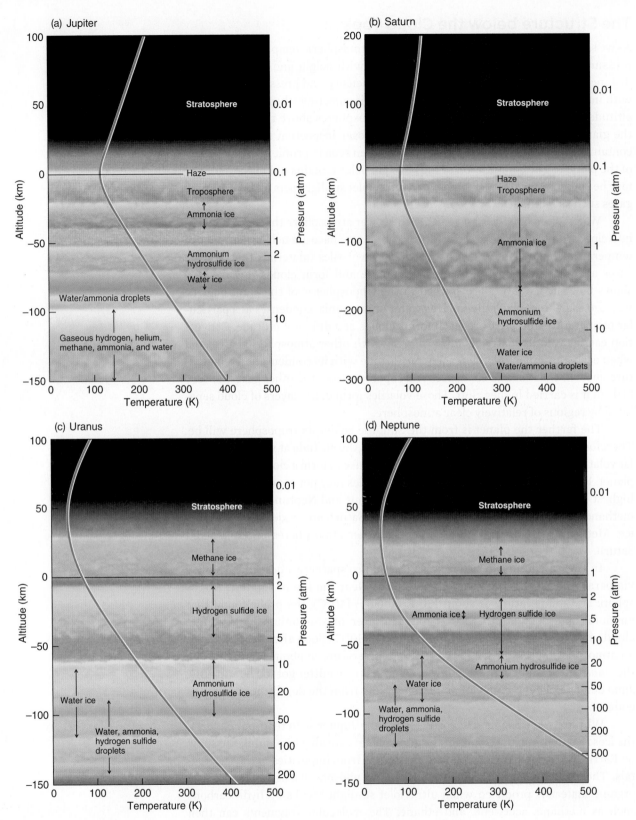

Figure 10.9 Volatile materials condense at different levels in the atmospheres of the giant planets, leading to chemically different types of clouds at different depths in the atmospheres. The red line in each diagram shows how atmospheric temperature changes with height. Because those planets have no solid surface, the zero point of altitude is arbitrary. Here, the arbitrary zero points of altitude are at 0.1 atmosphere (atm) for Jupiter (a) and Saturn (b) and at 1.0 atm for Uranus (c) and Neptune (d). The value of 1.0 atm corresponds to the atmospheric pressure at sea level on Earth. Saturn's altitude scale is compressed to better show the layered structure.

The atmospheric composition of Uranus and Neptune gives those planets their bluish green color. The upper tropospheres of Uranus and Neptune are relatively clear, with only a few white clouds that are probably composed of methane ice crystals. Methane gas is much more abundant in the atmospheres of Uranus and Neptune than in those of Jupiter and Saturn. Like water, methane gas tends to selectively absorb the longer wavelengths of light—yellow, orange, and red. Absorption of the longer wavelengths leaves only the shorter wavelengths—green and blue—to be scattered from the atmospheres of Uranus and Neptune.

Winds and Weather

On the giant planets, the thermal energy that drives convection comes both from the Sun and from the hot interiors of the planets themselves. Convection results from vertical temperature differences (Chapter 9). As heating drives air up and down, the Coriolis effect shapes that convection into atmospheric vortices, visible as isolated circular or oval cloud structures, such as the Great Red Spot on Jupiter and the Great Dark Spot on Neptune. As the atmosphere rises near the center of a vortex, it expands and cools. Cooling condenses certain volatile materials into liquid droplets, which then fall as rain. As they fall, the raindrops collide with surrounding molecules, stripping electrons from the molecules and thereby developing tiny electric charges in the air. The cumulative effect of countless falling raindrops can be an electric field so great that it ionizes the molecules in the atmosphere and creates a surge of current and a flash of lightning. A single observation of Jupiter's night side by *Voyager 1* revealed several dozen lightning bolts within an interval of 3 minutes. *Cassini* also has imaged lightning flashes in Saturn's atmosphere, and radio receivers on *Voyager 2* picked up lightning static in the atmospheres of both Uranus and Neptune.

The giant planets have much stronger zonal winds than those of the terrestrial planets. Because the giant planets are farther from the Sun, less thermal energy is available. They rotate rapidly, however, making the Coriolis effect very strong. In fact, the Coriolis effect is more important than atmospheric temperature patterns in determining the structure of the global winds. If we know the radius of the planet, we can find out how fast the features are moving, as shown in **Working It Out 10.1**, and find rotation speeds and wind speeds. **Figure 10.10** shows the wind speeds at various latitudes on the different planets.

JUPITER On Jupiter (Figure 10.10a), the strongest winds are equatorial, blowing from the west, at speeds up to 550 kilometers per hour (km/h). At higher latitudes, the winds alternate between blowing from the west or east in a pattern that might be related to Jupiter's banded structure. Near a latitude of 20° south, the Great Red Spot vortex appears to be caught between a pair of winds blowing from the west and east with opposing speeds of more than 200 km/h—indicating a relationship between zonal flow and vortices. Data from *Juno* suggest that the jet stream goes 3,000 km down into the atmosphere, much deeper than expected. The jet stream layer rotates differently according to latitude and contains about 1 percent of the mass of the planet.

SATURN The equatorial winds on Saturn (Figure 10.10b) also blow from the west but are stronger than those on Jupiter. The maximum wind speeds at any given time vary between 990 and 1,650 km/h. Saturn's winds appear to decrease with height in the atmosphere, so the apparent time variability of Saturn's equatorial winds may be nothing more than changes in the height of the cloud tops.

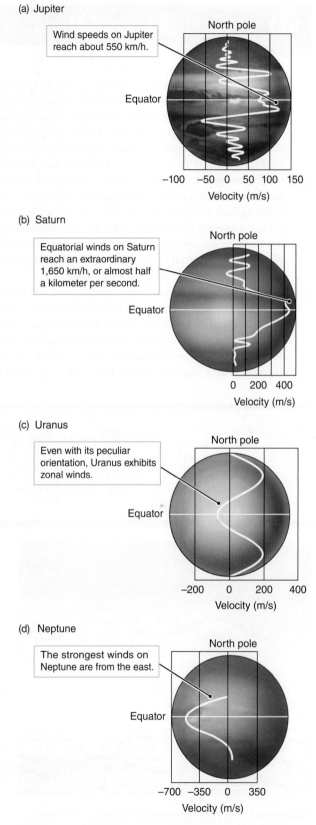

(a) Jupiter

Wind speeds on Jupiter reach about 550 km/h.

North pole

Equator

−100 −50 0 50 100 150
Velocity (m/s)

(b) Saturn

Equatorial winds on Saturn reach an extraordinary 1,650 km/h, or almost half a kilometer per second.

North pole

Equator

0 200 400
Velocity (m/s)

(c) Uranus

Even with its peculiar orientation, Uranus exhibits zonal winds.

North pole

Equator

−200 0 200 400
Velocity (m/s)

(d) Neptune

The strongest winds on Neptune are from the east.

North pole

Equator

−700 −350 0 350
Velocity (m/s)

Figure 10.10 The speed of the wind at various latitudes on the giant planets is shown by the white line. When the speed is positive (to the right of zero), the wind is from the west. When the speed is negative (to the left of zero), the wind is from the east.

10.1 Working It Out Measuring Wind Speeds on Different Planets

How do astronomers measure wind speeds on planets so far away? If we can see individual clouds in their atmospheres, we can measure their winds. As on Earth, clouds are carried by local winds. The local wind speed can be calculated by measuring the positions of individual clouds and noting how much they move during an interval of a day or so. We also need to know how fast the planet is rotating so we can measure the speed of the winds with respect to the planet's rotating surface. For the giant planets, no solid surface exists against which to measure the winds. Scientists must instead assume a hypothetical surface—one that rotates as though it were somehow "connected" to the planet's deep interior. Periodic bursts of radio energy caused by the rotation of a planet's magnetic field tell us how fast the interior of the planet is rotating.

If we know the radius of the planet, we can find the speed of the features. Consider a small white cloud in Neptune's atmosphere. The cloud, on Neptune's equator, is observed to be at longitude 73.0° west on a given day. (That longitude system is anchored in the planet's deep interior and already takes into account Neptune's rotation.) The spot is then seen at longitude 153.0° west exactly 24 hours later.

Neptune's equatorial winds have carried the white spot 80.0° in longitude in 24 hours.

The circumference (C) of a planet is given by $2\pi r$, where r is the equatorial radius. The equatorial radius of Neptune is 24,760 km, so Neptune's circumference is

$$C = 2\pi r = 2\pi \times 24{,}760 \text{ km} = 155{,}600 \text{ km}$$

The full circle represented by the circumference has 360° of longitude. The spot has moved 80°/360° of the circumference and thus has traveled

$$\frac{80}{360} \times 155{,}600 \text{ km} = 34{,}580 \text{ km}$$

in 24 hours. Therefore, the wind speed is given by

$$\text{Speed} = \frac{\text{Distance}}{\text{Time}} = \frac{34{,}580 \text{ km}}{24 \text{ h}} = 1{,}440 \text{ km/hr} = 400 \text{ m/s}$$

The equatorial winds are very strong and are blowing in a direction opposite to the planet's rotation. On Earth, the much slower equivalents of those winds are called *trade winds* (see Figure 9.10).

Alternating winds blowing from the east or west also occur at higher latitudes; but unlike on Jupiter, that alternation seems to bear no clear association with Saturn's atmospheric bands. That example is just one of the many unexplained differences among the giant planets.

Saturn's jet stream at latitude 45° north (**Figure 10.11a**) is a narrow meandering river of atmosphere with alternating crests and troughs. **Figure 10.11b** shows how the jet stream curves around regions of high and low pressure to create a wavelike structure. That behavior is similar to Earth's jet streams, in which high-speed winds blow generally from west to east but wander toward and away from the poles. Nested within the crests and troughs of Saturn's jet stream are anticyclonic and cyclonic vortices. They appear similar in both form and size to terrestrial high- and low-pressure systems, which bring alternating periods of fair and stormy weather.

URANUS Less is known about global winds on Uranus (Figure 10.10c) than about those on the other giant planets. When *Voyager 2* flew by Uranus in 1986, the few visible clouds were in the southern hemisphere because the northern hemisphere was in complete darkness at the time. The strongest winds observed were 650 km/h from the west in the middle to high southern latitudes, and no winds from the east were seen. Because Uranus's peculiar orientation makes its poles warmer than its equator, some astronomers had predicted that the global wind system of Uranus might differ greatly from that of the other giant planets. But *Voyager 2* observed that the Coriolis forces dominate on Uranus as they do on other planets, so the dominant winds on Uranus are zonal, just as on the other giant planets.

(a)

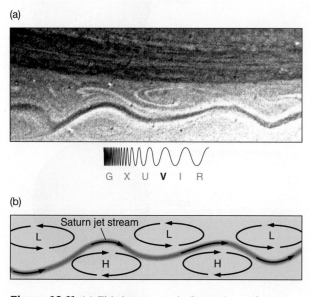

(b)

Figure 10.11 (a) This jet stream in Saturn's northern hemisphere, imaged by *Voyager*, is similar to jet streams in Earth's atmosphere. (b) The jet stream dips toward the equator below regions of low pressure and is forced toward the pole above regions of high pressure.

As Uranus moved along in its orbit, regions previously unseen by modern telescopes have become visible (**Figure 10.12**). Observations by HST and ground-based telescopes showed bright cloud bands in the far north extending more than 18,000 km and revealed wind speeds of up to 900 km/h. As Uranus approaches northern summer solstice in the year 2027, much more about its northern hemisphere will be learned.

NEPTUNE On Neptune (Figure 10.10d), the southern hemisphere's summer solstice occurred in 2005, so much of the north is still in darkness. Observers will have to wait until Neptune's equinox in 2045 to see its northern hemisphere. The strongest winds on Neptune occur in the tropics, similar to winds on Jupiter and Saturn. On Neptune, however, the winds are from the east rather than from the west, with speeds in excess of 2,000 km/h. Winds from the west with speeds higher than 900 km/h have been seen in Neptune's south polar regions. With wind speeds 5 times greater than those of the fiercest hurricanes on Earth, Neptune and Saturn are the windiest planets known.

CHECK YOUR UNDERSTANDING 10.2

Why does Jupiter appear reddish in color? (a) It is very hot; (b) its atmosphere consists of reddish chemical compounds; (c) it is moving very quickly; (d) it is rusty, like Mars.

10.3 The Interiors of the Giant Planets Are Hot and Dense

At the center of each giant planet is a dense, liquid core consisting of a very hot mixture of heavier materials such as water, molten rock, and metals. **Figure 10.13** illustrates the interior structure of the giant planets. As you can see, the gas giants differ from the ice giants in their amounts of hydrogen and ices. We look at each in turn.

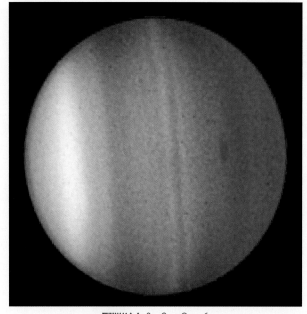

Figure 10.12 Uranus is approaching equinox in this 2006 Hubble Space Telescope image. Much of its northern hemisphere is becoming visible. The dark spot in the northern hemisphere (to the right) is similar to but smaller than the Great Dark Spot seen on Neptune in 1989.

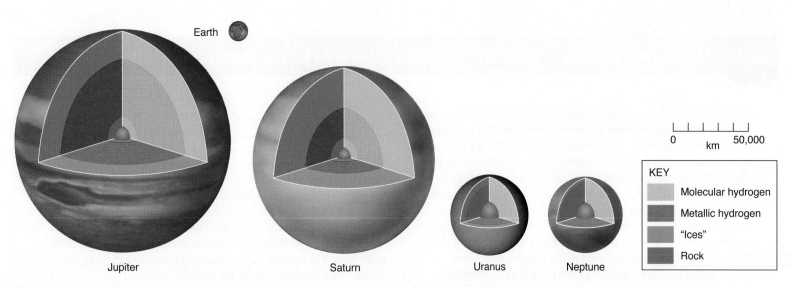

Figure 10.13 The interiors of the giant planets have central cores and outer liquid shells. Only Jupiter and Saturn have significant amounts of the molecular and metallic forms of liquid hydrogen surrounding their cores.

The Cores of Jupiter and Saturn

The overlying layers of Jupiter and Saturn press down on the liquid core, raising its temperature. For example, the pressure at Jupiter's core is about 45 million bars, and a pressure that high heats the fluid to 35,000 K. Central temperatures and pressures of the other, less massive giant planets are correspondingly lower than those of Jupiter. Water is still liquid at those enormously high temperatures because the extremely high pressures at the centers of the giant planets prevent water from turning to steam.

The internal energy that lies deep within the giant planets is left over from their formation. The giant planets are still contracting and converting their gravitational potential energy into thermal energy today as they did when they first formed, but they are doing it more slowly. The annual amount of contraction necessary to sustain their internal temperature is only a tiny fraction of their radius. Jupiter, for example, would need to contract by a bit less than 2 centimeters per year. The thermal energy from the core drives convection in the atmosphere and eventually escapes to space as radiation (**Working It Out 10.2**). For Saturn, moreover, and perhaps for Jupiter, liquid helium separates from a hydrogen-helium mixture under the right conditions and rains downward toward the core. As the droplets of liquid helium sink, they release their gravitational potential energy as thermal energy. Planetary physicists think that most of Saturn's internal energy and perhaps some of Jupiter's internal energy come from that separation of liquid helium. The continuous production of thermal energy is enough to replace the energy escaping from their interiors.

The pressure within the atmospheres of Jupiter and Saturn increases with depth because overlying layers of atmosphere press down on lower layers. At depths of a few thousand kilometers, the atmospheric gases of Jupiter and Saturn are so compressed by the weight of the overlying atmosphere that they turn to liquid. That depth roughly marks the lower boundary of the atmosphere. The difference between a liquid and a highly compressed, very dense gas is subtle, so Jupiter and Saturn have no clear boundary between the atmosphere and the ocean of liquid that lies below. Jupiter's atmosphere is about 20,000 km deep, and Saturn's atmosphere is about 30,000 km deep; at those depths, the pressure

---

10.2 Working It Out Internal Thermal Energy Heats the Giant Planets

The equilibrium between the absorption of sunlight and the radiation of infrared light into space was introduced in Chapter 5. How the resulting equilibrium temperature is modified by the greenhouse effect on Venus, Earth, and Mars was explained in Chapter 9. When that equilibrium is calculated for the giant planets, however, it doesn't match the measurements. According to those calculations, the equilibrium temperature for Jupiter should be 109 K, but the average temperature turns out to be 124 K. A difference of 15 K might not seem like much, but the energy radiated by an object depends on its temperature raised to the fourth power (the Stefan-Boltzmann law; Chapter 5). Applying that relationship to Jupiter, we get:

$$\left(\frac{T_{\text{actual}}}{T_{\text{expected}}}\right)^4 = \left(\frac{124\,\text{K}}{109\,\text{K}}\right)^4 = 1.67$$

That result implies that Jupiter is radiating roughly two-thirds more energy into space than it absorbs from sunlight. Similarly, the internal energy escaping from Saturn is observed to be about 1.8 times greater than the sunlight that it absorbs. Neptune emits 2.6 times more energy than it absorbs from the Sun. Those planets, then, are *not* in equilibrium; they are slowly contracting and thus generating more heat. The internal energy escaping from Uranus turns out to be smaller than the solar energy the planet absorbs.

climbs to 2 million bars and the temperature reaches 10,000 K (hotter than the surface of the Sun). Under those conditions, hydrogen molecules are battered so violently that their electrons are stripped free, and the hydrogen acts like a liquid metal. In that state, it is called metallic hydrogen. Those oceans of hydrogen and helium are tens of thousands of kilometers deep.

Differentiation has occurred and is still occurring in Saturn and perhaps in Jupiter. On Saturn, helium condenses out of the hydrogen-helium oceans. Helium also can be compressed to a metal, but it does not reach that metallic state under the physical conditions of the interiors of the giant planets. Because those droplets of helium are more dense than the hydrogen-helium liquid in which they condense, they sink toward the center of the planet, converting gravitational energy to thermal energy. That process heats the planet and enriches the helium concentration in the core while depleting it in the upper layers. In Jupiter's hotter interior, by contrast, the liquid helium is mostly dissolved together with the liquid hydrogen.

The heavy-element components of the cores of Jupiter and Saturn have masses of about 5–20 Earth masses. Jupiter and Saturn have total masses of 318 and 95 Earth masses, respectively. The heavy materials in their cores contribute little to their average chemical composition. Therefore, Jupiter and Saturn have approximately the same composition as the Sun and the rest of the universe: about 98 percent hydrogen and helium, leaving only 2 percent for everything else.

The Cores of Uranus and Neptune

Uranus and Neptune are less massive than Jupiter and Saturn, have lower interior pressures, and contain smaller fractions of hydrogen—their interiors probably contain only a small amount of liquid hydrogen, with little or none in a metallic state. Uranus and Neptune are made of denser material than Saturn and Jupiter. Neptune, the most dense of the giant planets, is about 1.6 times denser than water and only about half as dense as rock. Uranus is less dense than Neptune. Those observations tell us that water and other low-density ices, such as ammonia and methane, must be the major compositional components of Uranus and Neptune, along with lesser amounts of silicates and metals. The total amount of hydrogen and helium in those planets is probably limited to no more than 1 or 2 Earth masses, and most of those gases reside in the relatively shallow atmospheres of the planets. Computer simulations suggest that under their conditions of high pressure and temperature, the water that makes up so much of Uranus and Neptune might be *super-ionic*—a state between a liquid and a solid.

Why do Jupiter and Saturn have so much more hydrogen and helium than Uranus and Neptune? Although each giant planet formed around cores of rock and metal, they turned out differently. Those differences are an important clue to their origins. The variation may be due to the time those planets took to form and to the distribution of material from which they formed. The cores of Uranus and Neptune were smaller and formed much later than those of Jupiter and Saturn, when most of the gas in the protoplanetary disk had been blown away by the emerging Sun. The icy planetesimals from which they formed were more widely dispersed at their greater distances from the Sun. With more space between planetesimals, their cores would have taken longer to build up. Saturn may have captured less gas than Jupiter both because its core formed later and because less gas was available at its greater distance from the Sun. Some astronomers hypothesize that Uranus and Neptune might have formed in a location different from where they are now (see the Origins section).

Figure 10.14 The magnetic fields of the giant planets can be approximated by the fields from bar magnets offset and tilted with respect to the planets' axes. Compare those magnetic fields with Earth's, shown in Figure 8.12.

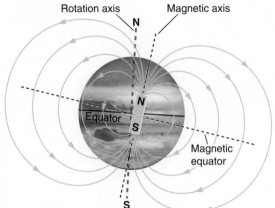

(a) **Jupiter**

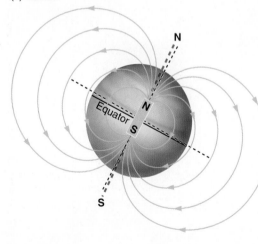

(b) **Saturn**

(c) **Uranus**

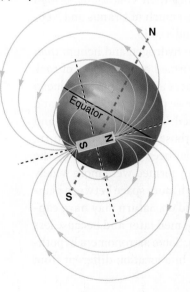

(d) **Neptune**

10.4 The Giant Planets Are Magnetic Powerhouses

All the giant planets have magnetic fields much stronger than Earth's: their field strengths range from 50 to 20,000 times stronger. Field strength falls off with distance, however, so the fields at the cloud tops of Saturn, Uranus, and Neptune have comparable strength to that of Earth's surface field. Even for Jupiter's exceptionally strong field, the field strength at the cloud tops is only about 15 times that of Earth's surface field. In Jupiter and Saturn, circulating currents within deep layers of metallic hydrogen generate magnetic fields. In Uranus and Neptune, magnetic fields arise within deep oceans of water and ammonia made electrically conductive by dissolved salts or ionized molecules. The magnetospheres of the giant planets are very large and interact with not only the solar wind (as Earth's does) but also the rings and moons that orbit the giant planets.

The Size and Shape of the Magnetospheres

Just as Earth's magnetic field traps energetic charged particles to form Earth's magnetosphere, the magnetic fields of the giant planets also trap energetic particles to form magnetospheres of their own. The most colossal of those by far is Jupiter's magnetosphere. Its radius is 100 times that of the planet itself, roughly 10 times the radius of the Sun. Even the relatively weak magnetic fields of Uranus and Neptune form magnetospheres that are comparable in size to the Sun. Evidence of the giant planets' magnetospheres comes from spacecraft in the outer Solar System, from telescopes orbiting Earth, and from radio emissions received on Earth.

Figure 10.14 illustrates the geometry of the magnetic fields of the giant planets as though they came from bar magnets. The differences in the orientations of the magnetic field axes are not well understood. Jupiter's magnetic axis is inclined 10° with respect to its rotation axis—an orientation similar to Earth's—but it is offset about a tenth of a radius from the planet's center. Saturn's magnetic axis is located almost precisely at the planet's center and is almost perfectly aligned with the rotation axis. The magnetic axis of Uranus is inclined nearly 60° to its rotation axis and is offset by a third of a radius from the planet's center. The orientation of Neptune's rotation axis is similar to that of Earth, Mars, and Saturn. Neptune's magnetic-field axis is inclined 47° to its rotation axis, however, and the center of that magnetic field is displaced from the planet's center by more than half the radius—an offset even greater than that of Uranus. The field is displaced primarily toward Neptune's southern hemisphere, thereby creating a field 20 times stronger at the southern cloud tops than at the northern cloud tops. The reason for the unusual geometries of the magnetic fields of Uranus and Neptune remains uncertain, but it is not related to the orientations of their rotation axes.

The magnetosphere also is influenced by the solar wind, which supplies some of the particles (Chapter 9). In addition, the pressure of the solar wind pushes on and compresses a magnetosphere, so the size and shape of a planet's magnetosphere depend on how the solar wind is blowing at any particular time. Planetary magnetic fields also divert the solar wind, which flows around magnetospheres the way a stream flows around boulders. Just as a rock in a river creates a wake that extends downstream (**Figure 10.15a**), the magnetosphere of a planet produces a wake that can extend for great distances.

Figure 10.15b shows that the wake of Jupiter's magnetosphere extends well past the orbit of Saturn. Jupiter's magnetosphere is the largest permanent structure in the Solar System, surpassed in size only by the tail of an occasional comet. If your eyes were sensitive to radio waves, the second-brightest object in the sky would be Jupiter's magnetosphere. The Sun would still be brighter, but even at a distance from Earth of 4.2–6.2 AU, Jupiter's magnetosphere would appear roughly twice as large as the Sun in the sky.

Saturn's magnetosphere also would be large enough to see, if we could see radio waves, but it would be much fainter than Jupiter's. Even though Saturn has a strong magnetic field, pieces of rock, ice, and dust in Saturn's spectacular rings act like sponges, soaking up magnetospheric particles soon after they enter the magnetosphere. With far fewer magnetospheric electrons, Saturn has much less radio emission. The magnetic tails of Uranus and Neptune have a curious structure. For both Uranus and Neptune, the tilt and the large displacement of the magnetic field from the center of each planet cause the magnetosphere to wobble as the planet rotates. That wobble causes the tail of the magnetosphere to twist like a corkscrew as it stretches away.

Rapidly moving electrons in planetary magnetospheres spiral around the magnetic field lines, and as they do so they emit a type of radiation, known as **synchrotron radiation**, concentrated in the low-energy radio part of the spectrum. Precise measurement of periodic variations in the radio signals "broadcast" by the giant planets indicates the planets' true rotation periods. The magnetic field of each planet is locked to the conducting liquid layers deep within the planet's interior, so the magnetic field rotates with the same period as the deep interior of the planet. With the fast and highly variable winds that

Figure 10.15 (a) Water flowing past a rock sweeps the algae against the rock and into a "tail" pointing in the direction of the water's flow. (b) The solar wind compresses Jupiter's (or any other planet's) magnetosphere on the side toward the Sun and draws it out into a magnetic tail away from the Sun. Jupiter's tail stretches beyond the orbit of Saturn. (Not to scale.)

(a)

(b)

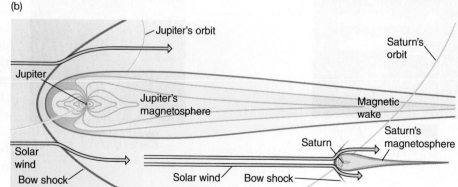

push around clouds in the atmospheres of the giant planets, measuring the radio emission is the only way to determine their true rotation periods.

Radiation Belts and Auroras of the Giant Planets

As a planet rotates on its axis, it drags its magnetosphere around with it, and charged particles are swept around at high speeds. Those fast-moving charged particles slam into neutral atoms, and the energy released in the resulting high-speed collisions heats the **plasma** to extreme temperatures. (A plasma is a gas consisting of electrically charged particles.) In 1979, while passing through Jupiter's magnetosphere, *Voyager 1* encountered a region of plasma with a temperature of more than 300 million K—20 times the temperature at the center of the Sun. *Voyager 1* did not melt when passing through that region because the plasma was so tenuous that the plasma's particles were very far apart in space. Each particle was extraordinarily energetic; too few of them were present to harm the probe.

Charged particles trapped in planetary magnetospheres are concentrated in *radiation belts*. Although Earth's radiation belts are severe enough to worry astronauts, the radiation belts that surround Jupiter are searing by comparison. In 1974, the *Pioneer 11* spacecraft passed through the radiation belts of Jupiter. Several of the instruments onboard were permanently damaged as a result, and the spacecraft itself barely survived to continue its journey to Saturn.

In addition to protons and electrons from the solar wind, the magnetospheres of the giant planets contain large amounts of various elements (some ionized), including sodium, sulfur, oxygen, nitrogen, and carbon. Those elements come from several sources, including the planets' extended atmospheres and the moons that orbit within them. The most intense radiation belt in the Solar System is a doughnut-shaped ring, or *torus*, of plasma associated with Io, the innermost of Jupiter's four Galilean moons. As we discuss in more detail in Chapter 11, Io has low surface gravity and violent volcanic activity. Some of the gases erupting from Io's interior escape and become part of Jupiter's radiation belt. As charged particles are slammed into Io by the rotation of Jupiter's magnetosphere, even more material is knocked free of its surface and ejected into space. Images of the region around Jupiter, taken in the light of emission lines from atoms of sulfur or sodium, show a faintly glowing ring of plasma supplied by Io. Other moons also influence the magnetospheres of the planets they orbit. *Cassini* found that Saturn's moon Enceladus leaks ionized molecules (including nitrogen), water vapor, and ice grains from icy geysers and provides most of the torus of plasma in Saturn's magnetosphere.

Charged particles spiral along the magnetic-field lines of the giant planets, bouncing back and forth between each planet's two magnetic poles, just as they do around Earth. As with Earth, those energetic particles collide with atoms and molecules in a planet's atmosphere, knocking them into excited energy states that decay and emit light. The results are bright auroral rings (**Figure 10.16**). Those auroral rings surround the magnetic poles of the giant planets, just as the aurora borealis and aurora australis ring the north and south magnetic poles, respectively, of Earth (see Figure 9.9a).

Figure 10.16 The Hubble Space Telescope took images of auroral rings around the poles of Jupiter (a) and Saturn (b). The auroral images (bright rings near the north pole) were taken in ultraviolet light and then superimposed on visible-light images.

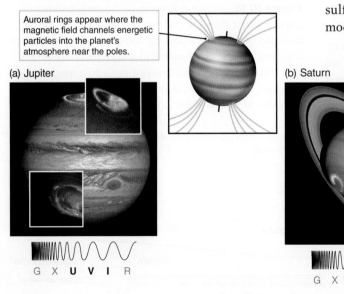

Auroral rings appear where the magnetic field channels energetic particles into the planet's atmosphere near the poles.

(a) Jupiter

G X **U** V I R

(b) Saturn

G X **U** V I R

Jupiter's auroras are the most powerful in the Solar System, and they have an added twist not seen on Earth. As Jupiter's magnetic field sweeps past Io, electrons spiral along Jupiter's magnetic-field lines. The result is a magnetic channel, called a **flux tube**, that connects Io with Jupiter's atmosphere near the planet's magnetic poles (**Figure 10.17**). Io's flux tube carries power roughly equivalent to the total power produced by all electrical generating stations on Earth. Much of the power generated within the flux tube is radiated away as radio energy. Those radio signals are received on Earth as intense bursts. However, a substantial fraction of the energy of the particles in the flux tube also is deposited into Jupiter's atmosphere. At the very location where Io's flux tube intercepts Jupiter's atmosphere is a spot of intense auroral activity. As Jupiter rotates, that spot leaves behind an auroral trail in Jupiter's atmosphere.

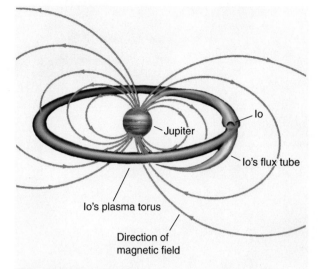

Figure 10.17 The geometry of Io's plasma torus and flux tube.

CHECK YOUR UNDERSTANDING 10.4

Why are the radiation belts around Jupiter much stronger than those around Earth? (a) Jupiter has larger storms than Earth; (b) Jupiter is colder than Earth; (c) Jupiter rotates faster than Earth; (d) Jupiter has a stronger magnetic field than Earth.

10.5 The Planets of Our Solar System Might Not Be Typical

In Chapters 8–10, we discussed the planets of our Solar System in detail. The categories of inner terrestrial rocky planets versus outer gas and ice giants were based on our eight planets—but how typical are those categories when compared with the many exoplanets that have been detected? Do other multiplanet systems resemble our own? As the number of confirmed exoplanets increases, astronomers can compare them statistically with those of our own system. Planetary scientists were surprised to find that exoplanets differ from the planets of our Sun. In the Kepler-20 system, for example, the large planets alternate with the smaller ones (**Figure 10.18**). In this section, we examine some of those differences.

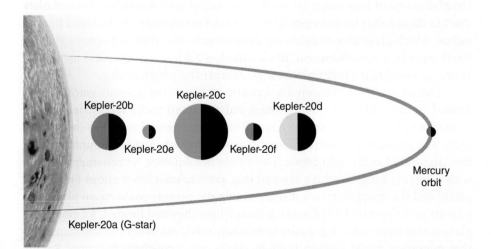

Figure 10.18 Illustration of the Kepler-20 system. Large and small planets alternate, but they are all closer to their star than Mercury is to the Sun.

Figure 10.19 Artist's depiction of the super-Jupiter planet Kappa Andromedae b.

Different Types of "Jupiters"

The first exoplanets were discovered using the radial velocity method (Chapter 7), which measures the wobble of a star caused by the gravity of its planet. That method is most successful at finding a massive planet located close to its star, where the planet's gravitational tug on the star is stronger than if the planet were of smaller mass or farther away. Those *hot Jupiters* are gas giants, within a few percent of an AU to half an AU from their respective stars, with correspondingly short and sometimes highly elliptical orbits. Current models of planetary formation suggest that not enough excess hydrogen would have existed that close to their stars for those planets to form there. Some of those planets may be *puffy Jupiters*, which have a larger radius and a lower density than Jupiter. Their densities are closer to that of Saturn. The larger radius is thought to come from a heated, and thus expanding, gaseous atmosphere.

Astronomers have identified several hundred exoplanets known as *super-Jupiters* because their masses are 2–13 times that of Jupiter. Some are also hot Jupiters, but most are not. Their higher mass gives them stronger gravitational contraction than that in planets with Jupiter's mass, so they shrink. As a result, most have a higher density than Jupiter. The gases can be compressed by self-gravity so much that the super-Jupiter could be denser than Earth. The stronger gravitational contraction would create hotter cores, so they might have more intense winds and weather than Jupiter has. An artist's depiction of the super-Jupiter planet Kappa Andromedae b, with about 13 times the mass of Jupiter, is shown in **Figure 10.19**.

Super-Earths to Mini-Neptunes

Currently, observations suggest that the most common exoplanet size is one with a radius between that of Earth and that of Neptune (4 times larger than Earth). However, no planet in our Solar System falls in that range. Planets with about 1.5–10 M_{Earth} are called *super-Earths*. Up to about 2 R_{Earth}, those planets get denser as they get larger, as expected for rocky planets. Above 2 R_{Earth}, however, most planets are puffier—a gaseous envelope surrounds the rocky core. **Figure 10.20** shows a plot of the size of planets versus their frequency. The gaseous planets at the higher end of that range are sometimes called *mini-Neptunes*. Some planets don't fit those rules; for example, KOI-314c has Earth's mass but 1.6 times Earth's radius, which gives that exoplanet a density more like that of Neptune than of Earth (that is, a mini-Neptune). Kepler-10c has 2.3 R_{Earth} and a mass as high as 17 M_{Earth}—making it a rocky planet with a surprisingly high mass.

In Chapter 7, we presented a scenario in which the gaseous giant planets formed in the cold outer Solar System and the small rocky terrestrial planets formed in the warm inner Solar System. That model seemed to make sense chemically and physically, so astronomers were shocked when the hot Jupiters were first discovered in the mid-1990s. To explain hot Jupiters, astronomers worked with computer models, which showed that gravitational interactions between a planet and the protoplanetary disk or among the planets could cause planets to migrate to different orbital distances from where they had formed. As more exoplanets that were *not* hot Jupiters were discovered, computer models suggested that migration might explain their locations, too, especially for super-Earths close in to their respective stars.

To date, the mix of planet types in our Solar System—outer giant gaseous planets and inner small rocky planets, all with nearly circular orbits—has not

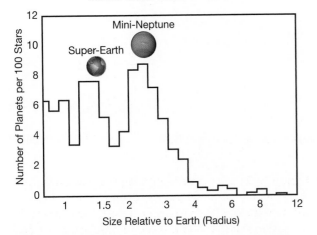

Figure 10.20 Smaller planets tend to be super-Earths or mini-Neptunes. Data on nearby super-Earths suggest that at about 2 R_{Earth}, planet density decreases with size because the planet has more gas accumulated on its rocky core.

been seen in several hundred extrasolar multiplanet systems. Observations and computer models show that many combinations of planetary sizes, masses, and compositions can exist within planetary systems. It is not yet known whether planetary systems like ours are rare or if they just haven't yet been discovered.

CHECK YOUR UNDERSTANDING 10.5

Place in order of increasing diameter the following types of exoplanets: (a) super-Earths; (b) puffy Jupiters; (c) super-Jupiters; (d) mini-Neptunes.

Origins Giant Planet Migration and the Inner Solar System

As long as only two objects are involved, the motions that result from Newton's law of gravitation are simple. Kepler's laws describe the regular, repeating elliptical orbits of planets around the Sun. When more than two objects are involved, however, the resulting motions may be anything but simple and regular. Each planet in the Solar System moves under the gravitational influence of the Sun combined with that of all the other planets. Although those extra influences are small, they are not negligible. Over millions of years, they lead to significant differences in the locations of planets in their orbits. In many possible exoplanetary systems, such interactions among planets might cause planets to dramatically change their orbits or even be ejected from the system entirely.

Computer models developed to understand the exoplanet systems have been applied to the Solar System as well, and the results are intriguing. Computer models of the formation of the Solar System show that the giant planets may not have formed in their current locations and could have migrated substantially in the early Solar System. A key point seems to be the gravitational influence of Saturn on Jupiter, especially the ratio of their orbital periods. **Figure 10.21** shows one type of migration model. When Saturn's orbital period became twice that of Jupiter, their respective orbits became more elongated. The result was an outward migration of Uranus and Neptune, whose orbits grew larger. Uranus and Neptune may actually have switched places. That shuffling cleared away nearby planetesimals, sending some to the inner Solar System, and made the planetary orbits more stable. In another set of models, Jupiter migrated inward to 1.5 AU—the current orbital distance of Mars. Then Saturn migrated inward even faster to a point at which its orbital period was 1.5 times that of Jupiter, and finally they both migrated outward, pushing Uranus and Neptune into larger orbits.

In some exoplanetary systems, gas giants are observed at the right distance from their respective stars to be in the habitable zone (see Chapter 7), but it is not known whether gas giants can support life. Super-Earths also are sometimes found in the habitable zone of their stars. In the Solar System, migrations of the giant planets could have helped confine and stabilize the orbits of the inner planets, so that they reside in or near the habitable zone. Migration of Jupiter may have destabilized the orbits of close-in hot planets, so that they crashed into the Sun. Planetesimals closer to 1.5 AU may have scattered, keeping Mars from growing so that it stayed small. A small Mars could not hold on to its atmosphere and its liquid water, making it a less likely starting point for life. In addition, the shuffling about of the outer planets may be responsible for the period of late heavy bombardment, which brought at least some of the atmospheric gases, water, and possibly organic molecules needed for life to form on Earth.

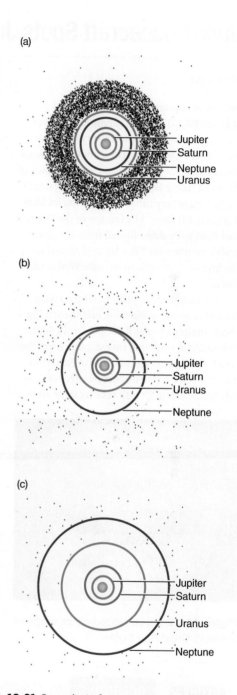

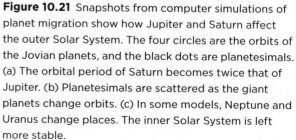

Figure 10.21 Snapshots from computer simulations of planet migration show how Jupiter and Saturn affect the outer Solar System. The four circles are the orbits of the Jovian planets, and the black dots are planetesimals. (a) The orbital period of Saturn becomes twice that of Jupiter. (b) Planetesimals are scattered as the giant planets change orbits. (c) In some models, Neptune and Uranus change places. The inner Solar System is left more stable.

Juno Spacecraft Spots Jupiter's Great Red Spot

jpl.nasa.gov

Images of Jupiter's Great Red Spot (**Figure 10.22**) reveal a tangle of dark, veinous clouds weaving their way through a massive crimson oval. The JunoCam imager aboard NASA's *Juno* mission snapped pics of the most iconic feature of the solar system's largest planetary inhabitant during its Monday (July 10) flyby. The images of the Great Red Spot were downlinked from the spacecraft's memory on Tuesday and placed on the mission's JunoCam website Wednesday morning.

"For hundreds of years scientists have been observing, wondering and theorizing about Jupiter's Great Red Spot," said Scott Bolton, *Juno* principal investigator from the Southwest Research Institute in San Antonio. "Now we have the best pictures ever of this iconic storm. It will take us some time

Figure 10.22 An enhanced-color image of Jupiter's Great Red Spot.

to analyze all the data from not only Juno-Cam, but *Juno*'s eight science instruments, to shed some new light on the past, present and future of the Great Red Spot."

As planned by the *Juno* team, citizen scientists took the raw images of the flyby from the JunoCam site and processed them, providing a higher level of detail than available in their raw form. The citizen-scientist images, as well as the raw images they used for image processing, can be found at: https://www.missionjuno.swri.edu/junocam/processing.

"I have been following the *Juno* mission since it launched," said Jason Major, a Juno-Cam citizen scientist and a graphic designer from Warwick, Rhode Island. "It is always exciting to see these new raw images of Jupiter as they arrive. But it is even more thrilling to take the raw images and turn them into something that people can appreciate. That is what I live for."

Measuring in at 10,159 miles (16,350 kilometers) in width (as of April 3, 2017) Jupiter's Great Red Spot is 1.3 times as wide as Earth. The storm has been monitored since 1830 and has possibly existed for more than 350 years. In modern times, the Great Red Spot has appeared to be shrinking.

All of *Juno*'s science instruments and the spacecraft's JunoCam were operating during the flyby, collecting data that are now being returned to Earth. *Juno*'s next close flyby of Jupiter will occur on Sept. 1.

Juno reached perijove (the point at which an orbit comes closest to Jupiter's center) on

July 10 at 6:55 p.m. PDT (9:55 p.m. EDT). At the time of perijove, *Juno* was about 2,200 miles (3,500 kilometers) above the planet's cloud tops. Eleven minutes and 33 seconds later, *Juno* had covered another 24,713 miles (39,771 kilometers), and was passing directly above the coiling, crimson cloud tops of the Great Red Spot. The spacecraft passed about 5,600 miles (9,000 kilometers) above the clouds of this iconic feature.

Juno launched on Aug. 5, 2011, from Cape Canaveral, Florida. During its mission of exploration, *Juno* soars low over the planet's cloud tops—as close as about 2,100 miles (3,400 kilometers). During these flybys, *Juno* is probing beneath the obscuring cloud cover of Jupiter and studying its auroras to learn more about the planet's origins, structure, atmosphere and magnetosphere.

Early science results from NASA's *Juno* mission portray the largest planet in our solar system as a turbulent world, with an intriguingly complex interior structure, energetic polar aurora, and huge polar cyclones.

"These highly-anticipated images of Jupiter's Great Red Spot are the 'perfect storm' of art and science. With data from *Voyager*, *Galileo*, *New Horizons*, Hubble and now *Juno*, we have a better understanding of the composition and evolution of this iconic feature," said Jim Green, NASA's director of planetary science. "We are pleased to share the beauty and excitement of space science with everyone."

ARTICLES   QUESTIONS

1. Why has the Great Red Spot been seen for only ~350 years?
2. Why is it difficult for astronomers to understand what is happening below the cloud tops?
3. Do a search to see whether the Great Red Spot is still shrinking.
4. Go to the website mentioned in the article. Why is *Juno* using "citizen scientists" to process the images?
5. What new observational capabilities does the *Juno* mission bring to understanding the Jovian system?

Summary

The giant planets are much larger and less dense than the terrestrial planets. They consist primarily of light elements rather than rock, and their outer atmospheres are much colder. Because of their rapid rotation and the Coriolis effect, zonal winds are very strong on those planets. Storms, such as those on Saturn, tend to be larger and longer-lived than storms on Earth. Volatiles become ices at various heights in those atmospheres, leading to a layered cloud structure. Jupiter, Saturn, and Neptune are still shrinking, and their gravitational energy is being converted to thermal energy, heating both the cores and the atmospheres from the inside. Uranus does not seem to have as large a heat source inside. Temperatures and pressures in the cores of the giant planets are very high, leading to unusual states of matter, such as metallic hydrogen and super-ionic water. The current locations of those planets might be different from where they formed: models suggest that their positions may have migrated.

LG 1 Differentiate the giant planets from one another and from the terrestrial planets. Uranus and Neptune were discovered by telescope, unlike all the other planets, which have been known since ancient times. Jupiter and Saturn are gas giants, whereas Uranus and Neptune are ice giants. Jupiter and Saturn are made up mostly of hydrogen and helium (a composition similar to that of the Sun), whereas Uranus and Neptune contain larger amounts of ices, such as water, ammonia, and methane, than are found in Jupiter and Saturn. Those compositions set them apart from the terrestrial planets. In addition, all four giant planets are much larger than Earth.

LG 2 Describe the atmosphere of each giant planet. We see only atmospheres on the giant planets because solid or liquid surfaces, if they exist, are deep below the cloud layers. Clouds on Jupiter and Saturn are composed of various kinds of ice crystals colored by impurities. Uranus and Neptune have relatively few clouds, and so their atmospheres appear more uniform. The most prominent atmospheric feature is the Great Red Spot in Jupiter's southern hemisphere.

LG 3 Explain the extreme conditions deep inside the giant planets. The ongoing collapse of the giant planets converts gravitational energy to thermal energy. That process heats most of the giant planets from within, producing convection. Powerful convection and the Coriolis effect drive high-speed winds in the upper atmospheres of all the giant planets. The interiors of the giant planets are very hot and very dense because of the high pressures exerted by their overlying atmospheres.

LG 4 Describe the magnetosphere of each giant planet. The giant planets have enormous magnetospheres that emit synchrotron radiation and interact with their moons. The rotation speed of a gas giant is found from periodic bursts of radio waves generated as the planet's magnetic field rotates.

LG 5 Compare the planets of our Solar System with those in exoplanetary systems. Systems of exoplanets found to date do not contain our Solar System's distribution of small, rocky inner planets and large, giant outer planets. Most planets found so far fall between Earth and Neptune in size. Our understanding of how solar systems form is incomplete.

? Unanswered Questions

- Did the Solar System start with more planets? In the same types of computer models of the early Solar System that we discussed in Origins, astronomers can run simulations with different initial configurations of planets to see which configurations evolve and then compare those with what is observed today. In one set of models, astronomers found that starting with five giant planets best reproduced the current outer Solar System. The fifth planet, which would have been kicked out of the Solar System after a close encounter with Jupiter, may still be wandering through the Milky Way.

- What are the mass and size of the core of each giant planet? Is a rocky core underneath the thick atmosphere? Is the core the size of a terrestrial planet or larger? The NASA *Juno* mission is measuring Jupiter's gravitational and magnetic fields to map the amount and distribution of mass in its core and atmosphere. Preliminary results from *Juno* suggest that Jupiter's core is larger and "fuzzier" than previously thought, so that question might be answered for Jupiter in a few years.

Questions and Problems

Test Your Understanding

1. The following steps lead to convection in the atmospheres of giant planets. After (a), place (b)–(f) in order.
   a. Gravity pulls particles toward the center.
   b. Warm material rises and expands.
   c. Particles fall toward the center, converting gravitational energy to kinetic energy.
   d. Expanding material cools.
   e. Thermal energy heats the material.
   f. Friction converts kinetic energy to thermal energy.

2. Deep inside the giant planets, water is still a liquid even though the temperatures are tens of thousands of degrees above the boiling point of water. That can happen because
   a. the density inside the giant planets is so high.
   b. the pressure inside the giant planets is so high.
   c. the outer Solar System is so cold.
   d. space has very low pressure.

3. Assume you want to deduce the radius of a Solar System planet as it occults a background star when the relative velocity between the planet and Earth is 30 km/s. If the star crosses through the middle of the planet and disappears for 26 minutes, what is the planet's radius?
   a. 3,000 km
   b. 23,000 km
   c. 15,000 km
   d. 5,000 km

4. Neptune's existence was predicted because
   a. Uranus did not seem to obey Newton's laws of motion.
   b. Uranus wobbled on its axis.
   c. Uranus became brighter and fainter in an unusual way.
   d. some of the solar nebula's mass was unaccounted for.

5. Which giant planet has the most extreme seasons?
   a. Jupiter
   b. Saturn
   c. Uranus
   d. Neptune

6. The magnetic fields of the giant planets
   a. align closely with the planet's rotation axis.
   b. extend far into space.
   c. are thousands of times stronger at the cloud tops than at Earth's surface field.
   d. have an axis that passes through the planet's center.

7. An occultation occurs when
   a. a star passes between Earth and a planet.
   b. a planet passes between Earth and a star.
   c. a planet passes between Earth and the Sun.
   d. Earth passes between the Sun and a planet.

8. Occultations directly determine a planet's
   a. diameter.
   b. mass.
   c. density.
   d. orbital speed.

9. The chemical compositions of Jupiter and Saturn are most similar to those of
   a. Uranus and Neptune.
   b. the terrestrial planets.
   c. their moons.
   d. the Sun.

10. Individual cloud layers in the giant planets have different compositions because
   a. the winds are all in the outermost layer.
   b. the Coriolis effect occurs only close to the "surface" of the inner core.
   c. convection does not occur on the giant planets.
   d. different volatiles freeze out at different temperatures.

11. The Great Red Spot on Jupiter is
   a. a surface feature.
   b. a storm that has been raging for more than 300 years.
   c. caused by the interaction between the magnetosphere and Io.
   d. currently about the size of North America.

12. How are Uranus and Neptune different from Jupiter and Saturn?
   a. Uranus and Neptune have a higher percentage of ices in their interiors.
   b. Uranus and Neptune have no rings.
   c. Uranus and Neptune have no magnetic field.
   d. Uranus and Neptune are closer to the Sun.

13. What could have caused the planets to migrate through the Solar System?
   a. gravitational pull from the Sun
   b. interaction with the solar wind
   c. accreting gas from the solar nebula
   d. gravitational pull from other planets

14. Zonal winds on the giant planets are stronger than those on the terrestrial planets because
   a. they have more thermal energy.
   b. the giant planets rotate faster.
   c. the moons of giant planets provide additional pull.
   d. the moons feed energy to the planet through the magnetosphere.

15. A "hot Jupiter" gets its name from the fact that
    a. its temperature has been measured to be higher than Jupiter's.
    b. it is located around a much hotter star than the Sun.
    c. it has very high density, so its temperature is high.
    d. it orbits very close to its central star.

Thinking about the Concepts

16. Describe how the giant planets differ from the terrestrial planets.

17. Jupiter's chemical composition is more like that of the Sun than Earth's is, yet both planets formed from the same protoplanetary disk. Explain why they are different today.

18. What can be learned about a Solar System object when it occults a star?

19. What drives the zonal winds in the atmospheres of the giant planets?

20. Compare the sequence of events in the Process of Science Figure in this chapter with the flowchart of the Process of Science Figure in Chapter 1. Redraw the flowchart, incorporating each event leading to the discovery of Uranus as examples in the appropriate boxes.

21. None of the giant planets are perfectly spherical. Explain why they have a flattened appearance.

22. What is the source of color in Jupiter's clouds? Uranus and Neptune, when viewed through a telescope, appear distinctly bluish green. What are the two reasons for their striking appearance?

23. Which giant planets have seasons similar to Earth's, and which one experiences extreme seasons?

24. Jupiter's core is thought to consist of rocky material and ices, all in a liquid state at a temperature of 35,000 K. How can materials such as water be liquid at such high temperatures?

25. Explain how astronomers measure wind speeds in the atmospheres of the giant planets.

26. What is the Great Red Spot? How has it changed over the centuries it has been observed?

27. Jupiter, Saturn, and Neptune radiate more energy into space than they receive from the Sun. What is the source of the additional energy?

28. When viewed by radio telescopes, Jupiter is the second-brightest object in the sky. What is the source of its radiation?

29. What creates auroras in the polar regions of Jupiter and Saturn?

30. How might migration of the outer giant planets affect the sizes and orbits of the inner planets?

Applying the Concepts

31. Figure 10.1 shows two sets of pictures of the outer planets. What is the difference between Figure 10.1a and Figure 10.1b?

32. Figure 10.10d shows the winds on Neptune. The graph, however, does not cover the full planet. Is that likely to mean that the wind speed is zero where no white line is present or that the wind speed is unknown where no white line is present? Explain your reasoning.

33. What creates metallic hydrogen in the interiors of Jupiter and Saturn, and why do we call it metallic?

34. Use Figure 10.17 to estimate the radius of Io's plasma torus in terms of the radius of Jupiter. Convert that value to kilometers, and then look up the answer on the Internet. How close did you get with your simple measurement?

35. The Sun appears 400,000 times brighter than the full Moon in Earth's sky. How far from the Sun (in astronomical units) would you have to go for the Sun to appear only as bright as the full Moon appears in Earth's nighttime sky? How does the distance you would have to travel compare with the semimajor axis of Neptune's orbit?

36. Uranus occults a star when the relative motion between Uranus and Earth is 23.0 km/s. An observer on Earth sees the star disappear for 37 minutes 2 seconds and notes that the center of Uranus passed directly in front of the star.
    a. On the basis of those observations, what value would the observer calculate for the diameter of Uranus?
    b. What could you conclude about the planet's diameter if its center did not pass directly in front of the star?

37. Jupiter's equatorial radius (R_{Jup}) is 71,500 km, and its oblateness is 0.065. What is Jupiter's polar radius (R_{polar})? Oblateness is given by $(R_{Jup} - R_{polar})/R_{Jup}$.

38. Ammonium hydrosulfide (NH_4HS) is a molecule in Jupiter's atmosphere responsible for many of its clouds. Using the periodic table in Appendix 3, calculate the molecular weight of an ammonium hydrosulfide molecule, where the atomic weight of a hydrogen atom is 1. (Recall from Working It Out 9.1 that the weight of a molecule is equal to the sum of the weights of its component atoms.)

39. Jupiter is an oblate planet with an average radius of 69,900 km. Earth's average radius is 6,370 km.
    a. Given that volume is proportional to the cube of the radius, how many Earth volumes could fit inside Jupiter?
    b. Jupiter is 318 times as massive as Earth. Show that Jupiter's average density is about one-fourth that of Earth.

40. The tilt of Uranus is 98°. From one of the planet's poles, how far from the zenith would the Sun appear on summer solstice?

41. A small cloud in Jupiter's equatorial region is observed to be at a longitude of 122.0° west in a coordinate system and to be rotating at the same rate as the deep interior of the planet. (West longitude is measured along a planet's equator toward the west.) Another observation, made exactly 10 Earth hours later, finds the cloud at a longitude of 118.0° west. Jupiter's equatorial radius is 71,500 km. What is the observed equatorial wind speed in kilometers per hour? Is that wind from the east or west?

42. The equilibrium temperature for Saturn should be 82 K, but the observed temperature is 95 K. How much more energy does Saturn radiate than it absorbs?

43. Neptune radiates 2.6 times as much energy into space as it absorbs from the Sun. Its equilibrium temperature (see Chapter 5) is 47 K. What is its true temperature?

44. Compare the graphs in Figures 10.9a and 10.9b. Does atmospheric pressure increase more rapidly with depth on Jupiter or on Saturn? Compare the graphs in Figures 10.9c and 10.9d. Does pressure increase more rapidly with depth on Uranus or on Neptune? Of the four giant planets, which has the fastest pressure rise with depth? The slowest?

45. Use Figure 10.9 to find the temperature at an altitude of 100 km on each of the four giant planets.

USING THE WEB

46. a. Go to websites for the NASA *Juno* mission (https://www.nasa.gov/mission_pages/juno/ and http://missionjuno.swri.edu), a spacecraft launched in 2011, which has been in orbit around Jupiter since 2016. Examine the mission's trajectory. Why did it loop around the Sun and pass Earth again in 2013 before heading to Jupiter? Why does the spacecraft carry a plaque dedicated to Galileo Galilei?
    b. What are the science goals of the mission? What is a recent discovery? Is the mission still active?

47. Go to the website for the *Voyager 1* and *2* missions (https://voyager.jpl.nasa.gov), which celebrated their 40th anniversary in 2017. *Voyager* visited and collected data on all four giant planets.
    a. Where are the spacecraft now? Click on "Galleries" and then "Images." Those are still the only close-up images of Uranus and Neptune. What was learned about those planets?
    b. Click on the "The Golden Record," and then on the right, look at scenes, greetings, music, and sounds from Earth. Suppose you were asked to make a new version of the Golden Record, a playlist to send on an upcoming space mission to outside the Solar System. What would you include in one or more of those categories?

48. Go to the *Cassini* website (https://saturn.jpl.nasa.gov). That mission ended in 2017, but data is still being analyzed. Why did NASA crash *Cassini* into Saturn? What are some of the main discoveries from that mission? Are new discoveries still being reported?

49. Go to the Extrasolar Planets encyclopedia (http://exoplanet.eu/catalog/).
    a. Under "Mass," look for a super-Jupiter planet with a mass significantly larger than that of Jupiter. How far is it from its star—is it a hot Jupiter? Click on the planet name—how was it discovered? If a radius is given, is it denser or less dense than Jupiter? Click twice on "Mass" to get a list in descending order—what is the most massive super-Jupiter in the catalog?
    b. Under "Mass," click on "M_{Jup}" so it changes to "M_{Earth}"; do the same under "Radius" so it shows "R_{Earth}." Look for a "Super-Earth." What is its radius? How was it detected? Is an estimated mass listed—if so, what is its density compared with that of Earth? Is it a hot or cold super-Earth?

50. Go to the website in Problem 49 or to the NASA Exoplanet Archive (https://exoplanetarchive.ipac.caltech.edu/ and click on "Multi-Planet Systems"). Look for a star that has multiple planets. Make a graph showing the distances of the planets from that star, and note the masses and sizes of the planets. Put the Solar System planets on the same axis. How does that exoplanet system differ from our Solar System?

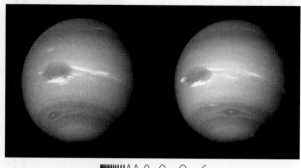

Figure 10.23 Images of Neptune, taken 17.6 hours apart by *Voyager 2*.

Study the two images of Neptune in **Figure 10.23**. The image on the left was taken first, whereas the image on the right was taken 17.6 hours later. During that time, the Great Dark Spot completed nearly one full rotation. The small storm at the bottom of the image completed slightly more than one rotation. You would be very surprised to see that result for locations on Earth.

1 What do those observations tell you about the rotation of visible cloud tops of Neptune?

_____

You can find the rotation period of the smaller storm by equating two ratios. First, use a ruler to find the distance (in millimeters) from the left edge of the planet to the small storm in each image (**Figure 10.24a**). The right edge of the planet is not illuminated, so you will have to estimate the radius of the circles traveled by the storms. You can do that by measuring from the edge of the planet to a line through the planet's center (**Figure 10.24b**). Because the small storm travels along a line of latitude close to a pole, the distance it travels is significantly less than the circumference of the planet.

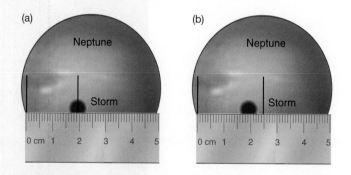

Figure 10.24 (a) How to measure the position of the storm. (b) How to measure the radius of the circle that the storm traveled.

2 Estimate the radius of the circle (in millimeters) that this small storm makes around Neptune by measuring from the edge of the disk to the line through the center of the planet.

_____

3 Find the circumference of that circle (in millimeters).

_____

Because the small storm rotated *more* than one time, the total distance it traveled is the circumference of the circle plus the distance between its locations in the two images.

4 Add the numbers you obtained in steps 2 and 3 to get the total distance traveled (in millimeters) between those images.

_____

Now take a ratio and find the rotation period. The ratio of the rotation period (T) to the time elapsed (t) must be equal to the ratio of the circumference of the circle around which it travels (C, in millimeters) to the total distance traveled (D, in millimeters): $T/t = C/D$.

5 You have all the numbers you need to solve for T. What value do you calculate for the small storm's rotation period? (To check your work, note that your answer should be less than 17.6 hours. Why?)

_____

Why does that calculation work at all? The actual distance the small storm traveled is *not* a few millimeters, nor is the circumference of the circle around which it travels. To find the actual distance or circumference, you would multiply both values by the same constant of proportionality. Because you are taking a ratio, however, that constant cancels out, so you might as well leave it out from the beginning.

6 Perform the corresponding measurements and calculations for the Great Dark Spot. What is its rotation period? Think carefully about how to find the total distance traveled because the Great Dark Spot has rotated around the planet *less* than once. (To check your work, note that your answer should be more than 17.6 hours. Why?)

_____

7 How similar are the rotation periods for those two storms?

_____

8 What does that comparison tell you about using that method to determine the rotation periods of the giant planets?

_____

9 What method do astronomers use instead?

_____

11

Planetary Moons and Rings

For centuries, Saturn's rings and the Galilean moons of Jupiter have delighted people who looked through telescopes. Since the dawn of the space age, robotic explorers traveling through the Solar System have revealed even more of the diverse collection of moons and rings orbiting other planets.

LEARNING GOALS

By the end of this chapter, you should be able to:

LG 1 Compare the orbits and formation of regular and irregular moons.

LG 2 Describe the evidence for geological activity and liquid oceans on some of the moons.

LG 3 Describe the composition, origin, and general structure of the rings of the giant planets.

LG 4 Explain the role gravity plays in the structure of the rings and the behavior of ring particles.

Why do some
planets have
rings?

11.1 Many Solar System Planets Have Moons

Most of the planets and some of the dwarf planets in our Solar System have moons orbiting them. New moons in the outer Solar System are still being discovered. As of 2018, the planets and dwarf planets of the Solar System have nearly 200 moons (some are provisional and need further confirmation). Many of those moons are unique worlds of their own, exhibiting geological processes similar to those on the terrestrial planets. Some moons have volcanic activity and atmospheres, and some are likely to contain liquid water under their icy surfaces. A few of those moons could have conditions suitable to some forms of life. Some of those planetary moons are listed in Appendix 4 (moons around asteroids are discussed in Chapter 12), and an updated list can be found through the "Using the Web" problems at the end of this chapter. In this section, we discuss the orbits and the formation of the moons.

The Distribution of the Moons

The moons of the Solar System are not distributed equally; most are among the giant planets. Only three moons are in the inner Solar System: Mars has two, Earth has one, and Mercury and Venus have none. Among the dwarf planets,

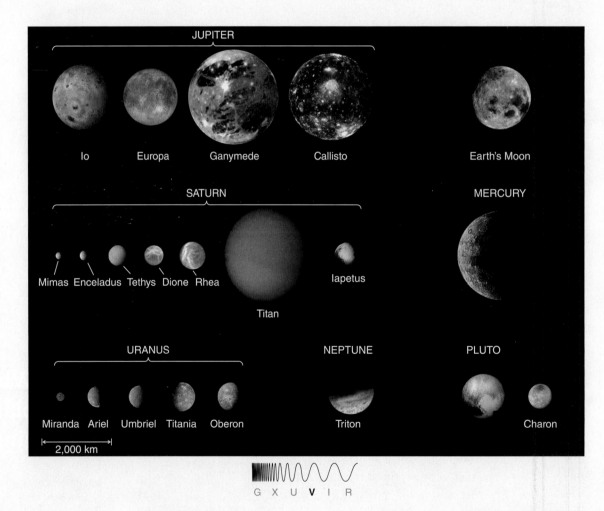

Figure 11.1 Various spacecraft have imaged the major moons of the Solar System. The images are to scale, with the planet Mercury and dwarf planet Pluto shown for comparison. At this scale, the martian moons, Phobos and Deimos, are too small to be shown.

Pluto has five known moons, Haumea has two, and Eris and Makemake each have one. The giant planets have the remaining moons. While the larger planets were forming, they had greater attracting mass and more debris around them; consequently, they have more moons.

Figure 11.1 shows the major moons in the Solar System. Some, like Earth's Moon, are made of rock. Others, especially in the outer Solar System, are mixtures of rock and water ice. A few are made almost entirely of ice. Two moons, Jupiter's Ganymede and Saturn's Titan, are larger in diameter than Mercury. The smallest known moons are only a kilometer in diameter. Although most moons have no atmosphere, Titan has an atmosphere denser than Earth's, and several have very low-density atmospheres. Moons probably accreted from smaller bodies in much the same way that planets accreted from planetesimals, although some may be the product of collisions.

The Orbits of the Moons

A moon can be classified according to its orbit as either *regular* or *irregular*. A **regular moon** lies near its planet's equatorial plane, is close to its planet, and has a nearly circular orbit (**Figure 11.2a**). A regular moon also moves in the same direction the planet rotates. About one-third of the moons in the Solar System are regular. Those moons probably formed from an accretion disk around a host planet at around the time the planet was forming. Our Moon, the Galilean moons of Jupiter, and Saturn's Titan are large, regular moons. Most regular moons are tidally locked to their parent planets. Recall from Chapter 4 that tidal locking causes a body to rotate synchronously with respect to its orbit, as Earth's Moon does. When a moon is tidally locked to its planet, the leading hemisphere permanently faces "forward," in the direction in which the moon is traveling in its orbit around the planet. The trailing hemisphere faces backward. The leading hemisphere is always flying directly into any local debris surrounding the planet, so that hemisphere may have more impact craters on its surface than the trailing hemisphere.

Compared with a regular moon, an **irregular moon** has a more elliptical and more inclined orbit and generally is farther away from its planet (**Figure 11.2b**). Most irregular moons orbit in a direction *opposite* to the rotation of their respective planets; that is, they have retrograde orbits. (Recall from Chapter 3 that apparent backward motion of a planet in the sky is called retrograde motion.) The largest irregular moons are Neptune's Triton and Saturn's Phoebe. Most of the recently discovered moons of the outer planets are irregular, and many are only a few kilometers across. Those moons are almost certainly bodies that formed elsewhere and were later captured by the planets.

Some of the regular moons have strange orbital characteristics. For example, Phobos, one of the two small moons of Mars (**Figure 11.3**), is so close to Mars that it actually orbits Mars faster than Mars rotates. As seen from Mars, Phobos rises in the west and sets in the east twice a day. Phobos is closer to its planet than any known moon. It is not known whether Phobos and the other moon of Mars, Deimos, were captured from the nearby asteroid belt or whether they evolved together with Mars, possibly after a collision early in the history of Mars.

Another strange regular moon is Pluto's Charon, which has a radius one-half that of Pluto. The two are in synchronous rotation, so each has one hemisphere that always faces the other body and another hemisphere that never faces the

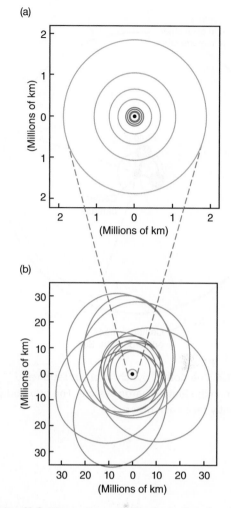

Figure 11.2 These diagrams illustrate the view from "above" the orbits of some of Jupiter's moons. (a) Closer moons, including the Galilean moons, are regular, with nearly circular orbits in Jupiter's equatorial plane. (b) Most of the more distant moons are irregular, with more elliptical, retrograde orbits not in the equatorial plane.

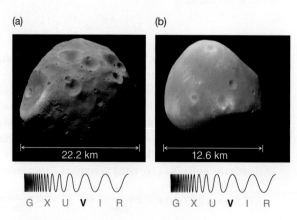
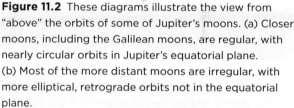

Figure 11.3 The *Mars Reconnaissance Orbiter* photographed the two tiny moons of Mars: (a) Phobos and (b) Deimos.

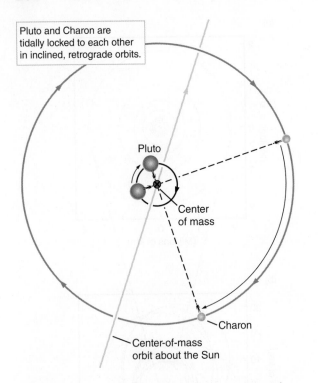

Pluto and Charon are tidally locked to each other in inclined, retrograde orbits.

Pluto

Center of mass

Charon

Center-of-mass orbit about the Sun

Figure 11.4 The doubly synchronous rotation and revolution in the Pluto-Charon system. The two bodies permanently face each other.

other body (**Figure 11.4**). Pluto and Charon are the only known pair in the Solar System in which both objects are tidally locked to each other in that way. Pluto's highly compact moon system may have been created by a massive collision between Pluto and another planetesimal, producing a cloud of debris that coalesced to form Charon and the four smaller moons of Pluto, perhaps similar to the way Earth's Moon formed (Chapter 8).

Some sets of moons are in synchronized orbits, called **orbital resonances**, in which the orbital period of one is a multiple of the orbital period of another. Jupiter's moons Ganymede, Europa, and Io are in a resonance of 1:2:4; for every one orbit of Ganymede, Europa orbits twice and Io orbits four times. When the moons are aligned, gravitational effects elongate Io's orbit, which varies from the tidal forces of Jupiter on Io. Pluto's five moons appear to be in a 1:3:4:5:6 sequence of near resonances. Pairs of some of Saturn's moons are in resonance, and resonances also exist between its moons and gaps in its rings.

The orbits of the moons not only indicate something about their origin. They also can be used to estimate the masses of their host planets by using Kepler's laws, just as the orbital properties of the planets can be used to find the mass of the Sun. An example of that calculation is shown in **Working It Out 11.1**.

CHECK YOUR UNDERSTANDING 11.1

Which of the following are characteristics of regular moons? (Choose all that apply.)
(a) They revolve around their planets in the same direction as the planets rotate.
(b) They have orbits that lie nearly in the equatorial planes of their planets. (c) They are usually tidally locked to their parent planets. (d) They are much smaller than all the known planets.

11.1 Working It Out Using Moons to Compute the Mass of a Planet

Recall from Working It Out 4.3 that Newton's version of Kepler's law for planets orbiting the Sun can be used to estimate the mass of the Sun. In Chapter 4, we used the following equation to calculate the mass (M) of the Sun:

$$M = \frac{4\pi^2}{G} \times \frac{A^3}{P^2}$$

where A is the semimajor axis of the orbit, P is the orbital period of a planet, and G is the universal gravitational constant.

For moons orbiting a planet, the same equation applies, as long as the moon is much less massive than the planet. Thus, we can use the orbital motion of the moons to estimate the mass of the planet. For example, Jupiter's moon Io has an orbital semimajor axis of $A = 422{,}000$ kilometers (km) and an orbital period of $P = 1.77$ days. To match the units in G, we need to convert P into seconds:

$$1.77 \text{ days} = 1.77 \text{ days} \times 24\frac{\text{h}}{\text{day}} \times 60\frac{\text{min}}{\text{h}} \times 60\frac{\text{s}}{\text{min}} = 152{,}928 \text{ s}$$

$G = 6.67 \times 10^{-20} \text{ km}^3/(\text{kg s}^2)$, so the mass of Jupiter (M_{Jup}) is given by

$$M_{\text{Jup}} = \frac{4\pi^2}{G} \times \frac{A^3}{P^2} = \frac{4\pi^2}{6.67 \times 10^{-20} \text{ km}/(\text{kg s})^2} \times \frac{(422{,}000 \text{ km})^3}{(152{,}92 \text{ s})^2}$$

$$M_{\text{Jup}} = 1.90 \times 10^{27} \text{ kg}$$

Using any other moon of Jupiter, you would get the same answer.

Back before Newton published his law of gravity and before any measured value of the gravitational constant G was possible, Galileo and Kepler showed that P^2/A^3 was the same for each of the four Galilean moons of Jupiter. That finding demonstrated that Kepler's law applied to systems other than planets orbiting the Sun.

11.2 Some Moons Have Geological Activity and Water

The moons of the Solar System can be grouped in several ways. Some groupings are based on the sequence of the moons in their orbits around their parent planets; others are based on the sizes or compositions of the moons. In this section, we organize our discussion of the moons by considering some of the same properties we discussed for the terrestrial planets: the history of the moons' geological activity and the presence of water and an atmosphere.

Some moons in the Solar System have been frozen in time since their formation during the early history of the Solar System, whereas others are even more geologically active than Earth. As with the terrestrial planets and Earth's Moon, surface features provide critical clues to a moon's geological history. For example, water ice is a common surface material among the moons of the outer Solar System, and the freshness of that ice indicates the age of those surfaces. Meteorite dust darkens the icy surfaces of moons just as dirt darkens snow in urban areas. A bright surface often means a fresh surface. The size and number of impact craters indicate the relative timing of events such as volcanism, and that timing enables scientists to gauge whether and when a moon may have been active in the past. Older surfaces have more craters. Observations of erupting volcanoes, which are found on Io and Enceladus, for example, are direct evidence that some moons are geologically active today.

Io, the Most Geologically Active Moon

One surprise in Solar System exploration was the discovery of active volcanoes on Io, the innermost of the four large moons of Jupiter. Because Io's orbit is elliptical, Jupiter's gravitational pull flexes Io's crust and generates enough thermal energy to melt parts of it. Did you ever take a piece of metal and bend it back and forth, eventually breaking it in half? Touching the crease line can burn your fingers. Just as bending metal in your hands creates heat, the continual flexing of Io generates enough energy to melt parts of its mantle. In that way, Jupiter's gravitational energy is converted into thermal energy, powering the most active volcanism in the Solar System.

Io is just slightly larger than our Moon. Its surface is covered with volcanic features, including vast lava flows, volcanoes, and volcanic craters (**Figure 11.5a**). Lava flows and volcanic ash bury impact craters as quickly as they form, so no impact craters have been observed on the surface. The *Voyager*, *Galileo*, and *New Horizons* spacecraft and the Keck telescope have observed hundreds of volcanic vents and active volcanoes on Io. The most vigorous eruptions spray sulfurous gases, ash, and lava hundreds of kilometers above the surface. Some of that material escapes from Io. Ash and other particles rain onto the surface as far as 600 km from the vents. The moon is so active that several huge eruptions often occur at the same time. **Figure 11.5b** shows the volcanic activity on Io—the source of the material supplying Io's plasma torus and flux tube (Figure 10.17).

The surface of Io displays pale shades of red, yellow, orange, and brown. Mixtures of sulfur, sulfur dioxide frost, and sulfurous salts of sodium and potassium probably cause that variety of colors. Bright patches may be fields of sulfur dioxide snow. Liquid sulfur dioxide flows beneath Io's surface, held at

(a)

G X U **V** I R

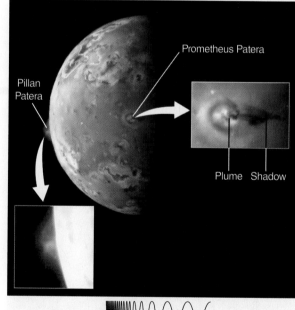

(b)

Pillan Patera

Prometheus Patera

Plume Shadow

G X U **V** I R

Figure 11.5 (a) Composite image of Jupiter's volcanically active moon Io constructed from pictures obtained by *Galileo*. (b) The plume from the crater Pillan Patera rises 140 km above the limb of the moon on the left, whereas the shadow of a 75-km-high plume can be seen to the right of the vent of Prometheus Patera, a volcanic crater near the moon's center.

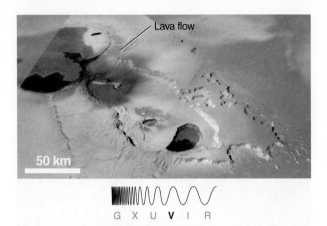

Figure 11.6 This *Galileo* image of Io shows regions where lava has erupted within a caldera. The molten lava flow is shown in false color to make it more visible.

Figure 11.7 This high-resolution *Galileo* image of Jupiter's moon Europa shows where the icy crust has been broken into slabs that have shifted into new positions. Those areas of chaotic terrain are characteristic of a thin, brittle crust of ice floating atop a liquid or slushy ocean.

Figure 11.8 Artist's conception of Europa, showing liquid bubbling up from the liquid ocean underneath the icy surface. Jupiter and Io are visible in the Europan sky.

high pressure by the weight of overlying material. Like water from a spring, that pressurized sulfur dioxide is pushed out through fractures in the crust, producing sprays of sulfur dioxide snow crystals that travel for up to hundreds of kilometers before settling back to the moon's surface. A similar process takes place with a carbon dioxide fire extinguisher—namely, liquid carbon dioxide contained at high pressure immediately turns to "dry ice" snow as it leaves the nozzle.

Spacecraft images reveal the plains, irregular volcanic craters, and flows, all related to the eruption of mostly silicate magmas onto the surface of Io. The images also show tall mountains, some nearly twice the height of Mount Everest. Huge structures have multiple summit craters showing a long history of repeated eruptions followed by the collapse of partially emptied magma chambers. Many of the chamber floors are very hot (**Figure 11.6**) and might still contain molten material similar to the magnesium-rich lavas that erupted on Earth more than 1.5 billion years ago. Volcanoes on Io are spread much more randomly than those on Earth, implying a lack of plate tectonics.

Because of its active volcanism, Io's mantle has turned inside-out more than once, leading to chemical differentiation. Volatiles such as water and carbon dioxide probably escaped into space long ago, whereas most of the heavier materials sank to the interior to form a core. Sulfur and various sulfur compounds, as well as silicate magmas, are constantly being recycled to form the complex surface we see today.

Evidence of Liquid Oceans on Europa and Enceladus

Jupiter's moon Europa is slightly smaller than our Moon, is made of rock and ice, and has an iron core. *Voyager* observed an outer shell of water ice with surface cracks and creases. Few impact craters are present, so the surface must be young. Regions of chaotic terrain (**Figure 11.7**) are places where the icy crust has been broken into slabs that have shifted into new positions. In other areas, the crust has split apart, and the gaps have filled in with new dark material rising from the interior. The young surface implies geological activity, which varies as Europa orbits Jupiter. As on Io, that activity is probably powered by continually changing tidal forces from Jupiter. However, the forces are not as strong on Europa as on Io because Europa is farther from Jupiter (**Working It Out 11.2**).

The *Galileo* spacecraft measured Europa's magnetic field and found that it is variable, indicating an internal electrically conducting fluid. Detailed computer models of the interior of Europa suggest that Europa has a global ocean 100 km deep that contains more water than held in all of Earth's oceans. That ocean might be salty as a result of dissolved minerals. The lightly cratered and thus geologically young surface indicates that energy exchange occurs between the icy crust and liquid water, but it is not known whether the icy crust is tens or hundreds of kilometers thick. It also is not yet known whether Europa has volcanic activity at the seafloor.

The Hubble Space Telescope may have detected two transient plumes of water vapor erupting from the icy surface, but they were not seen in later observations. The surface is too cold for liquid water, but lakes might lie a few kilometers underneath the surface. Scientists have reanalyzed observations of Europa in light of what has been learned about the many subsurface lakes in Antarctica, subglacial volcanoes in Iceland, and ice sheets at both poles of Earth. Many shallow "great lakes" may exist underneath the ice, and those would be prime targets for future exploration. **Figure 11.8** is an artist's schematic of Europa showing

11.2 Working it Out Tidal Forces on the Moons

Recall from Chapter 4 that the tidal force between a planet and its moon depends on the masses of the planet and the moon, and the size of the moon, divided by the cube of the distances between them:

$$F_{tidal} = \frac{2GM_{Jup}M_{moon}R_{moon}}{d_{Jup\text{-}moon}^3}$$

We can use that equation to compare the tidal forces between Jupiter and two of its moons by taking a ratio (as we did in Working It Out 4.4 to compare tidal forces between Earth and the Sun and Moon). For Io and Europa:

$$\frac{F_{tidal\text{-}Io}}{F_{tidal\text{-}Europa}} = \frac{\dfrac{2GM_{Jup}M_{Io}R_{Io}}{d_{Io}^3}}{\dfrac{2GM_{Jup}M_{Europa}R_{Europa}}{d_{Europa}^3}} = \frac{\dfrac{M_{Io}R_{Io}}{d_{Io}^3}}{\dfrac{M_{Europa}R_{Europa}}{d_{Europa}^3}}$$

After canceling out the mass of Jupiter (M_{Jup}) and the constant $2G$, and using data on Io and Europa from Appendix 4—namely, $M_{Io} = 8.9 \times 10^{22}$ kg, $M_{Europa} = 4.8 \times 10^{22}$ kg, $R_{Io} = 1,820$ km, $R_{Europa} = 1,560$ km, $d_{Io} = 422,000$ km, and $d_{Europa} = 671,000$ km—we have

$$\frac{F_{tidal\text{-}Io}}{F_{tidal\text{-}Europa}} = \frac{\dfrac{(8.9 \times 10^{22}) \times 1,820}{(422,000)^3}}{\dfrac{(4.8 \times 10^{22}) \times 1,560}{(671,000)^3}} = 8.7$$

Thus, the tidal forces on Io are much stronger than those on Europa.

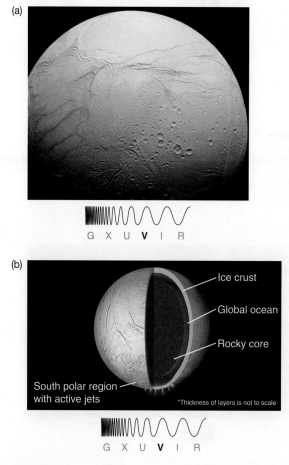

some of the water from the ocean leaking out at the surface, making it easier to study.

Enceladus, one of Saturn's icy moons, shows a variety of ridges, faults, and smooth plains. That evidence of tectonic processes is unexpected for a small (500 km) body. The activity on Enceladus is an example of **cryovolcanism**, which is similar to terrestrial volcanism but is driven by subsurface low-temperature liquids such as water and hydrogen rather than by molten rock. Some impact craters appear softened, perhaps by the viscous flow of ice, such as the flow that occurs in the bottom layers of glaciers on Earth. Parts of the moon have no craters, indicating recent resurfacing. Terrain near the south pole of Enceladus is cracked and twisted (**Figure 11.9a**). The cracks are warmer than their surroundings, suggesting that tidal heating and radioactive decay within the moon's rocky core heat the surrounding ice and drive it to the surface.

Enceladus has a liquid ocean buried beneath 30–40 km of ice crust and is 10 km deep (**Figure 11.9b**). Active cryovolcanic plumes expel water vapor, tiny ice crystals, and salts. Some of the crystals fall back onto the surface as an extremely fine, powdery snow. The rate of accumulation is very low—a fraction of a millimeter per year—but over time the snow builds up. *Cassini* scientists estimate that the snow may be 100 meters thick in one area near the south pole of Enceladus, indicating that the plume activity has continued on and off for at least tens of millions of years. Planetary scientists are very interested in sending a mission to Enceladus to explore this ocean.

Flexing from tidal forces is the likely source of the heat energy coming from Enceladus, as on Io and Europa. Enceladus has an orbital resonance with Saturn's moon Dione. A moon made completely of ice would be too stiff for tidal heating to be effective; tidal heating works more effectively with ice over liquid water or cracked ice with some liquid. It remains a mystery why Enceladus is so active, whereas Mimas, a neighboring moon of about the same size—but closer to Saturn and also subject to tidal heating—appears geologically dead.

Figure 11.9 Images of Enceladus taken by *Cassini*. (a) The deformed ice cracks (shown blue in false color) were found to be the sources of cryovolcanism. (b) Cryovolcanic plumes in the south polar region are seen spewing ice particles into space.

(a)

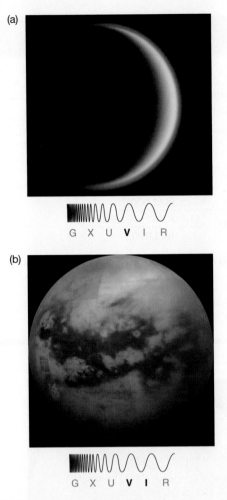

G X U V I R

(b)

G X U V I R

Figure 11.10 Images of Saturn's largest moon, Titan, taken by *Cassini*. (a) Titan's orange atmosphere is caused by organic, smoglike particles. (b) Infrared-light imaging penetrates Titan's smoggy atmosphere and reveals surface features.

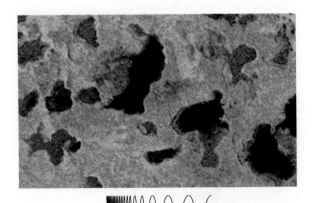

G X U V I R

Figure 11.11 Radar imaging (false color) near Titan's north pole shows lakes of liquid hydrocarbons covering 100,000 square kilometers (km²) of the moon's surface. Features such as islands, bays, and inlets are visible in many of those radar images.

Titan's Atmosphere, Lakes, and Ocean

Saturn's moon Titan is slightly larger than Mercury and has a composition of about 45 percent water ice and 55 percent rocky material. What makes Titan especially remarkable is its thick atmosphere and the presence of liquid methane on its surface. Whereas Mercury's secondary atmosphere has escaped to space due to proximity to the Sun, Titan's greater mass and distance from the Sun have allowed it to retain an atmosphere that is 30 percent denser than that of Earth. Titan's atmosphere, like Earth's, is mostly nitrogen. As Titan differentiated, various ices, including methane (CH_4) and ammonia (NH_3), emerged from the interior to form an early atmosphere. Ultraviolet photons from the Sun have enough energy to break apart ammonia and methane molecules—a process called **photodissociation**. Photodissociation of ammonia is the likely source of Titan's atmospheric nitrogen. Methane breaks into fragments that recombine to form organic compounds, including complex hydrocarbons such as ethane. Those compounds tend to cluster in tiny particles, creating organic smog much like the air over Los Angeles on a bad day; that gives Titan's atmosphere its characteristic orange hue (**Figure 11.10a**).

The *Cassini* spacecraft obtained close-up views of Titan's surface. Haze-penetrating infrared imaging showed broad regions of dark and bright terrain (**Figure 11.10b**). Radar imaging of Titan revealed irregularly shaped features in its northern hemisphere that appear to be widespread lakes and seas of methane, ethane, and other hydrocarbons (**Figure 11.11**). The photodissociative process by sunlight should have destroyed all atmospheric methane within a geologically brief period of about 50 million years, so a process must be renewing the methane being destroyed by solar radiation. That, along with the near absence of impact craters on the surface of Titan, suggests recent methane-producing activity. Radar views also indicate an active surface, showing features that resemble terrestrial sand dunes and channels. Heat supplied by radioactive decay could cause cryovolcanism that releases "new" methane from underground. The evidence of active cryovolcanism on Titan is indirect—the presence of abundant atmospheric methane and of methane lakes strongly suggests that Titan has some geological activity.

Titan has terrains reminiscent of those on Earth, with networks of channels, ridges, hills, and flat areas that may be dry lake basins. Those terrains suggest a methane cycle (analogous to Earth's water cycle) in which methane rain falls to the surface, washes the ridges free of the dark hydrocarbons, and then collects into drainage systems that empty into low-lying, liquid methane pools. Stubby, dark channels appear to be springs where liquid methane emerges from the subsurface; bright, curving streaks could be water ice that has oozed to the surface to feed glaciers. An infrared camera photographed a reflection of the Sun from such a lake surface. The type of reflection observed proves that the lake contains a liquid and is not frozen or dry. Recent observations might indicate waves on one of those lakes.

Titan is the only moon (besides Earth's) that has been landed upon. In 2005, *Cassini* released the *Huygens* probe, which plunged through Titan's atmosphere, taking pictures and measuring the moon's composition, temperature, pressure, and wind speeds. *Huygens* confirmed the presence of nitrogen-bearing organic compounds in the clouds. During its descent, *Huygens* encountered 120-m/s winds and temperatures as low as 88 kelvins (K). As it reached the surface, though, winds died down to less than 1 m/s and the temperature warmed to 112 K. The pictures taken by *Huygens* showed that the surface was wet with liquid methane, which evaporated as the probe—heated during its passage through the atmosphere—landed in the frigid soil. The surface also was rich with other organic (carbon-bearing) compounds, such as cyanogen and ethane. **Figure 11.12** shows that the surface around the landing site is relatively flat and littered with rounded "rocks" of water ice. The dark "soil" is probably a mixture of water and hydrocarbon ices.

As with Saturn's moon Enceladus and Jupiter's moon Europa, gravitational mapping provides indirect evidence that Titan also has an ocean buried beneath its surface. Some of Titan's surface features move by as much as 35 km, which suggests that the crust is sliding on an underlying liquid layer. The current model of Titan (**Figure 11.13**) is that its rigid ice shell varies in thickness and surrounds an ocean 100 km below the surface. That ocean would be made of water mixed with dissolved salts—possibly saltier than Earth's Dead Sea. In that model, methane outgassing would occur in hot spots.

Titan is the only moon with a significant atmosphere and the only Solar System body other than Earth that has standing liquid on the surface and a cycle of liquid rain and evaporation. In many ways, Titan resembles a primordial Earth, albeit at much lower temperatures. The presence of liquids and of organic compounds that could be biological precursors for life in the right environment makes Titan another high-priority target for continued exploration.

Figure 11.12 The two water-ice "rocks" just below the center of this *Huygens* image are about 85 centimeters (cm) from the camera and roughly 15 and 4 cm across, respectively.

G X U **V** I R

CHECK YOUR UNDERSTANDING 11.2A

Which of the following moons is *not* thought to have an ocean of water beneath its surface? (a) Io; (b) Europa; (c) Enceladus; (d) Titan

Cryovolcanism on Triton

Cryovolcanism also occurs on Triton, Neptune's largest moon. Triton is an irregular moon; its retrograde orbit suggests that Triton was captured by Neptune after the planet's formation. As Triton achieved its current circular, synchronous orbit, it experienced extreme tidal stresses from Neptune, generating large amounts of thermal energy. The interior may have melted, allowing Triton to become chemically differentiated.

Triton has a thin atmosphere and a surface composed mostly of ices and frosts of methane and nitrogen at a temperature of about 38 K. The relative absence of craters tells us the surface is geologically young. Part of Triton is covered with terrain that looks like the skin of a cantaloupe (**Figure 11.14**), with irregular pits and hills that may be caused by slushy ice emerging onto the surface from the interior. Veinlike features include grooves and ridges that could result from ice oozing out along fractures. The rest of Triton is covered with smooth volcanic plains. Irregularly shaped depressions as wide as 200 km formed when mixtures of water, methane, and nitrogen ice melted in the interior of Triton and erupted onto the surface, much as rocky magmas erupted onto the lunar surface and filled impact basins on Earth's Moon.

Colorless nitrogen ice creates a localized greenhouse effect, in which solar energy trapped beneath the ice raises the temperature at the base of the ice layer. A temperature increase of only 4 K vaporizes the nitrogen ice. As that gas is formed, the expanding vapor exerts very high pressures beneath the ice cap. Eventually, the ice ruptures and vents the gas explosively into the low-density atmosphere. *Voyager 2* found four of those active geyserlike cryovolcanoes on Triton. Each consisted of a plume of gas and dust as much as 1 km wide rising 8 km above the surface, where the plume was caught by upper atmospheric winds and carried for hundreds of kilometers downwind. Dark material, perhaps

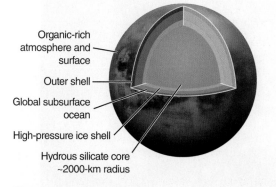

Organic-rich atmosphere and surface

Outer shell

Global subsurface ocean

High-pressure ice shell

Hydrous silicate core ~2000-km radius

Figure 11.13 Artist's conception of Titan's internal structure showing how Titan is differentiated, with a core of water-bearing rocks and a subsurface ocean of liquid water. A layer of high-pressure ice surrounds the core, with an outer ice shell on top of the subsurface ocean.

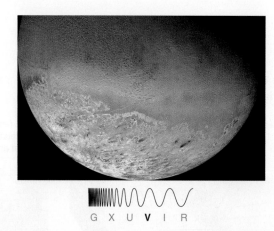

G X U **V** I R

Figure 11.14 *Voyager 2* mosaic showing various terrains on the Neptune-facing hemisphere of Triton. The lack of impact craters in the "cantaloupe terrain," visible at the top, indicates a geologically younger age than that of the bright, cratered terrain at the bottom.

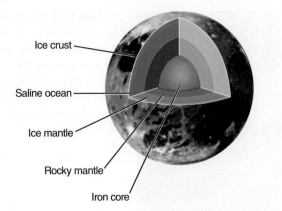

Figure 11.15 Artist's conception of the interior of Jupiter's moon Ganymede, showing that the ocean and ice may be stacked in multiple layers.

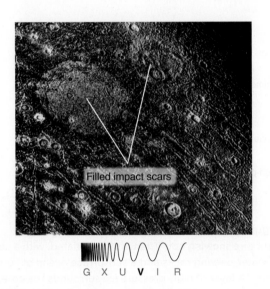

Figure 11.16 *Voyager* image showing filled impact scars on Jupiter's moon Ganymede.

silicate dust or radiation-darkened methane ice grains, is carried along with the expanding vapor into the atmosphere, from which it later settles to the surface, forming dark patches streaked out by local winds, as seen near the lower right of Figure 11.14.

Formerly Active Moons

Some moons show clear evidence of past ice volcanism and tectonic deformation, but no current geological activity. For example, Jupiter's moon Ganymede, the largest moon in the Solar System, is larger than the planet Mercury. When Jupiter was forming, low temperatures enabled grains of water ice to survive and coalesce along with dust grains into larger bodies at the distance of Ganymede's orbit. In less than half a million years, those bodies accreted to form Ganymede. Heating from accretion melted parts of Ganymede so that it is fully differentiated (**Figure 11.15**), with outer water layers, an inner silicate zone, and an iron-rich liquid core. As the moon cooled, much of the outer water layer froze, forming a dirty ice crust. Most of the denser materials sank to the central core, leaving an intermediate ice-silicate zone. Ganymede might also have a large, salty ocean underneath its icy surface, maybe 800 km deep, containing 25 times the volume of Earth's oceans.

Its surface is composed of two prominent terrains: a dark, heavily cratered (and therefore ancient) terrain, and a bright terrain characterized by ridges and grooves. The abundance of impact craters on Ganymede's dark terrain reflects the period of intense bombardment during the early history of the Solar System. The largest region of ancient dark terrain includes a semicircular area more than 3,200 km across on the leading hemisphere. Furrowlike depressions occurring in many dark areas are among Ganymede's oldest surface features. They may represent surface deformation from internal processes or may be relics of impact-cratering processes.

Impact craters on Ganymede range up to hundreds of kilometers in diameter, and the larger craters are proportionately shallower. The icy crater rims slowly slump, like a lump of soft clay. They are seen as bright, flat, circular patches found principally in the moon's dark terrain (**Figure 11.16**) and are thought to be scars left by early impacts onto a thin, icy crust overlying water or slush (**Figure 11.17**). In Chapter 8, we discussed how planetary surfaces can be fractured by faults or folded by compression resulting from movements initiated in the mantle. On Ganymede, the tectonic processes have been so intense that the fracturing and faulting have completely deformed the icy crust, destroying all

Figure 11.17 Filled impact scars form as viscous flow smooths out structures left by impacts on icy surfaces.

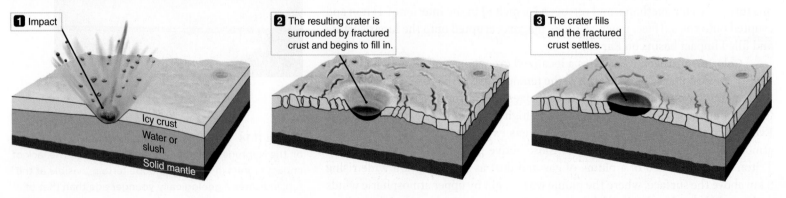

signs of older features, such as impact craters, and creating the bright terrain. The energy that powered Ganymede's early activity was liberated during a period of differentiation when the moon was very young. After differentiation was complete, that source of internal energy ran out, and geological activity ceased.

Many other moons show evidence that they experienced an early period of geological activity that resulted in a dazzling array of terrains. A 400-km impact crater scars Saturn's moon Tethys, covering 40 percent of its diameter, and an enormous canyonland wraps around at least three-fourths of the moon's equator. Saturn's moon Dione shows bright ice cliffs up to several hundred meters high, created by tectonic fracturing. The trailing hemisphere of Saturn's Iapetus is bright, reflecting half the light that falls on it, whereas much of the leading hemisphere is as black as tar. Those dark deposits appear *only* in the leading hemisphere of Iapetus, suggesting that they might be debris that was blasted off small retrograde moons of Saturn by micrometeoritic impacts and swept up by Iapetus as it moved along in its prograde orbit around Saturn.

Saturn's moon Mimas, no larger than the state of Ohio, is heavily cratered with deep, bowl-shaped depressions. The most striking feature on Mimas is a huge impact crater in the leading hemisphere (**Figure 11.18**). Named "Herschel" after astronomer Sir William Herschel, who discovered many of Saturn's moons, the crater is 130 km across—a third the size of Mimas itself. It is doubtful that Mimas could have survived the impact of a body much larger than the one that created Herschel. Some astronomers think that Mimas (and perhaps other small, icy moons as well) was hit many times in the past by objects so large as to fragment the moon into many small pieces. Each time that happened, the individual pieces still in Mimas's orbit would coalesce to re-form the moon, perhaps in much the same way that Earth's Moon coalesced from fragments that remained in orbit around Earth after a large planetesimal impacted Earth early in its history.

Areas on Uranus's small moon Miranda have been resurfaced by eruptions of icy slush or glacierlike flows. Other moons of Uranus—Oberon, Titania, and Ariel—are covered with faults and additional signs of early tectonism. On Ariel, in particular, very old, large craters appear to be missing, perhaps obliterated by earlier volcanism.

Geologically Dead Moons

Geologically dead moons, such as Jupiter's Callisto, Saturn's Hyperion, Uranus's Umbriel, and a large assortment of irregular moons, are moons for which little or no evidence exists of internal activity having occurred at any time since their formation. The surfaces of those moons are heavily cratered and show no modification other than the cumulative degradation caused by a long history of impacts.

Callisto is the third-largest moon in the Solar System, just slightly smaller than Mercury. Callisto also is the darkest of the Galilean moons of Jupiter, yet it is still twice as reflective as Earth's Moon. That brightness indicates that Callisto is rich in water ice, but with a mixture of dark, rocky materials. Except in areas that experienced large impact events, the surface is essentially uniform, consisting of relatively dark, heavily cratered terrain. Callisto's most prominent feature is a 2,000-km, multiringed structure of impact origin named Valhalla (the largest bright feature visible on Callisto's face in **Figure 11.19**). *Galileo* results suggest that a liquid ocean containing water or water mixed with ammonia could exist beneath the heavily cratered surface. Callisto may have partially differentiated, with rocky material separating from ices and sinking deeper into the interior.

G X U V I R

Figure 11.18 *Cassini* image showing Saturn's moon Mimas and the crater Herschel.

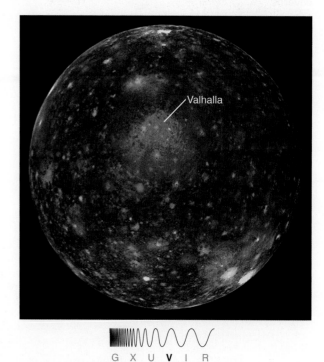

G X U V I R

Figure 11.19 *Galileo* image showing Jupiter's second-largest moon, Callisto. The moon's ancient surface is dominated by impact craters and shows no sign of early internal activity.

G X U V I R

Figure 11.20 Saturn's moon Hyperion rotates chaotically, with its rotation period and spin axis constantly changing. The 250-km moon's low density and spongelike texture, seen in this *Cassini* image, suggest that its interior houses a vast system of caverns.

Saturn's Hyperion (**Figure 11.20**) is one of the largest irregularly shaped moons and could be the remnant of an impact. The extensive craters look almost like sponges. Hyperion crosses Saturn's magnetosphere in its chaotic orbit, which seems to have left the moon with some electric charge. Umbriel, the darkest and third largest of Uranus's moons, appears uniform in color, reflectivity, and general surface features, indicating an ancient surface. The real puzzle posed by Umbriel is why it is geologically dead, whereas the surrounding large moons of Uranus have been active at least at some time in their past.

CHECK YOUR UNDERSTANDING 11.2B

Rank the following moons according to the density of impact craters (from most to least) that you would expect to observe on the surface. (a) Callisto; (b) Titan; (c) Io; (d) Ganymede

11.3 Rings Surround the Giant Planets

A planetary ring is a collection of particles—ranging in size from tiny grains to house-sized boulders—that orbit individually around a planet, forming a flat disk. Ring systems do not occur around the terrestrial planets but are found around each of the giant planets. **Figure 11.21** shows how the ring system of each giant planet varies in size and complexity: some systems extend for hundreds of thousands of kilometers, and some have detailed structure that includes many small rings. In this section, we discuss ring formation, composition, and evolution.

The Discovery of Planetary Rings

Saturn's rings have been observed for centuries. In 1610, Galileo observed two small objects next to Saturn and thought they might be similar to the four moons orbiting Jupiter. But Saturn's "moons" did not move, and 2 years later they disappeared. In 1655, Dutch instrument maker Christiaan Huygens (1629–1695) pointed a superior telescope of his own design at Saturn. Huygens observed that an apparently continuous flat ring surrounds the planet and that the ring's visibility changes with its apparent tilt as Saturn orbits the Sun. Over the next three centuries, astronomers discovered more rings around Saturn, but searches failed to detect rings around any other planet.

In 1977, a team of astronomers studying the atmosphere of Uranus during stellar occultations saw brief, minute changes in the brightness of a star as it first approached and then receded from the planet. The astronomers realized that meant that Uranus has rings. Over the next several years, stellar occultations revealed nine rings surrounding the planet. In 1986, *Voyager 2* imaged two additional rings of Uranus, and in 2005 the Hubble Space Telescope recorded two more, bringing the total to 13. In 1979, cameras on *Voyager 1* recorded a faint ring around Jupiter. The occultation technique also revealed arclike ring segments around Neptune, which were determined to be complete rings when *Voyager 2* reached Neptune in 1989.

The Orbits of Ring Particles

Ring particles follow Kepler's laws, so the speed and orbital period of each particle must vary with its distance from the planet. The closest particles move the fastest and have the shortest orbital periods (see Working It Out 11.1). The orbital

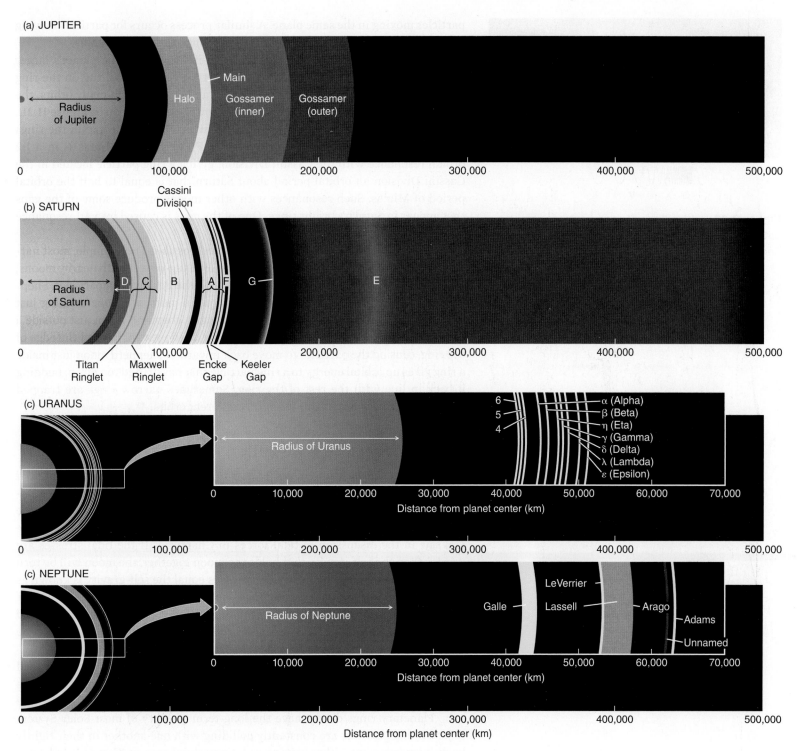

Figure 11.21 The ring systems of the four giant planets vary in size and complexity. Saturn's system, with its broad E Ring, is by far the largest and has the most complex structure in its inner rings.

periods of particles in Saturn's bright rings, for example, range from 5 hours 45 minutes at the inner edge of the innermost bright ring to 14 hours 20 minutes at the outer edge of the outermost bright ring. Ring particles have low speeds relative to one another because they are all orbiting in the same direction. A particle moving on an upward trajectory will bump into another particle on a downward trajectory, and the upward and downward motion will cancel—leaving the

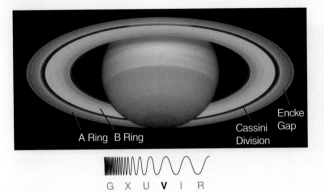

Figure 11.22 Hubble Space Telescope image showing Saturn and its A Ring, B Ring, Cassini Division, and Encke Gap. The C Ring is too dim to be seen clearly.

particles moving in the same plane. A similar process occurs for particles moving inward and outward, leaving the particles moving at a constant radius.

The orbits of ring particles can also be influenced by the planet's larger moons. If the moon is massive enough, it exerts a gravitational tug on the ring particles as it passes by. If that happens over and over through many orbits, the particles are pulled out of the area, leaving a lower-density gap (see Figure 11.21). Such is the case with Saturn's moon Mimas, which causes the gap in the rings around Saturn called the **Cassini Division** (**Figure 11.22**). Mimas is in a 1:2 orbital resonance with the Cassini Division, giving a ring particle located in the Cassini Division an orbital period about Saturn that is equal to half the orbital period of Mimas. Such resonances with other moons produce some of the gaps that appear in Saturn's bright rings. One of the gaps is caused by a 4:1 resonance between the ring particles and Mimas.

Other kinds of orbital resonances also are possible. For example, most narrow rings are caught up in a periodic gravitational tug-of-war with nearby moons, known as **shepherd moons** for how they "herd" the flock of ring particles. Shepherd moons are usually small and often come in pairs, with one orbiting just inside and the other just outside a narrow ring. A shepherd moon just outside a ring robs orbital energy from any particles that drift outward beyond the edge of the ring, causing the particles to move back inward. A shepherd moon just inside a ring gives up orbital energy to a ring particle that has drifted too far in, nudging it back in line with the rest of the ring. Sometimes narrow rings are trapped between two shepherd moons in slightly different orbits.

Ring Formation and Evolution

Much of the material found in planetary rings is thought to be the result of tidal stresses. If a moon (or other planetesimal) orbits a large planet, the force of gravity will be stronger on the side of the moon close to the planet and weaker on the side farther away. That difference in gravitational force stretches out the moon, as you saw in the discussion of tidal forces in Chapter 4. If the tidal stresses are greater than the self-gravity that holds the moon together, the moon will be torn apart. The distance at which the tidal stresses equal the self-gravity is known as the Roche limit. The Roche limit applies only to objects held together by their own gravity; it does not apply to objects, such as people or cars, that are held together by other forces. If a moon or planetesimal comes within a planet's Roche limit, the object is pulled apart by tidal stresses, leaving many small pieces orbiting the planet. Those pieces gradually spread out, and their orbits are circularized and flattened out by collisions. The fragmented pieces of the disrupted body are then distributed around the planet in the form of a ring.

Planetary rings do not have the long-term stability of most Solar System objects. Ring particles are constantly colliding with one another in their tightly packed environment, either gaining or losing orbital energy. That redistribution of orbital energy can cause particles at the ring edges to leave the rings and drift away, aided by nongravitational influences such as the pressure of sunlight. Although moons may help guide the orbits of ring particles and delay the dissipation of the rings themselves, at best that condition can be only temporary. Saturn's brightest rings might be nearly as old as Saturn, but most planetary rings eventually disperse.

Even Earth may have had several short-lived rings at various times during its long history. Many comets or asteroids must have passed within Earth's Roche limit (about 25,000 km for rocky bodies and more than twice that for icy bodies) to disintegrate into a swarm of small fragments to create a temporary ring. Unlike the giant planets, however, Earth lacks shepherd moons to provide orbital stability to rings.

The Composition of Ring Material

Because much of the material in the rings of the giant planets comes from their moons, the composition of the rings is similar to that of the moons. Saturn's bright rings probably formed when a moon or planetesimal came within the Roche limit of Saturn. Those rings reflect about 60 percent of the sunlight falling on them. They are made of water ice, though a slight reddish tint indicates that they must contain small amounts of other materials, such as silicates. The icy moons around Saturn or the frozen comets of the outer Solar System could easily supply that material.

Saturn's rings are the brightest in the Solar System and are the only ones known to be composed of water ice. In stark contrast, the rings of Uranus and Neptune are among the darkest objects known in the Solar System. Only 2 percent of the sunlight falling on them is reflected back into space, making the ring particles blacker than coal or soot. No silicates or similar rocky materials are that dark, so those rings are likely to be composed of organic materials and ices that have been radiation darkened by high-energy, charged particles in the magnetospheres of those planets. (Radiation blackens organic ices such as methane by releasing carbon from the ice molecules.) Jupiter's rings are of intermediate brightness, suggesting that they may be rich in silicate materials, like the innermost of Jupiter's small moons.

The jumble of fragments that make up Saturn's rings is understood to be a product of tidal disruption of a moon or planetesimal, but moons can contribute material to rings in other ways. The brightest of Jupiter's rings is a relatively narrow strand only 6,500 km across, consisting of material from the moons Metis and Adrastea. Those two moons orbit in Jupiter's equatorial plane, and the ring they form is narrow. Beyond that main ring, however, are the very different wispy rings called gossamer rings. The gossamer rings are supplied with dust by the moons Amalthea and Thebe. The innermost ring in Jupiter's system, called the halo ring, consists mostly of material from the main ring. As the dust particles in the main ring drift slowly inward toward the planet, they pick up an electric charge and are pulled into that thick torus by electromagnetic forces associated with Jupiter's powerful magnetic field.

Finally, moons may contribute ring material through volcanism. Volcanoes on Jupiter's moon Io continually eject sulfur particles into space, many of which are pushed inward by sunlight and find their way into a ring. The particles in Saturn's E Ring are ice crystals ejected from icy geysers on the moon Enceladus, in the very densest part of the E Ring (**Working It Out 11.3**). Ice particles ejected into space replace particles continually lost from Saturn's E Ring (**Figure 11.23**). The E Ring will survive for as long as Enceladus remains geologically active.

CHECK YOUR UNDERSTANDING 11.3

If rings are observed around a planet, that means: (a) a recent source of ring material is present; (b) the planet is newly formed; (c) the rings formed with the planet; (d) the rings are made of fine dust.

G X U V I R

Figure 11.23 Saturn's moon Enceladus (the large bright spot appearing to be on the ring) is the source of material in Saturn's E Ring. Note the distortion in the distribution of ring material around the moon. That distortion is caused by the moon's gravitational influence on the orbits of ring particles. Other bright objects in this *Cassini* image are distant stars.

11.3) Working It Out Feeding the Rings

The moons of the giant planets have a low surface gravity and a much lower escape velocity than the 11.2 km/s of Earth. Thus, volcanic emissions from some of those small moons can escape and supply material to a ring. Recall the equation from Working It Out 4.2 for the escape velocity from a spherical object of mass M and radius R:

$$v_{esc} = \sqrt{\frac{2GM}{R}}$$

Saturn's moon Enceladus has a mass of 1.08×10^{20} kg and a radius of 250 km. The escape velocity from Enceladus is given by

$$v_{esc} = \sqrt{\frac{2 \times [6.67 \times 10^{-20}\,\text{km}^3/(\text{kg s}^2)] \times (1.08 \times 10^{20}\,\text{kg})}{250\,\text{km}}}$$

$$v_{esc} = 0.24\,\text{km/s; or multiply by 3,600 s/h to get 864 km/h}$$

That escape velocity is much lower than the speed of the volcanic plumes on Enceladus, which is nearly 2,200 km/h. The icy particles from the plumes supply particles to Saturn's E Ring.

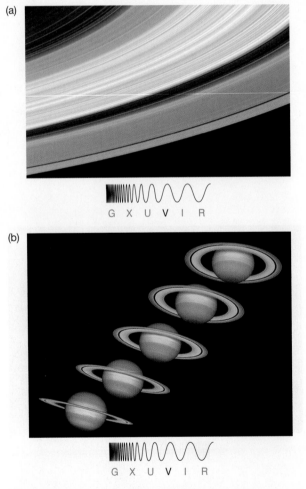

(a)

G X U V I R

(b)

G X U V I R

Figure 11.24 (a) *Cassini* image of the rings of Saturn showing many ringlets and minigaps. The cause of most of that structure has yet to be explained in detail. (b) Hubble Space Telescope took these images of Saturn over 6 years, showing how the rings appear to change their shape as Saturn moves through its orbit around the Sun. That effect is caused by changes in the position of Earth relative to the rings.

11.4 Ring Systems Have a Complex Structure

Huygens, with his mid-17th-century understanding of physics, thought that Saturn's ring was a solid disk surrounding the planet. Not until the mid-19th century did the brilliant Scottish mathematical physicist James Clerk Maxwell show that solid rings would be unstable and would quickly break apart. In this section, we examine the details of the rings of each outer planet.

Saturn's Magnificent Rings—A Closer Look

Saturn is adorned by a magnificent and complex system of rings, unmatched by any other planet in the Solar System. Figure 11.21b shows the components of Saturn's ring system and its major divisions and gaps. Among the four giant planets, Saturn's rings are the widest and brightest. The outermost bright ring, the A Ring, is the narrowest of the three bright rings. It has a sharp outer edge and contains several narrow gaps.

In 1675, the Italian-French astronomer Jean-Dominique Cassini (1625–1712) found a gap in the planet's seemingly solid ring. Saturn appeared to have two rings rather than one, and the gap that separated them became known as the *Cassini Division*. The Cassini Division is so wide (4,700 km) that the planet Mercury would almost fit within it. Astronomers once thought that the gap was empty, but images taken by *Voyager 1* show that the Cassini Division is filled with material, albeit less dense than the material in the bright rings.

The B Ring, whose width is roughly twice Earth's diameter, is the brightest of Saturn's rings and has no internal gaps on the scale of those seen in the other bright rings. The C Ring is so much fainter than neighboring rings that it often fails to show up in normally exposed photographs. Through the eyepiece of a telescope, that ring appears like delicate gauze. No known gap exists between the C Ring and either adjacent ring; only an abrupt change in brightness marks the boundary between them. The cause of that sharp change in the amount of ring material remains unknown. Too dim to be seen next to Saturn's bright disk, the D Ring is a fourth wide ring that was unknown until it was imaged by *Voyager 1*. The D Ring shows less structure than any of the bright rings, and it does not appear to have a definable inner edge. The D Ring may extend all the way down to the top of Saturn's atmosphere, where its ring particles would burn up as meteors.

Saturn's bright rings are not uniform. The A and C rings contain hundreds, and the B Ring, thousands, of individual ringlets, some only a few kilometers wide (**Figure 11.24a**). Each ringlet is a narrowly confined concentration of ring particles bounded on both sides by regions of relatively little material. About every 15 years, the plane of Saturn's rings lines up with Earth, and we view them edge on (**Figure 11.24b**). The rings are so thin that they almost vanish for a day or so in even the largest telescopes. While the glare of the rings is absent, astronomers search for undiscovered moons or other faint objects close to Saturn. In 1966, an astronomer looking for moons found weak but compelling evidence for a faint ring near the orbit of Saturn's moon Enceladus. In 1980, *Voyager 1* confirmed the existence of that faint ring, now called the E Ring, and found another, closer one known as the G Ring.

The E and G rings are examples of diffuse rings. In a diffuse ring, particles are far apart, and rare collisions between them can cause their individual orbits to become eccentric, inclined, or both. Because collisions are rare, the particles tend to remain in those disturbed orbits. Diffuse rings spread out horizontally and thicken vertically, sometimes without any obvious boundaries.

Diffuse rings contain tiny particles that show up best when the viewer is looking at those rings *into* the light—that is, in the direction of the Sun. In contrast, larger objects such as pebbles and boulders are easiest to see when the light illuminating them is coming from behind the viewer. For example, dust particles on your windshield appear brightest when you are driving toward the Sun. Photographers call that effect backlighting and often place their subjects in front of a bright light to highlight hair (**Figure 11.25a**). Backlighting happens when light falls on very small objects—those with dimensions a few times to several dozen times the wavelength of light. Cat hair and human hair are near the upper end of that range. Light falling on strands of hair tends to continue in the direction away from the source of illumination. Very little light is scattered off to the side, and almost none is scattered back toward the source.

Some of the dustier planetary rings are filled with particles just a few times larger than the wavelength of visible light. To a spacecraft approaching from the direction of the Sun, such rings may be difficult or even impossible to see. Those tiny ring particles scatter very little sunlight back toward the Sun and the approaching spacecraft. However, when the spacecraft passes by the planet and looks backward toward the Sun, those same dusty rings suddenly appear as a circular blaze of light, much like a halo surrounding the nighttime hemisphere of the planet. Many planetary rings are best seen with backlighting, and some, such as Saturn's G Ring (**Figure 11.25b**), have been observed only under those conditions. In 2009, astronomers using the infrared Spitzer Space Telescope discovered another diffuse ring around Saturn (**Figure 11.26**). That dusty ring is thicker than other rings, about 20 times larger than Saturn from top to bottom, and is tilted 27° with respect to the plane of the rest of the rings.

Although Saturn's bright rings are very wide—more than 62,000 km from the inner edge of the C Ring to the outer edge of the A Ring—they are extremely thin. Saturn's bright rings are no more than 100 meters thick and probably only a few tens of meters from their lower to upper surfaces. The diameter of Saturn's bright ring system is 10 million times the thickness of the rings. If the bright rings of Saturn were the thickness of a page in a book, their diameter would equal six football fields laid end to end.

Voyager 1 images showed that Saturn's F Ring is separated into several strands that appear to be intertwined as well as displaying what appear to be several knots and kinks. Saturn's F Ring is now understood to be a dramatic example of the action of a pair of shepherd moons. The F Ring is flanked by Prometheus,

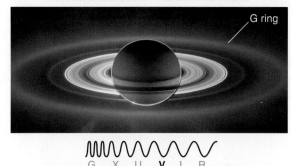

Figure 11.25 (a) Backlighting of hair creates a halo effect. (b) This backlit image of Saturn shows the G Ring.

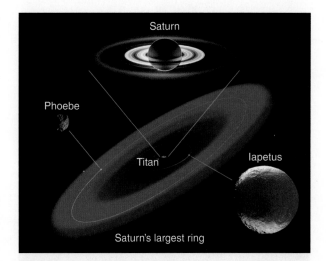

Figure 11.26 Artist's conception showing the highly inclined giant dust ring recently discovered around Saturn. That ring is so large that the rest of Saturn appears as a speck in the center (magnified in the inset).

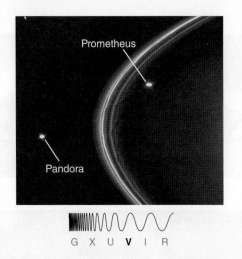

Figure 11.27 *Cassini* image showing Saturn's F Ring and its shepherd moons, Pandora and Prometheus. Also visible are some "kinks" in the inner ring.

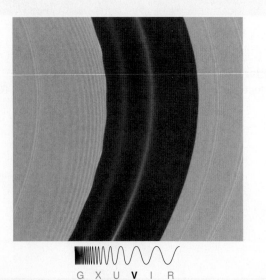

Figure 11.28 In this high-resolution view from *Cassini*, Saturn's Encke Gap reveals a scalloped pattern, caused by the moon Pan, along its inner edge.

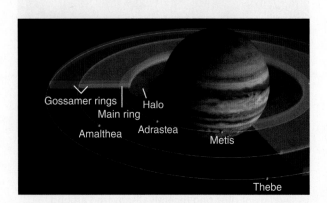

Figure 11.29 A diagram of the Jupiter ring system and the small moons that form the rings.

orbiting 860 km inside the ring, and Pandora, orbiting 1,490 km on the outside, as seen in a more recent *Cassini* image (**Figure 11.27**). Both moons are irregular in shape, with average diameters of 85 and 80 km, respectively. Because of their relatively large size and their proximity, the moons exert significant gravitational forces on nearby ring particles. The resulting tug-of-war between Prometheus pulling nearby ring particles into larger orbits and Pandora drawing its neighboring particles into smaller orbits causes the bizarre structure in the F Ring (see the **Process of Science Figure**).

The F Ring is not an isolated case. The 360-km-wide Encke Gap in the outer part of Saturn's A Ring contains two narrow rings that show bright knots and dark gaps—a structure that must be related to a 20-km moon named Pan that orbits within the gap. Small moons orbiting within ring gaps also can disturb ring particles along the edges of the gaps. **Figure 11.28** shows the scalloped pattern caused by Pan and that is found along the inner edge of the Encke Gap. Similarly, the 7-km moon Daphnis disrupts the inner and outer edges of Saturn's 35-km Keeler Gap, located near the outer edge of the A Ring.

Voyager 1 and then *Cassini* observed dozens of dark, spokelike features in the outer part of Saturn's B Ring. Those temporary features grow in a radial direction and are seasonal, lasting for less than half an orbit around Saturn, indicating that the particles in the spokes must be suspended above the ring plane, probably by electrostatic forces. One explanation is that when the charged particles interact with Saturn's magnetic field, the spokes rotate as the planet spins.

Rings around the Other Outer Planets

Ring structure among the other giant planets is not as diverse as Saturn's. Most rings other than Saturn's are narrow, although a few are diffuse. When *Voyager 1* scientists looked at Jupiter's ring system with the Sun behind the camera, they saw only a narrow, faint strand. But when *Voyager 2* looked back toward the Sun while in the shadow of the planet, Jupiter's rings suddenly blazed into prominence. **Figure 11.29** shows Jupiter's moons orbiting among that ring system. Most of the material in Jupiter's rings is made up of fine dust dislodged by meteoritic impacts on the surfaces of Jupiter's small inner moons.

Of the 13 rings of Uranus, nine are very narrow and widely spaced relative to their widths (see Figure 11.21c). Most are only a few kilometers wide, but they are many hundreds of kilometers apart. The two rings discovered by the Hubble Space Telescope in 2005 (**Figure 11.30**) are much wider and more distant than the narrow rings. The most prominent ring of Uranus, the Epsilon Ring, is eccentric and the widest of the planet's inner narrow rings, ranging between 20 and 100 km. The innermost ring is wide and diffuse, with an undefined inner edge. As with Saturn's D Ring, material in that ring may be spiraling into the top of the planetary atmosphere. When viewed under backlit conditions by *Voyager 2*, the space *between* the rings of Uranus turned out to be filled with dust, much as in Jupiter's ring system.

The rings in the Encke Gap are unusual but not unique. If shepherd moons are in eccentric or inclined orbits, they cause the confined ring also to be eccentric or inclined. That is the case for the Epsilon Ring of Uranus. Because shepherd moons can be so small, they often escape detection. According to current theories of ring dynamics, several still-unknown shepherd moons must be interspersed among the ring systems of the outer Solar System.

For a while, Neptune seemed to be the only giant planet devoid of rings. Then, in the early to mid-1980s, occultation searches by teams of astronomers began yielding confusing results. Several occultation events that appeared to be due to

Following Up on the Unexpected

Scientists expected to find dust particles in the rings of Saturn moving on undisturbed orbits. Instead, the F Ring particles seemed to disobey the laws of physics!

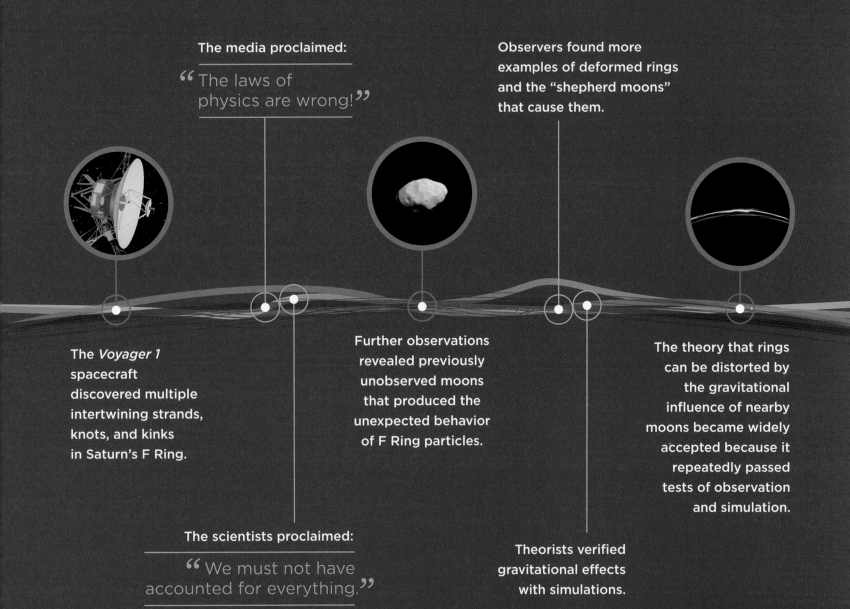

The media proclaimed:

" The laws of physics are wrong! "

Observers found more examples of deformed rings and the "shepherd moons" that cause them.

The *Voyager 1* spacecraft discovered multiple intertwining strands, knots, and kinks in Saturn's F Ring.

Further observations revealed previously unobserved moons that produced the unexpected behavior of F Ring particles.

The theory that rings can be distorted by the gravitational influence of nearby moons became widely accepted because it repeatedly passed tests of observation and simulation.

The scientists proclaimed:

" We must not have accounted for everything. "

Theorists verified gravitational effects with simulations.

Scientists are excited by apparent violations of well-supported theories because they may lead to new discoveries. One consequence is that unexpected or contradictory results often receive more attention than confirming results.

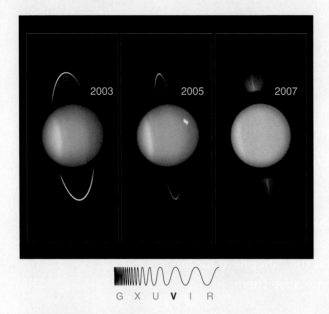

G X U **V** I R

Figure 11.30 The appearance of rings depends dramatically on lighting conditions and the angle from which they are seen. Earth's view of the rings of Uranus changes over several years, as shown.

Ring arc

G X U **V** I R

Figure 11.31 *Voyager 2* image showing the three brightest arcs in Neptune's Adams Ring. Neptune itself is greatly overexposed.

rings were seen on only one side of the planet. The astronomers concluded that Neptune was surrounded not by complete rings but rather by several arclike ring segments. When *Voyager 2* reached Neptune in 1989, it was determined that Neptune's rings were complete. The **ring arcs** are high-density segments within one of its narrow rings. All of Neptune's rings are faint and, except for the ring arcs, contain too little material to be detected through the stellar occultation technique.

Four of Neptune's six rings are very narrow, similar to the 13 narrow rings surrounding Uranus. The other two are a few thousand kilometers wide (see Figure 11.21d). Neptune's rings are named for 19th century astronomers who made major contributions to Neptune's discovery. Of those, the Adams Ring attracts the greatest attention. Much of the material in the Adams Ring is clumped into several ring arcs. Those high-density ring segments extend 4,000–10,000 km yet are only about 15 km wide. When first discovered, the ring arcs were a puzzle because mutual collisions among their particles should cause the particles to be spread more or less uniformly around their orbits. Most astronomers now attribute that clumping to orbital resonances with the moon Galatea, which orbits just inside the Adams Ring (**Figure 11.31**). Images obtained by the Hubble Space Telescope in 2004 and 2005 compared with those taken by *Voyager 2* in 1989 show that some parts of the Neptune ring arcs are unstable. Slow decay is evident in two of the arcs, Liberté and Courage, suggesting that they may disappear before the end of the century. Uranus's Lambda Ring and Saturn's G Ring also show ring arcs.

Moons and Rings around Exoplanets

Large and small planets, many in multiplanet systems, have been detected within our galaxy, demonstrating that exoplanets are common. If our Solar System is typical, we might expect that other planetary systems contain planets with rings and planets with large moons, perhaps with geological activity and water. However, identifying exoplanet moons and rings is at the limit of current astronomical instrumentation, and as of this writing no confirmed detections have been made.

The proposed methods for detecting *exomoons* are similar to those for detecting exoplanets. Recall that *Kepler* detects planets through the transit method (see Figures 7.19 and 7.20). The presence of a moon near the planet can slightly alter the depth and duration of the change in the light curve. The moon is probably in a different place in its orbit each time the planet orbits the star, so those alterations in the light curve will be different for each cycle. Over several cycles, the signature of a moon might be detected. It is also possible that astronomers could detect a large exomoon as the moon itself transits its star. Or a large moon could cause its planet's orbit to "wobble," which could be detected in the transit signal. NASA supercomputers are being used to analyze the large database from *Kepler* to look for signatures of exomoons.

Similarly, a large exoplanet with an extensive ring system, especially if the ring system has gaps, might be detectable through changes in the transit signal. The depth and length of the changes in the star's light curve could indicate the presence of such a system. **Figure 11.32** is an artist's conception of an exoplanet with rings and a large exomoon.

CHECK YOUR UNDERSTANDING 11.4

If you wanted to search for faint rings around a giant planet by sending a spacecraft on a flyby, you could make your best observations: (a) as the spacecraft approached the planet; (b) after the spacecraft passed the planet; (c) while the spacecraft is orbiting the planet; (d) during the closest flyby; (e) while the spacecraft is orbiting one of its moons.

Origins Extreme Environments

During the 1980s and 1990s, as the *Voyager* spacecraft were exploring the outer Solar System, biologists back on Earth were identifying strange forms of life. Off the coast of the Galápagos Islands, 2,500 meters beneath the ocean's surface, plates grind against each other, creating friction, high temperatures, and seafloor volcanism. Mineral-rich, superheated water pours out of hydrothermal vents. The surrounding water contains very little dissolved oxygen. No sunlight reaches those depths, yet in the total darkness of the ocean bottom, life abounds. From tiny bacteria to shrimp to giant clams and tube worms, sea life thrives in that severe environment. Being without sunlight, the small, single-celled organisms at the bottom of the local food chain get their energy from *chemosynthesis*, a process by which inorganic materials are converted into food by using chemical energy. Biologists call such life-forms *extremophiles*.

Similarly, robust types of bacteria are found flourishing in the scalding waters of Yellowstone's hot springs; in the bone-dry oxidizing environment of Chile's Atacama Desert; and in the Dead Sea, where salt concentrations run as high as 33 percent. Bacteria have even been found in core samples of ancient ice 3,600 meters below the East Antarctic ice sheet. When it comes to harsh habitats, life is amazingly adaptable. If life can exist under such extreme conditions on Earth, might it also exist on those moons of the giant planets that have the ingredients necessary for life on Earth: liquid water, an energy source, and the presence of organic compounds?

When scientists realized that Mars and Venus were not as Earth-like as once imagined, prospects for finding life elsewhere in the Solar System seemed dim. Now astrobiologists are turning their attention to some of the small worlds around the giant planets far from the Sun. Those moons may supply clues about the history of life in the Solar System. Their environments may be similar to some of the ecological niches on Earth that support extremophiles. The conditions necessary to create and support life on Earth—liquid water, heat, and organic material—could all be present in oceans on some of those moons.

Enceladus, Saturn's geologically active ice moon, spews salty water-ice grains, indicating liquid water below the surface. Perhaps its south polar region is habitable. The fractured ice floes on Jupiter's Europa may cover an ocean warmed and enriched by geothermal vents similar to those that dot the floors of Earth's oceans. Like Europa, Callisto also shows magnetic variability, possibly indicating a salty ocean. Saturn's Titan has an atmosphere, liquid methane lakes, and a possible subsurface ocean. Titan is similar in several ways to a much earlier Earth.

The presence of comet-borne organic material in large bodies of water on Europa, Callisto, Enceladus, or Titan cannot yet be confirmed. Methane can arise from biological processes or can come from chemical or geochemical processes, so its detection on Enceladus and Titan is tantalizing but is not definitive evidence of life. In addition to the presence of methane, spectroscopy reveals organic gases in Titan's massive atmosphere. Titan's nitrogen atmosphere contains compounds of biological interest. For example, five molecules of hydrogen cyanide (HCN) will spontaneously combine to form adenine, one of the four primary components of DNA and RNA. HCN also is a building block of amino acids, which combine, in turn, to form proteins. Photodissociation and recombination of those various gases produce complex organic molecules that then rain out onto Titan's surface as a frozen tarry sludge. Biochemists think that many of those substances are biological precursors, similar to the organic molecules that preceded the development of life on Earth.

Astronomers anticipate future exploration of those moons, which may yield fascinating clues to the origins of terrestrial life.

Figure 11.32 Artist's conception showing a hypothetical Earth-like moon in the foreground, orbiting around a Saturn-like exoplanet.

ARTICLES   QUESTIONS

Spacecraft's 13 years in orbit have produced suggestions of potentially habitable worlds.

NASA killed *Cassini* to avoid contaminating Saturn's moons

By **NICOLE MORTILLARO**, CBC News

NASA scientists killed the hard-working *Cassini* spacecraft to avoid contaminating Saturn's moons with Earth microbes because they may have the potential to support life.

On Friday at 7:55 a.m. ET, the world said goodbye to *Cassini* after 20 years in space and 13 years orbiting Saturn and its moons, providing incredibly detailed, high-resolution photos of what many consider the jewel of our solar system.

Though *Cassini*'s life was extended twice with new missions, the spacecraft had a limited fuel supply. So NASA announced in April that it would carry out one final mission, dubbed The Grand Finale.

Scientists and engineers altered the orbit of the spacecraft to run into Saturn, where it would safely burn up in its atmosphere.

Cassini's fate arose directly from its success in shedding a light on the possibility of life on other worlds.

On Wednesday, NASA's planetary science division director Jim Green spoke about Enceladus, a small, icy moon, spewing organic material that likely originates from a subsurface ocean.

"What we thought was an icy ball, when we observed the southern hemisphere, and geysers of water spewing out into the Saturn system, it amazed us," he told reporters.

"And it began changing the way we view the habitability or potential habitability of moons in the outer part of our solar system."

Allowing *Cassini* to run out of fuel would leave NASA no way to control the spacecraft. One day it could crash into Enceladus or even Titan, Saturn's largest moon, which is also being studied for potential habitability.

While there is no guarantee that there is microbial life on any of these moons, there is a chance of contaminating their surface with Earth microbes.

"Now because of the importance of Enceladus that *Cassini* has shown us, and of Titan, another potential world that could be habitable for life, perhaps not like we know it, but perhaps completely different than ours, we had to make decisions on how to dispose of the spacecraft," Green said.

"And that led us, inevitably, to the plan of taking *Cassini* and plunging into Saturn."

On the plus side, *Cassini*'s Grande Finale has given us unprecedented views of the planet's north pole as well as its intricate ring system. For the first time in its mission,

the spacecraft flew between the planet and the rings.

And, with its success and valuable insight into the potential habitability of other worlds, it's likely there could be new missions.

Missions have been suggested to explore Enceladus, including its interior and the subsurface liquid body, Larry Soderblom, an interdisciplinary scientist on the Cassini mission, told CBC News.

"We might find in fact a higher probability than I'm willing to admit right now that life exists in the interior of Enceladus."

While astronomers and planetary enthusiasts may miss the data and photos gathered by Saturn, NASA's *Juno* spacecraft is also orbiting Jupiter. And there are plans for the *Europa Clipper*, which will conduct flybys of another icy, potentially habitable moon, Europa.

If there is a chance for life to be found in the Saturn system, NASA is eager to protect it.

"Because of planetary protection and our desire to go back to Enceladus and go back to Titan and go back to the Saturn system, we must protect those bodies for future exploration," Green said.

ARTICLES   QUESTIONS

1. How could *Cassini* "contaminate" a moon?
2. Do a search on YouTube to see the end of *Cassini*. What did the final dives reveal about Saturn's rings?
3. Do a search to see whether new missions to Saturn's moons have been approved. Which moons will be explored?
4. Go to the Web page for the NASA Office of Planetary Protection (https://planetaryprotection.nasa.gov/overview). Which moons are affected by that policy? How might that policy affect future plans for a lander on Europa? What was done to protect Mars from contamination during the NASA Mars lander missions?

Source: Nicole Mortillaro, "NASA killed Cassini to avoid contaminating Saturn's moons," CBC News, September 15, 2017. Reprinted by permission of CBC Licensing.

Summary

The moons of the outer Solar System are composed of rock and ice. A few moons are geologically active, but most are dead. All four giant planets have ring systems, which are temporary, created from and maintained by moons also in orbit around those planets. Scientists are excited about the evidence for oceans on several of the moons and the possibility that some form of life may exist in those oceans.

LG 1 **Compare the orbits and formation of regular and irregular moons.** Most of the regular moons were formed along with their parent planets and have short and nearly circular orbits. Irregular moons were captured later, have more elongated orbits, and often orbit in the opposite direction of the planet's rotation. Observations of the orbits of moons can be used to find the masses of their host planets.

LG 2 **Describe the evidence for geological activity and liquid oceans on some of the moons.** Observations indicate that Jupiter's Io is the most volcanically active body in the Solar System. Jupiter's moon Europa contains an enormous subsurface ocean and probably has some geological activity. Saturn's moon Titan has lakes of liquid methane and perhaps a deep, salty ocean. Saturn's moon Enceladus and Neptune's moon Triton have cryovolcanoes. The large Galilean moons of Jupiter, Ganymede and Callisto, also may have subsurface oceans. Some moons were geologically active in the past, as indicated by crater scars and areas smoothed by flowing fluids. Moons that have always been geologically dead show nothing but impact craters on their surfaces.

LG 3 **Describe the composition, origin, and general structure of the rings of the giant planets.** Rings are formed of countless numbers of particles all in the same plane, held to the host planet by gravity. Some rings form when moons cross a planet's Roche limit. The composition of those moons determines the composition of the rings that form from them. Shepherd moons often maintain and shape rings by gravitationally pulling and pushing those particles as they pass by. Ring particles also interact gravitationally with one another. Some rings may be temporary features held in place by moons. Saturn's bright rings and its E Ring are made primarily of water ice: the rings of the other planets are composed of darker materials.

LG 4 **Explain the role gravity plays in the structure of the rings and the behavior of ring particles.** Saturn's ring system is the most complex, and it is the best laboratory for understanding gravitational interactions between moons and rings, interactions between rings, and ring formation and dissipation. Gravity holds the ring particles in orbit around the planet, and gravitational interactions with moons determine the size and shape of rings.

? Unanswered Questions

- What is the source of Titan's nitrogen atmosphere? One group of experimenters studied that by blasting a laser at water-ammonia (H_2O-NH_3) ice to simulate cometary impacts and see whether nitrogen gas (N_2) forms. They concluded that the observed amount of N_2 in Titan's atmosphere could have been created from ammonia ice in that way. Another idea is that those gases were accreted during the formation of the moon. Using data from the *Huygens* probe, other researchers conclude that if Titan had differentiated like Ganymede, hydrothermal activity released the gases from a hot core. Astronomers want to understand why Titan has an atmosphere and the other larger moons do not.

- What is the origin of Saturn's brightest rings? One early hypothesis is that the rings come from a moon that approached too close to the planet, but Saturn's moons are composed of rock and ice, and the rings are solely ice. Some recent computer models start with a differentiated moon the size of Titan that had a rock and iron core and a large, icy mantle. As the moon in the model slowly migrates toward Saturn and crosses the Roche limit, tidal forces rip away its water ice, but not the rocky core. According to that model, the core might have continued migrating inward until it fell into Saturn, with the ice forming Saturn's ring. As time went on, the ring spread, and as material crossed the Roche limit outward, Saturn's small moonlets formed. Computers are only now getting fast enough to test those models, which suggest a unified origin for many of the moons and rings.

Questions and Problems

Test Your Understanding

1. Categorizing moons by geological activity is helpful because
   a. comparing them reveals underlying physical processes.
   b. geological activity levels drop with distance from the Sun.
   c. the size and composition of the moons depend on geological activity.
   d. most moons are very similar to one another.

2. Why are Ganymede and Callisto geologically dead, whereas the other two Galilean moons of Jupiter are active?
   a. They are larger.
   b. They are farther from Jupiter.
   c. They are more massive.
   d. They have retrograde orbits.

3. Moons of outer planets may serve as a home for life because
   a. some have liquid water.
   b. some have organic molecules.
   c. some have an interior source of energy.
   d. all of the above

4. Io has the most volcanic activity in the Solar System because
   a. it is continually being bombarded with material in Saturn's E Ring.
   b. it is one of the largest moons and its interior is heated by radioactive decays.
   c. of gravitational friction caused by the moon Enceladus.
   d. its interior is tidally heated as it orbits around Jupiter.
   e. the ice on the surface creates a large pressure on the water below.

5. Gravitational interactions with moons produce
   a. fine structure within rings.
   b. short-lived rings.
   c. smoothed-out rings.
   d. rings with spokes.

6. Saturn's bright rings are located within the planet's Roche limit, supporting the theory that those rings (select all that apply)
   a. formed of moons torn apart by tidal stresses.
   b. formed when Saturn formed.
   c. are relatively recent.
   d. are temporary.

7. The story of the F Ring of Saturn is an example of
   a. an unexplained phenomenon.
   b. media bias.
   c. the self-correcting nature of science.
   d. a violation of causality.

8. If a moon revolves opposite to its planet's rotation, it probably
   a. was captured after the planet formed.
   b. had its orbit altered by a collision.
   c. has a different composition from that of other moons.
   d. formed very recently in the Solar System's history.

9. Under what lighting conditions are the tiny dust particles found in some planetary rings best observed?
   a. viewed from the shadowed side of the planet, looking toward deep space
   b. viewed from the shadowed side of the planet, looking toward the Sun
   c. viewed from the near side of the planet, looking toward deep space
   d. viewed from the near side of the planet, looking toward the Sun

10. Planets in the outer Solar System have more moons than those in the inner Solar System because
    a. the solar wind was weaker there.
    b. more debris was present around the outer planets when they were forming.
    c. the outer planets captured most of their moons.
    d. more planetesimal collisions occurred far from the Sun.

11. The difference between a moon and a planet is that
    a. moons orbit planets, whereas planets orbit stars.
    b. moons are smaller.
    c. moons and planets have different compositions.
    d. moons and planets formed differently.

12. Scientists determine the geological history of the moons of the outer planets from
    a. seismic probing.
    b. radioactive dating.
    c. surface features.
    d. time-lapse photography.

13. The energy that keeps Io's core molten comes from
    a. the Sun.
    b. radioactivity in the core.
    c. residual heat from the collapse.
    d. Jupiter's gravity.

14. We classify moons as formerly active if they
    a. are covered in craters.
    b. have no young craters.
    c. have regions with few craters.
    d. have regions with no craters.

15. Volcanoes on Enceladus affect the E Ring of Saturn by
    a. pushing the ring around.
    b. stirring the ring particles.
    c. supplying ring particles.
    d. dissipating the ring.

Thinking about the Concepts

16. Explain the process that drives volcanism on Jupiter's moon Io.

17. Describe cryovolcanism and explain its similarities and differences with respect to terrestrial volcanism. Which moons show evidence of cryovolcanism?

18. Discuss evidence supporting the idea that Europa has a subsurface ocean of liquid water.

19. Titan contains abundant amounts of methane. What process destroys methane in that moon's atmosphere?

20. In certain ways, Titan resembles a frigid version of the early Earth. Explain the similarities.

21. Some moons display signs of geological activity in the past. Identify some of the evidence for past activity.

22. Why do the outer planets but not the inner planets have rings? Describe a ground-based technique that led to the discovery of rings around the outer planets.

23. What are ring arcs, and where are they found?

24. Identify and explain two possible mechanisms that can produce planetary ring material.

25. Explain two mechanisms that create gaps in Saturn's bright-ring system.

26. Describe ways in which diffuse rings differ from other planetary rings.

27. In Chapter 1, we stated that "all scientific theories are provisional." Explain how the discovery of the detailed structure of Saturn's F Ring (as described in the Process of Science Figure) challenged a scientific theory and how that apparent conflict was ultimately resolved.

28. Astronomers think that most planetary rings eventually dissipate. Explain why the rings do not last forever. Describe and explain a mechanism that keeps planetary rings from dissipating.

29. Name one ring that might continue to exist indefinitely, and explain why it could survive when others might not.

30. Make a case for sending a space mission to one of the moons. Which moon would you choose, and what observations would you try to obtain?

Applying the Concepts

31. Io has a mass of 8.9×10^{22} kg and a radius of 1,820 km.
    a. Using the formula provided in Working It Out 11.3, calculate Io's escape velocity.
    b. How does Io's escape velocity compare with the vent velocities of 1 km/s from its volcanoes?

32. Use the value of P^2/A^3 for Europa, as in Working It Out 11.1, to estimate the mass of Jupiter.

33. Follow Working It Out 11.1 to use one of Saturn's moons to estimate the mass of that planet

34. Study Figure 11.2.
    a. Are the scales on Figure 11.2a and 11.2b linear or logarithmic?
    b. About how much larger is the space shown in Figure 11.2b than in Figure 11.2a?

35. Planetary scientists have estimated that Io's extensive volcanism could be covering the moon's surface with lava and ash to an average depth of up to 3 millimeters (mm) per year.
    a. If Io is a sphere with a radius of 1,820 km, what are its surface area and volume?
    b. What is the volume of volcanic material deposited on Io's surface each year?
    c. How many years would it take for volcanism to perform the equivalent of depositing Io's entire volume on its surface?
    d. How many times might Io have "turned inside-out" over the age of the Solar System?

36. Use the formula for tidal forces in Working It Out 11.2 to answer the following questions:
    a. If the radius of the moon increases but its mass stays the same, what happens to the tidal force?
    b. If the radius of the moon's orbit decreases, what happens to the tidal force?
    c. If the mass of the central planet increases, what happens to the tidal force?

37. Use Working It Out 11.2 to compare the tidal force between Jupiter and Io with the tidal force between Earth and its Moon.

38. A 60-kg astronaut is spacewalking outside the International Space Station, 380 km above Earth. Use Working It Out 11.2 to find the tidal force on the astronaut. Assume that her feet point toward the center of Earth.

39. Assuming that all other numbers are held constant, make a graph of the tidal force versus the distance between a planet and its moon. On the same graph, plot the gravitational force (which falls off as $1/d^2$). Compare the two graphs to determine the relative importance of tidal forces and gravitational forces at various distances.

40. Particles at the very outer edge of Saturn's A Ring are in a 7:6 orbital resonance with the moon Janus. If the orbital period of Janus is 16 hours 41 minutes (16^h41^m), what is the orbital period of the outer edge of Ring A?

41. Use Working It Out 11.3 to find the escape velocity from Saturn's moon Janus.

42. The inner and outer diameters of Saturn's B Ring are 184,000 and 235,000 km, respectively. If the average thickness of the ring is 10 meters and the average density is 150 kilograms per cubic meter (kg/m^3), what is the mass of Saturn's B Ring?

43. The mass of Saturn's small, icy moon Mimas is 3.8×10^{19} kg. How does that mass compare with the mass of Saturn's B Ring, as calculated in Question 42? Why is that comparison meaningful?

44. The inner and outer diameters of Saturn's B Ring are 184,000 and 235,000 km, respectively. Use that information to find the ratio of the periods of particles at those two diameters. Does the B Ring orbit like a solid disk or like a collection of separate particles?

45. Use the escape velocity equation in Working It Out 11.3 to answer the following questions
    a. For more massive planets, is the escape velocity higher or lower?
    b. For larger planets, is the escape velocity higher or lower?
    c. If you know the escape velocity of a planet, what other piece of information do you need to find the planet's mass?

USING THE WEB

46. Go to *Sky & Telescope*'s "Jupiter's Moons" Web page (http://www.skyandtelescope.com/wp-content/observing-tools/jupiter_moons/jupiter.html). Enter your date and time. Where are the four Galilean moons? Keep clicking on "+1 hour" to see how their positions change. Which moon passes in front of (transits) Jupiter? If possible, observe those moons for a couple of nights through a small telescope, binoculars, or telephoto camera lens. Sketch the positions of the moons.

47. Look at the updated lists of giant planet moons on NASA's "Our Solar System" website (https://solarsystem.nasa.gov/planets; click on "Moons") or on the Carnegie Institution of Washington Department of Terrestrial Magnetism's (DTM) "Jupiter Satellite Page" (http://home.dtm.ciw.edu/users/sheppard/satellites). What are two of the more recently discovered moons of one of the planets? Are the orbits retrograde? What are the eccentricities and inclinations of the orbits? Where would those new moons fit on the graph of orbits on the DTM's website? Where did the names come from? Why are some of the more recent moons labeled "provisional"?

48. Go to the website for the former *Cassini* mission (https://saturn.jpl.nasa.gov). Watch the video at https://saturn.jpl.nasa.gov/resources/732/ to listen to the "hiss" from the aurora. What moon causes Saturn's aurora?

49. Missions to the moons:
    a. Go to the website for NASA's *Clipper* mission to Europa (https://www.nasa.gov/europa). What is the science plan for the mission? Why will the spacecraft orbit Jupiter and not Europa? What is its status? When will it get to Europa? Will it include a lander?
    b. Go to the website for the European Space Agency's *JUICE* (*JUpiter ICy moons Explorer*) mission (http://sci.esa.int/juice/). Which moons will that mission study? What are the goals of the mission? What is the status of the project? Why will *JUICE* take longer to get to Jupiter than the *Clipper* mission?
    c. Do a search to see whether a new mission to the one of the moons of Saturn has been approved.

50. Do a search to see whether moons or rings have been confirmed on exoplanets. Why is that topic of interest to astronomers?

EXPLORATION

digital.wwnorton.com/astro6

Part A: Finding the Image Scale

Finding the scale of an image is like finding the scale on a map. On a map, each inch or centimeter represents miles or kilometers of actual space. The same thing is true in an image. If you take a picture of a meter stick and then measure the meter stick in the picture to be 10 cm long, you know that 10 cm in the picture represents 1 meter of actual space, and 1 cm in the picture represents 10 cm of actual space.

To find the scale, you must compare the size of something in the image with its actual size in space. In **Figure 11.33**, the moon Io is the known object.

1 Use a ruler to measure the diameter of that image of Io in millimeters.

_____

2 Estimate the error in your measurement. (How far off could your measurement be?)

_____

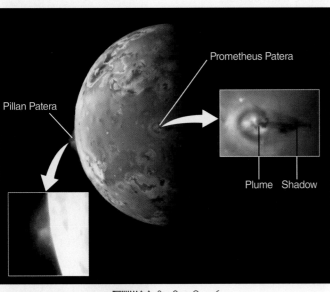

Figure 11.33 A *Galileo* image of Jupiter's moon Io, a very active moon. This image, which appeared previously in the chapter as Figure 11.5b, offers an opportunity to make some measurements.

3 Find the radius from that diameter.

_____

4 Look up the actual radius of Io (in kilometers) in Appendix 4 or online.

_____

5 Find the image scale (*s*) as follows:

$$s = \frac{\text{Actual size of object}}{\text{Size of object on the image}}$$

_____

6 What are the units of that image scale?

_____

Part B: Finding the Sizes of Features on Io

7 Near the center of Io is a geyser, surrounded by a black circle and a white ring. That is Prometheus Patera. What is the diameter of the black ring around Prometheus in that image (in millimeters)? (Do not use the inset image.)

_____

8 Multiply the measured diameter by the image scale to find the actual size of the circle around Prometheus. Identify something on Earth that is about the same size.

_____

_____

9 On the limb of the moon is a purple plume from an erupting sulfur geyser. Follow steps 7 and 8 to find the height of that plume. Identify something on Earth that is about the same height.

_____

_____

That measurement has at least two sources of error. One is the error of measurement—how well you use a ruler and how accurately you determined the top and the bottom of the plume. The other source of error is uncertainty about whether the plume is *exactly* at the limb of the moon. If the eruption occurred on the far side of Io, would your calculation be an overestimate or underestimate of the plume height?

12

Dwarf Planets and Small Solar System Bodies

Here in Chapter 12, we explore the small bodies remaining from the formation of the Solar System (see Chapter 7), including dwarf planets, irregular moons, asteroids, and comets. Those remaining planetesimals, and the fragments that some of them continually create, have revealed much about the physical and chemical conditions of the earliest moments in the history of the Solar System and how the Solar System formed and evolved. In addition, those planetesimals are important because some fraction of the water, gases, and organic material found on Earth and in the inner Solar System came from comets and asteroids.

....

LEARNING GOALS

By the end of this chapter, you should be able to:

LG 1 List the categories of small bodies and identify their locations in the Solar System.

LG 2 Describe the defining characteristics of the dwarf planets in the Solar System.

LG 3 Describe the origin of the types of asteroids, comets, and meteorites.

LG 4 Explain how asteroids, comets, and meteoroids provide important clues about the history and formation of the Solar System.

LG 5 Describe what has been learned from observations of recent impacts in the Solar System.

The European Space Agency landed a probe on Comet 67P/Churyumov-Gerasimenko in 2014. ▶▶▶

Why land a
spacecraft on
a comet?

(a)

(b)

(c)

G X U **V I R**

Figure 12.1 All images are from the *New Horizons* spacecraft flyby in 2015. (a) Enhanced color image of Pluto. (b) Dark highlands shown in the lower right border a section of icy plains. (c) Pluto's largest moon, Charon.

12.1 Dwarf Planets May Outnumber Planets

Recall from Chapter 7 that very early in the history of the Solar System—when the Sun was becoming a star—tiny grains of primitive material stuck together to produce swarms of small bodies called *planetesimals*. Those that formed in the hotter, inner part of the Solar System were composed mostly of rock and metal, whereas those in the colder, outer part were composed of ice, organic compounds, and rock. Some of the objects collided to become planets and moons. Many are still present, however, and they remain a scientifically important component of the present-day Solar System.

Dwarf planets, asteroids, Kuiper Belt objects, comets, and meteoroids are smaller than planets and orbit the Sun. Dwarf planets are found in the asteroid belt and in the **Kuiper Belt**. The asteroid belt in the region between the orbits of Mars and Jupiter contains most of the asteroids in the Solar System. The Kuiper Belt is a disk-shaped population of comet nuclei extending from Neptune's orbit (30 astronomical units [AU]) to about 50 AU.

The dwarf planets orbit the Sun and have round shapes, but because they have relatively small mass, they have not cleared the area around their orbits. As of this writing, the Solar System has five officially recognized dwarf planets: Pluto, Eris, Haumea, Makemake, and Ceres (their properties are tabulated in Appendix 4). Ceres is a large object in the main asteroid belt, whereas the others are found in the Kuiper Belt. Hundreds of dwarf planet candidates exist, but their shapes have not yet been measured well enough to classify them definitively.

Pluto

Throughout the 19th century, discrepancies were observed between the observed and predicted orbital positions of Uranus and Neptune. Early in the 20th century, astronomers hypothesized that an unseen body was perturbing the orbits of those planets. Astronomers called that body Planet X and estimated that it had 6 times Earth's mass and was located beyond Neptune's orbit. Astronomer Clyde W. Tombaugh (1906–1997) discovered Planet X in 1930, not far from its predicted position. It became the Solar System's ninth planet and was named Pluto for the Roman god of the underworld. However, observational evidence soon indicated that the mass of Pluto was far too small to have produced the perturbations in the orbits of Uranus and Neptune. When astronomers reanalyzed the 19th century observations, they found that the orbital "discrepancies" were a mistake. Pluto's discovery thus turned out to be a coincidence.

Pluto's orbital period is 248 Earth years. Its orbit is elliptical and is tilted with respect to the plane of the Solar System. Pluto's orbit periodically crosses inside Neptune's nearly circular orbit—from 1979 to 1999, Pluto was closer to the Sun than Neptune. Pluto has only two-thirds the diameter of our Moon. It has five known moons, the largest of which is Charon, about half the size of Pluto. The total mass of the Pluto-Charon system is 1/400 that of Earth, or 1/5 the mass of the Moon. Similar to Uranus, Pluto rotates nearly on its side—that is, its equatorial plane is nearly perpendicular to its orbital plane. Pluto and Charon are a tidally locked pair: each has one hemisphere that always faces the other.

As they discovered more about Pluto and other objects beyond Neptune's orbit, some astronomers questioned Pluto's classification as a planet, and a debate ensued. In 2005, astronomers identified an object more distant than Pluto, later named Eris, and then Eris's moon, Dysnomia. Observations of Dysnomia's orbit yielded a mass for Eris, which turned out to be about 28 percent greater than

Pluto's mass. Pluto and Eris have similar nitrogen and methane abundances and a relatively large moon. Then the inevitable question emerged: Should astronomers consider Eris the Solar System's tenth planet? Or should neither Pluto nor Eris be called a planet? The members of the International Astronomical Union (IAU) decided in August 2006: Pluto is round like the classical planets, but it cannot clear its neighborhood, so it was reclassified as a dwarf planet. (see the **Process of Science Figure**).

Pluto and Charon were not on *Voyager*'s route through the Solar System, so until recently only limited information was available about their surface properties or geological history. That changed when NASA's *New Horizons* flyby spacecraft passed within 12,500 km of Pluto in July 2015. Astronomers were surprised to find that Pluto and Charon had geological activity. *New Horizons* images of Pluto showed varied surface features, including a large bright region whose western half is a basin containing nitrogen, methane, and carbon monoxide ices, possibly from a large impact (**Figure 12.1a**). **Figure 12.1b** shows highlands and icy plains. Pluto's surface contains an icy mixture of frozen water, carbon dioxide, methane, and carbon monoxide, with flowing nitrogen ice. Pluto has a thin atmosphere of nitrogen, methane, ethane, and carbon monoxide: those gases freeze out of the atmosphere when Pluto is more distant from the Sun and therefore colder. Astronomers suspect, but have not confirmed, that Pluto has cryovolcanoes (Section 2 in Chapter 11) and a subsurface ocean of liquid water. Charon has no atmosphere. Its surface has deep canyons, which might have formed as an ancient ocean froze and pushed the surface outward (**Figure 12.1c**).

Ceres

In 1801, Sicilian astronomer Giuseppe Piazzi found a bright object between the orbits of Mars and Jupiter. He named the new object Ceres. Piazzi thought he might have found a hypothetical "missing planet," but as more objects were discovered orbiting the region between Mars and Jupiter, astronomers classified Ceres as belonging to a new category of Solar System objects called asteroids. Ceres is the largest body in the main asteroid belt. It also is now called a dwarf planet because although it is round (**Figure 12.2a**), it has not cleared its surroundings. With a diameter of about 940 km, Ceres is larger than most moons but smaller than any planet. It contains about a third of the total mass in the asteroid belt but only about 1.3 percent of the mass of Earth's Moon. Ceres rotates on its axis with a period of about 9 hours, typical of many asteroids.

Ceres is being explored by the *Dawn* mission, which went into orbit around Ceres in 2015 and is expected to remain until its fuel runs out in 2018. The spacecraft found many features similar to those seen on the inner planets. For example, the 34-km-wide Haulani Crater (**Figure 12.2b**) has landslides on the crater's limb and blue streaks radiating outward from the central crater peak. The color in that image is enhanced, and blue indicates young surfaces on Ceres. Those color variations indicate that the surface of Ceres has a different composition from that of the subsurface layer. Ceres has a large mountain, Ahuna Mons, that is 4–5 km (13,000–16,000 feet) high. It seems to be a salty-mud cryovolcano that may have been active in the last few hundred million years. Evidence also indicates that Ceres may have had cryovolcanoes that were flattened out by gravity over hundreds of millions of years. Because those volcanoes are made of ice rather than rock, they flow downward slowly, much like a blob of honey gradually spreads out in the bottom of a mug.

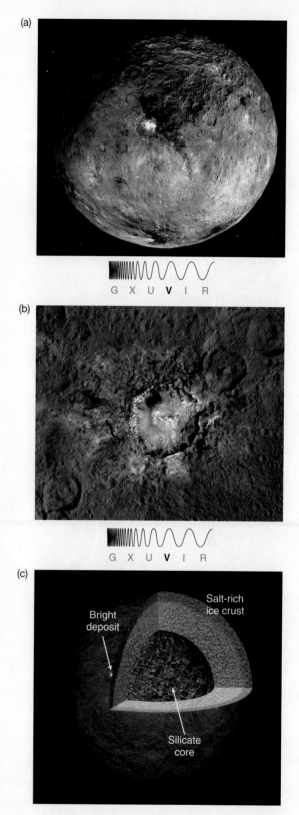

Figure 12.2 (a) Dwarf planet Ceres photographed by the *Dawn* mission. (b) The 34-km-wide Haulani Crater shows landslides on the crater's limb, as well as smooth material and a central peak on the crater floor. Blue indicates a younger surface on Ceres. The color in this image is enhanced. (c) Diagram of the interior of Ceres.

12.1 Working It Out Eccentric Orbits

Many of the objects discussed in this chapter are far from the Sun. Their complete orbits take many years, but observing a complete orbit is not necessary to determine an object's semimajor axis and eccentricity: those values can be obtained from watching how the object moves in just a fraction of its orbit. Astronomers can calculate the orbits of distant objects as they approach the Sun in a highly elliptical orbit and determine whether they will come near Earth.

Kepler's third law for objects orbiting the Sun is $P^2 = A^3$, where P is the period of the orbit and A is the semimajor axis (Chapter 3). **Figure 12.4** shows how eccentricity (e) is defined mathematically as the distance from the center of the orbit to one focus (the Sun) divided by the semimajor axis (A). (The eccentricities of the orbits of the planets in the Solar System range from 0.007 for Venus, which is nearly circular, to 0.2 for Mercury.) The types of objects discussed here in Chapter 12 generally have higher eccentricities.

The eccentricity can be related to the closest approach and the farthest distance in the orbit (Figure 12.4). That is, the object's closest approach to the Sun, its **perihelion**, equals $A \times (1 - e)$, and the object's farthest distance from the Sun, its **aphelion**, equals $A \times (1 + e)$. So, if we know the semimajor axis and eccentricity of an orbit, we can calculate how close to and how far away from the Sun an object's orbit takes it.

What are perihelion and aphelion for the dwarf planet Eris? The eccentricity of Eris's orbit is 0.44, and the semimajor axis of its orbit is 67.7 AU. Therefore,

Perihelion $= A \times (1 - e) = 67.7 \times (1 - 0.44) = 67.7 \times 0.56 = 37.9$ AU

Aphelion $= A \times (1 + e) = 67.7 \times (1 + 0.44) = 67.7 \times 1.44 = 97.5$ AU

Eris is now close to aphelion. When approaching perihelion, though, Eris will cross the orbit of Pluto, whose distance varies from 29.7 to 48.9 AU.

What are perihelion and aphelion for Apollo asteroid 2005 YU55, which has a semimajor axis of 1.14 AU and an orbital eccentricity of 0.43?

Perihelion $= A \times (1 - e) = 1.14 \times (1 - 0.43) = 1.14 \times 0.57 = 0.65$ AU

Aphelion $= A \times (1 + e) = 1.14 \times (1 + 0.43) = 1.14 \times 1.43 = 1.63$ AU

Those results indicate that the orbit of 2005 YU55 crosses the orbits of Earth and Mars. In November 2011, that asteroid passed 324,900 km from Earth—about 85 percent of the distance to the Moon.

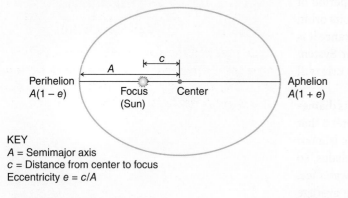

KEY
A = Semimajor axis
c = Distance from center to focus
Eccentricity $e = c/A$

Figure 12.4 The relationships between eccentricity, aphelion, and perihelion in an elliptical orbit.

12.2 Asteroids Are Pieces of the Past

After Piazzi found Ceres in 1801, several similar objects were discovered between the orbits of Mars and Jupiter. Because those new objects appeared in astronomers' eyepieces as nothing more than faint points of light, William and Caroline Herschel (the brother-sister pair of astronomers who discovered Uranus) named them *asteroids*, a Greek word meaning "starlike." As the years went by, more asteroids were discovered. Now, an estimated 1 million to 2 million asteroids larger than 1 km exist, as well as many more that are smaller. Because those objects are in the Solar System, many of them move quickly enough across the sky that their motion is noticeable over a few hours. Both professional and amateur astronomers have discovered asteroids.

An asteroid is a primitive planetesimal that did not become part of the accretion process that formed planets. The planetesimals that formed our Solar System's planets and moons have been so severely modified by planetary processes that nearly all information about their original physical condition and chemical composition has been lost. By contrast, asteroids and comet nuclei constitute an ancient and far more pristine record of what the early Solar System was like. Asteroids are composed of the same types of rocky and metallic materials that became the inner planets, and comets are composed of the same types of icy materials that became the outer planets. Thus, planetary scientists study asteroids to learn about the inner planets and their formation. In this section, we study the orbits and composition of the asteroids.

The Distribution of Asteroids

Asteroids are found throughout the Solar System. Most orbit the Sun in several distinct zones, with many residing between the orbits of Mars and Jupiter in the **main asteroid belt**. The main belt contains at least 1,000 objects larger than 30 km, of which about 200 are larger than 100 km. Although a great many asteroids exist, they account for only a tiny fraction of the matter in the Solar System. Some of the asteroids are bound to another asteroid in a double system, and more than 200 asteroids have moons, some similar in size to the asteroids themselves. At least one asteroid has a ring.

Asteroids are not distributed randomly throughout the main asteroid belt: several empty regions exist. **Figure 12.5** shows that very few asteroids orbit at specific distances from the Sun. Those "gaps" in the asteroid belt are called **Kirkwood gaps**, after Daniel Kirkwood (1814–1895), the astronomer who first recognized them. The orbital periods of some moons around their planets are numerically related—that idea of orbital resonances was introduced in Chapter 11. Similarly, all the Kirkwood gaps in the asteroid belt correspond to resonances: asteroid orbits related to the orbital period of Jupiter by the ratio of two small integers. The boundaries of the asteroid belt are set by some of those resonances. The inner boundary of the asteroid belt, at 1.8 AU, corresponds to the 5:1 orbital resonance of Jupiter, wherein the asteroids make five orbits for every orbit of Jupiter. The outer boundary, at 3.3 AU, corresponds to the 2:1 orbital resonance of Jupiter.

Those gaps are caused by gravitational interactions with Jupiter. Imagine an asteroid orbiting the Sun. While crossing the line between Jupiter and the Sun, that asteroid passes closest to Jupiter; the asteroid will receive an outward tug from Jupiter's gravity. The gravitational force of the Sun on the asteroid is more than 360 times stronger than the gravitational force that Jupiter exerts on the asteroid there, so a single close pass between Jupiter and the asteroid does very little to the asteroid's orbit. For an asteroid *not* in orbital resonance with Jupiter, the tiny gravitational tugs from Jupiter come at a different place in its orbit each time. The effects of those random tugs average out, and as a result even multiple passes close to Jupiter have little overall effect.

Now consider an asteroid starting with an orbital period exactly half that of Jupiter, a 2:1 orbital resonance. After 11.86 years—two complete asteroid orbits and one Jupiter orbit—the asteroid, Jupiter, and the Sun line up in the same location. The repeated tugs from Jupiter at the same location in the asteroid's orbit add together and move the asteroid out of that orbit, so a gap in the distribution of asteroids forms there. That is why no asteroids are in a 2:1 resonance with Jupiter. Asteroids in other orbital resonances, such as a 3:1 resonance, are similarly moved from their orbits. Asteroids are less likely to be found in the Kirkwood gaps because their gravitational interaction with Jupiter prevents them from staying there.

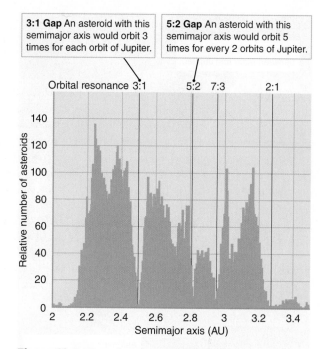

Figure 12.5 The relative number of asteroids in the main belt with a given orbital semimajor axis. The gaps in the distribution of asteroids, called *Kirkwood gaps*, are caused by orbital resonances with Jupiter.

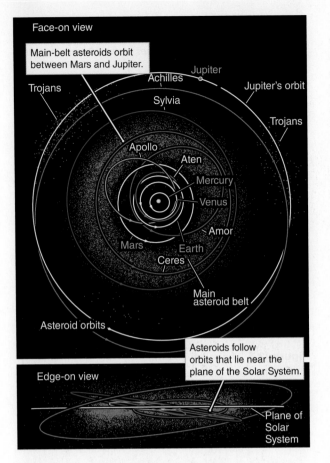

Face-on view

Main-belt asteroids orbit between Mars and Jupiter.

Trojans

Achilles Jupiter

Sylvia Jupiter's orbit

Trojans

Apollo

Aten

Mercury

Venus

Amor

Mars Earth

Ceres

Main asteroid belt

Asteroid orbits

Asteroids follow orbits that lie near the plane of the Solar System.

Edge-on view

Plane of Solar System

Figure 12.6 Face-on and edge-on views of asteroid orbits. Blue dots show the locations of known asteroids at a single point in time. The orbits of Aten, Amor, and Apollo (prototype members of some groups of asteroids) are shown. Most asteroids, such as Sylvia, are main-belt asteroids. Achilles was the first Trojan asteroid discovered.

Several groups of asteroids exist outside the main asteroid belt. They are divided according to their orbital characteristics (**Figure 12.6**). **Trojan** asteroids share Jupiter's orbit and are held in place by interactions with Jupiter's gravitational field. Three other groups are defined by their relationship to the orbits of Earth and Mars: **Apollo** asteroids cross the orbits of Earth and Mars, **Aten** asteroids cross Earth's orbit but not that of Mars, and **Amor** asteroids cross the orbit of Mars but not Earth's. All three of those groups are named for a prototype asteroid that represents the group.

Asteroids whose orbits bring them within 1.3 AU of the Sun are called **near-Earth asteroids** because their orbits bring them close to Earth's orbit, at 1 AU. Those asteroids, along with a few comet nuclei, are known collectively as **near-Earth objects** (**NEOs**). NEOs occasionally collide with Earth or the Moon. Astronomers estimate that 500–1,000 NEOs have diameters larger than a kilometer. Collisions with NEOs are geologically important and have dramatically altered the history of Earth and life on Earth (see Chapter 8).

Part of NASA's mission is to identify and track NEOs. NASA's Wide-field Infrared Survey Explorer (WISE), an infrared telescope in space, surveyed the entire sky during 2010. The data suggest that about 20,000 mid-sized asteroids (100 meters to 1 km) exist near Earth. WISE also observed more than 150,000 asteroids in the main belt, including 33,000 new ones, as well as 2,000 Jovian Trojans. The spacecraft was reactivated in late 2013 as NEOWISE; it has discovered asteroids and comets and now is searching for NEOs.

The Composition and Classification of Asteroids

Most asteroids are relics of rocky or metallic planetesimals that originated between the orbits of Mars and Jupiter. Although early collisions between those planetesimals created several bodies large enough to differentiate, Jupiter's tidal disruption and possible orbital migration prevented them from forming a single Moon-sized planet. As they orbit the Sun, asteroids continue to collide with one another, producing small fragments of rock and metal. Most meteorites are pieces of those asteroidal fragments that have found their way to Earth and crashed to its surface.

Studies of meteorites led to a scheme for classifying asteroids by composition. Meteorites found on Earth come from asteroids, which come from planetesimals, as shown in **Figure 12.7**. As larger planetesimals accreted smaller objects, thermal energy from impacts and the decay of radioactive elements heated them. Despite that heating, some planetesimals never reached the high temperatures needed to melt their interiors: they simply cooled. The planetesimals look like rubble piles, pretty much as they were when they formed. Those planetesimals, the most common type of asteroid in the main belt, are called **C-type** (carbon-type) asteroids. They are composed of primitive material that has largely been unmodified since the origin of the Solar System almost 4.6 billion years ago.

In contrast, some planetesimals were heated enough by impacts and radioactive decay to cause them to melt and differentiate, with denser matter such as iron sinking to their centers. Lower-density material—such as compounds of calcium, silicon, and oxygen—floated toward the surfaces of those planetesimals and formed mantles and crusts of silicate rocks. **S-type** (stony) asteroids may be pieces of the mantles and crusts of such differentiated planetesimals and are chemically similar to volcanic rocks found on Earth. S-type asteroids were hot enough at some point to lose their carbon compounds and other volatile materials

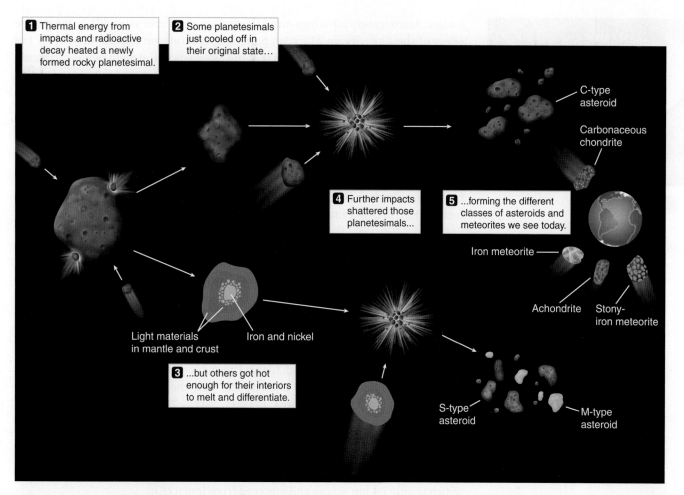

1 Thermal energy from impacts and radioactive decay heated a newly formed rocky planetesimal.

2 Some planetesimals just cooled off in their original state...

C-type asteroid

Carbonaceous chondrite

4 Further impacts shattered those planetesimals...

5 ...forming the different classes of asteroids and meteorites we see today.

Iron meteorite

Light materials in mantle and crust

Iron and nickel

Achondrite   Stony-iron meteorite

3 ...but others got hot enough for their interiors to melt and differentiate.

S-type asteroid

M-type asteroid

Figure 12.7 The fate of a rocky planetesimal in the young Solar System depends on whether it gets large and hot enough to melt and differentiate, as well as on the impacts it experiences. Different histories led to the varieties of asteroids and meteorites found today. (Images not to scale.)

to space. **M-type** (metal) asteroids are fragments of the iron- and nickel-rich cores of one or more differentiated planetesimals that shattered into small pieces during collisions with other planetesimals.

Recently, some asteroids have been shown to have ice on their surface. Using a ground-based infrared telescope, astronomers found ice on 24 Themis, one of the largest main-belt asteroids (diameter 200 km) that orbits the Sun at the outer edge of the asteroid belt. Water ice covers its surface, and organic molecules also were found there. Hydrated minerals have been found on meteorites thought to have come from outer-main-belt asteroids, but that was the first direct detection of water ice on an asteroid. The discovery may indicate that a continuum rather than a strict boundary exists between icy comets and rocky asteroids. The observations support the idea that both asteroids and comets brought water and organic material to the early Earth.

With a few exceptions, the mass of an asteroid is too small for self-gravity to have pulled it into a spherical shape. Some asteroids have highly elongated irregular shapes, like potatoes, suggesting objects that either are fragments of larger bodies or were created haphazardly from collisions between smaller

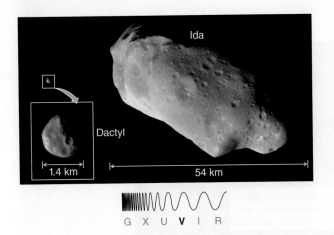

G X U V I R

Figure 12.8 This *Galileo* spacecraft image shows the asteroid Ida with its tiny moon, Dactyl (shown enlarged in the inset).

bodies. Astronomers have measured the masses of several asteroids by noting the effect of their gravity on Mars, on passing spacecraft, on one another, or by the orbits of their moons. The total mass of the asteroids in the main belt is estimated to be about 3 times the mass of Ceres, or 4 percent of the mass of the Moon. Their densities can be found from their mass and size and range between 1.3 and 3.5 times the density of water. The lower-density asteroids are shattered heaps of rubble, with large voids between the fragments.

Asteroids rotate just as planets and moons do, although irregularly shaped asteroids can wobble a lot as they spin. Their rotation periods range from 2 hours to longer than 40 Earth days. Rotation periods for asteroids are measured by watching changes in their brightness as they alternately present their broad and narrow faces to Earth. Different groups of asteroids have different average rotations.

Asteroids Viewed Up Close

Spacecraft have visited several asteroids. In 1991, the *Galileo* mission passed by two S-type asteroids while on its way to Jupiter. The small asteroid Gaspra is cratered and irregular in shape, about 18 × 11 × 9 km. Faint groovelike patterns may be fractures from the impact that chipped Gaspra from a larger planetesimal. Distinctive colors imply that Gaspra is covered with a variety of rock types. Later, *Galileo* passed close to asteroid Ida in the outer part of the main asteroid belt (**Figure 12.8**). *Galileo* flew so close to Ida that the probe's cameras could see details as small as 10 meters across. Ida is 60 × 19 × 25 km, and its surface is about a billion years old, twice the age estimated for Gaspra. Like Gaspra, Ida contains fractures, indicating that those asteroids must be made of relatively solid rock. That finding supports the idea that some asteroids are chips from larger, solid objects. The *Galileo* images also revealed a tiny moon orbiting Ida, called Dactyl, which is only 1.4 km across and cratered from impacts.

The first spacecraft to land on an asteroid was *NEAR Shoemaker*, which was gently crash-landed onto asteroid Eros in 2002 after a year of taking observations. Chemical analyses confirmed that the composition of Eros is like that of primitive meteorites. In November 2005, the Japanese spacecraft *Hayabusa* made contact with the small (less than 0.5 km) S-type asteroid Itokawa. *Hayabusa* collected small samples of dust that were returned to Earth in 2010—the first sample-return mission from an asteroid. Chemical analysis showed that such S-type asteroids are the parents of a type of meteorite found on Earth. The findings also suggested that Itokawa had been much larger, more than 20 km, when it formed.

In 2011, NASA's *Dawn* spacecraft went into orbit around Vesta (**Figure 12.9**), the second-most-massive body in the asteroid belt (after Ceres). Vesta is smaller (diameter, 525 km) than the terrestrial planets but larger than the other visited asteroids. The data from *Dawn* indicate that Vesta is a leftover intact protoplanet that formed within the first 2 million years of the aggregation of the first solid bodies in the Solar System. Vesta has an iron core and is differentiated, so it is more like the planets than it is like other asteroids.

Vesta's spectrum matches the reflection spectrum of a peculiar group of meteorites that look like rocks taken from iron-rich lava flows on Earth and the Moon. A collision—or two—that created the two large impact basins in the south polar region of Vesta (**Figure 12.10a**) blasted material into space that then landed on Earth as those meteorites. Those basins are only 1 billion to 2 billion years old. The younger basin is 500 km across and 19 km deep—a depth greater than the height of Mauna Kea in Hawaii (more than 10 km, measured

G X U V I R

Figure 12.9 This image of Vesta was taken by *Dawn* in 2012. Its north pole is in the middle of the image.

from the ocean floor). Smaller adjacent impact craters in the northern hemisphere (**Figure 12.10b**) were nicknamed "Snowman."

Two spacecraft are visiting asteroids, with the intent to return samples to Earth for analysis. The Japanese space agency mission *Hayabusa 2* has landed on asteroid 162173 Ryugu. After studying the asteroid for 1.5 years, *Hayabusa 2* will return a sample to Earth in late 2020. NASA's *OSIRIS-REx* (Origins, Spectral Interpretation, Resource Identification, Security-Regolith Explorer) mission is in transit to near-Earth asteroid 101955 Bennu. The probe will arrive in late 2018 and bring a sample to Earth in 2023. For both missions, the goal is to study those relics from the early Solar System to look for water and organic compounds.

CHECK YOUR UNDERSTANDING 12.2

Remnants of volcanic activity on the asteroid Vesta indicate that members of the asteroid belt: (a) were once part of a single protoplanet that was shattered by collisions; (b) have all undergone significant chemical evolution since formation; (c) occasionally grow large enough to become differentiated and geologically active; (d) used to be volcanic moons orbiting other planets.

12.3 Comets Are Clumps of Ice

Comets are icy planetesimals that formed from primordial material. They spend most of their time adrift in the frigid outer reaches of the Solar System. Comet nuclei put on a show only when their orbit brings them deep enough into the inner Solar System to undergo destructive heating from the Sun—they emit streams of dust and gas. In this section, we examine the orbits and composition of comets.

Early cultures viewed the sudden and unexpected appearance of a bright comet as an omen. Comets were often seen as dire warnings of disease, destruction, and death, or sometimes as portents of victory in battle or as heavenly messengers announcing the impending birth of a great leader. The earliest records of comets date from as long ago as the 23rd century BCE. Until the end of the Middle Ages, comets were regarded as mysterious temporary atmospheric phenomena rather than as astronomical objects. In the 16th century, Tycho Brahe reasoned that if comets were atmospheric phenomena like clouds, their appearance and location in the sky should be very different to observers located many miles apart. But when Tycho compared sightings of comets made by observers at several sites, he found no evidence of such differences and concluded that comets must be at least as far away as the Moon.

The Homes of the Comets

A *comet* is a complex object consisting of a small, solid, icy nucleus; an atmospheric halo; and a tail of dust and gas: a comet nucleus is the "heart" of the comet and contains most of the comet's mass. When very distant from the Sun, the comet consists entirely of frozen nucleus. As a comet comes near enough to the Sun to heat up, we can see visible changes. When a comet is near enough to the Sun to show the effects of solar heating, it is called an **active comet**, or often simply *comet*. Most comet nuclei are much too small and far away to be seen, so no one really knows how many exist. Estimates for our Solar System range as high as a trillion (10^{12}) comet nuclei—more than the number of stars in the Milky Way Galaxy—but astronomers have seen only several thousand.

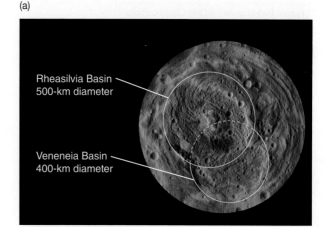

(a)

Rheasilvia Basin
500-km diameter

Veneneia Basin
400-km diameter

(b)

G X U **V** I R

Figure 12.10 (a) Impact basins at the south pole of Vesta. By counting the craters on top of it, astronomers estimate Rheasilvia to be 1 billion years old. Veneneia is partly beneath Rheasilvia and is estimated to be 2 billion years old. Red indicates higher elevation. (b) Three craters in the northern hemisphere—60, 50, and 22 km across, respectively. The feature was nicknamed "Snowman."

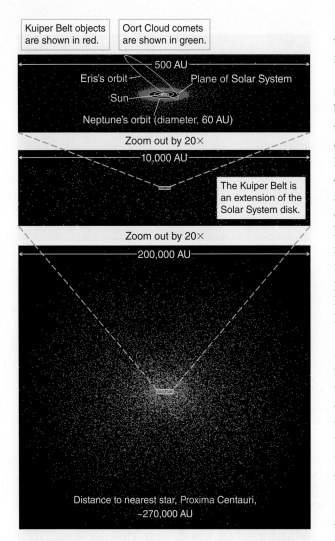

Figure 12.11 The top image shows that most comets near the inner Solar System populate an extension to the disk of the Solar System called the Kuiper Belt (red). The orbit of dwarf planet Eris is highly inclined with respect to the disk. The middle image zooms out to show the expanded Kuiper Belt. The bottom image zooms out to illustrate the spherical Oort Cloud, which is far larger and contains many more comet nuclei (green).

We can extrapolate where comets come from by observing their orbits as they pass through the inner Solar System. Comets fall into two distinct groups named for scientists Gerard Kuiper (1905–1973) and Jan Oort (1900–1992).

KUIPER BELT The Kuiper Belt is a disk-shaped population of comet nuclei that begins about 30 AU from the Sun, near the orbit of Neptune, and extends outward to about 55 AU (**Figure 12.11**). Comets from the Kuiper Belt orbit the Sun in a disk-shaped region aligned with the Solar System. The innermost part of the Kuiper Belt contains tens of thousands of icy planetesimals known as Kuiper Belt objects (KBOs) or sometimes as *trans-Neptunian objects* (*TNOs*). The largest KBOs are similar in size to Pluto and Eris. With a few exceptions, the sizes of KBOs are difficult to determine because although brightness and approximate distance are known, their albedos are uncertain. Reasonable limits for the albedos can set maximum and minimum values for their size. Like asteroids, some KBOs have moons, and at least one has three moons. We know very little of the chemical and physical properties of KBOs because of their great distance. After encountering Pluto in 2015, the *New Horizons* spacecraft continued outward into the Kuiper Belt, where it will fly close to the 40-km object 2014 MU69 in early 2019.

One of the larger known KBOs, called Quaoar (pronounced "kwa-whar"), is also one of the few whose size astronomers have independently measured—about 900 km. From its apparent brightness, distance, and size, astronomers calculate Quaoar's albedo to be 0.20, making it more reflective than the nuclei of those comets that have entered the inner Solar System but far less reflective than Pluto. Quaoar's remote location and pristine condition have allowed some volatile ices to survive on its surface, including crystalline water ice, methane, and ethane. Quaoar has a nearly circular orbit about the Sun and has a small moon, which enables astronomers to estimate the larger body's mass.

The icy planetesimals in the Kuiper Belt are packed closely enough to interact gravitationally from time to time. When that happens, one object gains energy, whereas the other loses it. The "winner" may gain enough energy to be sent into an orbit that reaches far beyond the boundary of the Kuiper Belt. The "loser" may fall inward toward the Sun.

OORT CLOUD Unlike the flat disk of the Kuiper Belt, the **Oort Cloud** is a spherical distribution of planetesimals (Figure 12.11) much too distant to be seen by even the most powerful telescopes. Astronomers determine the size and shape of the Oort Cloud from the orbits of its comets, which approach the Sun from all directions and from as far away as 100,000 AU—nearly halfway to the nearest stars.

Sedna is an object in the inner Oort Cloud whose highly elliptical orbit around the Sun takes it from 76 AU out to 937 AU. With such an extended orbit, Sedna requires more than 11,000 years to make a single trip around the Sun. When discovered in 2003, Sedna was about 90 AU from the Sun and getting closer. It will reach its perihelion in 2076. Herschel Space Observatory data suggest an albedo of 0.30 and a size of 1,000 km. Sedna has no known moon, so its mass is hard to estimate. Water and methane ices have been detected in its spectrum. Like dwarf planet Eris, Sedna has a highly eccentric orbit. A second object in the inner Oort Cloud, 2012 VP113, was recently detected. Its distance ranges from 80 to 452 AU from the Sun, and it is thought to be about half the size of Sedna.

Inner Solar System objects are close enough to the Sun that disturbances external to the Solar System never exert more than a tiny fraction of the gravitational force of the Sun on them. In the distant Oort Cloud, however, comet nuclei are so far from the Sun, and the Sun's gravitational force on them is so weak, that they are barely bound to the Sun at all. The tug of a slowly passing star or interstellar cloud can compete with the Sun's gravity, significantly stirring up the Oort Cloud and changing the orbits of its objects. If the interaction adds to the orbital energy of a comet nucleus, the comet may move outward to an even more distant orbit or perhaps escape from the Sun. A comet nucleus that loses orbital energy as a result of that type of interaction will fall inward. Some of those comet nuclei come all the way into the inner Solar System, where they may appear briefly in Earth's skies before returning once again to the Oort Cloud.

The Orbits of Comets

The lifetime of a comet nucleus depends on how often it passes by the Sun and how close it gets. About 400 **short-period comets** are known, which by definition have periods less than 200 years. In addition, each year astronomers discover about six new **long-period comets**, whose orbital periods are longer than 200 years. The total number of long-period comets observed to date is about 3,000.

Figure 12.12 shows the orbits of several comets, nearly all highly elliptical, with one end of the orbit close to the Sun and the other in the distant parts of the

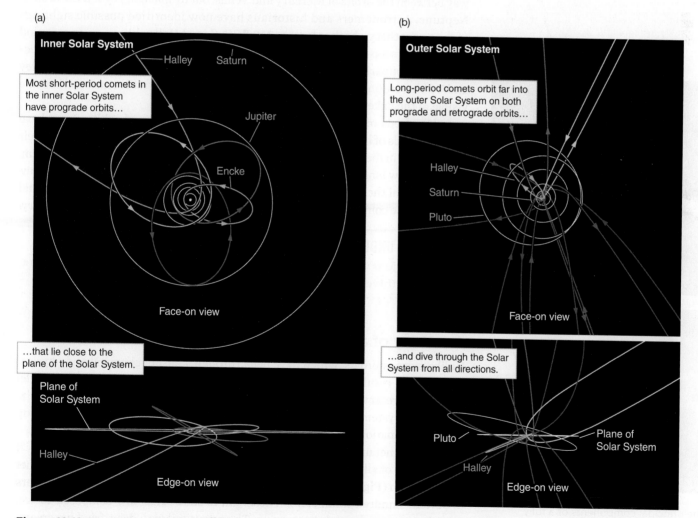

Figure 12.12 The orbits of several comets in face-on and edge-on views of the Solar System. Populations of (a) short-period comets and (b) long-period comets have very different orbital properties. Comet Halley, which appears in both diagrams for comparison, is a short-period comet.

Solar System. Most comets passing through the inner Solar System have long orbital periods that carry them back to the Oort Cloud or the Kuiper Belt. Long-period comets were scattered to the outer Solar System by gravitational interactions, so they come into the inner Solar System from all directions. Some orbit the Sun in the same direction that the planets orbit (prograde), whereas others orbit in the opposite direction (retrograde).

Conversely, short-period comets tend to be prograde and to have orbits in the ecliptic plane, often passing close enough to a planet for its gravity to change the comet's orbit about the Sun. Short-period comets likely originated in the Kuiper Belt, but as they fell in toward the Sun, gravitational encounters with Jupiter forced them into their current short-period orbits relatively close to the Sun.

Comet Halley is the most famous short-period comet. In 1705, Edmund Halley, using the gravitational laws of his colleague Isaac Newton, noted that a bright comet from 1682 had an orbit remarkably similar to those of comets seen in 1531 and 1607. He concluded that all three were the same comet and predicted that it would return in 1758. When it reappeared, astronomers quickly named it Halley's Comet and heralded it as a triumph for the genius of both Newton and Halley. Comet Halley's highly elongated orbit takes it from perihelion, about halfway between the orbits of Mercury and Venus, out to aphelion beyond the orbit of Neptune. Astronomers and historians have now identified possible sightings of the comet that go back at least to 240 BCE. Comet Halley has an average period of 76 years. Its most recent appearance was in 1986, and it was not especially spectacular in comparison with its appearance in 1910 because in 1986, Comet Halley and Earth were on opposite sides of the Sun. Comet Halley will return and become visible to the naked eye once again in summer 2061.

Hundreds of long-period comets have well-determined orbits. Some have orbital periods of hundreds of thousands or even millions of years. Almost all their time is spent in the Oort Cloud in the frigid, outermost regions of the Solar System. Orbits of a few long-period comets are shown in Figure 12.12b. Those are the comets that reveal the existence of the Oort Cloud. Because of their very long orbital periods, those comets have been observed only once throughout recorded history.

CHECK YOUR UNDERSTANDING 12.3A

Oort Cloud objects will pass close to the Sun, and become comets, only if their orbits are (a) very large; (b) very small; (c) highly elliptical; (d) highly inclined.

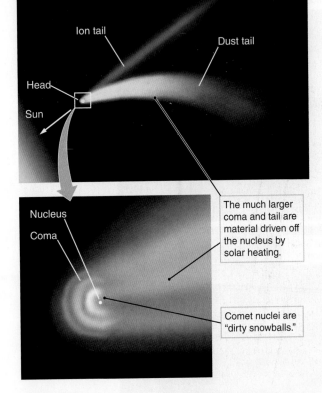

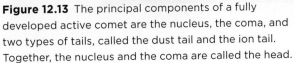

Figure 12.13 The principal components of a fully developed active comet are the nucleus, the coma, and two types of tails, called the dust tail and the ion tail. Together, the nucleus and the coma are called the head.

Anatomy of an Active Comet

Unlike most asteroids, which have been through a host of chemical and physical changes as a result of collisions, heating, and differentiation, most comet nuclei have been preserved over the past 4.6 billion years by the "deep freeze" of the outer Solar System. Comet nuclei are made of nearly pristine material remaining from the formation of the Solar System.

The comet nucleus at the center is the smallest component of a comet, but it is the source of all the material that we see stretched across the sky as the comet nears the Sun (**Figure 12.13**). Comet nuclei range in size from a few dozen meters to several hundred kilometers. Those "dirty snowballs" are composed of ice, organic compounds, and dust grains. They are similar to deep-fried ice cream, with a soft porous interior surrounded by a crunchy crust of hardened water-ice crystals, topped off with sooty dust and organic molecules.

As a comet nucleus nears the Sun, sunlight heats its surface, vaporizing ices that stream away, carrying dust particles along. That change from solid to gas is called **sublimation**. For example, dry ice (frozen carbon dioxide) does not melt like water ice but instead sublimates directly into carbon dioxide gas—that is why it is called "dry." Set a piece of dry ice out in the Sun on a summer day, and you will get a pretty good idea of what happens to a comet. The gases and dust driven from the nucleus of an active comet form a nearly spherical atmospheric cloud around the nucleus called the **coma**. The nucleus and the inner part of the coma form the comet's **head**. Pointing from the head of the comet in a direction more or less away from the Sun are long streamers of dust, gas, and ions called **tails**.

The tails are the largest and most spectacular part of a comet. The tails also are the "hair" for which comets are named. (*Comet* comes from the Greek word *kometes*, which means "hairy one.") Active comets have two types of tails, called the **ion tail** and the **dust tail** (Figure 12.13). Many of the atoms and molecules that make up a comet's coma are ions. Being electrically charged, ions in the coma experience the effect of the solar wind—the stream of charged particles that blows continually away from the Sun. The solar wind pushes on those ions, rapidly accelerating them to speeds of more than 100 kilometers per second (km/s)—far greater than the orbital velocity of the comet itself—and sweeps them out into a long wispy structure. Because the particles that make up the ion tail are so quickly picked up by the solar wind, an ion tail is usually very straight: beginning at the head of the comet, an ion tail points directly away from the Sun.

Dust particles in the coma can also have a net electric charge and are subject to the force of the solar wind. Sunlight also exerts a force on cometary dust. But dust particles are much more massive than individual ions, so they are accelerated more gently and do not reach such high relative speeds as those of the ions. As a result, the dust particles cannot keep up with the comet, and the dust tail often curves away from the head of the comet as the dust particles are gradually pushed from the comet's orbit in the direction away from the Sun (Figure 12.13).

Figure 12.14 shows the tails of a comet at various points in its orbit. Both types of tails always point *away* from the Sun, regardless of which direction the comet is moving. As the comet approaches the Sun, then, its two tails trail *behind* its nucleus, but the tails extend *ahead* of the nucleus as the comet moves away from the Sun. Tails vary greatly from one comet to another. Some comets display both types of tails simultaneously; others, for reasons not yet understood, produce no tails at all. A tail often forms as a comet crosses the orbit of Mars, where the increase in solar heating drives gas and dust away from the nucleus.

The gas in a comet's tail is even more tenuous than the gas in its coma, with densities of no more than a few hundred particles per cubic centimeter. That is much, much less than the density of Earth's atmosphere, which at sea level contains more than 10^{19} molecules per cubic centimeter. Dust particles in the tail are typically about 1 micron (μm) in diameter, roughly the size of smoke particles.

The nuclei of short-period comets have been badly worn out by repeated exposure to heating by the Sun. As the volatile ices are driven from a nucleus, some of the dust and organics are left behind on the surface. (Similarly, as a pile of dirty snow melts, the dirt left behind is concentrated on the surface of the snow.) The buildup of that covering slows down cometary activity. In contrast, long-period comets are usually relatively pristine. More of their supply of volatile ices still remains near the surface of the nucleus, and they can produce a truly magnificent show.

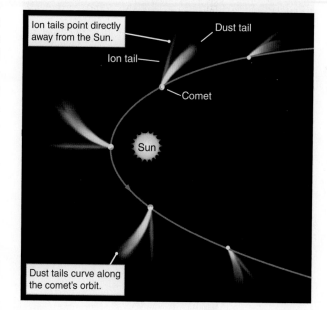

Figure 12.14 The orientation of the dust and ion tails at several points in a comet's orbit. The ion tail points directly away from the Sun, whereas the dust tail curves along the comet's orbit.

GXUVIR

Figure 12.15 Comet McNaught in 2007 was the brightest comet to appear in decades, but its true splendor was visible only to observers in the Southern Hemisphere.

GXUVIR

Figure 12.16 Comet Hale-Bopp was a great comet in 1997. The ion tail is blue in this image, and the dust tail is white.

Most naked-eye comets first develop a coma and then an extended tail as they approach the inner Solar System. Comet McNaught in 2007 was such a comet, the brightest to appear in nearly 50 years. The comet's nucleus and coma were visible in broad daylight as its orbit carried it within 25 million km of the Sun. When Comet McNaught had passed behind the Sun and next appeared in the evening skies to observers in the Southern Hemisphere, its tail had grown to more than 160 million km and stretched 35° across the sky (**Figure 12.15**). Comet McNaught came into the inner Solar System from the Oort Cloud, but it left on a path that will carry it out of the Solar System.

Comet Hale-Bopp in 1997 was a spectacular long-period comet with a long, beautiful tail (**Figure 12.16**). Hale-Bopp was a large comet, with a nucleus estimated at 60 km in diameter. It was discovered far from the Sun, near Jupiter's orbit, 2 years before its perihelion passage. That early discovery extended the total time available to study its development and plan observations as it approached the Sun. Warmed by the Sun, the nucleus produced large quantities of gas and dust and as much as 300 tons of water per second, with lesser amounts of carbon monoxide, sulfur dioxide, cyanogen, and other gases. Comet Hale-Bopp will continue its outward journey for more than 1,000 years, and it will not return to the inner Solar System until sometime around the year 4530.

Comet Ikeya-Seki is a member of a family of comets called **sungrazers**, comets whose perihelia are located very close to the surface of the Sun. Many sungrazers fail to survive even a single orbit of the Sun. Ikeya-Seki became so bright as it neared perihelion in 1965 that it was visible in broad daylight, close to the Sun in the sky. Sungrazers generally come in groups, with successive comets following in nearly identical orbits. Each member of such a group started as part of a single larger nucleus that broke into pieces during an earlier perihelion passage.

A half dozen or so long-period comets arrive each year. Most pass through the inner Solar System at relatively large distances from Earth or the Sun and never become bright enough to attract much public attention. On average, a spectacular comet appears about once per decade.

Visits to Comets

Comets pose an engineering challenge for spacecraft designers. Researchers seldom have enough advance knowledge of a comet's visit or its orbit to mount a successful mission to intercept it. The relative speed between an Earth-launched spacecraft and a comet can be extremely high. Observations must be made very quickly, and danger exists of high-speed collisions with debris from the nucleus. About a dozen spacecraft have been sent to rendezvous with comets, including five spacecraft sent to Comet Halley by the Soviet, European, and Japanese space agencies in 1986. Much of what we know about comet nuclei and the innermost parts of the coma comes from data sent back by those missions. The spacecraft observed gas and dust jets, impact craters, and ice and dust on the comet nuclei.

The Soviet *Vega 1* and *Vega 2* and the European *Giotto* spacecraft entered the coma of Comet Halley when they were still nearly 300,000 km from its nucleus. The dust from Comet Halley was a mixture of light organic substances and heavier rocky material, and the gas was about 80 percent water and 10 percent carbon monoxide, with smaller amounts of other organic molecules. The surface of Comet Halley's nucleus is among the darkest known objects in the Solar System. That means it is rich in complex organic matter that must have been present as dust in the disk around the young Sun—perhaps even in the interstellar cloud from which the Solar System formed. As the three spacecraft passed near Halley's nucleus, they observed jets of gas and dust moving away from its surface at speeds of up to 1 km/s, far above the escape velocity. By observing the jets of material streaming away from the nucleus of Halley, planetary scientists estimated that Comet Halley must have lost one-tenth of 1 percent of its mass as it went around the Sun.

Several space missions have visited short-period comets. In 2004, NASA's *Stardust* spacecraft flew within 235 km of the nucleus of Comet Wild 2 (**Figure 12.17**). Comet Wild 2 had previously resided between the orbits of Jupiter and Uranus, but a close encounter with Jupiter in 1974 had perturbed its orbit, bringing the relatively pristine body closer to the Sun as it traveled between the orbits of Jupiter and Earth. At the time of *Stardust*'s encounter with Wild 2, the comet had made only five trips around the Sun in its new orbit. Wild 2's nearly spherical nucleus is about 5 km across. At least 10 gas jets were active, some of which carried large chunks of surface material. The surface of Wild 2 is covered with features that may be impact craters modified by ice sublimation, small landslides, and erosion by jetting gas. Some craters show flat floors, suggesting a relatively solid interior beneath a porous surface layer.

The *Stardust* mission collected dust samples from Wild 2, which were returned to Earth in 2006. It found new kinds of organic materials unlike any seen before in materials from space. They are more primitive than those observed in meteorites and may have formed before the Solar System itself. Those grains can be used to investigate the conditions under which the Sun and planets formed. Minerals that form at high temperature also have been found, supporting the idea that the solar wind blew material out of the inner Solar System very early in the system's history. Scientists will be studying the particles from that mission for many years.

In 2005, NASA's *Deep Impact* spacecraft launched a 370-kilogram (kg) impacting projectile into the nucleus of Comet Tempel 1 at a speed of more than 10 km/s (**Figure 12.18**). The impact sent 10,000 tons of water and dust flying off into space at speeds of 50 m/s. A camera mounted on the projectile snapped

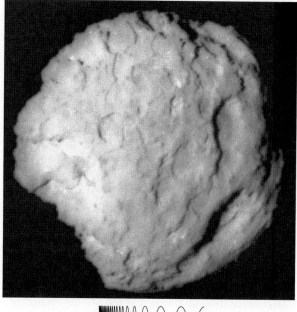

Figure 12.17 The nucleus of Comet Wild 2 was imaged by the *Stardust* spacecraft, which also sampled the comet's tail.

(a)

(b)

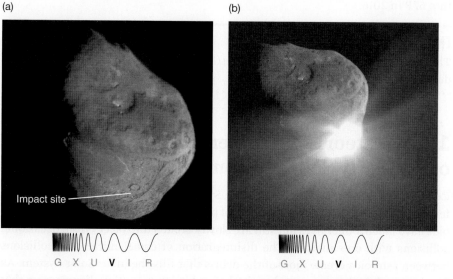
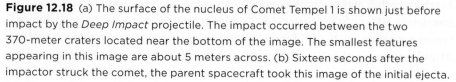

Figure 12.18 (a) The surface of the nucleus of Comet Tempel 1 is shown just before impact by the *Deep Impact* projectile. The impact occurred between the two 370-meter craters located near the bottom of the image. The smallest features appearing in this image are about 5 meters across. (b) Sixteen seconds after the impactor struck the comet, the parent spacecraft took this image of the initial ejecta.

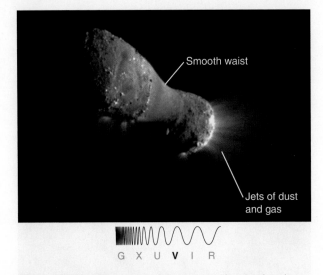

G X U V I R

Figure 12.19 This image of Comet Hartley 2 taken by the *EPOXI* spacecraft reveals two distinct surface types. Water seeps through the dust at the smooth "waist" of the comet's nucleus, whereas carbon dioxide jets shoot gas, dust, and chunks of ice from the rough areas.

photos of its target until the projectile was vaporized by the impact. Observations of the event were made both locally by *Deep Impact* and back on Earth by orbiting and ground-based telescopes. Water, carbon dioxide, hydrogen cyanide, iron-bearing minerals, and a host of complex organic molecules were identified in the Comet Tempel 1 impact. The comet's outer layer is composed of fine dust with the consistency of talcum powder. Beneath the dust are layers made up of water ice and organic materials. Well-formed impact craters, which had been absent in previous images of Comet Wild 2, also were seen.

In 2010, the *EPOXI* spacecraft flew past Comet Hartley 2 (**Figure 12.19**), imaging not only jets of dust and gas, indicating a remarkably active surface, but also an unusual separation of rough and smooth areas. Carbon dioxide jets shoot out from the rough areas. The narrow part at the middle of Figure 12.19 is a smooth, inactive area where ejected material has fallen back onto the cometary nucleus. Further observations by the Herschel Space Observatory showed that the water on Hartley 2 has the same ratio of hydrogen isotopes as that of the water in Earth's oceans. That finding suggests that some of Earth's water could have originated in the Kuiper Belt. Measurements of water on comets from the Oort Cloud have a different ratio, and so they have been ruled out as the source of Earth's water.

The European Space Agency spacecraft *Rosetta* visited Comet 67P/Churyumov-Gerasimenko in 2014 (see the chapter-opening figure). The comet is "binary"—it consists of two pieces that bumped together billions of years ago. Organic compounds such as glycine (an amino acid), ethanol, and complex carbons were identified in the comet's dust. A small, separate spacecraft landed on the comet and sent back data for 2.5 days before running out of power. The landing site was dust-covered solid ice, too thick to drill. The coma is composed of water vapor, carbon monoxide (CO), carbon dioxide (CO_2), hydrogen sulfide, and ammonia. *Rosetta* orbited the comet as it approached the Sun and observed changes as the comet heated up. Dust and gas were released from the comet, including a bright jet. About half that dust is made up of organic molecules. The water on that comet did not have the same isotope ratio as Earth's water (see Reading Astronomy News). After 2 years of observations, *Rosetta* was crashed into 67P in 2016.

CHECK YOUR UNDERSTANDING 12.3B

The nucleus of a comet is mostly: (a) solid ice; (b) solid rock; (c) a porous mix of ice and dust; (d) frozen carbon dioxide.

12.4 Meteorites Are Remnants of the Early Solar System

Comet nuclei that enter the inner Solar System generally disintegrate within a few hundred thousand years as a result of passing repeatedly near the Sun. Asteroids have much longer lives but still are slowly broken into pieces from occasional collisions with one another. The disintegration of comet nuclei and collisions between asteroids create most of the debris that fills the inner Solar System. As Earth and other planets move along in their orbits, they continually sweep up that fine debris. Most of the meteoroids Earth encounters come from the disintegration of comets or asteroids. Meteoroids are small solid bodies ranging in size from 10 millimeters (mm) to 100 meters. When a meteoroid enters Earth's atmosphere, frictional heat causes the air to glow, producing an atmospheric phenomenon called a *meteor*. If a meteoroid survives to reach the planet's surface, we call it a *meteorite*. Earth sweeps up about 100,000 kg of meteoritic debris every day, and particles smaller than 100 mm eventually settle to the ground as fine dust. In this section, we look more closely at meteorites and what can be learned about the early Solar System from them.

Observations of Meteors

If you stand outside for a few minutes on a moonless, starry night, away from bright city lights, you will almost certainly see a meteor, commonly known as a *shooting star*. The larger pieces that survive the plunge through Earth's atmosphere are usually fragments of asteroids. Most of the smaller pieces that burn up in the atmosphere before reaching the ground are cometary fragments typically less than a centimeter across and having about the same density as cigarette ash.

A 1-gram meteoroid (about half the mass of a dime) entering Earth's atmosphere at 50 km/s has a kinetic energy comparable to that of an automobile cruising along at the fastest highway speeds. Scientists measuring meteor altitudes with radar find that the altitudes of meteors are between 50 and 150 km. Most meteoroids are so small and fragile that they burn up before reaching Earth's surface. A meteor may streak across 100 km of Earth's atmosphere and last at most a few seconds. Meteorites probably litter the surfaces of all solid planets and moons.

Nearly all ancient cultures were fascinated by those rocks from the sky. Iron from meteorites was used to make the earliest tools. Early Egyptians preserved meteorites along with the remains of their pharaohs, Japanese placed them in Shinto shrines, and ancient Greeks worshiped them. Despite many eyewitness accounts of meteorite falls, however, many people were slow to accept that those peculiar rocks actually come from far beyond Earth. By the early 1800s, scientists had documented so many meteorite falls that their true origin was indisputable. Today, hardly a year passes without a recorded meteorite fall, including some that have caused damage.

Fragments of asteroids are much denser than cometary meteoroids. If an asteroid fragment is large enough—about the size of your fist—it can survive all the way to the ground to become a meteorite. The fall of a 10-kg meteoroid can produce a fireball so bright that it lights up the night sky more brilliantly than the full Moon. Such a large meteoroid, traveling many times faster than the speed of sound, may create a sonic boom heard hundreds of kilometers away. The meteoroid may even explode into multiple fragments as it nears the end of its flight. Some fireballs glow with a brilliant green color, caused by elements in the meteoroid that created them.

Meteor showers occur when Earth's orbit crosses the orbit of a comet or asteroid and passes through a concentration of cometary or asteroid debris. During a shower, many meteors can be observed in just a few hours. More than a dozen comets and at least two asteroids have orbits that come close enough to Earth's orbit to produce annual meteor showers—see the list in **Table 12.1**. Because the meteoroids in a shower are in similar orbits, they all enter Earth's atmosphere moving in the same direction—the paths through the

TABLE 12.1	Selected Meteor Showers	
Shower	Approximate Date	Parent Object
Quadrantids	January 3–4	Asteroid 2003 EH1
Lyrids	April 21–22	Comet Thatcher
Eta Aquariids	May 5–6	Comet Halley
Perseids	August 12–13	Comet Swift-Tuttle
Draconids	October 8–9	Comet Giacobini-Zinner
Orionids	October 21–22	Comet Halley
Taurids	November 5–6	Comet Encke
Leonids	November 17–18	Comet Tempel-Tuttle
Geminids	December 13–14	Asteroid Phaethon
Ursids	December 22–23	Comet Tuttle

(a)

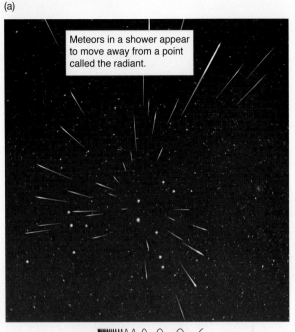

Meteors in a shower appear to move away from a point called the radiant.

G X U V I R

(b)

However, the meteor paths are actually parallel.

Figure 12.20 (a) Meteors appear to stream away from the radiant of the Leonid meteor shower. (b) Such streaks are actually parallel paths that appear to emerge from a vanishing point, as in our view of these railroad tracks.

sky are parallel to one another. Therefore, all the meteors appear to originate from the same point in the sky (**Figure 12.20a**), just as the parallel rails of a railroad track appear to vanish to a single point in the distance (**Figure 12.20b**). That point is called the shower's **radiant**.

The Perseid shower occurs in August because that's when Earth crosses the orbit of Comet Swift-Tuttle. Although spread out along the comet's orbit, the debris is more concentrated near the comet itself. In 1992, Comet Swift-Tuttle returned to the inner Solar System for the first time since its discovery in 1862, resulting in an exceptional Perseid meteor shower with counts of up to 500 meteors per hour.

In mid-November of each year, Earth passes almost directly through the orbit of Comet Tempel-Tuttle, a short-period comet with an orbital period of 33.2 years. That crossing produces the Leonid meteor shower, which is usually weak because Comet Tempel-Tuttle distributes little of its debris around its orbit in most years. In 1833 and 1866, however, Tempel-Tuttle was not far away when Earth passed through its orbit, and the Leonid showers were so intense that meteors filled the sky with as many as 100,000 meteors per hour. Further perturbations of the comet's orbit caused a spectacular Leonid shower in 1966 that may have produced as many as a half-million meteors per hour. The Leonid shower put on less spectacular but still impressive shows between 1999 and 2003, when several thousand meteors per hour were seen.

Types of Meteoroids

As asteroids orbit the Sun, they occasionally collide with one another, chipping off smaller rocks and bits of dust. Sometimes, one of those fragments is captured by Earth's gravity and survives its fiery descent through Earth's atmosphere as a meteor. Thousands of meteorites reach the surface of Earth every day, but only a tiny fraction of those are ever found and identified. Antarctica offers the best meteorite hunting in the world because in many places, the only stones to be found on the ice are meteorites. Because Antarctica has little precipitation, Antarctic meteorites also tend to show little weathering or contamination from terrestrial dust or organic compounds, making them excellent specimens for study.

Meteorites are categorized as stony, iron, or stony-iron on the basis of their materials and the degree of differentiation they experienced within their parent bodies (**Figure 12.21**). More than 90 percent of meteorites are **stony meteorites**, which are similar to terrestrial silicate rocks. A stony meteorite is characterized by the thin coating of melted rock that forms as it passes through the atmosphere. Many stony meteorites contain small, round spherules called **chondrules**, once-molten droplets that rapidly cooled to form crystallized spheres ranging in size from that of sand grains to that of marbles. Stony meteorites containing chondrules are called **chondrites** (Figure 12.21a); those without chondrules are known as **achondrites** (Figure 12.21b). The most primitive meteorites are called **carbonaceous chondrites** because they are chondrites rich in carbon. Indirect measurements suggest that those meteorites are about 4.56 billion years old—consistent with all other measurements of the time that has passed since the Solar System formed.

Iron meteorites (Figure 12.21c) come from M-type asteroids. Iron meteorites can be recognized by their melted and pitted appearance generated by frictional heating as they streaked through the atmosphere. Many iron meteorites are never found, either because they land in water or because they are not recognized as meteorites. The Mars *Opportunity* rover discovered a few iron meteorites on the martian surface (**Figure 12.22**). Both their appearance—typical of iron meteorites found on Earth—and their position on the smooth, featureless plains made them instantly recognizable.

Stony-iron meteorites consist of a mixture of rocky material and iron-nickel alloys (Figure 12.21d). They are relatively rare.

Meteorites and the History of the Solar System

Meteorites are extremely valuable because they are samples of the same relatively pristine material that makes up asteroids. Astronomers can take meteorites into the laboratory and thus study them more easily than sending spacecraft to comets or asteroids. Scientists compare meteorites to rocks found on Earth and the Moon and contrast their structure and chemical makeup with those of rocks studied by spacecraft that have landed on Mars and Venus. Comparing the spectra of meteorites with those of asteroids and planets reveals their origin.

Meteorites come from asteroids, which originate from stony-iron planetesimals. A few planetesimals between the orbits of Mars and Jupiter evolved toward becoming tiny planets before being shattered by collisions. Some became volcanically active, with lava erupting onto their surfaces. But instead of forming planets, those planetesimals broke into pieces in collisions with other planetesimals.

Some types of meteorites fail to follow the patterns discussed so far. Whereas most achondrites have ages of 4.5 billion to 4.6 billion years, some are less than 1.3 billion years old. Other achondrites are chemically and physically similar to the soil and the atmospheric gases that NASA's lander instruments have measured on Mars. The similarities are so strong that most planetary scientists think those meteorites are pieces of Mars knocked into space by large asteroidal impacts—as a result, researchers can study pieces of Mars in laboratories here on Earth. In 1996, a NASA research team announced that the meteorite ALH84001, found in Antarctica, showed possible physical and chemical evidence of past life on Mars, but the claim is still debated (see Chapter 24).

Another group of meteorites bear striking similarities to samples returned from the Moon. Like the meteorites from Mars, they are chunks of the Moon blasted into space by impacts and later fell to Earth. Therefore, meteorites from Earth may have fallen on the Moon. If they are ever collected from the unchanging Moon, they could tell us about what conditions were like on the early Earth.

Zodiacal Dust

Like meteoroids, **zodiacal dust** is a mixture of cometary debris and ground-up asteroid material. Just as you can "see" sunlight streaming through an open window by observing its reflection from dust drifting in the air, you can see the sunlight reflected off tiny zodiacal dust particles that fill the inner Solar System close to the plane of the ecliptic. On a clear, moonless night, not long after the western sky has grown dark, that dust is visible as a faint column of light slanting upward from the western horizon along the path of the ecliptic. That band, called the

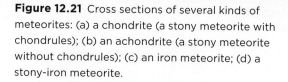

Figure 12.21 Cross sections of several kinds of meteorites: (a) a chondrite (a stony meteorite with chondrules); (b) an achondrite (a stony meteorite without chondrules); (c) an iron meteorite; (d) a stony-iron meteorite.

Figure 12.22 This 2 meter long iron meteorite lying on the surface of Mars was imaged by the Mars exploration rover *Curiosity*.

Figure 12.23 Zodiacal light shines in the western sky after sunset as seen from the La Silla Observatory in Chile.

zodiacal light, also can be seen in the eastern sky just before dawn (**Figure 12.23**). With good eyes and an especially dark night, you may be able to follow the zodiacal dust band all the way across the sky. In its brightest parts, the zodiacal light can be several times brighter than the Milky Way, for which it is sometimes mistaken.

The dust grains are roughly a millionth of a meter in diameter—the size of smoke particles. Near Earth, each cubic kilometer of space contains only a few particles of zodiacal dust. The total amount of zodiacal dust in the entire Solar System is estimated to be 10^{16} kg, equivalent to a solid body about 25 km across, or roughly the size of a large comet nucleus. Grains of zodiacal dust are constantly being lost as they are swept up by planets or pushed out of the Solar System by the pressure of sunlight. Such interplanetary dust grains have been recovered from Earth's upper atmosphere by aircraft flying very high. If not replaced by new dust from comets, all zodiacal dust would be gone within the brief span of 50,000 years.

In the infrared region of the spectrum, thermal emission from the band of warm zodiacal dust makes it one of the brightest features in the sky. It is so bright that astronomers wanting to observe faint infrared sources are often hindered by its foreground glow.

CHECK YOUR UNDERSTANDING 12.4

Meteorites contain clues to which of the following? (Choose all that apply.) (a) the age of the Solar System; (b) the temperature in the early solar nebula; (c) changes in the composition of the primitive Solar System; (d) changes in the rate of cratering in the early Solar System; (e) the physical processes that controlled the formation of the Solar System.

12.5 Comet and Asteroid Collisions Still Happen Today

Almost all hard-surfaced objects in the Solar System still bear the scars of a time when tremendous impact events were common. In previous chapters, we explained how impacts from long ago may have been responsible for the formation of our Moon, the variation in hemispheric features on Mars, and the death of the dinosaurs. Although such impacts are far less frequent today than they once were, they still happen. In this section, we examine a few examples of recent collisions.

Comet Shoemaker-Levy 9 Collided with Jupiter

Early in the 20th century, the orbit of a comet nucleus called Shoemaker-Levy 9 from the Kuiper Belt was perturbed, and the comet's new orbit carried it close to Jupiter. Eventually, it was captured by Jupiter and orbited the planet. In 1992, that comet passed so close to Jupiter that tidal stresses broke it into two dozen major fragments, which then spread out along its orbit. The fragments took one more 2-year orbit around the planet, and throughout a week in 1994, the entire string of fragments crashed into Jupiter. The impacts occurred just behind the limb of the planet, so they were not visible on Earth until Jupiter's rotation put the impact points in view. Astronomers using ground-based telescopes and the HST could see immense plumes rising from the impacts to heights of more than

3,000 km above the cloud tops at the limb. The debris in those plumes then rained back onto Jupiter's stratosphere, causing ripples like pebbles thrown into a pond. **Figure 12.24** shows some HST images of the impact features. Sulfur and carbon compounds released by the impacts formed Earth-sized scars in the atmosphere that persisted for months.

Collisions with Earth

In summer 1908, a remote region of western Siberia was blasted with the energy equivalent of 2,000 times the energy of the atomic bomb dropped on Hiroshima in 1945. Eyewitness accounts detailed the destruction of dwellings, the incineration of reindeer (including one herd of 700), and the deaths of at least five people. Although trees were burned or flattened over more than 2,150 square kilometers (km²)—an area greater than metropolitan New York City—no crater was left behind. The Tunguska event (named for the nearby river) was the result of a tremendous high-altitude explosion that occurred when a small body hit Earth's atmosphere, ripped apart, and formed a fireball before reaching Earth's surface. Recent expeditions to the Tunguska area have recovered resin from the trees blasted by the event. Chemical traces in the resin suggest that the impacting object may have been a stony asteroid.

In February 2013, a known near-Earth object about half the size of an American football field passed so close to Earth that it came within the orbit of artificial satellites. That near miss was uneventful, and the object simply continued on its way. In an unrelated event on the same day, however, a previously unknown meteoroid estimated to have a radius of about 20 meters exploded over Chelyabinsk, Russia. The shock waves from that explosion damaged thousands of buildings in six cities and injured more than 1,000 people. That was likely to have been the largest impact on Earth since the Tunguska event, and many observations were recorded of the effect on that less remote location.

From car dashboard, cell phone, and security camera video and images (**Figure 12.25a**), as well as reports of the time between the brighter-than-the-Sun flash and the sonic boom that followed, scientists determined the trajectory and speed of the incoming object as it traveled through the atmosphere. They estimate a preimpact orbit of the object in the inner asteroid belt and think it originally broke off from a known 2-km asteroid. From small pieces collected

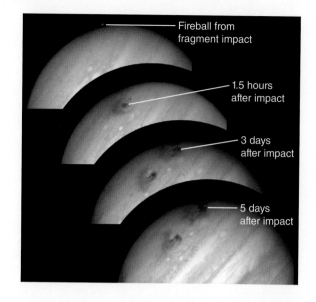

Figure 12.24 HST images of the evolution of the scar left by one fragment of Comet Shoemaker-Levy 9 when it impacted Jupiter in 1994.

Figure 12.25 (a) In February 2013, a meteoroid entered the atmosphere over Russia, creating a fireball that eyewitnesses said was brighter than the Sun. (b) A 600-kg piece of the meteorite on display.

(a)

(b)

over a wide area and from a large, 600-kg chunk found in a frozen-over lake (**Figure 12.25b**), scientists could analyze the object's composition and density. It seems to be similar in composition to that of Itokawa, the S-type asteroid whose dust was collected and returned to Earth. It is estimated that only a few hundredths of 1 percent of the original 10-million-kg mass has been found on the surface of Earth. The energy of the explosion was about 30 times the energy released by the Hiroshima atomic bomb. Most of that energy went into the atmosphere, however, heating and breaking up the meteoroid at a much higher altitude than that at which bombs are detonated, so the effects on the ground were less than those of a bomb.

Those impacts are sobering events. The distribution of relatively large asteroids in the inner Solar System indicates it is highly improbable that an asteroid will impact a populated area on Earth within your lifetime. However, many comets and smaller asteroids have unknown orbits, and several previously unknown long-period comets enter the inner Solar System each year. If on a collision course with Earth, even a large comet might not be noticed until just a few weeks or months before impact. For example, Comet Hyakutake was discovered only 2 months before it passed near Earth, and a potentially destructive asteroid that just missed Earth in 2002 was not discovered until 3 days *after* its closest approach. The Chelyabinsk meteoroid was in an orbit such that it wasn't detectable at all. Earth's geological and historical record suggests that actual impacts by large bodies are infrequent events.

As many as 10 million asteroids larger than a kilometer across may exist, but only about 130,000 have well-known orbits, and most of the unknown asteroids are too small to see until they come very close to Earth. The U.S. government—along with the governments of several other nations—is aware of the risk posed by near-Earth objects. Although the probability of a collision between a small asteroid and Earth is small, the consequences could be catastrophic (**Working It Out 12.2**), so NASA has been given a congressional mandate to catalog all NEOs and to scan the skies for those that remain undiscovered.

12.2 Working It Out Impact Energy

How much energy can be released from the impact of a comet nucleus? The kinetic energy of a moving object is given by

$$E_K = \frac{1}{2}mv^2$$

where E_K is the kinetic energy in joules (J), m is the mass in kilograms, and v is the speed in meters per second (m/s).

Suppose an asteroid or comet nucleus that is 10 km in diameter with a mass of 5×10^{14} kg hits Earth at a speed of 20 km/s = 20×10^3 m/s. Putting those values into the preceding equation gives us

$$E_K = \frac{1}{2} \times (5 \times 10^{14}\,\text{kg}) \times (20 \times 10^3\,\text{m/s})^2$$

$$E_K = 1.0 \times 10^{23}\,\text{J}$$

How much energy is that? A 1-megaton hydrogen bomb (67 times as energetic as the Hiroshima atomic bomb) releases 4.2×10^{15} J. If we divide the energy from our comet impact by that number, we have

$$\frac{1.0 \times 10^{23}\,\text{J}}{4.2 \times 10^{15}\,\text{J/megaton H-bomb}} = 2.4 \times 10^7$$

$$= 24\,\text{million 1-megaton H-bombs}$$

That's a lot of energy, and it shows why impacts have been so important in the history of the Solar System.

CHECK YOUR UNDERSTANDING 12.5
How do astronomers determine the origin of a meteorite that reaches Earth?

Origins   Comets, Asteroids, Meteoroids, and Life

Water is essential to life on Earth, so where did it all come from? Scientists have thought that some of Earth's water was contributed during impacts of icy planetesimals early in the history of the Solar System. The icy planetesimals condensed from the protoplanetary disk surrounding the young Sun and grew to their current size near the orbits of the giant planets. Those planetesimals later interacted with the giant planets, which may themselves have been migrating to and from the inner Solar System. In such interactions, some of the planetesimals were flung outward to form the Kuiper Belt and Oort Cloud, and some were thrown inward toward the Sun, possibly hitting Earth. Because much of the mass in comet nuclei and some in asteroids appears to be in the form of water ice, some of Earth's current water supply may have come from that early bombardment. Spacecraft have measured the type of water in several comets and asteroids, and the water on Earth best matches that of the primitive carbonaceous chondrites and a few, but not most, comets.

Comets and asteroids can threaten life on Earth. Occasional collisions of comet nuclei and asteroids with Earth have probably resulted in widespread devastation of Earth's ecosystem and in the extinction of many species. Passing stars or the periodic passage of the Sun through giant gas clouds located in denser regions of the Milky Way Galaxy may have resulted in showers of comet nuclei into the inner Solar System, possibly changing the climate and contributing to mass extinctions. Although certainly qualifying as global disasters for the plants and animals alive at the time, such events also represent global opportunities for new life-forms to evolve and fill the niches left by species that did not survive. As noted in Chapter 8, such a collision with an asteroid or comet probably played a central role in ending the 180-million-year reign of dinosaurs and provided an opportunity for the evolution of mammals.

In studying comets, astronomers may also have found a key to the chemical origins of life on Earth. Comets are rich in complex organic material—the chemical basis for terrestrial life—and cometary impacts on the young Earth may have played a role in chemically seeding the planet. If comets are pristine samples of the material from which the Sun and planets formed, organic material must be widely distributed throughout interstellar space. Radio telescope observations of vast interstellar clouds throughout the Milky Way confirm the presence of organic material. The fact that asteroid belts and storms of comets have been observed in distant solar systems could have significant implications as astronomers consider the possibility of life elsewhere in the universe.

Scientists report on the water observed on Comet 67P/Churyumov-Gerasimenko.

Rosetta Spacecraft Finds Water on Earth Didn't Come from Comets

By **REBECCA JACOBSON**, *PBS NewsHour*

It's a mystery that has baffled scientists for decades: Where did Earth's water come from?

Some scientists believed comets might have been the original source of the Earth's oceans. But a study published this week in the journal *Science* is sending scientists back to the drawing board. In its first published scientific data, the ROSINA mass spectrometer on board the *Rosetta* probe found that water on Comet 67P/Churyumov–Gerasimenko doesn't match the water on Earth.

The result is surprising, says Kathrin Altwegg, principal investigator for ROSINA at the University of Bern and one of the authors of the study. For decades, scientists had ruled out comets from the Oort Cloud at the very edge of our Solar System as the source of Earth's water.

But three years ago, an analysis of water on the Hartley 2 comet near Jupiter found a perfect match to the Earth's oceans. That finding led scientists to believe that Earth's water could have come from much closer comets, either near Jupiter or in the Kuiper Belt just beyond Neptune. Comet 67P/Churyumov–Gerasimenko is one of those Jupiter family comets, which scientists believe originated in the Kuiper Belt.

"That was a big surprise, but now we are back to what I expected," she said. "I think it's very nice to see the diversity we have in Kuiper Belt, to see that not everything is as simple as it seemed."

To find the origin of Earth's water, scientists analyze the water's "fingerprint," says Claudia Alexander, project scientist of the U.S. *Rosetta* Project at NASA's Jet Propulsion Laboratory. Water has a chemical isotopic signature, which works just like a fingerprint. Planets, comets, even minerals all have a fingerprint, Alexander says, and scientists are looking for a match to Earth's.

On Earth, water is mostly two parts hydrogen and one part oxygen—H_2O. But there's also "heavy" water, Alexander explained, which is made with deuterium—a hydrogen atom with a neutron. That heavy water is what the *Rosetta* spacecraft found on Comet 67P/Churyumov–Gerasimenko. It's also a closer match to the water scientists have found on other comets, ruling them out as Earth's water source, Alexander said.

"The clues don't quite all add up," she said.

Altwegg agrees, saying that it's not likely the other Kuiper Belt comets have a match to Earth's water, but further studies would be helpful.

"You would have to assume that 67P is the exception in the Kuiper Belt," she said. "We need more missions to Kuiper Belt comets, which would be fabulous."

There are several ideas to explain the origin of Earth's water, Alexander said. Some believe that water has been on Earth since its formation, that it was beaten out of other minerals as the planet formed. Others think "wet planetesimals" near Jupiter—which were like planetary Silly Putty, loose sticky blobs of rock and ice, Alexander said—collided with Earth in the early formation of the Solar System.

Alexander believes the answer could be a combination of any of these ideas. The *Dawn* mission in 2015 [is studying] the water on the asteroid Ceres, near Jupiter. If it's a match for Earth's water, it may be another clue, Alexander said. But the *Rosetta* finding is a huge step in solving the mystery, she said.

"I think this is a big deal. . . . For me, I've not always been a believer in the story that comets brought the water," Alexander said. "In some respects, I'm somewhat relieved (this finding) doesn't confirm it. It's more complicated than that. I think we need more forensic evidence to settle the score."

1. This article uses the word *fingerprint* when discussing isotope ratios. Earlier in this book, we used that word in our discussion of spectral lines. How are the two phenomena similar?
2. What might explain why the water on Comet Hartley 2 matched Earth's water, but the water on Comet 67P/Churyumov-Gerasimenko did not?
3. Is the water on other comets more like the water on Comet Hartley 2 or on Comet 67P/Churyumov-Gerasimenko?
4. What are other possible sources of Earth's water?
5. Do an Internet search to see whether any results from the *Dawn* mission indicate the type of water observed on Ceres.

Source: "Rosetta spacecraft finds water on Earth didn't come from comets," by Rebecca Jacobson. *PBS NewsHour*, December 10, 2014. © 2014 NewsHour Productions LLC. Reprinted by permission.

Summary

The story of how planetesimals, asteroids, and meteorites are related is a great success of planetary science. Scientists have assembled a wealth of information about that diverse collection of objects to piece together a picture of how planetesimals grow, differentiate, and then shatter in collisions. That story fits well with the even larger story of how most planetesimals were accreted into the planets and their moons (Chapter 7). Comets, asteroids, and meteoroids may have supplied Earth with water, volatiles, and organic material in the early history of the Solar System. Impacts with large asteroids or comet nuclei may have led to mass extinctions on Earth that eventually enabled mammals (for example) to evolve. Recent spacecraft missions to comets and asteroids have begun to reveal details about the composition of those bodies.

LG 1 **List the categories of small bodies and identify their locations in the Solar System.** Small bodies in the Solar System that orbit the Sun include dwarf planets, asteroids, comets, Kuiper Belt objects, and meteoroids. Most asteroids are located in the main asteroid belt between the orbits of Mars and Jupiter. Comets are small, icy planetesimals that reside in the frigid regions of the Kuiper Belt and the Oort Cloud, beyond the planets. The orbits of nearly all comets are highly elliptical, with one end of the orbit close to the Sun and the other in the distant parts of the Solar System.

LG 2 **Describe the defining characteristics of the dwarf planets in the Solar System.** Pluto, Eris, Haumea, Makemake, and Ceres are massive enough to have pulled themselves into spherical shapes. But they are classified as dwarf planets, not planets, because they are not massive enough to have cleared their surroundings of other bodies.

LG 3 **Describe the origin of the types of asteroids, comets, and meteorites.** Asteroids are small Solar System bodies made of rock and metal. Although early collisions between those planetesimals created several bodies large enough to differentiate, Jupiter's tidal disruption (and possible migration) prevented them from forming a single planet. Comets that venture into the inner Solar System are warmed by the Sun, producing an atmospheric coma and a tail. Meteoroids are small fragments of asteroids and comets. When a meteoroid enters Earth's atmosphere, frictional heat causes the air to glow, producing a meteor. Meteor showers occur when Earth passes through a trail of cometary debris. A meteoroid that survives to a planet's surface is called a meteorite. The various types of meteoroids that are formed depend on the differentiation of the parent body.

LG 4 **Explain how asteroids, comets, and meteoroids provide important clues about the history and formation of the Solar System.** Asteroids, comets, and meteoroids are leftover debris from the formation of the Solar System. Asteroids are composed of the same type of material that became the inner planets, and comets are composed of the same type of material that became the outer planets. They provide samples of the initial composition and properties of the Solar System and furnish samples of material from its entire history.

LG 5 **Describe what has been learned from observations of recent impacts in the Solar System.** Impacts by comets and meteoroids have been observed recently on Jupiter and Earth. The debris from those impacts helps astronomers understand the conditions in the early Solar System when impacts were much more frequent.

? Unanswered Questions

• What can asteroids and comets reveal about the migration of the giant planets in the early Solar System? It is not accidental that the main asteroid belt and the Kuiper Belt are at the boundaries of the orbits of the giant planets. As noted in Chapter 10, the giant planets may have moved around quite a bit in the early Solar System. The migration of the giant planets may explain the spread in the orbits of the objects in the main asteroid belt. Migration may also have brought icy objects—such as comet nuclei—out of the Kuiper Belt and into the main asteroid belt. The Kuiper Belt would have been closer to the Sun originally, and Jupiter and Saturn would have pushed it outward. And, finally, the migration might have sent objects from both belts into the inner Solar System, creating the heavy bombardment of 4 billion years ago.

• Does a "Planet Nine" exist? That is a hypothetical planet in the far Solar System that was proposed to explain some irregularities in the orbits of objects in the Kuiper belt and beyond. The planet would be about 10 times the mass of Earth, about 2–4 times the size of Earth—similar to Uranus and Neptune. It would have an elongated orbit, with a distance from the Sun of about 200–1,200 AU, and would take 10,000–20,000 years to orbit the Sun. Originally it might have migrated outward in the early Solar System or have been captured from another star. Scientists have extensively searched through existing data and made new observations, but nothing has been found. Because the planet would be cold, it would radiate in the infrared and might be detected by the James Webb Space Telescope.

• Does a small, dim star exist in the neighborhood of the Sun that comes close periodically and stirs up the Oort Cloud, sending a much larger than average number of comets into the inner Solar System? Some scientists think the fossil data show periodic mass extinctions on Earth and have investigated astronomical causes of the extinctions. One hypothesis is that a distant companion to the Sun has a wide orbit and periodically has come closer to the Sun as they both have

traveled around the Milky Way Galaxy. That companion may have stirred up the Oort Cloud as a result. That Oort Cloud disturbance sent many comets to the inner Solar System, which could have caused many impacts on Earth, leading to a mass extinction similar to the one that wiped out the dinosaurs. But some scientists dispute that Earth's extinctions have occurred at regular intervals, and surveys with infrared telescopes have found no evidence of a small companion star.

Questions and Problems

Test Your Understanding

1. This chapter has focused on leftover planetesimals. What became of most of the other planetesimals?
   a. They evaporated.
   b. They left the Solar System.
   c. They became part of larger bodies.
   d. They fragmented into smaller pieces.

2. The three types of meteorites come from different parts of their parent bodies. Stony-iron meteorites are rare because
   a. they are hard to find.
   b. the volume of a differentiated body that has both stone and iron is small.
   c. the Solar System has very little iron.
   d. the magnetic field of the Sun attracts the iron.

3. As a comet leaves the inner Solar System, the ion tail points
   a. back along the orbit.
   b. forward along the orbit.
   c. toward the Sun.
   d. away from the Sun.

4. Congress tasked NASA with searching for near-Earth objects because
   a. they might impact Earth, as others have.
   b. they are close by and easy to study.
   c. they are moving fast.
   d. they are scientifically interesting.

5. Meteorites can provide information about all of the following except
   a. the early composition of the Solar System.
   b. the composition of asteroids.
   c. the composition of comets.
   d. the Oort Cloud.

6. Perihelion is the point in an orbit _____ the Sun; aphelion is the point in an orbit _____ the Sun.
   a. closest to; farthest from
   b. farthest from; closest to
   c. at one focus of; at the other focus of

7. Kuiper Belt objects (KBOs) are mostly comet nuclei. Why don't they display comae and tails?
   a. Most of the material has already been stripped from the objects.
   b. They are too far from the Sun.
   c. They are too close to the Sun.
   d. The comae and tails are pointing away from Earth, behind the object.

8. Asteroids are small
   a. rock and metal objects orbiting the Sun.
   b. icy objects orbiting the Sun.
   c. rock and metal objects found only between the orbits of Mars and Jupiter.
   d. icy bodies found only in the outer Solar System.

9. Aside from their periods, short-period and long-period comets differ in that
   a. short-period comets orbit prograde, whereas long-period comets orbit retrograde.
   b. short-period comets contain less ice, whereas long-period comets contain more.
   c. short-period comets do not develop ion tails, whereas long-period comets do.
   d. short-period comets come closer to the Sun at closest approach than long-period comets.

10. On average, a bright comet appears about once each decade. Statistically, that means that
    a. one will definitely be observed every tenth year.
    b. one will definitely be observed in each 10-year period.
    c. exactly 10 comets will be observed in a century.
    d. about 10 comets will be observed in a century.

11. Most asteroids are located between the orbits of
    a. Earth and Mars.
    b. Mars and Jupiter.
    c. Jupiter and Saturn.
    d. the Kuiper Belt and the Oort Cloud.

12. Comets, asteroids, and meteoroids may be responsible for delivering a significant fraction of the current supply of _____ to Earth.
    a. mass
    b. water
    c. oxygen
    d. carbon

13. An iron meteorite most likely came from
    a. an undifferentiated asteroid.
    b. a differentiated asteroid.
    c. a planet.
    d. a comet.

14. Meteor showers occur because Earth passes through the path of
    a. another planet.
    b. a planetesimal.
    c. a comet.
    d. the Moon.

15. Dwarf planets differ from the other planets in that they
    a. have no atmosphere.
    b. have no moons.
    c. are all very far from the Sun.
    d. have lower mass.
    e. are covered in ice.

Thinking about the Concepts

16. Describe ways in which Pluto differs significantly from the Solar System planets.

17. By what criteria did Pluto fail to be considered a planet under the new IAU definition? Explain how that decision demonstrates the self-correcting nature of science.

18. How does the composition of an asteroid differ from that of a comet nucleus?

19. Define *meteoroid*, *meteor*, and *meteorite*.

20. What are the differences between a comet and a meteor in size, distance, and how long they remain visible?

21. Most meteorites are 4.54 billion years old. Carbonaceous chondrites, however, are 4.56 billion years old (20 million years older). What determines the time of "birth" of those pieces of rock? What does that information tell you about the history of their parent bodies?

22. Most asteroids are found between the orbits of Mars and Jupiter, but astronomers are especially interested in the relative few whose orbits cross that of Earth. Why?

23. How could you and a friend, armed only with your cell phones and knowledge of the night sky, prove conclusively that meteors are an atmospheric phenomenon?

24. Suppose you find a rock that has all the characteristics of a meteorite. You take it to a physicist friend, who confirms that it is a meteorite but says that radioisotope dating indicates an age of only a billion years. What might be the origin of that meteorite?

25. Describe differences between the Kuiper Belt and the Oort Cloud as sources of comets. What is the ultimate fate of a comet from each of those reservoirs?

26. What are the three parts of a comet? Which part is the smallest in radius? Which is the most massive?

27. In 1910, Earth passed directly through the tail of Comet Halley. Among the various gases in the tail was hydrogen cyanide (HCN). HCN is deadly to humans, yet nobody became ill from that event. Why?

28. Comets have two types of tails. Describe them and explain why they sometimes point in different directions.

29. What is zodiacal light, and what is its source?

30. How might comets and asteroids have contributed to the origin of life on Earth?

Applying the Concepts

31. Comet 67P/Churyumov-Gerasimenko has a diameter of 4 km and a mass of 10^{13} kg.
    a. What is the density of the comet? How does that value compare with the density of water?
    b. What is the escape velocity from the surface of that comet?

32. Ceres has a diameter of 975 km and a period of about 9 hours. What is the rotational speed of a point on the surface of that dwarf planet?

33. Figure 12.11 shows the scale of the Solar System out to the Oort Cloud. Judging from that figure, what fraction of the distance between the Sun and Proxima Centauri is occupied by the Oort Cloud, a part of the Solar System?

34. What are the perihelion and aphelion distances for Pluto and Eris? (Hint: Use Working It Out 12.1 and the information in Appendix 4.)

35. Use Working It Out 12.2 to find the impact energy (in joules) of an asteroid with a mass of 4.6×10^{11} kg traveling at 40 km/s. Does that energy depend on the "target" of impact? What is the equivalent in 1-megaton hydrogen bombs?

36. Earth's Moon has a diameter of 3,474 km and orbits at an average distance of 384,400 km. At that distance, it subtends an angle just slightly larger than half a degree in Earth's sky. Pluto's moon Charon has a diameter of 1,186 km and orbits at a distance of 19,600 km from the dwarf planet.
    a. Compare the appearance of Charon in Pluto's skies with the Moon in Earth's skies.
    b. Describe where in the sky Charon would appear as seen from various locations on Pluto.

37. One recent estimate concludes that nearly 800 meteorites with mass greater than 100 grams (massive enough to cause personal injury) strike Earth each day. If you present a target of 0.25 square meter (m^2) to a falling meteorite, what is the probability that you will be struck by a meteorite during your 100-year lifetime? (The surface area of Earth is approximately 5×10^{14} m^2.)

38. Electra is a 182-km asteroid accompanied by a small moon orbiting at a distance of 1,350 km in a circular orbit with a period of 3.92 days.
    a. What is the mass of Electra?
    b. What is Electra's density?

39. Calculate the orbital radius of the Kirkwood gap that is in a 3:1 orbital resonance with Jupiter.

40. The orbital periods of Comets Encke, Halley, and Hale-Bopp are 3.3, 76, and 2,530 years, respectively. Their orbital eccentricities are 0.847, 0.967, and 0.995, respectively.
    a. What are the semimajor axes (in astronomical units) of the orbits of those comets?
    b. What are the minimum and maximum distances from the Sun (in astronomical units) reached by Comets Halley and Hale-Bopp in their respective orbits?
    c. Which region of the Solar System did each probably come from?
    d. Which would you guess is the most pristine comet among the three? Which is the least? Explain your reasoning.

41. Comet Halley has a mass of approximately 2.2×10^{14} kg. It loses about 3×10^{11} kg each time it passes the Sun.
    a. The first confirmed observation of the comet was made in 240 BCE. If you assume a constant period of 76.4 years, how many times has it reappeared since that early sighting?
    b. How much mass has the comet lost since 240 BCE?
    c. What percentage of the comet's total mass today does that amount represent?

42. If Comet Halley is approximated as a sphere 5 km in radius, what is its density if it has a mass of 2.2×10^{14} kg? How does that density compare with that of water ($1,000$ kg/m³)?

43. A cubic centimeter of the air you breathe contains about 10^{19} molecules. A cubic centimeter of a comet's tail may typically contain 200 molecules. Calculate the volume of comet tail material that would hold 10^{19} molecules.

44. Some near-Earth objects are in binary systems, so it is possible to estimate their mass. How much energy would be released if a near-Earth asteroid with mass $m = 4.6 \times 10^{11}$ kg hit Earth at a velocity $v = 5$ km/s?

45. The estimated amount of zodiacal dust in the Solar System remains constant at approximately 10^{16} kg, yet zodiacal dust is constantly being swept up by planets or removed by the pressure of sunlight.
    a. If all the dust disappeared (at a constant rate) over a span of 30,000 years, what would the average production rate have to be, in kilograms per second, to maintain the current content?
    b. Is that an example of static or dynamic equilibrium? Explain your answer.

USING THE WEB

46. Dwarf planets:
    a. Go to planetary astronomer Mike Brown's website of dwarf planets (http://gps.caltech.edu/~mbrown/dps.html). How many dwarf planets does he think are in the Solar System? Why is it difficult officially to certify an object as a dwarf planet?
    b. Go to the website for the *New Horizons* mission (http://pluto.jhuapl.edu), which reached Pluto in 2015 and is scheduled to pass Kuiper Belt Object (KBO) 2014 MU69 in early 2019. Click on "News Center." What has been learned from that mission?

47. Citizen Science: Go to Backyard Worlds: Planet 9 (https://www.zooniverse.org/projects/marckuchner/backyard-worlds-planet-9). In that project you help search for new objects in the far reaches of the Solar System. Read all the sections in "About," and then click on "Classify" and look at some images.

48. Go to NASA's Asteroid Watch website (https://www.jpl.nasa.gov/asteroidwatch). What is new? Has a new discovery or a recent flyby occurred? Was the asteroid studied with a spacecraft, an orbiting telescope, or a ground-based telescope? What has been learned about the object?

49. Go to the Space Weather website (http://spaceweather.com). Are any comets now visible with the naked eye? Scroll down to "Near Earth Asteroids." Are any "close encounters" coming up in the next few months? Click on a few asteroid names to access the JPL Small-Body Database. Note the values of e and a in the table under the orbit. Calculate the NEO's closest and farthest distances from the Sun. In each case, how close will the NEO be to Earth when it is at its closest? How large is the object?

50. Do a search for the Japanese *Hayabusa 2* and the NASA *OSIRIS-REx* missions, each of which is visiting an asteroid in 2018 and is scheduled to bring back a sample to Earth after making observations. Did they arrive at their asteroid? What has been discovered from the observations? Has a sample been returned to Earth—if so, what has been learned?

digital.wwnorton.com/astro6

Astronomers often discover asteroids and other small Solar System objects by comparing two (or more) images of the same star field and looking for bright spots that have moved between the images. The four images in **Figure 12.26** are all "negative images": every dark spot would actually be bright in the sky, and all the white space is dark sky. A negative image sometimes helps the observer pick out faint details, and it is preferable for printing and photocopying. Study those four images. Can you find an asteroid that moves across the field in those images? That's the hard way to do it. A much easier method is to use a "blink comparison," which lets you look at one image and then another very quickly. Make three photocopies of each image, cut out each one, and align them all carefully on top of one another in sequence, so that the stars overlap. You should have 12 pieces of paper, in this order: Image 1, Image 2, Image 3, Image 4; Image 1, Image 2, . . . , and so on. Staple the top edge and flip the pages with your thumb, looking carefully at the images. Can you find the asteroid now?

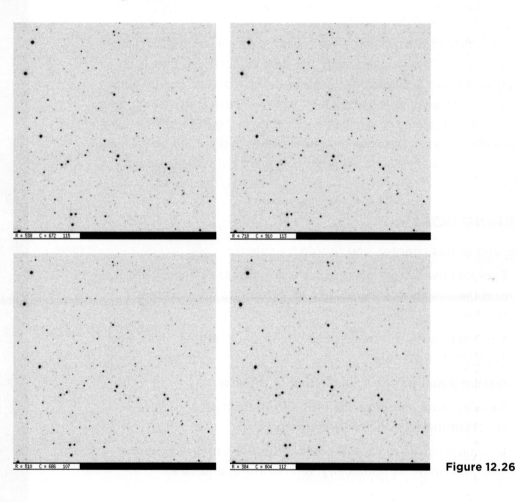

R = 538   C = 672   115

R = 718   C = 910   113

R = 510   C = 686   107

R = 384   C = 804   112

Figure 12.26

1 Circle the asteroid in each image.

The "blink comparison" method takes advantage of a feature of the human eye-brain connection. Humans are much better at noticing things that move than things that are stationary.

2 Why might that feature be a helpful evolutionary adaptation?

_____

_____

A third method is available, which sometimes makes things easier to see but requires high-quality digital images. In that method, one image is subtracted from another.

3 If you used the subtraction method with two of those images, what would you expect to see in the resulting image?

_____

_____

13

Taking the Measure of Stars

To all but the largest telescopes, even nearby stars are just points of light in the night sky. Astronomers study the stars by observing their light, by using the laws of physics discussed in earlier chapters, and by finding patterns in subgroups of stars that are then applied to other stars. Astronomers use knowledge of geometry, radiation, and orbits to begin to answer basic questions about stars, such as how they are similar to or different from the Sun, and whether they might have planets orbiting around them as the Sun does.

LEARNING GOALS

By the end of this chapter, you should be able to:

LG 1 Explain how astronomers use the brightness of nearby stars and their distances from Earth to determine how luminous they are.

LG 2 Explain how astronomers obtain the temperatures, sizes, and composition of stars.

LG 3 Describe how astronomers estimate the masses of stars.

LG 4 Classify stars and organize that information on a Hertzsprung-Russell (H-R) diagram.

LG 5 Explain how the mass and composition of a main-sequence star determine its luminosity, temperature, and size.

The constellation Orion. In addition to stars we can see numerous nebula of gas and dust. ▶▶▶

Why do stars
have different
colors?

353

NASA's Hubble Extends Stellar Tape Measure 10 Times Farther into Space

NASA Press Release

Using NASA's Hubble Space Telescope, astronomers now can precisely measure the distance of stars up to 10,000 light-years away—10 times farther than previously possible.

Astronomers have developed yet another novel way to use the 24-year-old space telescope by employing a technique called spatial scanning, which dramatically improves Hubble's accuracy for making angular measurements. The technique, when applied to the age-old method for gauging distances called astronomical parallax, extends Hubble's tape measure 10 times farther into space (**Figure 13.21**).

"This new capability is expected to yield new insight into the nature of dark energy, a mysterious component of space that is pushing the universe apart at an ever-faster rate," said Nobel laureate Adam Riess of the Space Telescope Science Institute (STScI) in Baltimore, Maryland.

Parallax, a trigonometric technique, is the most reliable method for making astronomical distance measurements, and a practice long employed by land surveyors here on Earth. The diameter of Earth's orbit is the base of a triangle, and the star is the apex where the triangle's sides meet. The lengths of the sides are calculated by accurately measuring the three angles of the resulting triangle.

Astronomical parallax works reliably well for stars within a few hundred light-years of Earth. For example, measurements of the distance to Alpha Centauri, the star system closest to our Sun, vary only by 1 arcsec. This variance in distance is equal to the apparent width of a dime seen from 2 miles away.

Stars farther out have much smaller angles of apparent back-and-forth motion that are extremely difficult to measure. Astronomers have pushed to extend the parallax yardstick ever deeper into our galaxy by measuring smaller angles more accurately.

This new long-range precision was proven when scientists successfully used Hubble to measure the distance of a special class of bright stars called Cepheid variables, approximately 7,500 light-years away in the northern constellation Auriga. The technique worked so well, they are now using Hubble to measure the distances of other far-flung Cepheids.

Such measurements will be used to provide firmer footing for the so-called cosmic "distance ladder." This ladder's "bottom rung" is built on measurements to Cepheid variable stars that, because of their known brightness, have been used for more than a century to gauge the size of the observable universe. They are the first step in calibrating far more distant extra-galactic milepost markers such as Type Ia supernovae.

Riess and the Johns Hopkins University in Baltimore, Maryland, in collaboration with Stefano Casertano of STScI, developed a technique to use Hubble to make measurements as small as five-billionths of a degree.

To make a distance measurement, two exposures of the target Cepheid star were taken 6 months apart, when Earth was on opposite sides of the Sun. A very subtle shift in the star's position was measured to an accuracy of

1/1,000 the width of a single image pixel in Hubble's Wide Field Camera 3, which has 16.8 megapixels total. A third exposure was taken after another 6 months to allow for the team to subtract the effects of the subtle space motion of stars, with additional exposures used to remove other sources of error.

Riess shares the 2011 Nobel Prize in Physics with another team for his leadership in the 1998 discovery that the expansion rate of the universe is accelerating—a phenomenon widely attributed to a mysterious, unexplained dark energy filling the universe. This new high-precision distance measurement technique is enabling Riess to gauge just how much the universe is stretching. His goal is to refine estimates of the universe's expansion rate to the point where dark energy can be better characterized.

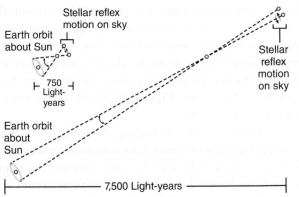

Figure 13.21 By applying a technique called spatial scanning to an age-old method for gauging distances called astronomical parallax, scientists now can use NASA's Hubble Space Telescope to make precision distance measurements 10 times farther into our galaxy than was previously possible.

1. How many parsecs are in 7,500 light-years? In 10,000 light-years?
2. What is the parallax angle of a star 7,500 light-years away? Ten thousand light-years away?
3. Why were the exposures taken 6 months apart?
4. Why is making parallax measurements from space better than doing so from the ground on Earth?
5. How does improving the accuracy of the distances to nearby stars from trigonometric parallax affect astronomers' estimates of farther stars' distances, measured using spectroscopic parallax?

Summary

Finding the distances to stars is a difficult but important task for astronomers. Parallax and spectroscopic parallax are two methods that astronomers use to determine distances to stars. Brightness and distance can be used to obtain the luminosity. Careful study of the light from a star, including its spectral lines, gives the temperature, size, and composition of the star. Study of binary systems gives the masses of stars of various spectral types, which we can extend to all stars of the same spectral type. The H-R diagram shows the relationship among the various physical properties of stars. The mass of a star is the major determining factor in the changes it undergoes. The habitable zone is the distance from a star in which a planet could have the right temperature for liquid water to exist on its surface. Stars of different luminosities and temperatures have habitable zones of different widths at different distances from the star.

LG 1 **Explain how astronomers use the brightness of nearby stars and their distances from Earth to determine how luminous they are.** The distance to a nearby star is measured by finding its parallax—by measuring how the star's apparent position changes in the sky over a year. The nearest star (other than the Sun) is about 4 light-years (1.3 parsecs) away. The brightness of a star in the sky can be measured directly, and brightness and distance can be used to obtain the star's luminosity—how much light the star emits.

LG 2 **Explain how astronomers obtain the temperatures, sizes, and composition of stars.** The color of a star depends on its temperature; blue stars are hotter and red stars are cooler. The radius can be computed from the temperature and luminosity of the star. Small, cool stars greatly outnumber large, hot stars. Spectral lines carry a great deal of information about a star, including what chemical elements and molecules are present in the star.

LG 3 **Describe how astronomers estimate the masses of stars.** Masses of stars are measured in binary star systems by observing the effects of the gravitational pull between the stars. Newton's universal law of gravitation and Kepler's laws connect the motion of the star to the forces they experience and thus to their masses.

LG 4 **Classify stars and organize that information on a Hertzsprung-Russell (H-R) diagram.** The H-R diagram shows the relationship among the various physical properties of stars. Temperature increases to the left, so that hotter stars lie on the left side of the diagram, whereas cooler stars lie on the right. Luminosity increases vertically, so that the most luminous stars lie near the top of the diagram. A star's luminosity class and temperature indicate its size. The mass and composition of a main-sequence star determine its luminosity, temperature, and size. Ninety percent of stars lie along the main sequence.

LG 5 **Explain how the mass and composition of a main-sequence star determine its luminosity, temperature, and size.** The mass and composition of a main-sequence star determine its original position on the H-R diagram. The main sequence on the H-R diagram is actually a sequence of masses. That position connects its other properties such as its luminosity, temperature, and size.

? Unanswered Questions

- What is the *upper* limit for stellar mass? Both theory and observation have shown that the *lower* limit for stellar mass is approximately 0.08 M_{Sun} with temperature about 2000 K. However, neither theory nor observation has yielded a definitive value for the upper limit. Many astronomers believe that the upper limit lies somewhere around 150–200 M_{Sun}. The very first stars that formed in the universe might have been even larger.

- Are planets with life likely to be orbiting around main-sequence stars of type M? Those low-luminosity stars are the most common type in the Milky Way, and the Kepler telescope has detected many planets orbiting such stars. However, the habitable zone of an M star is very close to the star, so that the planet may be strongly affected by streams of charged particles blowing off the star. Unless the planet has a strong protective magnetic field, that radiation may decrease or completely strip a planet of its atmosphere. Another complication is that a close-in planet is likely to be tidally locked to the star, so that one hemisphere of the planet receives light and the other hemisphere is permanently dark. That imbalance of light and heat might make the planet uninhabitable.

Questions and Problems

Test Your Understanding

1. Star A and star B are nearly the same distance from Earth. If star A is half as bright as star B, which of the following statements must be true?
   a. Star B is farther away than star A.
   b. Star B is twice as luminous as star A.
   c. Star B is hotter than star A.
   d. Star B is larger than star A.

2. Star A and star B are two nearby stars. If star A is blue and star B is red, which of the following statements must be true?
   a. Star A is hotter than star B.
   b. Star A is cooler than star B.
   c. Star A is farther away than star B.
   d. Star A is more luminous than star B.

3. Star A and star B are two stars nearly the same distance from Earth. If star A is blue and star B is red, but they have equal brightness, which of the following statements is true?
   a. Star A is more luminous than star B.
   b. Star A is larger than star B.
   c. Star A is smaller than star B.
   d. Star A is less luminous than star B.

4. What does it most likely mean when a star has very weak hydrogen lines and is blue?
   a. The star is too hot for hydrogen lines to form.
   b. The star has no hydrogen.
   c. The star is too cold for hydrogen lines to form.
   d. The star is moving too fast to measure the lines.

5. Star A and star B are a binary system. If the Doppler shift of star A's absorption lines is 3 times the Doppler shift of star B's absorption lines, which of the following statements is true?
   a. Star A is 3 times as massive as star B.
   b. Star A is one-third as massive as star B.
   c. Star A is closer than star B.
   d. The binary pair is moving toward Earth, but star A is farther away.

6. Star A and star B are two red stars at nearly the same distance from Earth. If star A is many times brighter than star B, which of the following statements is true?
   a. Star A is a main-sequence star, whereas star B is a red giant.
   b. Star A is a red giant, whereas star B is a main-sequence star.
   c. Star A is hotter than star B.
   d. Star A is a white dwarf, whereas star B is a red giant.

7. Star A and star B are two blue stars at nearly the same distance from Earth. If star A is many times brighter than star B, which of the following statements is true?
   a. Star A is a main-sequence star, whereas star B is a red giant.
   b. Star A is a main-sequence star, whereas star B is a blue giant.
   c. Star A is a white dwarf, whereas star B is a blue giant.
   d. Star A is a blue giant, whereas star B is a white dwarf.

8. In which region of an H-R diagram would you find the main-sequence stars with the widest habitable zones?
   a. upper left
   b. upper right
   c. center
   d. lower left
   e. lower right

9. If star A is more massive than star B, and both are main-sequence stars, star A is _____ than star B. (Choose all that apply.)
   a. more luminous
   b. less luminous
   c. hotter
   d. colder
   e. larger
   f. smaller

10. A telescope on Mars could measure the distances to more stars than can be measured from Earth because
    a. the resolution of the telescope would be better.
    b. Mars has a thin atmosphere.
    c. it would be closer to the stars.
    d. the parallax "baseline" would be longer.

11. Star A and star B are two nearby stars. If star A has a parallactic angle 4 times as large as star B's, which of the following statements is true?
    a. Star A is one-quarter as far away as star B.
    b. Star A is 4 times as far away as star B.
    c. Star A has moved through space one-quarter as far as star B.
    d. Star A has moved through space 4 times as far as star B.

12. If star A appears twice as bright as star B but is also twice as far away, star A is _____ as luminous as star B.
    a. 8 times   c. twice
    b. 4 times   d. half

13. In Table 13.1, the percentage of hydrogen in the Sun decreases when changing from percentage by number of atoms to percentage by mass, but the percentage of helium increases. Why?
    a. Hydrogen is more massive than helium.
    b. Helium is more massive than hydrogen.
    c. Hydrogen is located in a different part of the Sun.
    d. The mass of hydrogen is hard to measure.

14. Capella (in the constellation Auriga) is the sixth-brightest star in the sky. When viewed with a high-power telescope, it turns out that Capella is actually two pairs of binary stars: the first pair are G-type giants, whereas the second pair are M-type main-sequence stars. What color does Capella appear to be?
    a. red
    b. yellow
    c. blue
    d. The color cannot be determined from that information.

15. An eclipsing binary system has a primary eclipse (star A is eclipsed by star B) that is deeper (more light is removed from the light curve) than the secondary eclipse (star B is eclipsed by star A). What does that information tell you about stars A and B?
    a. Star A is hotter than star B.
    b. Star B is hotter than star A.
    c. Star B is larger than star A.
    d. Star B is moving faster than star A.

Thinking about the Concepts

16. The distances of nearby stars are determined by their parallaxes. Why are the distances of stars farther from Earth more uncertain?

17. To know certain properties of a star, you must first determine the star's distance. For other properties, knowing its distance is unnecessary. Explain why an astronomer does or does not need to know a star's distance to determine each of the following properties: size, mass, temperature, color, spectral type, and chemical composition.

18. Albireo, in the constellation Cygnus, is a visual binary system whose two components can easily be seen with even a small, amateur telescope. Viewers describe the brighter star as "golden" and the fainter one as "sapphire blue."
    a. What does that description tell you about the relative temperatures of the two stars?
    b. What does that description tell you about their respective sizes?

19. Very cool stars have temperatures around 2500 K and emit Planck spectra with peak wavelengths in the red part of the spectrum. Do those stars emit any blue light? Explain your answer.

20. The stars Betelgeuse and Rigel are both in the constellation Orion. Betelgeuse appears red, whereas Rigel is bluish white. To the eye, the two stars seem equally bright. If you can compare the temperature, luminosity, or size from just that information, do so. If not, explain why.

21. Explain why the stellar spectral types (O, B, A, F, G, K, M) are not in alphabetical order. What sequence of temperatures is defined by those spectral types?

22. Other than the Sun, the only stars whose mass astronomers can measure *directly* are those in eclipsing or visual binary systems. Why? How do astronomers estimate the masses of stars that are not in eclipsing or visual binary systems?

23. Once the mass of a certain spectral type of star located in a binary system has been determined, it can be assumed that all other stars of the same spectral type and luminosity class have the same mass. Why is that a reasonable assumption?

24. Explain why the Kepler Mission is finding eclipsing binary stars while it is searching for extrasolar planets by using the transit method.

25. Scientific advances often require the participation of scientists from all over the world, working on the same problem over many decades, even centuries. Compare that mode of "collaboration" with collaborations in your courses (perhaps on final projects or papers). What mechanisms must be in place to allow scientists to collaborate across space and time in that way?

26. What would happen to our ability to measure stellar parallax if we were on Mars? What if we were on Venus or Jupiter?

27. Figure 13.7 has an absorption line at about 410 nm that is weak for O stars and weak for G stars but very strong in A stars. That particular line comes from the transition from the second excited state of hydrogen up to the sixth excited state. Why is that line weak in O stars? Why is it weak in G stars? Why is it strongest in the middle of the range of spectral types?

28. Which kinds of binary systems are best observed edge-on? Which kinds are best observed face-on?

29. In Figure 13.10, two stars orbit a common center of mass.
    a. Explain why star 2 has a smaller orbit than star 1.
    b. Re-sketch that picture for the case in which star 1 has a very low mass, perhaps close to that of a planet.
    c. Re-sketch that picture for the case in which star 1 and star 2 have the same mass.

30. If our Sun were a blue main-sequence star, and Earth was still 1 AU from the Sun, would you expect Earth to be in the habitable zone? What about if our Sun were a red main-sequence star?

Applying the Concepts

31. Suppose that Figure 13.1b included a third star, located 4 times as far away as star A. How much less than star A would the third star appear to move each year? How much less than star B?

32. Suppose you see an object jump from side to side by half a degree as you blink back and forth between your eyes. How much farther away is an object that moves only one-third of a degree?

33. Logarithmic (log) plots show major steps along an axis scaled to represent equal factors, most often factors of 10. Why do astronomers sometimes use a log plot instead of the more conventional linear plot? Is the horizontal axis of the H-R diagram in Figure 13.15 logarithmic or linear?

34. Figure 13.5 is plotted logarithmically on both axes. The luminosities are in units of solar luminosities.
    a. How much more luminous than the Sun is a star on the far right side of the plot?
    b. How much less luminous than the Sun is a star on the far left side of the plot?

35. Compared with the Sun, how luminous, large, and hot is a star that has 10 times the mass of the Sun? Use Figure 13.17 to answer that question.

36. Sirius, the brightest star in the sky, has a parallax of 0.379 arcsec. What is its distance in parsecs? In light-years? How long does the light take to reach Earth?

37. Sirius is actually a binary pair of two A-type stars. The brighter star is called the "Dog Star" and the fainter is called the "Pup Star" because Sirius is in the constellation Canis Major (meaning "big dog"). The Dog Star appears about 6,800 times brighter than the Pup Star, even though both stars are at the same distance from Earth. Compare the temperatures, luminosities, and sizes of those two stars.

38. Sirius and its companion orbit around a common center of mass with a period of 50 years. The mass of Sirius is 2 times the mass of the Sun.
    a. If the orbital velocity of the companion is 2.35 times greater than that of Sirius, what is the mass of the companion?
    b. What is the semimajor axis of the orbit?

39. Sirius is 25 times more luminous than the Sun, and Polaris (the "North Pole Star") is 2,500 times more luminous than the Sun. Sirius appears 24 times brighter than Polaris. How much farther away is Polaris than Sirius? Use your answer from Problem 36 to find the distance of Polaris in light-years.

40. Betelgeuse (in Orion) has a parallax of 0.00451 ± 0.00080 arcsec, as measured by the Hipparcos satellite. What is the distance to Betelgeuse, and what is the uncertainty in that measurement?

41. The star Achernar has a Hipparchus parallax of 0.02339 arcsec and appears about as bright as Betelgeuese (previous problem) in the sky. Which star is actually more luminous? Knowing that Betelgeuse appears reddish, whereas Achenar appears bluish, which star would you say is hotter, and why?

42. The Sun is about 16 trillion (1.6×10^{13}) times brighter than the faintest stars visible to the naked eye.
    a. How far away (in astronomical units) would an identical solar-type star be if it were just barely visible to the naked eye?
    b. What would be its distance in light-years?

43. If $m_1 = m_2$ in Figure 13.9, where would the center of mass be located? If $m_1 = 2m_2$, where would the center of mass be located?

44. Find the peak wavelength of blackbody emission for a star with a temperature of about 10,000 K. In what region of the spectrum does that wavelength fall? What color is that star?

45. About 1,470 watts (W) of solar energy hits each square meter of Earth's surface. Use that value and the distance to the Sun to calculate the Sun's luminosity.

USING THE WEB

46. Go to the European Space Agency's *Gaia* mission website (https://www.esa.int/Our_Activities/Space_Science/Gaia). How has it helped astronomers determine the distances to more stars? Why is making parallax measurements from space better than measuring from the ground on Earth? Do a search for *Gaia* news (for example, at http://sci.esa.int/gaia/)—what has been learned from the released data?

47. Go to the Kepler Eclipsing Binary Catalog (http://keplerebs.villanova.edu) to see what new observations look like. Pick a few stars to study. Look at the last two columns (figures, labeled "LC Figs" and "SC Figs"). The "raw" and "dtr" figures are rough, but the "pf" figure shows a familiar light curve. How deep is the eclipse; that is, how much lower is the "normalized flux" during maximum eclipse?

48. Do a search for a photograph of your favorite constellation (or go outside and take a picture yourself). Can you see different colors in the stars? What do the colors tell you about the surface temperatures of the stars? From your photograph, can you tell which are the three brightest stars in the constellation? Those stars will be named "alpha" (α), "beta" (β), and "gamma" (γ) for that constellation. Look up the constellation online and see whether you chose the right stars. What are their temperatures and luminosities? What are their distances?

49. Citizen science: Go to the website for the Stellar Classification Online Public Exploration (SCOPE) project (http://scope.pari.edu/takepart.php). The project uses crowdsourcing to classify stars seen on old photographic plates of pictures taken in the Southern Hemisphere. Create an account, review the science and the FAQs, and then click on "To Take Part" to see some practice examples. Then go to "Classify," choose a photographic plate, and classify a few stars.

50. Watch the short NASA video about planets in binary star systems: https://nasaviz.gsfc.nasa.gov/11106. Why do astronomers think these systems are not very common? Do a search on "circumbinary planets"—have many more been found?

Visit the Digital Resources Page and on the Student Site open the "H-R Diagram" Interactive Simulation for Chapter 13. This simulation enables you to compare stars on the H-R diagram in two ways. You can compare an individual star (marked by a red X) to the Sun by varying its properties in the box in the left half of the window. Or you can compare groups of the nearest and brightest stars. Play around with the controls for a few minutes to familiarize yourself with the simulation.

Begin by exploring how changes to the properties of the individual star change its location on the H-R diagram. First, press the "Reset" button at the top right of the window.

Decrease the temperature of the star by dragging the temperature slider to the left. Notice that the luminosity remains the same. Because the temperature has decreased, each square meter of star surface must be emitting less light. What other property of the star changes to keep the total luminosity of the star constant?

Predict what will happen when you move the temperature slider all the way to the right. Now do it. Did the star behave as you expected?

1 As you move to the left across the H-R diagram, what happens to the radius?

_____

_____

_____

2 What happens as you move to the right?

_____

_____

_____

Press "Reset" and experiment with the luminosity slider.

3 As you move up on the H-R diagram, what happens to the radius?

_____

_____

_____

4 What happens as you move down?

_____

_____

_____

Press "Reset" again and predict how you would have to move the slider bars to move your star into the red giant portion of the H-R diagram (upper right). Adjust the slider bars until the star is in that area. Were you correct?

5 How would you adjust the slider bars to move the star into the white dwarf area of the H-R diagram?

_____

_____

Press the "Reset" button and explore the right-hand side of the window. Add the nearest stars to the graph by clicking their radio button under "Plotted Stars." Using what you learned above, compare the temperatures and luminosities of those stars with the Sun (marked by the X).

6 Are the nearest stars generally hotter or cooler than the Sun?

_____

7 Are the nearest stars generally more or less luminous than the Sun?

_____

Press the radio button for the brightest stars. That action will remove the nearest stars and add the brightest stars in the sky to the plot. Compare those stars with the Sun.

8 Are the brightest stars generally hotter or cooler than the Sun?

_____

9 Are the brightest stars generally more or less luminous than the Sun?

_____

10 How do the temperatures and luminosities of the brightest stars in the sky compare with the temperatures and luminosities of the nearest stars? Does that information support the claim in the chapter that more low-luminosity stars than high-luminosity stars exist? Explain.

_____

_____

_____

14 Our Star—The Sun

Because the Sun is the only star close to Earth, much of our detailed knowledge about stars has come from studying the Sun. In Chapter 13, we looked at the physical properties of distant stars, including their mass, luminosity, size, temperature, and chemical composition. Here in Chapter 14, we ask fundamental questions about Earth's local star: How does the Sun work? Where does it get its energy? How has its luminosity changed over the billions of years since the Solar System formed?

LEARNING GOALS

By the end of this chapter, you should be able to:

LG 1 Describe the balance between the forces that determine the structure of the Sun.

LG 2 Explain how mass is converted to energy in the Sun's core and how long the Sun will take to use up its fuel.

LG 3 Sketch a physical model of the Sun's interior, and list the ways that energy moves outward from the Sun's core toward its surface.

LG 4 Describe how observations of solar neutrinos and seismic vibrations on the surface of the Sun test astronomers' models of the Sun.

LG 5 Describe the solar activity cycles of 11 and 22 years, and explain how those cycles are related to the Sun's changing magnetic field.

This image is a combination of several extreme ultraviolet images of the Sun from the Solar Dynamics Observatory. ▶▶▶

What makes
the Sun shine?

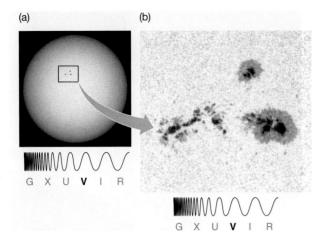

(a)   (b)

G X U **V** I R

G X U **V** I R

Figure 14.17 (a) This image from the Solar Dynamics Observatory (SDO), taken in 2010, shows a large sunspot group. Sunspots are magnetically active regions cooler than the surrounding surface of the Sun. (b) This high-resolution view shows the sunspots in this group.

cools, convection cannot carry it downward, so it makes a cooler (and therefore darker) spot on the surface of the Sun. **Figure 14.17** shows a large sunspot group. Sunspots appear dark, but only in contrast to the brighter surface of the Sun (**Working It Out 14.2**).

Early telescopic observations of sunspots made during the 17th century led to the discovery of the Sun's rotation, which has an average period of about 27 days as seen from Earth and 25 days relative to the stars. Because Earth orbits the Sun in the same direction that the Sun rotates, observers on Earth see a slightly longer rotation period. Observations of sunspots also show that the Sun's rotation period is shorter at its equator than at higher latitudes, an effect called **differential rotation**. Differential rotation is possible only because the Sun is a large ball of gas rather than a solid object.

Figure 14.18 shows the structure of a sunspot on the surface of the Sun. A sunspot consists of an inner dark core called the **umbra**, surrounded by a less dark region called the **penumbra**. The penumbra shows an intricate radial pattern, reminiscent of the petals of a flower. Sunspots are caused by magnetic fields thousands of times greater than the magnetic field at Earth's surface. They occur in pairs connected by loops in the magnetic field. Sunspots range in size from a few tens of kilometers across up to complex groups that may contain several dozen individual spots and span as much as 150,000 km. The largest sunspot groups can be seen without a telescope.

Although sunspots occasionally last 100 days or longer, half of all sunspots come and go in about 2 days, and 90 percent are gone within 11 days. The number and distribution of sunspots change in a pattern averaging 11 years called the **sunspot cycle**. **Figure 14.19a** shows data for several recent cycles. At the beginning of a cycle, sunspots appear at solar latitudes of about 30° north and south of

---

14.2   Working It Out Sunspots and Temperature

Sunspots are about 1500 K cooler than their surroundings. What does that lower temperature tell us about their luminosity? Think back to the Stefan-Boltzmann law in Chapter 5. The flux, $\mathcal{F}$, from a blackbody is proportional to the fourth power of the temperature, T:

$$\mathcal{F} = \sigma T^4$$

The constant of proportionality is the Stefan-Boltzmann constant, σ, which has a value of 5.67×10^{-8} W/(m² K⁴).

The flux is the amount of energy coming from a square meter of surface every second. How much less energy comes out of a sunspot than out of the rest of the Sun? Using 4500 and 6000 K for the temperatures of a typical sunspot and the surrounding photosphere, respectively, we can set up two equations:

$$\mathcal{F}_{spot} = \sigma T^4_{spot} \quad \text{and} \quad \mathcal{F}_{surface} = \sigma T^4_{surface}$$

We could solve each of those separately and then divide the value of $\mathcal{F}_{spot}$ by $\mathcal{F}_{surface}$ to find out how much fainter the sunspot is, but solving for the ratio of the fluxes eliminates σ:

$$\frac{\mathcal{F}_{spot}}{\mathcal{F}_{surface}} = \frac{\sigma T^4_{spot}}{\sigma T^4_{surface}} = \frac{T^4_{spot}}{T^4_{surface}} = \left(\frac{T_{spot}}{T_{surface}}\right)^4$$

Plugging in our values for T_{spot} and $T_{surface}$ gives

$$\frac{\mathcal{F}_{spot}}{\mathcal{F}_{surface}} = \left(\frac{4500 \text{ K}}{6000 \text{ K}}\right)^4 = 0.32$$

and multiplying both sides by $F_{surface}$ gives

$$\mathcal{F}_{spot} = 0.32\mathcal{F}_{surface}$$

So, the amount of energy coming from a square meter of sunspot every second is about one-third as much as the amount of energy coming from a square meter of surrounding surface every second. In other words, the sunspot is about one-third as bright as the surrounding photosphere. If you could cut out the sunspot and place it elsewhere in the sky, it would be brighter than the full Moon.

the solar equator. Over the following years, sunspots are found closer to the equator as their number increases to a maximum and then declines. As the last few sunspots approach the equator, new sunspots again begin appearing at middle latitudes, and the next cycle begins. **Figure 14.19b** shows the number of sunspots at a given latitude plotted against time: that diagram of opposing diagonal bands is called the sunspot "butterfly diagram."

In the early 20th century, solar astronomer George Ellery Hale (1868–1938) was the first to show that the 11-year sunspot cycle is actually half of a 22-year magnetic cycle during which the direction of the Sun's magnetic field reverses after each 11-year sunspot cycle. **Figure 14.19c** shows how the average strength of the magnetic field at every latitude has changed over more than 35 years. The direction of the Sun's magnetic field flips at the maximum of each sunspot cycle. Sunspots come in pairs, with one spot (the leading sunspot) in front of the other with respect to the Sun's rotation. In one 11-year sunspot cycle, the leading sunspot in each pair tends to be a north magnetic pole, whereas the trailing sunspot tends to be a south magnetic pole. In the next 11-year sunspot cycle, that polarity is reversed: the leading sunspot in each pair is a south magnetic pole, whereas the trailing sunspot tends to be a north magnetic pole. The

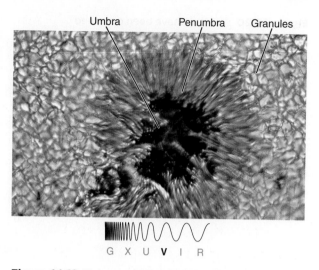

Figure 14.18 This very high-resolution view of a sunspot shows the dark umbra surrounded by the lighter penumbra. The solar surface around the sunspot bubbles with separate cells of hot gas called *granules*. The smallest features are about 100 km across.

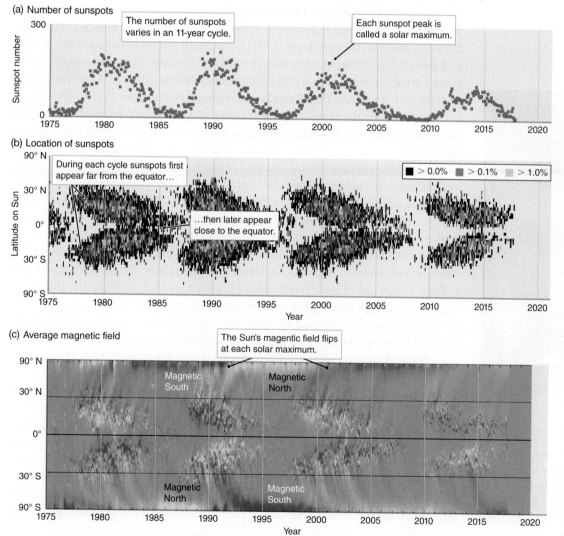

Figure 14.19 (a) The number of sunspots varies, as shown in this graph of the past few solar cycles. (b) The "solar butterfly" diagram shows the fraction of the Sun covered by sunspots at each latitude. The data are color coded to show the percentage of the strip at that latitude that is covered in sunspots at that time: black, 0 to less than 0.1 percent; red, 0.1–1.0 percent; yellow, greater than 1.0 percent. (c) The Sun's magnetic poles flip every 11 years. Yellow indicates magnetic north, whereas blue indicates magnetic south.

Figure 14.20 Sunspots have been observed for hundreds of years. In this plot, the 11-year cycle in the number of sunspots (half of the 22-year solar magnetic cycle) is clearly visible. Sunspot activity varies greatly. The period from the middle of the 17th century to the early 18th century, when almost no sunspots were seen, is called the *Maunder minimum*.

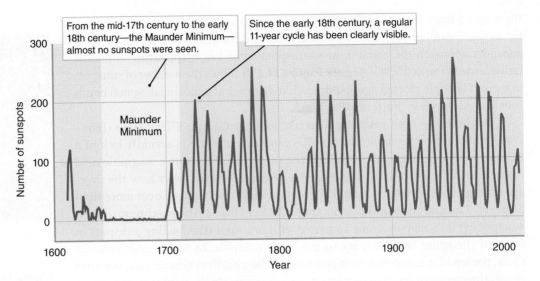

From the mid-17th century to the early 18th century—the Maunder Minimum—almost no sunspots were seen.

Since the early 18th century, a regular 11-year cycle has been clearly visible.

Maunder Minimum

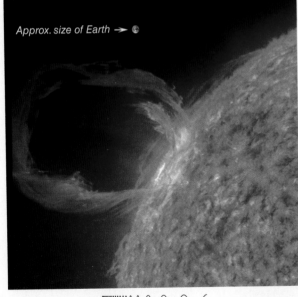

Approx. size of Earth →

G X **U** V I R

Figure 14.21 Solar prominences are magnetically supported arches of hot gas that rise high above active regions on the Sun. Here, you can see a close-up view at the base of a large prominence. An image of Earth is included for scale (it is not actually that close to the Sun).

transition between those two magnetic polarities occurs near the peak of each sunspot cycle. Magnetic activity on the Sun affects the photosphere, chromosphere, and corona.

The graph in **Figure 14.20** shows 400 years of sunspot observations. The 11-year cycle is neither perfectly periodic nor especially reliable. The time between peaks in the number of sunspots actually varies between about 9.7 and 11.8 years. The number of spots seen during a given cycle fluctuates as well, and in some periods sunspot activity has disappeared almost entirely. An extended lull in solar activity, called the **Maunder minimum**, lasted from 1645 to 1715. Typically, about six peaks of solar activity occur in 70 years, but virtually no sunspots were seen during the Maunder minimum, and auroral displays were less frequent than usual.

Sunspots are only one of several phenomena that follow the Sun's 22-year cycle of magnetic activity. The peaks of the cycle, called **solar maxima**, are times of intense activity. Sunspots are often accompanied by a brightening of the solar chromosphere that is seen most clearly in emission lines such as Hα. Those bright regions are known as solar active regions. The magnificent loops arching through the solar corona, shown in **Figure 14.21**, are solar **prominences**, magnetic flux tubes of relatively cool (5000–10,000 K) but dense gas extending through the million-kelvin gas of the corona. Those prominences are anchored in the active regions. Although most prominences are relatively quiet, others can erupt out through the corona, towering a million kilometers or more over the surface of the Sun and ejecting material into the corona at velocities of 1,000 km/s.

Solar flares are the most energetic form of solar activity, violent eruptions in which enormous amounts of magnetic energy are released over a few minutes to a few hours. **Figure 14.22** shows solar flares erupting from two sunspot groups. The images in Figures 14.22a and b, taken in ultraviolet light, show material at very high temperatures. The spots in the visible-light image (Figure 14.22c) are at the base of the activity seen in Figures 14.22a and b. Solar flares can heat gas to temperatures of 20 million K, and they are the source of intense X-rays and gamma rays. Hot **plasma** (consisting of atoms stripped of some or all of their electrons) moves outward from flares at speeds that can reach 1,500 km/s. Magnetic effects can then accelerate subatomic particles to almost the speed of light. Such

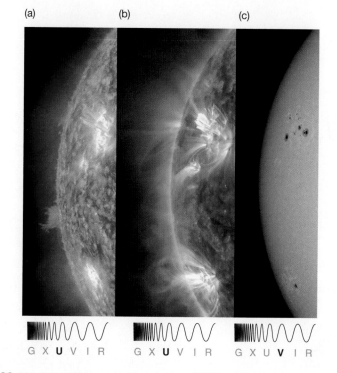

(a)   (b)   (c)

G X **U** V I R   G X **U** V I R   G X U **V** I R

Figure 14.22 The Solar Dynamics Observatory (SDO) observed these active regions of the Sun that produced solar flares in August 2011. (a) Activity near the surface at 60,000 K is visible in extreme ultraviolet light (along with a prominence rising up from the Sun's edge). (b) Viewed at other ultraviolet wavelengths, many looping arcs and plasma heated to about 1 million K become visible. (c) The dark spots in this image are the magnetically intense sunspots that are the sources of all the activity.

events, called **coronal mass ejections** (**CMEs**) (**Figure 14.23**), send powerful bursts of energetic particles outward through the Solar System. CMEs occur about once per week during the minimum of the sunspot cycle and as often as several times per day near the maximum of the cycle.

Solar Activity Affects Earth

The amount of solar radiation received at the distance of Earth from the Sun has been measured to be, on average, 1,361 watts per square meter (W/m^2). Satellite measurements of the amount of radiation coming from the Sun (**Figure 14.24**) show that this value varies by as much as 0.2 percent over periods of a few weeks, as dark sunspots in the photosphere and bright spots in the chromosphere move across the disk. Overall, however, the increased radiation from active regions on the Sun more than makes up for the reduction in radiation from sunspots. On average, the Sun seems to be about 0.1 percent brighter during the peak of a solar cycle than it is at its minimum.

Solar activity affects Earth in many ways. Solar active regions are the source of most of the Sun's extreme ultraviolet and X-ray emissions—energetic radiation that heats Earth's upper atmosphere and, during periods of increased solar activity, causes Earth's upper atmosphere to expand. When that happens, the swollen upper atmosphere can significantly increase the atmospheric drag on spacecraft

Coronal mass ejection

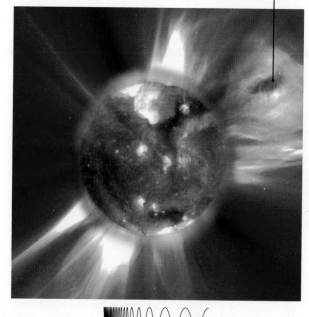

G X **U** V I R

Figure 14.23 This Solar and Heliospheric Observatory (*SOHO*) image shows a coronal mass ejection (upper right); a simultaneously recorded ultraviolet image of the solar disk is superimposed.

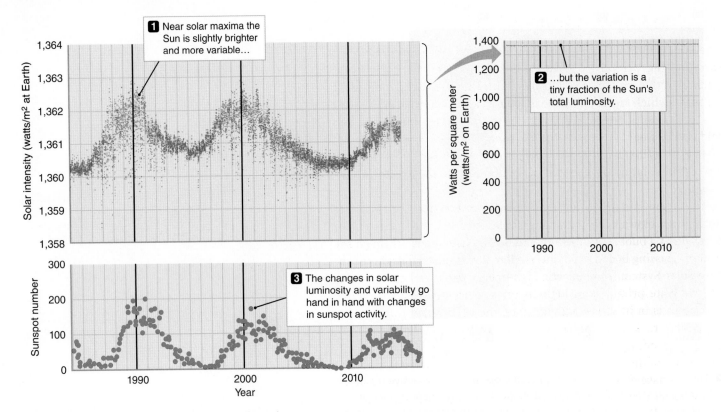

Figure 14.24 Measurements taken by satellites show that the amount of light from the Sun changes slightly.

orbiting at relatively low altitudes, such as that of the Hubble Space Telescope, causing their orbits to decay. Periodic boosts have been necessary to keep the Hubble Space Telescope in its orbit.

Earth's magnetosphere is the result of the interaction between Earth's magnetic field and the solar wind. Increases in the solar wind accompanying solar activity, especially CMEs directed at Earth, can disrupt Earth's magnetosphere. Spectacular auroras can accompany such events, as can magnetic storms that have disrupted electric power grids and caused blackouts across large regions. CMEs emitted in the direction of Earth also hinder radio communication and navigation, and they can damage sensitive satellite electronics, including communication satellites. In addition, energetic particles accelerated in solar flares pose one of the greatest dangers to human exploration of space.

Detailed observations from the ground and from space help astronomers understand the complex nature of the solar atmosphere. The Solar and Heliospheric Observatory (*SOHO*) spacecraft is a joint mission between NASA and the European Space Agency (ESA). *SOHO* moves in lockstep with Earth at a location approximately 1,500,000 km from Earth that is almost directly in line between Earth and the Sun. *SOHO* carries 12 scientific instruments that monitor the Sun and measure the solar wind upstream of Earth. In addition, NASA's Solar Dynamics Observatory (SDO) studies the solar magnetic field to predict when major solar events will occur, instead of simply responding after they happen.

CHECK YOUR UNDERSTANDING 14.4

Sunspots appear dark because: (a) they have very low density; (b) magnetic fields absorb most of the light that falls on them; (c) they are regions of very high pressure; (d) they are cooler than their surroundings.

Origins The Solar Wind and Life

Energetic particles accelerated in solar flares need to be considered when astronauts are orbiting Earth in a space station or, someday, traveling to the Moon or farther. Earth's magnetic field protects life on the surface from those energetic particles, which travel along the magnetic-field lines to Earth's poles, creating the auroras. But the Moon does not have that protection because its magnetic field is very weak. Astronauts on the lunar surface would be exposed to as much radiation as astronauts traveling in space. The strength of the solar wind varies with the solar cycle, as noted in Section 14.4, so the exposure danger varies as well.

The Solar System is surrounded by the **heliosphere** (**Figure 14.25**), in which the solar wind blows against the interstellar medium and clears out an area like the inside of a bubble. As the Sun and Solar System move through the Milky Way Galaxy, passing in and out of interstellar clouds, that heliosphere protects the entire Solar System from galactic high-energy particles known as cosmic rays that originate primarily in high-energy explosions of massive dying stars. When the Sun is in its lower-activity state, the heliosphere is weaker, so more galactic cosmic rays enter the Solar System. In addition, the intensity of those cosmic rays depends on where the Sun and Solar System are located in their orbit about the center of the Milky Way Galaxy.

Some scientists have theorized that at times when the Sun was quiet and the heliosphere was weaker than average, and the Solar System was passing through a particular part of the galaxy, the cosmic-ray flux in the Solar System—and on Earth—increased. That increased flux possibly led to a disruption in Earth's ozone layer and possibly contributed to a mass extinction in which many species died out on Earth.

Thus, in addition to heat and light on Earth, the extension of the Sun through the solar wind may have affected the evolution of life on Earth—and may affect the ability of humans to live and work in space.

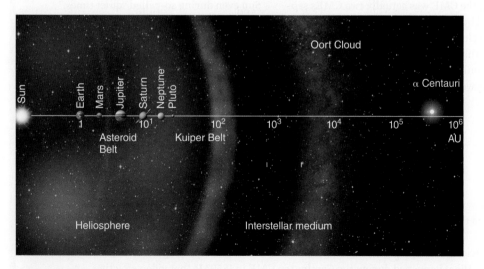

Figure 14.25 The heliosphere of the Sun is a bubble of charged particles covering the Solar System. The heliosphere is formed by the solar wind blowing against the interstellar medium. Both the *Voyager* spacecraft are now past 100 AU. The scale, indicating how far objects are from the Sun, is logarithmic.

Carrington-Class CME Narrowly Misses Earth

By DR. TONY PHILLIPS, Science@NASA

Last month (April 8–11, 2014), scientists, government officials, emergency planners, and others converged on Boulder, Colorado, for NOAA's Space Weather Workshop—an annual gathering to discuss the perils and probabilities of solar storms.

The current solar cycle is weaker than usual, so you might expect a correspondingly low-key meeting. On the contrary, the halls and meeting rooms were abuzz with excitement about an intense solar storm that narrowly missed Earth.

"If it had hit, we would still be picking up the pieces," says Daniel Baker of the University of Colorado, who presented a talk entitled *The Major Solar Eruptive Event in July 2012: Defining Extreme Space Weather Scenarios.*

The close shave happened almost two years ago. On July 23, 2012, a plasma cloud or "CME" rocketed away from the Sun as fast as 3,000 km/s, more than 4 times faster than a typical eruption. The storm tore through the Earth's orbit, but fortunately Earth wasn't there. Instead it hit the *STEREO-A* spacecraft. Researchers have been analyzing the data ever since, and they have concluded that the storm

was one of the strongest in recorded history. "It might have been stronger than the Carrington Event itself," says Baker.

The Carrington Event of September 1859 was a series of powerful CMEs that hit Earth head-on, sparking Northern Lights as far south as Tahiti. Intense geomagnetic storms caused global telegraph lines to spark, setting fire to some telegraph offices and disabling the "Victorian Internet." A similar storm today could have a catastrophic effect on modern power grids and telecommunication networks. According to a study by the National Academy of Sciences, the total economic impact could exceed $2 trillion or 20 times greater than the costs of a Hurricane Katrina. Multi-ton transformers fried by such a storm could take years to repair and impact national security.

A recent paper in *Nature Communications* authored by UC Berkeley space physicist Janet G. Luhmann and former postdoc Ying D. Liu describes what gave the July 2012 storm Carrington-like potency. For one thing, the CME was actually *two* CMEs separated by only 10 to 15 minutes. This double storm cloud traveled through a region of space that had been cleared out by another CME only four days earlier. As a result, the CMEs were not decelerated as much as

usual by their transit through the interplanetary medium.

Had the eruption occurred just one week earlier, the blast site would have been facing Earth, rather than off to the side, so it was a relatively narrow escape.

When the Carrington Event enveloped Earth in the 19th century, technologies of the day were hardly sensitive to electromagnetic disturbances. Modern society, on the other hand, is deeply dependent on Sun-sensitive technologies such as GPS, satellite communications, and the Internet.

"The effect of such a storm on our modern technologies would be tremendous," says Luhmann.

During informal discussions at the workshop, Nat Gopalswamy of the Goddard Space Flight Center noted that "without NASA's *STEREO* probes, we might never have known the severity of the 2012 superstorm. This shows the value of having 'space weather buoys' located all around the Sun."

It also highlights the potency of the Sun even during so-called "quiet times." Many observers have noted that the current solar cycle is weak, perhaps the weakest in 100 years. Clearly, even a weak solar cycle can produce a very strong storm. Says Baker, "We need to be prepared."

1. What is a coronal mass ejection (CME)?
2. Why would a CME cause disruptions on Earth?
3. Explain how the 2012 storm missed Earth by 1 week.
4. More sensationalistic headlines for this story claimed that Earth almost "was sent back to the Dark Ages." What did they mean by that exaggeration?
5. Go to the NASA press release for this event (https://science.nasa.gov/science-news/science-at-nasa/2014/23jul_superstorm/), and click to watch the 4-minute "ScienceCast" video. What happened during the Carrington CME in 1859? Is that video effective at communicating the science information to the nonspecialist?

Summary

The forces due to pressure and gravity balance each other in hydrostatic equilibrium, maintaining the Sun's structure. Nuclear reactions converting hydrogen to helium are the source of the Sun's energy. Energy created in the Sun's core moves outward to the surface, first by radiation and then by convection. The solar wind may adversely affect astronauts located in space or on planets that lack a protective magnetic field, but it also has protected the Solar System from galactic, high-energy, cosmic-ray particles.

LG 1 **Describe the balance between the forces that determine the structure of the Sun.** The outward radiation pressure of the hot gas inside the Sun balances the inward pull of gravity at every point. That balance is dynamically maintained. An energy balance is also maintained, with the energy produced in the core of the Sun balancing the energy lost from the surface.

LG 2 **Explain how mass is converted to energy in the Sun's core and how long the Sun will take to use up its fuel.** In the core of the Sun, mass is converted to energy via the proton-proton chain. When four hydrogen atoms fix to one helium atom, some mass is lost. That mass is released as energy, nearly all of which leaves the Sun either as photons or as neutrinos. Neutrinos are elusive, almost massless particles that interact only very weakly with other matter. Observations of neutrinos confirm that nuclear fusion is the Sun's primary energy source.

LG 3 **Sketch a physical model of the Sun's interior, and list the ways that energy moves outward from the Sun's core toward its surface.** The interior of the Sun is divided into zones defined by how energy is transported in that region. Energy moves outward through the Sun by radiation and by convection.

LG 4 **Describe how observations of solar neutrinos and seismic vibrations on the surface of the Sun test astronomers' models of the Sun.** The Sun has multiple layers, each with a characteristic pressure, density, and temperature. Neutrinos directly probe the interior of the Sun. That model of the interior of the Sun has been tested by helioseismology, in much the same way that the model of Earth's interior has been tested by seismology.

LG 5 **Describe the solar activity cycles of 11 and 22 years, and explain how those cycles are related to the Sun's changing magnetic field.** Activity on the Sun follows a cycle that peaks every 11 years but takes 22 full years for the magnetic field to reverse. Sunspots are photospheric regions cooler than their surroundings, and they reveal the cycles in solar activity. Material streaming away from the Sun's corona creates the solar wind, which moves outward through the Solar System until it meets the interstellar medium. Solar storms, including ejections of mass from the corona, produce auroras and can disrupt power grids and damage satellites.

? Unanswered Questions

- Will nuclear fusion become a major source of energy production on Earth? Scientists have been working on controlled nuclear fusion for more than 60 years, since the first hydrogen bombs were developed. So far, however, replicating the conditions inside the Sun has posed too many difficulties. Nuclear fusion requires that we have hydrogen isotopes at very high temperature, density, and pressure, just as when a hydrogen bomb explodes. However, controlled nuclear fusion requires that we confine that material long enough to get more energy out than we put in. Several major experiments have attempted to fuse isotopes of hydrogen. An alternative approach is to fuse an isotope of helium, ^{3}He, which has only three particles in the nucleus (two protons and one neutron). On Earth, ^{3}He is found in very limited supply. But ^{3}He is in much greater abundance on the Moon, so some people propose setting up mining colonies on the Moon to extract ^{3}He for use in fusion reactions on Earth or possibly even on the Moon (see question 49b at the end of the chapter).

- Are large timescale variations in Earth's climate—ice ages—related to solar activity? Solar activity affects Earth's upper atmosphere and may affect weather patterns as well. Variations in the amount of radiation from the Sun might be responsible for past variations in Earth's climate. Current models indicate that observed variations in the Sun's luminosity could account for only about 0.1-K differences in Earth's average temperature—much less than the effects due to the ongoing buildup of carbon dioxide in Earth's atmosphere. Triggering the onset of an ice age may require a sustained drop in global temperatures of only 0.2–0.5 K, so astronomers are continuing to investigate a possible link between solar variability and long-timescale changes in Earth's climate.

Questions and Problems

Test Your Understanding

1. The physical model of the Sun's interior has been confirmed by observations of
   a. neutrinos and seismic vibrations.
   b. sunspots and solar flares.
   c. neutrinos and positrons.
   d. sample returns from spacecraft.
   e. sunspots and seismic vibrations.

2. Place in order the following steps in the fusion of hydrogen into helium. If two or more steps happen simultaneously, use an equals sign (=).
   a. A positron is emitted.
   b. One gamma ray is emitted.
   c. Two hydrogen nuclei are emitted.
   d. Two ^{3}He collide and become ^{4}He.
   e. Two hydrogen nuclei collide and become ^{2}H.
   f. Two gamma rays are emitted.
   g. A neutrino is emitted.
   h. One deuterium nucleus and one hydrogen nucleus collide and become ^{3}He.

3. Sunspots, flares, prominences, and coronal mass ejections are all caused by
   a. magnetic activity on the Sun.
   b. electrical activity on the Sun.
   c. the interaction of the Sun's magnetic field and the interstellar medium.
   d. the interaction of the solar wind and Earth's magnetic field.
   e. the interaction of the solar wind and the Sun's magnetic field.

4. The structure of the Sun is determined by not only the balance between the forces due to _____ and gravity but also the balance between energy generation and energy _____.
   a. pressure; production   c. ions; loss
   b. pressure; loss   d. solar wind; production

5. In the proton-proton chain, four hydrogen nuclei are converted to a helium nucleus. That does not happen spontaneously on Earth because the process requires
   a. vast amounts of hydrogen.
   b. very high temperatures and densities.
   c. hydrostatic equilibrium.
   d. very strong magnetic fields.

6. The solar neutrino problem pointed to a fundamental gap in our knowledge of
   a. nuclear fusion.
   b. neutrinos.
   c. hydrostatic equilibrium.
   d. magnetic fields.

7. Sunspots change in number and location during the solar cycle. That phenomenon is connected to
   a. the rotation rate of the Sun.
   b. the temperature of the Sun.
   c. the magnetic field of the Sun.
   d. the tilt of the axis of the Sun.

8. Suppose an abnormally large amount of hydrogen suddenly fused in the core of the Sun. Which of the following would be observed first?
   a. The Sun would become brighter.
   b. The Sun would swell and become larger.
   c. The Sun would become bluer.
   d. The Sun would emit more neutrinos.

9. The solar corona has a temperature of 1 million to 2 million K; the photosphere has a temperature of only about 6000 K. Why isn't the corona much, much brighter than the photosphere?
   a. The magnetic field traps the light.
   b. The corona emits only X-rays.
   c. The photosphere is closer to us.
   d. The corona has a much lower density.

10. The Sun rotates once every 25 days relative to the stars. The Sun rotates once every 27 days as seen from Earth. Why are those two numbers different?
    a. The stars are farther away.
    b. Earth is smaller.
    c. Earth moves in its orbit as the Sun rotates.
    d. The Sun moves relative to the stars.

11. Place the following regions of the Sun in order of increasing radius.
    a. corona   e. chromosphere
    b. core   f. photosphere
    c. radiative zone   g. a sunspot
    d. convective zone

12. Coronal mass ejections
    a. carry away 1 percent of the mass of the Sun each year.
    b. are caused by breaking magnetic fields.
    c. are always emitted in the direction of Earth.
    d. are unimportant to life on Earth.

13. As energy moves out from the Sun's core toward its surface, it first travels by _____, then by _____, and then by _____.
    a. radiation; conduction; radiation
    b. conduction; radiation; convection
    c. radiation; convection; radiation
    d. radiation; convection; conduction

14. Energy is produced primarily in the center of the Sun because
    a. the strong nuclear force is too weak elsewhere.
    b. that's where neutrinos are created.
    c. that's where most of the helium is.
    d. the temperature and density are high enough in the core.

15. The solar wind pushes on the magnetosphere of Earth, changing its shape, because
    a. the solar wind is so dense.
    b. the magnetosphere is so weak.
    c. the solar wind contains charged particles.
    d. the solar wind is so fast.

Thinking about the Concepts

16. Explain how hydrostatic equilibrium acts as a safety valve to keep the Sun at its constant size, temperature, and luminosity.

17. Two of the three atoms in a molecule of water (H_2O) are hydrogen. Why are Earth's oceans not fusing hydrogen into helium and setting Earth ablaze?

18. Why are neutrinos so difficult to detect?

19. Explain the proton-proton chain through which the Sun generates energy by converting hydrogen to helium.

20. On Earth, nuclear power plants use *fission* to generate electricity. In fission, a heavy element such as uranium is broken into many atoms, where the total mass of the fragments is less than that of the original atom. Explain why fission could not be powering the Sun today.

21. If an abnormally large amount of hydrogen suddenly fused in the core of the Sun, what would happen to the rest of the Sun? Would the Sun change as seen from Earth?

22. Study the Process of Science Figure. If the follow-up experiments did not detect the other types of neutrinos, what would have been the next step for scientists at that point?

23. What is the solar neutrino problem, and how was it solved?

24. The Sun's visible "surface" is not a true surface, but a feature called the photosphere. Explain why the photosphere is not a true surface.

25. How are orbiting satellites and telescopes affected by the Sun?

26. Describe the solar corona. Under what circumstances can it be seen without special instruments?

27. In the proton-proton chain, the mass of four protons is slightly greater than the mass of a helium nucleus. Explain what happens to that "lost" mass.

28. What have sunspots revealed about the Sun's rotation?

29. Why are different parts of the Sun best studied at different wavelengths? Which parts are best studied from space?

30. Why is studying the interaction of the solar wind with the interstellar medium important?

Applying the Concepts

31. In Figure 14.10, density and temperature are both graphed versus height.
    a. Is the height axis linear or logarithmic? How do you know?
    b. Is the density axis linear or logarithmic? How do you know?
    c. Is the temperature axis linear or logarithmic? How do you know?

32. Using the data in Figures 14.19b and c, present an argument that sunspots occur in regions of strong magnetic field.

33. Study Figure 14.17a and Figure 14.20.
    a. Estimate the fraction of the Sun's surface covered by the large sunspot group in Figure 14.17a. (Remember that you are seeing only one hemisphere of the Sun.)
    b. From the graph in Figure 14.20, estimate the average number of sunspots that occurs at solar maximum.
    c. On average, what fraction of the Sun could be covered by sunspots at solar maximum? Is that a large fraction?
    d. Compare your conclusion with the graph of intensity in Figure 14.24. Does that graph make sense to you?

34. The Sun has a radius equal to about 2.3 light-seconds. Explain why a gamma ray produced in the Sun's core does not emerge from the Sun's surface 2.3 seconds later.

35. Assume that the Sun's mass is about 300,000 Earth masses and that its radius is about 100 times that of Earth. The density of Earth is about 5,500 kg/m^3.
    a. What is the average density of the Sun?
    b. How does that value compare with the density of Earth? With the density of water?

36. The Sun shines by converting mass into energy according to $E = mc^2$. Show that if the Sun produces 3.85×10^{26} J of energy per second, it must convert 4.3 million metric tons (4.3×10^9 kg) of mass per second into energy.

37. Assume that the Sun has been producing energy at a constant rate over its lifetime of 4.6 billion years (1.4×10^{17} seconds).
    a. How much mass has it lost creating energy over its lifetime?
    b. The current mass of the Sun is 2×10^{30} kg. What fraction of its current mass has been converted into energy over the lifetime of the Sun?

38. Suppose our Sun was an A5 main-sequence star, with twice the mass and 12 times the luminosity of the Sun, a G2 star. How long would that A5 star fuse hydrogen to helium? What would that mean for Earth?

39. Imagine that the source of energy inside the Sun changed abruptly.
    a. How long would it take before a neutrino telescope detected the event?
    b. When would a visible-light telescope see evidence of the change?

40. On average, how long do particles in the solar wind take to reach Earth from the Sun if they are traveling at an average speed of 400 km/s?

41. If a sunspot appears only 70 percent as bright as the surrounding photosphere, and the photosphere has a temperature of approximately 5780 K, what is the temperature of the sunspot?

42. The hydrogen bomb represents an effort to create a process similar to what takes place in the core of the Sun. The energy released by a 5-megaton hydrogen bomb is 2×10^{16} J.
    a. This textbook, *21st Century Astronomy*, has a mass of about 1.6 kg. If all its mass were converted to energy, how many 5-megaton bombs would it take to equal that energy?
    b. How much mass did Earth lose each time a 5-megaton hydrogen bomb was exploded?

43. Verify the claim made at the start of this chapter that the Sun produces more energy per second than all the electric power plants on Earth could generate in a half-million years. Estimate or look up how many power plants are on the planet and how much energy an average power plant produces. Be sure to account for different kinds of power, such as coal, nuclear, and wind.

44. Let's examine the reason that the Sun cannot power itself by chemical reactions. Using Working It Out 14.1 and the fact that an average chemical reaction between two atoms releases 1.6×10^{-19} J, estimate how long the Sun could emit energy at its current luminosity. Compare that estimate with the known age of Earth.

45. The Sun could get energy from gravitational contraction for a time period of $(GM^2_{Sun}/R_{Sun}L_{Sun})$. How long would the Sun last at its current luminosity? (Be careful with units!)

USING THE WEB

46. a. Go to the SDO website (http://sdo.gsfc.nasa.gov). Under "Data," select "The Sun Now" and view the Sun at many wavelengths. What activity do you observe in the images at the location of any sunspots seen in the "HMI Intensity-gram" images? (You can download a free SDO app by Astra to get real-time images on your mobile device.) Look at a recent news story from the SDO website. What was observed, and why is it newsworthy?
    b. Go to the *STEREO* mission's website (https://stereo.gsfc.nasa.gov). What is *STEREO*? Where are the spacecraft located? How does that configuration enable observations of the entire Sun at once? (You can download the app "3-D Sun" to get the latest images on your mobile device.)

47. a. What are the science goals of NASA's *Interface Region Imaging Spectrograph* (*IRIS*) mission (https://www.nasa.gov/mission_pages/iris/)? What has it discovered?
    b. An older space mission, *SOHO* (Solar and Heliospheric Observatory; https://sohowww.nascom.nasa.gov), was launched in 1995 by NASA and ESA. Click on "The Sun Now" to see today's images. The Extreme Ultraviolet imaging Telescope (EIT) images are in the far ultraviolet and show violent activity. How do those images differ from the ones of SDO in question 46b?
    c. Go to the Daniel K. Inouye Solar Telescope (DKIST) website (https://atst.nso.edu). That adaptive-optics telescope under construction on Haleakala, Maui, will be the largest solar telescope. Why is studying the magnetic field of the Sun important? What are some advantages of studying the Sun from a ground-based telescope instead of a space-based telescope? What wavelengths does the DKIST observe? Why is Maui a good location? When is the telescope scheduled to be completed?

48. a. Go to the Space Weather website (http://spaceweather.com). Are any solar flares happening today? What is the sunspot number? Is it about what you would expect for this year? (In the column at left, click on "What is the sunspot number?" to see a current graph.) Are any coronal holes present today?
    b. Citizen science: Go to the Solar Stormwatch II website (https://www.zooniverse.org/projects/shannon-/solar-stormwatch-ii), a Zooniverse project from the Royal Observatory in Greenwich, England. Create an account for Zooniverse if you don't already have one. Log in and click on "Get Started" and "Project Tutorial." Mark up two sets of images (and save a screen shot for your homework).

49. a. Go to the National Ignition Facility (NIF) & Photon Science website (https://lasers.llnl.gov/about/). Under "Science," click on "How to Make a Star." How are lasers used in experiments to develop controlled nuclear fusion on Earth? How does the fusion reaction here differ from that in the Sun?
    b. An alternative approach is to fuse ^{3}He + ^{3}He instead of the hydrogen isotopes. On Earth, however, ^{3}He is in limited supply. Helium-3 is in much greater abundance on the Moon, so some people propose setting up mining colonies on the Moon to extract ^{3}He for fusion reactions on Earth. Do an Internet search on "helium 3 Moon." Which countries are talking about going to the Moon for that purpose? What is the timeline for when that might happen? What are the difficulties?

50. a. Go to https://voyager.jpl.nasa.gov/mission/status/. Where are the *Voyager* spacecraft now? Has *Voyager 2* crossed into interstellar space?
    b. Go to the website for the Interstellar Boundary Explorer (*IBEX*) (https://www.nasa.gov/mission_pages/ibex/). What has *IBEX* learned about the solar wind and the interstellar medium?

digital.wwnorton.com/astro6

The proton-proton chain powers the Sun by fusing hydrogen into helium. That fusion process produces several types of particles as by-products, as well as energy. In this Exploration, we study the steps of the proton-proton chain in detail, with the intent of helping you keep them straight.

Visit the Digital Resources Page on the Student Site and open the "Proton-Proton Chain" Interactive Simulation in Chapter 14.

Watch the animation all the way through once.

Play the animation again, pausing after the first collision. Two hydrogen nuclei (both positively charged) have collided to produce a new nucleus with only one positive charge.

1 Which particle carried away the other positive charge?

_____

_____

2 What is a neutrino? Did the neutrino enter the reaction, or was the neutrino produced in the reaction?

_____

_____

_____

Compare the interaction on the top with the interaction on the bottom.

3 Did the same reaction occur in each instance?

_____

_____

Resume playing the animation, pausing it after the second collision.

4 What two types of nuclei entered the collision? What type of nucleus resulted?

_____

_____

5 Was charge conserved in that reaction, or did a particle have to carry charge away?

_____

_____

6 What is a gamma ray? Did the gamma ray enter the reaction, or was it produced by the reaction?

_____

_____

_____

Resume the animation again, and allow it to run to the end.

7 What nuclei enter the final collision? What nuclei are produced?

_____

_____

_____

8 In chemistry, a catalyst is a species that facilitates a reaction but is not used up in the process. Do any nuclei act like catalysts in the proton-proton chain?

_____

_____

Make a table of inputs and outputs. Which particles in the final frame of the animation were inputs to the reaction? Which were outputs? Fill in your table with those inputs and outputs.

9 Which outputs are converted into energy that leaves the Sun as light?

_____

_____

10 Which outputs could become involved in another reaction immediately?

_____

_____

11 Which output is likely to stay in that form for a very long time?

_____

_____

_____

15

The Interstellar Medium and Star Formation

Stars form from clouds of dust and gas, which collapse until nuclear fusion begins. The formation process is faster for more massive stars. The most massive stars take tens of thousands of years to form, whereas the least massive stars take hundreds of millions of years. To understand star formation, astronomers have observed many stars at various stages of development. In this chapter, we will describe the interstellar environment from which stars both large and small form. Then we will focus on the forming star—the *protostar*—and explain how it becomes a star.

LEARNING GOALS

By the end of the chapter, you should be able to:

LG 1 Describe the types and states of material that exist in the space between the stars and how that material is detected.

LG 2 Explain the conditions under which a cloud of gas can contract into a stellar system and the role that gravity and angular momentum play in forming stars and planets.

LG 3 List the steps in the evolution of a protostar, and explain how the mass of the protostar affects its evolution.

LG 4 Describe the track of a protostar as it evolves to a main-sequence star on the Hertzsprung-Russell (H-R) diagram.

Stars are forming in this giant molecular cloud. ▶▶▶

How do
stars form?

15.1 The Interstellar Medium Fills the Space between the Stars

The space between the stars contains thin gas, interrupted by giant clouds of cool gas and dust. Those clouds can be observed in the visible part of the spectrum in three ways: as they absorb light from objects behind them, emit light from excited atoms, or reflect starlight from nearby stars. Together, the clouds and the thin gas are known as the **interstellar medium**.

Stars interact with the interstellar medium. Stars form from the interstellar medium and return much of their material back to it when they die. Outflows from dying stars pile up material in their path like snow in front of a snowplow, clearing out vast hot bubbles. Those hot bubbles of high-pressure gas compress clouds, driving up their densities and triggering the formation of new stars. Interstellar clouds are destroyed by those violent events. In turn, new clouds are formed from the swept-up gas. Those clouds go on to produce new stars. In addition, energy from stars heats and stirs the interstellar medium. For example, ultraviolet radiation from massive, hot stars warms gas that then pushes outward into its surroundings. In this section, we survey the varied states of the interstellar medium.

The Composition and Density of the Interstellar Medium

The Sun formed from the interstellar medium, so the chemical composition of the interstellar medium near the Sun is similar to the chemical composition of the Sun (see Table 13.1). In the interstellar medium, hydrogen accounts for about 90 percent of the number of atomic nuclei, and the remaining 10 percent is almost all helium. The more massive elements account for only 0.1 percent of the atomic nuclei, or about 2 percent of the mass in the interstellar medium (because those atoms are more massive than hydrogen or helium, they account for a larger fraction of the mass in the interstellar medium than they do of the number of individual atoms). Roughly 99 percent of interstellar matter is a gas, consisting of individual atoms or molecules moving about freely, as the molecules in the air do.

The air that you breathe has a density 100 billion billion times greater than the average density of the interstellar medium. Each cubic centimeter (cm^3) of the air around you contains about 2.7×10^{19} molecules. The local interstellar medium has an average density of about 0.1 atom/cm^3. Stated another way, imagine a tube with the diameter of your fist that stretches from the Solar System to the center of our galaxy, 26,000 light-years away. Now imagine a tube with the same diameter that stretches between your eye and the floor on Earth. The amount of material in both tubes is about the same.

Interstellar Dust and Its Effects on Light

About 1 percent of the mass in the interstellar medium is in the form of solid grains, called **interstellar dust**. Ranging in size from little more than large molecules up to particles about 1 micron (μm) across, those solid grains are more like the particles of soot from a candle flame than the dust that collects on a windowsill. It would take several hundred average interstellar grains to span the thickness of a single human hair. Interstellar dust begins to form when larger atoms such as iron, silicon, and carbon stick together to form grains in dense, relatively cool environments such as the outer atmospheres and "stellar winds" of cool, red giant stars or in dense material thrown into space by stellar explosions. Once those grains are in the interstellar medium, other atoms and molecules stick to

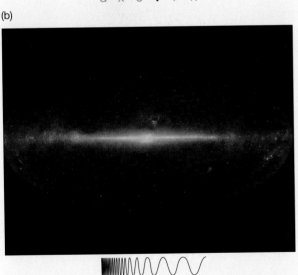

(a)

G X U V I R

(b)

G X U V I R

Figure 15.1 (a) This all-sky picture of the Milky Way was taken in visible light. The dark splotches blocking the view are dusty interstellar clouds. The center of this and all the other all-sky images in this chapter is the center of the Milky Way. (b) This all-sky picture was taken in the near infrared. Infrared radiation penetrates the interstellar dust, providing a clearer view of the stars in the disk of the Milky Way.

them. About half of the interstellar matter that is more massive than helium is found in interstellar grains.

Interstellar dust is extremely effective at absorbing and diverting light, so the view of distant objects is affected even by the low-density interstellar medium. That effect is called **interstellar extinction**. Go out on a dark summer night in the Northern Hemisphere (or on a dark winter night in the Southern Hemisphere) and look closely at the Milky Way, visible as a band of diffuse light running through the constellation Sagittarius. A dark "lane" is running roughly down the middle of that bright band, splitting it in two, as shown in **Figure 15.1a**. That dark band is caused by interstellar extinction, dust that dims the light from distant stars. Recall the comparison of the air between you and the floor with the interstellar medium stretching to the center of the galaxy. If the interstellar medium were compressed to the same density as air, it would be so full of dust that you could not easily see your hand 10 centimeters (cm) in front of your face. When spread out between Earth and the distant stars, the interstellar dust dims light just as effectively. Although interstellar extinction is most noticeable in the dark lanes in the Milky Way, it has a lesser but still important effect on starlight coming from all directions in the galaxy.

Figure 15.1 shows two images of the Milky Way Galaxy: one taken in visible light, the other taken in the infrared (IR). Those two images are *all-sky images*: they portray the entire sky surrounding Earth. They have been oriented so that the disk of the Milky Way, in which the Solar System is embedded, runs horizontally across the center of the image. The dark clouds that block the shorter-wavelength visible light (Figure 15.1a) seem to have vanished in the longer-wavelength infrared image (**Figure 15.1b**). That observation shows that interstellar extinction affects different regions of the electromagnetic spectrum differently. When astronomers observe in long-wavelength regions of the spectrum, such as infrared or radio wavelengths, they can make observations through the clouds to the center of the galaxy and beyond.

To understand why dust obscures short-wavelength radiation but not long-wavelength radiation, think about waves on the ocean, as shown in **Figure 15.2a**. Imagine you are on the ocean in a boat. If the ocean waves have a wavelength much longer than your boat, the waves cause you to move slowly up and down. But that is all that happens: no energy is transferred from the wave to the boat, so the wave travels on unaffected by the interaction. The situation is different if the wavelength of the waves is close in size to the length of your boat. For example, if the waves are roughly half the size of your boat, the front of the boat may be on a wave crest while the back of the boat is in a trough or vice versa. The boat will tip wildly back and forth as the waves pass by, and energy will be transferred from the wave to the boat; the wave loses energy in the interaction. If the size of the boat and the wavelength of the waves are the right match, even low waves will rock the boat. Comparing those two situations shows that the wave loses very little energy when it is much bigger than the boat. But the wave loses a lot of energy when it is of similar size or smaller than the boat, and the wave can be blocked by the boat if the boat is large enough in comparison.

The interaction of light with matter is more involved than that of a boat rocking on the ocean, but the same basic idea applies, as shown in **Figure 15.2b**. Tiny interstellar dust grains effectively block the transmission of short-wavelength ultraviolet light and blue light because they have wavelengths comparable to or smaller than the typical size of dust grains. Our view of the Milky Way in those wavelengths is limited. In contrast, longer-wavelength infrared and radio radiation interact less strongly with the tiny interstellar dust grains, so in those portions of the spectrum we get a more complete view of the Milky Way.

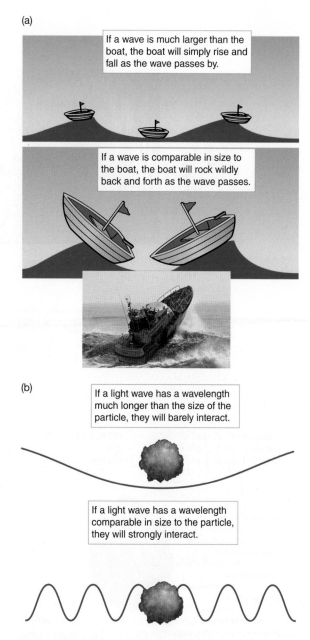

(a)

If a wave is much larger than the boat, the boat will simply rise and fall as the wave passes by.

If a wave is comparable in size to the boat, the boat will rock wildly back and forth as the wave passes.

(b)

If a light wave has a wavelength much longer than the size of the particle, they will barely interact.

If a light wave has a wavelength comparable in size to the particle, they will strongly interact.

Figure 15.2 (a) Just as boats interact most strongly with ocean waves that have a wavelength similar to the length of the boat, (b) particles interact most strongly with light that has wavelengths near the same size as the particle.

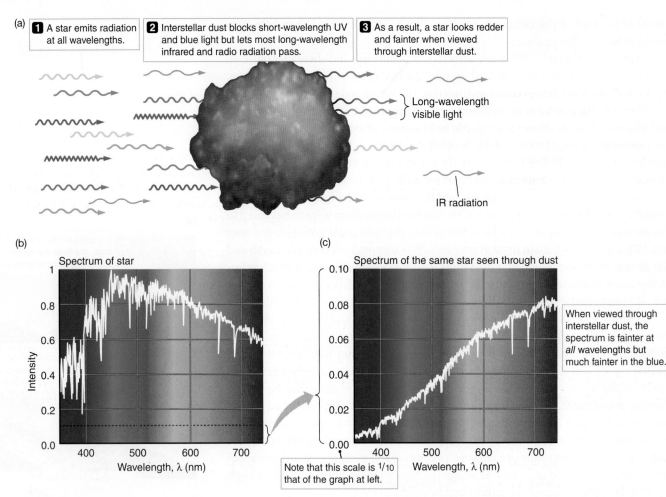

(a)
1 A star emits radiation at all wavelengths.

2 Interstellar dust blocks short-wavelength UV and blue light but lets most long-wavelength infrared and radio radiation pass.

3 As a result, a star looks redder and fainter when viewed through interstellar dust.

Long-wavelength visible light

IR radiation

(b) Spectrum of star

Intensity / Wavelength, λ (nm)

(c) Spectrum of the same star seen through dust

When viewed through interstellar dust, the spectrum is fainter at *all* wavelengths but much fainter in the blue.

Note that this scale is ¹/₁₀ that of the graph at left.

Wavelength, λ (nm)

Figure 15.3 (a) The wavelengths of ultraviolet and blue light are close to the size of interstellar grains, so the grains effectively block that light. Grains are less effective at blocking longer-wavelength light. Therefore, the spectrum of a star (b) when seen through an interstellar cloud (c) appears fainter and redder.

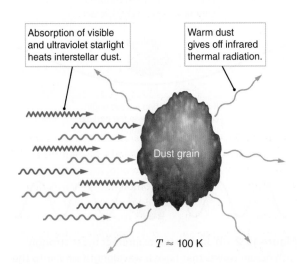

Absorption of visible and ultraviolet starlight heats interstellar dust.

Warm dust gives off infrared thermal radiation.

Dust grain

$T \approx 100$ K

Figure 15.4 The temperature of interstellar grains depends on the equilibrium between absorbed and emitted radiation.

The presence of dust significantly affects the spectrum of an object, as shown in **Figure 15.3**. Extinction occurs at all wavelengths (Figure 15.3a), causing an object viewed through dust to be fainter at all wavelengths than it would be otherwise. Extinction also affects short-wavelength blue light more than it affects long-wavelength red light, so the object appears less blue than it really is (Figure 15.3b and c). Removing the blue light causes an object to appear more red, so that effect is called **reddening**. Correcting for the reddening effect of dust can be one of the most difficult parts of interpreting astronomical observations, often adding to uncertainty in the measurement of an object's properties.

At longer infrared wavelengths, interstellar extinction from small gas particles is less effective, but larger dust particles still play an important role. In Chapter 5, we discussed how Wein's law can be used to relate the temperature of an object to the wavelength at which it shines most brightly, and how the equilibrium between absorbed sunlight and emitted thermal radiation determines the temperatures of the terrestrial planets (see Figure 5.19). As **Figure 15.4** illustrates, a similar equilibrium is at work in interstellar space, where dust is heated by starlight and surrounding gas to temperatures of tens to hundreds of kelvins. According to Wien's law, dust with a temperature of 100 K shines most strongly at a wavelength of 29 μm, whereas cooler dust—at 10 K—shines most strongly at a

15.1 Working It Out Dust Glows in the Infrared

The temperature of interstellar dust can be found from its spectrum. Wien's law, discussed in Chapter 5, relates the temperature of an object to the peak wavelength (λ_{peak}) of its emitted radiation. For warm dust at a temperature of 100 K (recall that 1 μm = 10^{-6} meter = 1,000 nanometers [nm]):

$$\lambda_{peak} = \frac{2{,}900 \ \mu m \ K}{T} = \frac{2{,}900 \ \mu m \ K}{100 \ K} = 29 \ \mu m$$

For cooler dust, at 10 K:

$$\lambda_{peak} = \frac{2{,}900 \ \mu m \ K}{T} = \frac{2{,}900 \ \mu m \ K}{10 \ K} = 290 \ \mu m$$

The temperature and the peak wavelength are inversely proportional, so if the temperature drops, the peak wavelength gets longer. For the temperatures common for dust in the interstellar medium, the peak wavelength is in the far-infrared region of the electromagnetic spectrum.

longer wavelength (**Working It Out 15.1**). Such warm dust produces much of the diffuse light present in infrared observations. **Figure 15.5a**, from NASA's Wide-field Infrared Survey Explorer (WISE), shows the sky at combined wavelengths of 3.4, 12, and 22 μm. That light is produced by dust hotter than 100 K. In **Figure 15.5b**, a far-infrared image of the sky shows the Milky Way's dark, dusty, cooler clouds shining at a wavelength of 100 μm.

Temperatures and Densities of Interstellar Gas

The gas and dust in interstellar space are roughly evenly divided between dense regions called **interstellar clouds** and the **intercloud gas** spread between them. The clouds fill about 2 percent of the volume of interstellar space, whereas the remaining 98 percent is filled with thin gas.

The properties of intercloud gas vary from place to place (**Table 15.1**). About half of the volume of interstellar space is filled with an intercloud gas

(a)

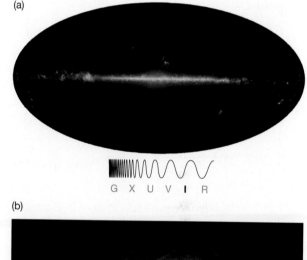

G X U V I R

(b)

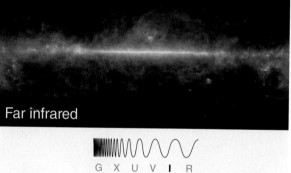

Far infrared

G X U V I R

Figure 15.5 (a) This WISE infrared image shows the plane of the Milky Way at combined wavelengths of 3.4, 12, and 22 μm. (b) This all-sky image, in the far-infrared wavelength of 100 μm, is a combination of two images from two space telescopes—the Infrared Astronomical Satellite (IRAS) and the Cosmic Background Explorer (COBE). The image shows dust throughout the Milky Way Galaxy.

TABLE 15.1 Typical Properties of Components of the Interstellar Medium

Component	Temperature (K)	Number Density (atoms/cm³)	Size of Cube per Gram* (km)	State of Hydrogen
Hot intercloud gas	~1 million	~0.005	~8,000	Ionized
Warm intercloud gas	~8000	0.01–1	~800	Ionized or neutral
Cold intercloud gas	~100	1–100	~80	Neutral
Interstellar clouds	~10	100–1,000	~8	Molecular or neutral

*This is the length of one side of the cube of space you would need to search to find 1 gram of the material. It is another way of thinking about density.

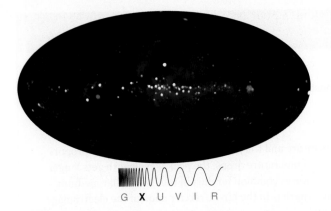

Figure 15.6 Bright spots in this image are distant X-ray sources, including objects such as bubbles of very hot, high-pressure gas surrounding the sites of recent explosions of supernovae. Supernovae are more common in the disk of the Milky Way, where most stars are located.

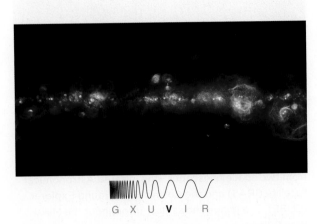

Figure 15.7 Warm interstellar gas (about 8000 K) glows in the Hα line of hydrogen. This image of the Hα emission from much of the northern sky reveals the complex structure of the interstellar medium.

that is extremely hot—millions of kelvins—heated primarily by the energy of tremendous stellar explosions called *supernovae* (**Figure 15.6**). Because the temperature is so high, the atoms in the gas are moving very rapidly. The gas has an extremely low density; typically, you would have to search a liter (1,000 cm³) or more of hot intercloud gas to find a single atom. Therefore, if you were floating in that million-kelvin intercloud gas, it would do little to keep you warm. Because so few atoms are present, you would radiate energy away and cool off much faster than the very hot gas around you could replace the lost energy.

Hot intercloud gas glows faintly in the X-ray portion of the electromagnetic spectrum, and orbiting X-ray telescopes observe the entire sky aglow with faint X-rays. That observation indicates that the Solar System is passing through a bubble of hot intercloud gas that may be the remnant of a supernova explosion 300,000 years ago. The bubble has a density of about 0.005 hydrogen atom/cm³ and is at least 650 light-years across.

Not all intercloud gas is as hot as that around our Solar System. Most other intercloud gas is "warm," with a temperature of about 8000 K and a density ranging from about 0.01 to 1 atom/cm³. Ultraviolet starlight with wavelengths shorter than 91.2 nm has enough energy to **ionize** hydrogen, meaning to strip the electron away from the nucleus. About half of the volume of warm intercloud gas is kept ionized by ultraviolet starlight. That ionized gas "uses up" the ultraviolet light from stars, so the remaining half of the volume of warm intercloud gas is protected and remains in an un-ionized state. Similarly, the ozone layer in Earth's upper atmosphere shields the planet surface from ultraviolet light from the Sun.

One way to look for both warm and hot interstellar gas is to study the spectra of distant stars. Most commonly, that gas produces absorption lines in the spectra of distant stars when atoms in the gas absorb starlight at particular wavelengths. Those absorption lines can be used to find the temperature, density, and chemical composition of the gas. Regions of warm, ionized intercloud gas, as in supernova remnants, also can produce emission lines when protons and electrons constantly recombine into hydrogen atoms. When a proton and an electron combine to form a neutral hydrogen atom, energy is emitted in the form of electromagnetic radiation. Typically, the resulting hydrogen atom is left in an excited state (see Chapter 5). The atom then drops down to lower and lower energy states, emitting a photon at each step. Therefore, warm, ionized interstellar gas glows in emission lines characteristic of hydrogen, with the Hα (hydrogen alpha) line often the strongest. That line occurs in the red part of the spectrum at a wavelength of 656.3 nm. Other elements undergo a similar process.

The faint, diffuse emission in **Figure 15.7** comes mostly from warm (about 8000 K), ionized intercloud gas glowing in Hα. The bright spots are called **H II regions** ("H two regions") because the hydrogen atoms are ionized; H I atoms are neutral, and H II atoms are ionized. H II regions form a rough sphere around hot luminous O and B stars because those stars produce enough ultraviolet radiation to ionize even relatively dense interstellar clouds around them. H II regions indicate areas of active star formation because O stars do not live long enough to move very far from where they formed. H II regions around O stars are the very clouds from which those stars were born.

One of the closest H II regions to the Sun is the Orion Nebula, located 1,340 light-years from the Sun in the constellation Orion (**Figure 15.8**). Almost all the ultraviolet light that powers that nebula comes from a single hot star, and only a few hundred stars are forming in its immediate vicinity. In contrast, a dense star

(a)

(b)

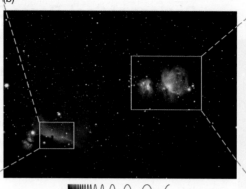

(c)

Figure 15.8 The Orion Nebula is only a small part of the larger Orion star-forming region. The dark Horsehead Nebula is seen at the lower left of image (b) and in (a). The circular halos around the bright stars in (a) are a photographic artifact. (c) The Orion Nebula is seen as a glowing region of interstellar gas surrounding a cluster of young, hot stars. New stars are still forming in the dense clouds surrounding the nebula.

G X U **V** I R   G X U **V** I R   G X U **V** I R

cluster containing thousands of hot, luminous stars powers a giant H II region called 30 Doradus, located in the Large Magellanic Cloud, a small companion galaxy to the Milky Way located 160,000 light-years away. If 30 Doradus were as close as the Orion Nebula, it would be bright enough in the nighttime sky to cast shadows.

Warm, *neutral* hydrogen gas also produces an emission line, although it gives off radiation in a way different from that of warm, *ionized* hydrogen. Many subatomic particles, including protons and electrons, have a property called *spin* that causes them to behave as though each particle has a bar magnet, with a north and a south pole, built into it. As demonstrated in **Figure 15.9**, a hydrogen atom can exist in only two configurations: either the magnetic "poles" of the proton and electron point in opposite directions or they are aligned. Those configurations have different energies. When the two "magnets" point in the same direction, the atom has slightly less energy than when they point in the opposite direction. If left undisturbed long enough, a hydrogen atom in the higher energy unaligned state will spontaneously jump to the lower energy state, emitting a photon in the process. The energy difference between the two magnetic spin states of a hydrogen atom is extremely small, so the emitted photon has a wavelength of 21 cm, in the radio region of the spectrum. Interactions between atoms in the gas will later bump the hydrogen atoms back to the higher energy state, refreshing the supply of atoms that can produce the 21-cm line.

The tendency for hydrogen atoms to emit 21-cm radiation is extremely weak. On average, you would have to wait about 11 million years for an individual hydrogen atom in the higher energy state to jump spontaneously to the lower energy state and give off a photon. But the universe has a lot of hydrogen, so at any given time, many atoms are making that transition. In **Figure 15.10**, the sky is aglow with 21-cm radiation from neutral hydrogen. Because of its long wavelength, 21-cm radiation freely penetrates dust in the interstellar medium, enabling astronomers to observe neutral hydrogen throughout the galaxy. Measurements

(a)

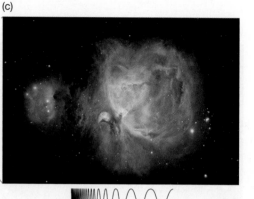

Poles opposite
(higher energy state)

N   S

Proton / Electron

S   N

Poles aligned
(lower energy state)

N   N

Proton / Electron

S   S

A 21-cm photon is emitted when poles go from being opposite to aligned (a spin flip).

(b)

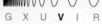

Energy

E_2 (poles opposite)

$E_{photon} = E_2 - E_1$

$\lambda_{2\text{-}1} = 21$ cm

E_1 (poles aligned)

Figure 15.9 (a) A slight difference in energy occurs when the poles of the proton and electron are aligned compared with when they are opposite. (b) This energy difference corresponds to a photon of 21 cm.

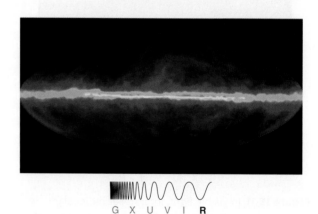

G X U V I **R**

Figure 15.10 This radio image of the sky shows the distribution of neutral hydrogen gas throughout the galaxy. Red indicates directions of the highest hydrogen density, and blue and black show areas with little hydrogen. Radio waves penetrate interstellar dust, allowing astronomers to probe the structure of the galaxy.

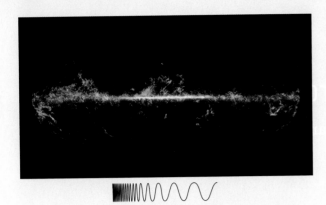

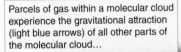

G X U V I **R**

Figure 15.12 This all-sky image from the Planck observatory, a spacecraft that observed in the microwave region of the spectrum, shows the distribution of carbon monoxide (CO), which traces molecular clouds where stars are born.

smallest possible molecule. Cold molecular hydrogen is very difficult to observe directly because it has no emission lines in the long-wavelength regions of the spectrum.

In addition to molecular hydrogen, approximately 150 other molecules have been observed in interstellar space. They range from very simple structures such as carbon monoxide (CO), to complex organic compounds such as methanol (CH_3OH), to some molecules with large carbon chains. Very large carbon molecules, made of hundreds of individual atoms, have sizes between those of large interstellar molecules and small interstellar grains. Among the more important molecules is CO. The ratio of CO to H_2 is relatively constant, as far as has been tested, and interstellar CO (**Figure 15.12**) is often used to estimate the amounts and distribution of interstellar H_2, which is more difficult to observe directly.

CHECK YOUR UNDERSTANDING 15.1

When visible light from an object passes through the interstellar medium (choose all that apply): (a) the object appears dimmer; (b) the object appears bluer; (c) the object appears brighter; (d) the object appears redder.

15.2 Molecular Clouds Are the Cradles of Star Formation

Stars and planets form from large clouds of dust and gas in the interstellar medium. In this section, we explore the first steps of that process as a cloud begins to contract and fragment to form stars.

Self-Gravity in the Molecular Cloud

As shown in **Figure 15.13**, each part of an interstellar cloud experiences a gravitational attraction from every other part of the cloud. The sum of all those forces acting on a particular parcel of gas will always point toward the center of mass of the cloud. That sum is the *net force* of gravity and indicates the direction in which the parcel will begin to move. The gravitational attraction between all the parts of a cloud is called *self-gravity*, which acts to hold the cloud together.

Recall from our discussion of the Sun in Chapter 14 that hydrostatic equilibrium is the balance between gravity and pressure in a stable object. Interstellar clouds are not always in hydrostatic equilibrium—most interstellar clouds are large and very thin, so their self-gravity is weak. The internal pressure pushing out is much stronger than the self-gravity, so the cloud should expand. But the much hotter gas surrounding the clouds also exerts a pressure inward on a cloud. That external hot gas helps hold a cloud together and often serves as a trigger for collapse. If a cloud is massive enough and dense enough (or becomes so after a triggering event), self-gravity becomes stronger than the internal pressure, so the clouds collapse under their own weight, beginning a chain of events that will form a new generation of stars.

If self-gravity in a molecular cloud is much greater than internal pressure, gravity should win outright, and the cloud should rapidly collapse toward its

Parcels of gas within a molecular cloud experience the gravitational attraction (light blue arrows) of all other parts of the molecular cloud...

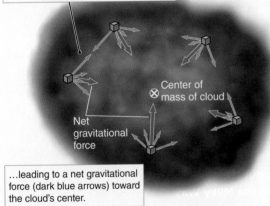

Center of mass of cloud

Net gravitational force

...leading to a net gravitational force (dark blue arrows) toward the cloud's center.

Figure 15.13 Self-gravity causes a molecular cloud to collapse, drawing parcels of gas toward a single point inside the cloud. The lighter blue arrows are examples of forces on the parcel due to other parcels of gas. The darker blue arrows show the sum of all those forces. That net force always points toward the center of mass of the cloud.

center. In practice, the process goes very slowly because several other effects stand in the way of the collapse. One effect that slows the collapse of a cloud is conservation of angular momentum, described in Chapter 7. Others are turbulence, the effects of magnetic fields, and thermal pressure. Even though those effects may slow the collapse of a molecular cloud, in the end gravity will dominate. One part of the cloud can lose angular momentum to another part of the cloud, allowing the part of the cloud with less angular momentum to collapse further. Neutral matter crosses magnetic field lines, gradually increasing the gravitational pull toward the center until the force on the charged particles is large enough to drag the magnetic field toward the center as well. Turbulence fades away. The details of those processes are complex and are the subject of much current research. The important point is that the effects that slow the collapse of a molecular cloud are temporary, and gravity is persistent. As the forces that oppose the cloud's self-gravity gradually fade, the cloud shrinks.

Astronomy in Action: Angular Momentum

AstroTour: Star Formation

Molecular Clouds Fragment as They Collapse

Some regions within a molecular cloud are denser and collapse more rapidly than surrounding regions. **Figure 15.14** shows the process of collapse in a molecular cloud. Slight variations in the density of the cloud become very dense, localized concentrations of gas. Instead of collapsing into a single object, the molecular cloud fragments into very dense **molecular-cloud cores**. A single molecular cloud may form hundreds or thousands of molecular-cloud cores, each of which is typically a few light-months in size. Some of those dense cores will eventually form stars.

As a molecular-cloud core collapses, the cloud's self-gravity grows stronger because the force of gravity is inversely proportional to the square of the radius. Suppose a cloud is 4 light-years across. When the cloud has collapsed to 2 light-years across, the different parts of the cloud are, on average, only half as far apart as when the collapse started. As a result, the self-gravity is 4 times stronger. When the cloud is one-fourth as large as it was at the beginning of the

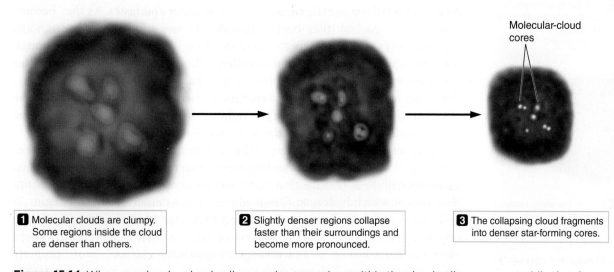

1 Molecular clouds are clumpy. Some regions inside the cloud are denser than others.

2 Slightly denser regions collapse faster than their surroundings and become more pronounced.

3 The collapsing cloud fragments into denser star-forming cores.

Molecular-cloud cores

Figure 15.14 When a molecular cloud collapses, denser regions within the cloud collapse more rapidly than less dense regions. As that process continues, the cloud fragments into several very dense molecular-cloud cores embedded within the large cloud. Those cloud cores may go on to form stars.

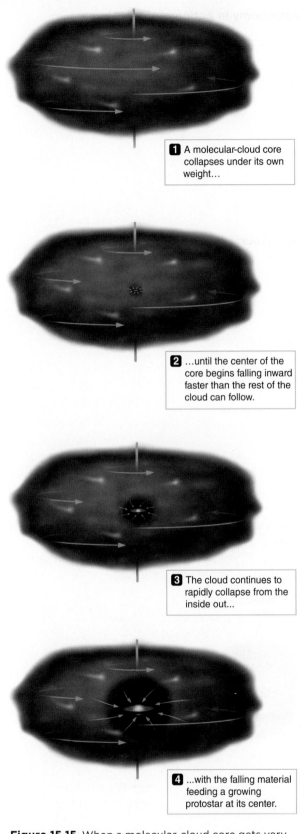

1 A molecular-cloud core collapses under its own weight…

2 …until the center of the core begins falling inward faster than the rest of the cloud can follow.

3 The cloud continues to rapidly collapse from the inside out…

4 …with the falling material feeding a growing protostar at its center.

Figure 15.15 When a molecular-cloud core gets very dense, it collapses from the inside out. Conservation of angular momentum causes the infalling material to form an accretion disk that feeds the growing protostar.

collapse, the self-gravity is 16 times stronger. As a core collapses, the self-gravity increases; as self-gravity increases, the collapse speeds up; as the collapse speeds up, the self-gravity increases even faster.

Eventually, gravity overwhelms the opposing forces due to pressure, magnetic fields, and turbulence. That happens first near the center of the molecular-cloud core where the cloud material is densest. The inner parts of the molecular-cloud core start to fall rapidly inward. Those inner layers supported the weight of the layers farther out. Without the support of that inner material, the more distant material begins to fall freely toward the center. The molecular-cloud core collapses from the inside out, as shown in **Figure 15.15**.

CHECK YOUR UNDERSTANDING 15.2

Molecular clouds fragment as they collapse because: (a) the rotation of the cloud throws some mass to the outer regions; (b) the density increases fastest in the center of the cloud; (c) density variations from place to place grow larger as the cloud collapses; (d) the interstellar wind is stronger in some places than others.

15.3 Formation and Evolution of Protostars

Because of conservation of angular momentum, material that falls inward in a collapsing molecular-cloud core accumulates in a flat, rotating accretion disk. Most of that material eventually finds its way inward to the center of the disk. The object forming there, which will eventually become a star, is called a **protostar**. Just as an ice skater spins faster when she draws her arms in (recall Figure 7.5), the protostar spins faster than the original cloud. In this section, we follow the evolution of the protostar as it becomes a star.

A Protostar Forms

As particles fall toward the center, they move faster and faster. As they become more densely packed, they begin to crash into one another, causing random motions and raising the temperature of the core. When the particles are hotter, they are moving faster, in random directions. Those random motions of particles are collectively known as the *thermal energy*. Thus, the collapse of the molecular-cloud core converts gravitational energy that was stored in the large, diffuse cloud into thermal energy, and the gas in the outer layers of the protostar is heated to thousands of kelvins, causing the protostar to shine.

Because of the accumulation of thermal energy, the core gradually reaches more than 1 million K. At that temperature, deuterium can fuse with hydrogen to create helium-3. (Notice that this reaction occurs at a lower temperature than that at which hydrogen fusion occurs.) Once deuterium fusion begins, it drives convection in the core. Recall from Chapter 14 that convection is the transport of energy by moving packets of gas. Convection temporarily keeps more gas from falling in and creates an apparent "surface"—more properly called a photosphere. The photosphere radiates away energy from the protostar. The hotter it gets, the more energy it radiates and the bluer that radiation becomes (see Working It Out 5.3). Because the protostar has not yet finished collapsing, the photosphere of a protostar is tens of thousands of times larger than the photosphere of the Sun today. Each square meter radiates away energy, so despite being thousands of times more luminous than the Sun, the protostar has a lower temperature.

Although the protostar is extremely luminous, astronomers often cannot observe it in visible light for two reasons: First, the photosphere of the protostar is relatively cool, so most of its radiation is in the infrared part of the spectrum. Second, and even more important, the protostar is buried deep in the heart of a dense and dusty molecular cloud. Instead, astronomers view protostars in the infrared part of the spectrum because much of the longer-wavelength infrared light from a protostar can escape through the cloud. Sometimes, as the dust absorbs the visible light, it warms up, and that heated dust also glows in the infrared. Even when astronomers cannot view a protostar directly, they can sometimes view that heated dust and know that a collapsing molecular-cloud core is hidden inside.

Sensitive infrared instruments developed since the 1980s have revolutionized the study of protostars and other young stellar objects. Clouds that appeared dark in the visible region of the spectrum, when viewed in the infrared, have revealed themselves to be clusters of dense cloud cores, young stellar objects, and glowing dust. Nearby hot stars sometimes blow away concealing dust and gas, so that molecular-cloud cores can be viewed more directly. For example, stars are forming in nodules attached to the tops of the columns of dust and gas in the Eagle Nebula, shown in **Figure 15.16**.

The Evolving Protostar

At any instant, the protostar is in balance: the forces from hot gas pushing outward and the force of gravity pulling inward exactly oppose each other. However, that balance is constantly changing. Once the core switches from convection to radiation, the deuterium in the core becomes depleted because it is no longer being replaced by material from the outer layers brought in by convection. The

Figure 15.16 The Eagle Nebula contains dense columns of molecular gas and dust at the edge of an H II region. The yellow box in the left image identifies the region magnified in the second and third images.

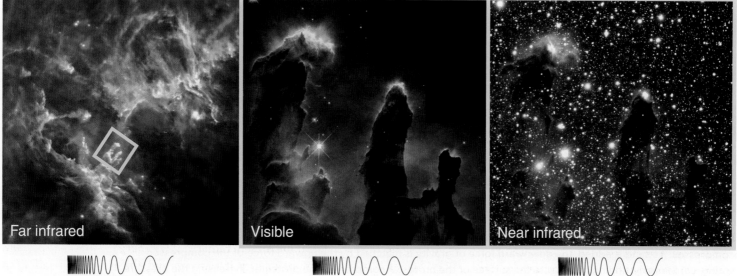

Far infrared

G X U V I R

Visible

G X U V I R

Near infrared

G X U V I R

nuclear reactions slow down, reducing the pressure from the hot gas and allowing material to resume falling onto the protostar. That infalling material adds to the mass and self-gravity of the protostar and therefore increases the weight that inner layers of the protostar must support. The protostar also slowly loses its internal thermal energy by radiating it away.

How can an object be in perfect balance and yet be changing at the same time? Consider an Earth-bound example. **Figure 15.17a** shows a simple spring balance, which works on the principle that the more a spring is compressed, the harder it pushes back. You can measure the weight of an object by determining the point at which the pull of gravity and the push of the spring are equal.

When sand is poured slowly onto the spring balance, at any instant the downward weight of the sand is balanced by the upward force of the spring. As the weight of the sand increases, the spring is slowly compressed. The spring and the weight of the sand are always in balance, but that balance is changing as more sand is added. Turning to **Figure 15.17b**, you can see that the situation is analogous to that of the protostar, in which the outward pressure of the gas behaves like the spring. The self-gravity is always matched by the internal pressure pushing out.

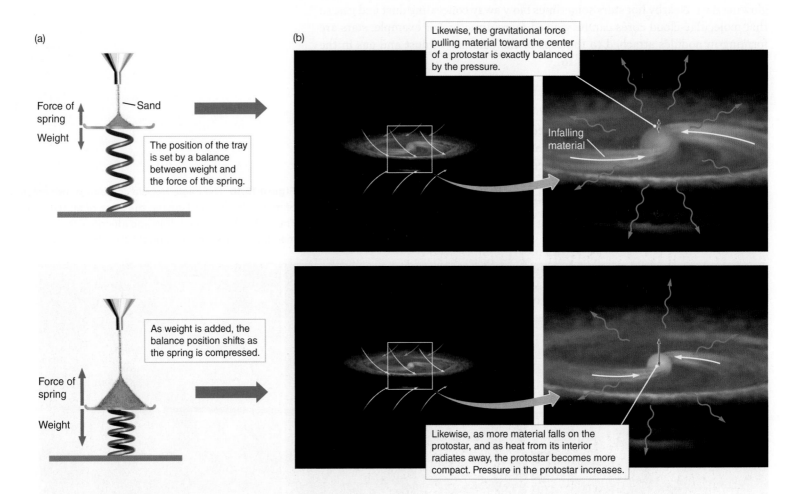

(a)

Force of spring

Weight

Sand

The position of the tray is set by a balance between weight and the force of the spring.

As weight is added, the balance position shifts as the spring is compressed.

Force of spring

Weight

(b)

Likewise, the gravitational force pulling material toward the center of a protostar is exactly balanced by the pressure.

Infalling material

Likewise, as more material falls on the protostar, and as heat from its interior radiates away, the protostar becomes more compact. Pressure in the protostar increases.

Figure 15.17 (a) As weight is added to a pan on top of a spring, the pressure on the spring increases, and the spring compresses. As sand is added, the downward force of gravity is matched by the upward force of the compressed spring. (b) Similarly, adding material to the surface of the protostar compresses the protostar, increasing the pressure inside. That balance between gravity and pressure determines the structure of a protostar.

Material falls onto the protostar, adding to its self-gravity. While the protostar slowly loses internal thermal energy by radiating it away, the material that has fallen onto the protostar compresses the protostar and heats it up. The interior becomes denser and hotter, and the pressure rises—just enough to balance the increased weight of the material above it. Dynamic balance is always maintained as the protostar slowly contracts.

Figure 15.18 illustrates that chain of events as the protostar shrinks. Gravitational energy is converted to thermal energy, which heats the core, raising the pressure to oppose gravity. That process continues, with the protostar becoming smaller and smaller and its interior growing hotter and hotter. If the protostar is massive enough, its interior will eventually become so hot that nuclear fusion of hydrogen to helium can begin. That is when the transition from protostar to star takes place. The distinction between the two is that a protostar draws its energy from gravitational collapse, whereas a star draws its energy from thermonuclear reactions in its interior.

Whether the protostar will actually become a star depends on its mass. As the protostar slowly collapses, the temperature at its center rises. If the protostar's mass is greater than about 0.08 times the mass of the Sun (0.08 M_{Sun}), the

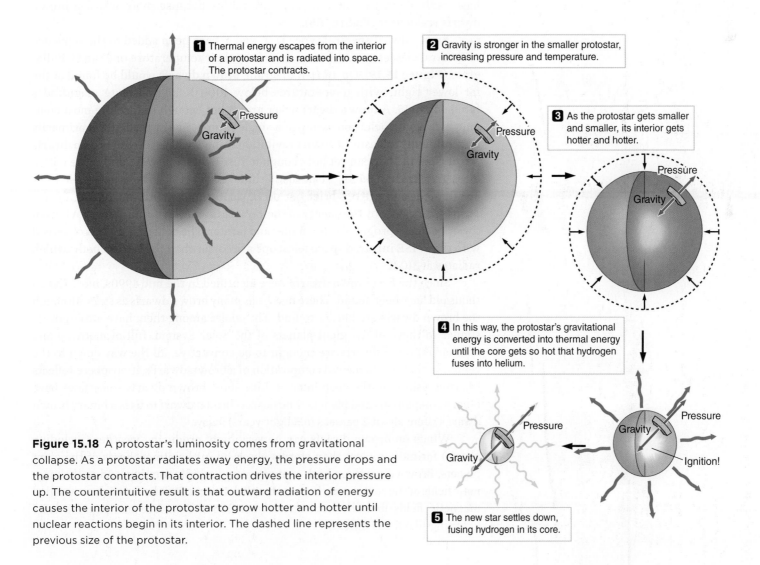

Figure 15.18 A protostar's luminosity comes from gravitational collapse. As a protostar radiates away energy, the pressure drops and the protostar contracts. That contraction drives the interior pressure up. The counterintuitive result is that outward radiation of energy causes the interior of the protostar to grow hotter and hotter until nuclear reactions begin in its interior. The dashed line represents the previous size of the protostar.

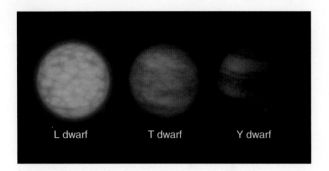

Figure 15.19 This artist's conception shows the three types of brown dwarf stars: L dwarfs (T ~1700 K); T dwarfs (T ~1200 K); and Y dwarfs (T ~500 K).

temperature in its core will eventually reach 10 million K and fusion of hydrogen into helium will begin. The newly born star will once again adjust its structure until it is radiating energy away from its surface at just the rate that energy is being liberated in its interior. As it does so, it achieves hydrostatic and thermal equilibrium and "settles" onto the main sequence of the Hertzsprung-Russell (H-R) diagram, where it will spend most of its life.

Brown Dwarfs

If the mass of the protostar is less than 0.08 M_{Sun}, it will never reach the point at which sustained nuclear fusion takes place. An object of roughly that mass is called a **brown dwarf** or sometimes a *substellar object*. Though forming in the same way that a star forms, in many respects a brown dwarf is more like a giant planet Jupiter than like a star. The International Astronomical Union (IAU) has set the smallest mass for a brown dwarf at 13 Jupiter masses (13 M_{Jup}), although some astronomers think that 10 M_{Jup} is more likely the lower limit at which brief nuclear fusion can occur. If the mass of the object is greater than 13 M_{Jup}, it must be a brown dwarf, not a supermassive planet. The upper limit of brown dwarf masses is about 70 M_{Jup}. Despite that range of masses, brown dwarfs all have radii about the same as Jupiter's radius because more massive brown dwarfs are denser (Figure 7.15).

Brown dwarf spectral types L, T, and Y have been added to the sequence of spectral classes to sort brown dwarfs by their temperature onto an H-R diagram (**Figure 15.19**). On an H-R diagram, brown dwarfs would be found at the far lower right, with temperatures below 1000 K and absolute magnitudes fainter than 15. A brown dwarf never grows hot enough to fuse the most common hydrogen nuclei consisting of a single proton, but instead glows primarily by continually releasing its own gravitational energy. The cores of brown dwarfs larger than 13 M_{Jup} can get hot enough to fuse deuterium (^{2}H), and those with a mass greater than 65 M_{Jup} can fuse lithium. But both of those energy sources are limited, and after a brief period of deuterium or lithium fusion, brown dwarfs shine only by the energy of their own gravitational contraction. A brown dwarf becomes progressively smaller and fainter. The coldest Y dwarfs observed with the WISE infrared space telescope are colder than the human body, which radiates at 310 K.

Since the first brown dwarfs were identified in the mid-1990s, more than a thousand have been found. There may be as many brown dwarfs as stars, although the brown dwarfs are harder to find. The cooler among them have atmospheres similar to those of the giant planets of the Solar System, full of methane and ammonia. Brown dwarfs are thought to be convective, all the way down to the center, so that the elemental composition of a brown dwarf's atmosphere reflects the composition in the deep interior. Like stars, brown dwarfs sometimes have binary companions and planets. The nearest brown dwarf to us is a binary brown dwarf system about 2 parsecs (6.5 light-years) away.

Winds on brown dwarfs can be very high, producing weather (including clouds) far more violent than storms observed in the atmospheres of the giant planets. Brown dwarfs are both convective and rapidly rotating, and so the magnetic fields at the surface can become tangled. Interactions between those tangled magnetic fields and the hot material beneath the surface can cause a brief lightning-like X-ray flare.

CHECK YOUR UNDERSTANDING 15.3

The energy required to begin nuclear fusion in a protostar originally came from: (a) the gravitational potential energy of the protostar; (b) the kinetic energy of the protostar; (c) the wind from nearby stars; (d) the pressure from the interstellar medium.

15.4 Evolution before the Main Sequence

Protostars and young evolving stars change their location on the H-R diagram as they settle into the main sequence, where they will spend the bulk of their lives. In this section, we examine some of the early stages in the life of the new stars.

The Evolutionary Track of an Evolving Star

In Chapter 13, we introduced the H-R diagram and used it to help explain how the properties of stars differ. As stars evolve through their lifetimes, their position on the H-R diagram changes, creating a path known as an **evolutionary track**. Because a protostar is so large, it is more luminous than a star of the same temperature on the main sequence, so a protostar's evolutionary track is located above the main sequence on the H-R diagram. **Figure 15.20** shows the evolutionary tracks of protostars of several masses. In a higher-mass protostar, a radiative zone develops around the core, causing the surface temperature to increase. Because of that increase in surface temperature, the luminosity of a higher-mass protostar stays roughly constant even as it contracts; on the H-R diagram, its track moves left horizontally to the main sequence. Lower-mass protostars are convective all the way to the core, so the surface temperature stays about the same as they collapse. As those protostars collapse, the luminosity decreases. On the H-R diagram, the track moves vertically down to the main sequence; that path is called a **Hayashi track** (Figure 15.20b).

In the 1960s, the theoretical physicist Chushiro Hayashi (1920–2010) explained the difference between the surface temperature of a star or protostar and the temperature deep in its interior. Hayashi showed that the atmospheres of stars and protostars contain a natural thermostat: the H^- ion. (An H^-, or "H minus," ion is a hydrogen atom that has acquired an extra electron and therefore has a negative charge.) The amount of H^- in the atmosphere of a protostar is highly sensitive to the temperature at the protostar's surface. The cooler the atmosphere of a star, the more slowly atoms and electrons are moving and the easier it is for a hydrogen atom to hold on to an extra electron. As a result, the cooler the atmosphere of the star, the more H^- is present.

The H^- ion, in turn, helps control how much energy a star or protostar radiates away. The more H^- the atmosphere of the star or protostar has, the more opaque the atmosphere is and the more effectively the thermal energy of the protostar is trapped in its interior. Imagine that the surface of the protostar is "too cool," meaning that extra H^- forms in the atmosphere and makes the atmosphere of the protostar more opaque. The atmosphere thus traps more of the radiation trying to escape, and the trapped energy heats up the star. As the temperature climbs, the H^- ions are changed to neutral H atoms. Now imagine the

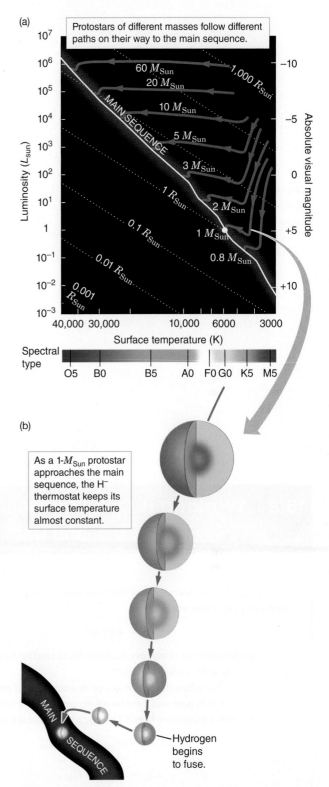

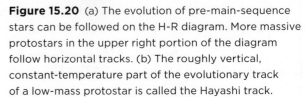

Figure 15.20 (a) The evolution of pre-main-sequence stars can be followed on the H-R diagram. More massive protostars in the upper right portion of the diagram follow horizontal tracks. (b) The roughly vertical, constant-temperature part of the evolutionary track of a low-mass protostar is called the Hayashi track.

other possibility—that the protostar is too hot. Then H⁻ in the protostar's atmosphere is destroyed, so the atmosphere becomes more transparent, allowing radiation to escape more freely from the interior. Because the protostar cannot hold on to enough of its energy to stay warm, the surface cools. In either case—too cold or too hot—H⁻ is formed or destroyed, respectively, until the star's atmosphere once again traps just the right amount of escaping radiation. The H⁻ ion is basically doing the same thing that you do with your bedcovers at night. If you get too cold, you pile on extra covers to trap your body's thermal energy and keep warm (corresponding to more H⁻ ions with the star). If you get too hot, you kick off some covers to cool off (fewer H⁻ ions).

The amount of H⁻ in the atmosphere keeps the surface temperature of the lower-mass protostar somewhere between about 3000 and 5000 K, depending on the protostar's mass and age. Because the surface temperature of the protostar is not changing much, the amount of energy per unit time (power) radiated away by each square meter of the surface of the protostar does not change much, either. Recall the Stefan-Boltzmann law from Chapter 5, which says that the amount radiated by each square meter of an object's surface depends on its temperature. As the protostar shrinks, the area of its surface shrinks as well. With fewer square meters of surface from which to radiate, the luminosity of the protostar drops. As viewed from outside, the protostar stays at nearly the same temperature and color but gradually gets fainter as it evolves toward its eventual life as a main-sequence star. The relationship among surface temperature, luminosity, and radius of protostars is further explored in **Working It Out 15.2**.

15.2 Working It Out Luminosity, Surface Temperature, and Radius of Protostars

In Chapter 13, you learned how the luminosity, surface temperature, and radius of a star are related:

$$L = 4\pi R^2 \sigma T^4$$

What can that equation reveal about the changing properties of the protostar as its radius shrinks?

Suppose that when the Sun was a protostar, it had a radius 10 times what it is now and a surface temperature of 3300 K. What would its luminosity have been? The equations for each are

$$L_{\text{protostar}} = 4\pi R_{\text{protostar}}^2 \sigma T_{\text{protostar}}^4$$

and

$$L_{\text{Sun}} = 4\pi R_{\text{Sun}}^2 \sigma T_{\text{Sun}}^4$$

We can set that relationship up as a ratio, comparing the luminosity of the protostar Sun with its luminosity now, L_{Sun}:

$$\frac{L_{\text{protostar}}}{L_{\text{Sun}}} = \frac{4\pi R_{\text{protostar}}^2 \sigma T_{\text{protostar}}^4}{4\pi R_{\text{Sun}}^2 \sigma T_{\text{Sun}}^4}$$

We rewrite that expression as follows, grouping like terms together:

$$\frac{L_{\text{protostar}}}{L_{\text{Sun}}} = \frac{4\pi\sigma}{4\pi\sigma} \times \left(\frac{R_{\text{protostar}}}{R_{\text{Sun}}}\right)^2 \times \left(\frac{T_{\text{protostar}}}{T_{\text{Sun}}}\right)^4$$

Then we cancel out the constants, $4\pi\sigma$, and use the value for $T_{\text{Sun}} = 5780$ K from Chapter 14. We know that the protostar's radius is 10 times that of the Sun, so $R_{\text{protostar}}/R_{\text{Sun}} = 10$. Then the equation becomes

$$\frac{L_{\text{protostar}}}{L_{\text{Sun}}} = \left(\frac{10}{1}\right)^2 \times \left(\frac{3,300}{5,780}\right)^4 = 10^2 \times (0.57)^4 = 10.6$$

So the Sun was about 10.6 times more luminous as a protostar than it is now. We see that on the H-R diagram of protostars (see Figure 15.20). As a 1-M_{Sun} star approaches the main sequence on the diagram, it moves down (toward lower luminosity) and to the left (toward higher surface temperature).

(a)

Material moves in toward the protostar in the accretion disk.

Bipolar jets and outflows flow away from the young star and disk.

Wind

Jets

This structure is seen in images of forming stars.

(b)

Figure 15.21 (a) Material falls onto an accretion disk around a protostar and then moves inward, eventually falling onto the protostar. In the process, some of that material is driven away in powerful jets that stream perpendicular to the disk. (b) This infrared Spitzer Space Telescope image shows jets streaming outward from a young, developing star. Note the nearly edge-on, dark accretion disk surrounding the young star.

Bipolar Outflow

As shown in **Figure 15.21a**, material falls onto the accretion disk around a young stellar object and moves inward toward the equator of the star. Meanwhile, other material is blown away from the protostar and disk in two opposite directions from the plane of the disk. The resulting stream of material away from the protostar is called a **bipolar outflow**. Powerful outflows can disrupt the molecular-cloud core and accretion disk from which the protostar formed, stopping material from falling onto the protostar.

Some bipolar outflows from young stellar objects are slow and fairly disordered, but others produce remarkable **jets** of material moving at hundreds of kilometers per second (**Figure 15.21b**). The material in those jets flows out into the interstellar medium, where it heats, compresses, and pushes away surrounding interstellar gas. Knots of glowing gas accelerated by jets are called **Herbig-Haro objects** (or **HH objects** for short), named after the two astronomers who first identified them and associated them with star formation. An example is shown in **Figure 15.22**.

The origin of outflows from protostars is not well understood, but current models suggest that outflows are the result of magnetic interactions between the protostar and the disk. The interior of a lower-mass protostar on its Hayashi track is convective. That convection, coupled with the protostar's rapid rotation, forms a dynamo, similar to the dynamo that drives the Sun's magnetic field but much

G X U V I R

Figure 15.22 In this image of Herbig-Haro object 212, taken by ESO's Infrared Spectrometer And Array Camera (ISAAC), the accretion disk of the protostar is visible as a fuzzy dark band cutting between the two jets. The protostar is concealed inside that accretion disk and is only a few thousand years old. The jets are remarkably symmetric, with several knots appearing at intervals. That finding suggests that the jet pulses regularly and over a short timescale—possibly as short as 30 years. At the ends of the jets, ejected gas collides with interstellar dust and gas at several hundred kilometers per second, creating bow shocks.

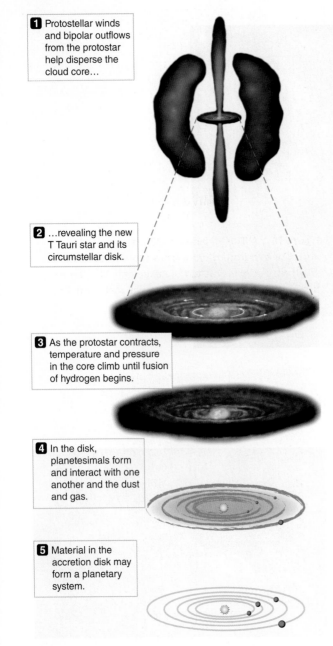

1 Protostellar winds and bipolar outflows from the protostar help disperse the cloud core...

2 ...revealing the new T Tauri star and its circumstellar disk.

3 As the protostar contracts, temperature and pressure in the core climb until fusion of hydrogen begins.

4 In the disk, planetesimals form and interact with one another and the dust and gas.

5 Material in the accretion disk may form a planetary system.

Figure 15.23 An overview of how stars like the Sun form, beginning with the onset of stellar winds and ending with the ignition of a star sitting at the center of a revolving system of planets.

more powerful. The resulting strong magnetic field might cause the protostar to begin blowing a powerful wind. As the rotation of the protostar drags the magnetic field lines around, the lines wind up like the fibers in a rope. Those tightly wound magnetic field lines collimate the outflow into jets.

As the wind from the protostar disperses the remains of the dusty molecular-cloud core from which the protostar formed, the first direct, visible-light view of the protostar emerges. Some of those protostars of lower mass are called **T Tauri stars**. The name comes from the first recognized member of that class of objects, the star labeled T in the constellation Taurus. Higher-mass protostars of spectral type B or A are called Herbig Be or Ae stars. **Figure 15.23** summarizes the evolution of a protostar from the onset of winds to the formation of planets (recall that steps 4 and 5 were covered in Chapter 7).

The Influence of Mass

Astronomers are interested in how and why molecular clouds fragment to form stars with a range of masses. Astronomers do not understand why some cloud cores become 1-M_{Sun} stars, whereas others become 5- or 10-M_{Sun} stars. The details of that division—specifically, what fraction of newly formed stars will have which masses—are crucial to understand how observations of the stars near the Sun today relate to the history of star formation in our galaxy.

A look around the disk of our galaxy reveals a variety of stars—some very old and others very young. If those were the only stars available to study, scientists could not easily learn much about how stars evolve. But astronomers have long known that stars are often found close together in collections called **star clusters**. Star clusters are collections of stars that all formed in the same place, from the same material, and at about the same time. **Figure 15.24** shows one such star cluster: a group called the Pleiades, or "Seven Sisters." Clusters such as that one serve as extremely useful samples for studying star formation. Even though the few brightest and most massive stars in a cluster dominate any observation of a cluster, most of the stars in a cluster are less massive than the Sun. In fact, some star-forming regions seem to form no high-mass stars at all.

After a molecular-cloud core collapses, how a protostar evolves depends almost entirely on its mass. A star like the Sun takes about 10 million years or so to descend its Hayashi track and become a star on the main sequence. The total time for such a star to form—from initial fragmentation up to the ignition of hydrogen fusion—might be more like 30 million years. Because the self-gravity of a more massive core is stronger, more massive cores collapse to form stars more quickly. A 10-M_{Sun} star will form in only 100,000 years. A 100-M_{Sun} star might take less than 10,000 years. By comparison, a 0.1-M_{Sun} star might take 100 million years finally to reach the main sequence.

The 30 million years that the Sun took to form is a long time, yet still just a tiny fraction of the 10 billion years during which the Sun will steadily fuse hydrogen into helium as a main-sequence star. Because stars spend so much longer on the main sequence than they do as protostars, it is no wonder that so few among the many stars visible in the sky are protostars. But every star was once a protostar, including the Sun.

CHECK YOUR UNDERSTANDING 15.4

What causes the wind from a protostar to form jets? (a) The accretion disk can supply material to the jet only along the equator of the star. (b) Magnetic fields wind up and direct the outflow. (c) The star is larger at the equator, so material leaves the poles more easily. (d) The star spins faster at the equator, so material leaves the poles more easily.

Origins Star Formation, Planets, and Life

When astronomers consider the possibility of other life in the universe, one of the first things they think about is the formation of stars and planets. Life probably needs planets, and planets form along with stars. The conditions under which a star is born, and the mass and chemical composition that it has when it begins its nuclear fusion, set the stage for the rest of its life. In Chapter 7, we noted that most stars have planets that form at about the same time as the star. If the star is going to have rocky planets with hard surfaces (such as the planets of the inner Solar System) or gaseous planets with molten rocky cores and rocky moons (such as the planets of the outer Solar System), the material from which the star and planets form must be "enriched" with the heavy elements that make up those rocky surfaces.

Those enriched clouds would also provide elements essential to life on Earth. In addition to the presence of organic molecules mentioned earlier in the chapter, astronomers have detected water in star-forming regions such as W3 IRS5 (**Figure 15.25**). Water exists as ice mixed with dust grains in the cool molecular clouds or exists as vapor when it is closer to a protostar and the dust grains and ice evaporate. In 2011, the Herschel Space Observatory detected oxygen molecules (O_2, the type we breathe) in a star-forming complex in Orion. Oxygen is the third-most-common element in the universe, yet it had not been decisively observed before in molecular form. That oxygen also may have come from the melting and evaporation of water ice on the tiny dust grains.

As noted in Chapter 13, astronomers had doubted that planets could exist in stable orbits in binary star systems, but now a few such systems have been found. Planets that form within associations of O and B stars may be too unstable to last very long. Isolated planets unattached to any star are moving through the Milky Way. Perhaps those rogue planets were gravitationally ejected soon after they formed in a multiple system. But those planets do not have a source of energy like Earth's Sun. Astronomers theorize that only planets that orbit stars can support life. So when trying to estimate the possibility of life in the galaxy, astronomers include estimations of the rate of formation of stars in the galaxy, along with the fraction of stars that have planets. Advances in the study of star formation and planet detection help astronomers understand better the conditions under which life might develop elsewhere.

G X U V I R

Figure 15.24 The Pleiades, or "Seven Sisters," is a cluster of hundreds of young stars, about seven of which are easily visible to the naked eye. The diffuse blue light around the stars is starlight scattered by interstellar dust.

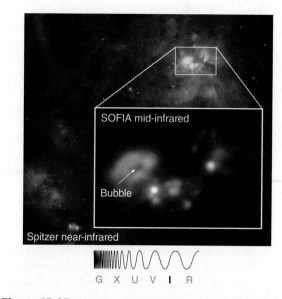

G X U V I R

Figure 15.25 Water was detected in the W3 IRS5 star-forming complex. Here, W3 IRS5 is observed with the Spitzer Space Telescope in near infrared and with the Stratospheric Observatory for Infrared Astronomy (SOFIA) telescope in mid-infrared. A massive star has cleared the dust and gas from a small bubble, sweeping it into a dense shell (green).

Astronomers report on their search for dust from outside the Solar System.

Interstellar Dust Discovered inside NASA Spacecraft

By IRENE KLOTZ, *Discovery News*

Thanks to a massive effort by 30,716 volunteers, scientists have pinpointed what appear to be seven precious specks of dust from outside the Solar System, each bearing unique stories of exploded stars, cold interstellar clouds, and other past cosmic lives.

The Herculean effort began eight years ago after NASA's *Stardust* robotic probe flew by Earth to deposit a capsule containing samples from a comet and dust grains from what scientists hoped would be interstellar space. The spacecraft was outfitted with panels containing a smoke-like substance called aerogel that could trap and preserve fast-moving particles.

Stardust twice put itself into position to fish for interstellar grains, which are so small that a trillion of them would fit in a teaspoon. The only way scientists back on Earth would be able to find them was by the microscopic trails the grains made as they plowed into the aerogel.

"When we did the math we realized it would take us decades to do the search ourselves," physicist Andrew Westphal, with the University of California, Berkeley, told *Discovery News*.

The team used an automated microscope to scan the collector and put out a call for volunteers.

"This whole approach was treated with pretty justifiable criticism by people in my community. They said, 'How can you trust total strangers to take on this project?'" Westphal said.

"We really didn't know how else to do it. We still don't," he added.

Recruits were trained and had to pass a test before they were given digital scans to peruse. Scientists sometimes inserted images with known trails just to see if the volunteers, known as "dusters," would spot them.

"We were very pleased to see that people are really good at finding these tracks, even really, really difficult things to find," Westphal said.

More than 50 candidate dust motes turned out to be bits of the spacecraft itself, but scientists found seven specks that bear chemical signs of interstellar origin and travel.

The grains are surprisingly diverse in shape, size, and chemical composition. The larger ones, for example, have a fluffy, snowflake-like structure.

Additional tests are needed to verify the grains' interstellar origins and ferret out their histories. But the grains are so tiny that with currently available technology, additional analysis would mean their demise.

"It'll probably be years before we can do a lot more with these samples," said space scientist Mike Zolensky, who oversees NASA's collection of cosmic dust, moon rocks, and other extraterrestrial samples at the Johnson Space Center in Houston.

"But we've got them safely tucked away and we can hang on to them until those techniques come along," Zolensky said.

1. Why do scientists want to identify interstellar dust grains?
2. How were those dust grains distinguished from Solar System grains?
3. Why can't the scientists do a complete analysis on those particles?
4. How did volunteer "citizen scientists" assist with that project?
5. Visit the website noted in question 49 in the "Questions and Problems" section later in this chapter. Is that project continuing? What else has been discovered?

Source: Irene Klotz, "Interstellar dust discovered inside NASA spacecraft." Originally published by Seeker, August 14, 2014. Reprinted by permission of Group Nine Media.

Summary

Astronomers have never watched the full process of a single star forming from beginning to end. Instead, they have observed many stars at different stages in their formation and evolution at different wavelengths, and they have used their knowledge of physical laws to tie those observations together into a coherent, consistent description of how, why, and where stars form. Stars form from material in the interstellar medium, under the influence of gravity, often triggered by external factors such as the winds of hot stars nearby. The conditions under which a star is born determine whether it will have planets along with the chemical elements required by life as it exists on Earth.

LG 1 **Describe the types and states of material that exist in the space between the stars and how that material is detected.** The interstellar medium ranges from cold, relatively dense molecular clouds to hot, tenuous intercloud gas heated and ionized by energy from stars and stellar explosions. Dust and gas in the interstellar medium blocks much visible light but becomes more transparent at longer, infrared wavelengths. Different phases of the interstellar medium emit various types of radiation and can be observed at different wavelengths, ranging from radio waves to X-rays. Neutral hydrogen cannot be detected at visible and infrared wavelengths, but its presence is revealed by its 21-cm emission.

LG 2 **Explain the conditions under which a cloud of gas can contract into a stellar system and the role that gravity and angular momentum play in forming stars and planets.** Star formation begins when the self-gravity of dense clouds exceeds outward pressure. The clouds collapse, heat up, and fragment to form stars. The conservation of angular momentum is important to the formation of disks during the collapse. Forming stars are detected directly from their infrared emission and indirectly from the dust around them, which is heated by the forming star.

LG 3 **List the steps in the evolution of a protostar, and explain how the mass of the protostar affects its evolution.** Protostars collapse, radiating away their gravitational energy until fusion starts in their cores. When protostars reach hydrostatic and thermal equilibrium, they settle onto the main sequence. Stars form in clusters from dense cores buried within giant molecular clouds. A protostar must have a mass of at least 0.08 M_{Sun} to become a true star. Brown dwarfs are neither stars nor planets but have masses in between.

LG 4 **Describe the track of a protostar as it evolves to a main-sequence star on the Hertzsprung-Russell (H-R) diagram.** Because star formation takes tens of thousands to millions of years, astronomers learn about the formation of stars from observations of many protostars at various stages of development. Higher-mass protostars move horizontally across the H-R diagram, at roughly constant luminosity, to the main sequence. Lower-mass protostars drop nearly straight down, at constant temperature, toward the main sequence. Once hydrogen fusion begins, the star moves nearly horizontally until it reaches the main sequence.

? Unanswered Questions

- Many questions about star formation remain. For example, how must theories be modified to explain how binary stars or other multiple-star systems form? At what point during star formation is it determined that a collapsing cloud core will form several stars instead of just one? Some models suggest that this split may happen early in the process, during the fragmentation and collapse of the molecular cloud. The advantage of those ideas is that they provide a natural way of dealing with much of the cloud core's angular momentum: it goes into the orbital angular momentum of the stars around each other. Other models suggest that additional stars may form from the accretion disk around an initially single protostar.

- Do high-mass and low-mass stars form very differently? The smallest stars, spectral type M, are most likely to form as single stars, but a high fraction of medium-mass stars are formed in binary pairs. According to one theory, those binaries start out as triple systems, from which the smallest star is ejected, leading to a remaining pair and a single star. The highest-mass stars are less likely to form alone; many form in larger groups of massive stars in which the formation of one large star may trigger the formation of another nearby in the molecular cloud.

Questions and Problems

Test Your Understanding

1. Phases of the interstellar medium include (choose all that apply)
   a. hot, low-density gas.
   b. cold, high-density gas.
   c. hot, high-density gas.
   d. cold, low-density gas.

2. Dust in the interstellar medium can be observed in
   a. visible light.
   b. infrared radiation.
   c. radio waves.
   d. X-rays.

3. The interstellar medium in the Sun's region of the galaxy is closest in composition to that of
   a. the Sun.
   b. Jupiter.
   c. Earth.
   d. comets in the Oort Cloud.

4. Interstellar dust is effective at blocking visible light because
   a. the dust is so dense.
   b. dust grains are so few.
   c. dust grains are so small.
   d. dust grains are so large.

5. Hot intercloud gas is heated primarily by
   a. starlight.
   b. protostars.
   c. neutrinos.
   d. supernova explosions.

6. Astronomers determined the composition of the interstellar medium from
   a. observing its emission and absorption lines.
   b. measuring the composition of the planets.
   c. return samples from spacecraft.
   d. composition of meteorites.

7. In astronomy, the term *bipolar* refers to outflows that
   a. point in opposite directions.
   b. alternate between expanding and collapsing.
   c. rotate about a polar axis.
   d. show spiral structure.

8. Which of the following has contributed most to our understanding of star formation?
   a. Astronomers have observed star formation as it happens for a few stars.
   b. Astronomers have observed star formation as it happens for many stars.
   c. Astronomers have observed many stars at various steps of the formation process.
   d. Theoretical models predict how stars form.

9. Cold neutral hydrogen can be detected because
   a. it emits light when electrons drop through energy levels.
   b. it blocks the light from more distant stars.
   c. it is always hot enough to glow in the radio and infrared wavelengths.
   d. the atoms in the gas change spin states.

10. The Hayashi track is a nearly vertical evolutionary track on the H-R diagram for low-mass protostars. Which of the following would you expect from a protostar moving along a vertical track?
    a. The star remains the same brightness.
    b. The star remains the same luminosity.
    c. The star remains the same color.
    d. The star remains the same size.

11. Which two forces establish hydrostatic equilibrium in an evolving protostar?
    a. the force from pressure and gravity
    b. the force from pressure and the strong nuclear force
    c. gravity and the strong nuclear force
    d. energy emitted and energy produced

12. Suppose you are studying a visible-light image of a distant galaxy, and you see a dark lane cutting across the bright disk. That dark line is most likely caused by
    a. gravitational instabilities that clear the area of stars.
    b. dust in the Milky Way blocking the view of the distant galaxy.
    c. dust in the distant galaxy blocking the view of stars in the disk.
    d. a flaw in the instrumentation.

13. What causes a hydrogen atom to radiate a photon of 21-cm radio emission?
    a. The electron drops down one energy level.
    b. The formerly free electron is captured by the proton.
    c. The electron flips to an aligned spin state.
    d. The electron flips to an unaligned spin state.

14. Astronomers know that dusty accretion disks exist around protostars because
    a. a dark band often appears across the protostar.
    b. a bright band often appears across the protostar.
    c. accretion disks should be there, according to theory.
    d. planets are in the Solar System.

15. What is the single most important property of a star that will determine its evolution?
    a. mass
    b. composition
    c. temperature
    d. radius

Thinking about the Concepts

16. The interstellar medium is approximately 99 percent gas and 1 percent dust. Why does dust and not gas block a visible-light view of the galactic center?

17. Explain why observations in the infrared are necessary for astronomers to study the detailed processes of star formation.

18. How does the material in interstellar clouds and intercloud gas differ in density and distribution?

19. When a star forms inside a molecular cloud, what happens to the cloud? Can a molecular cloud remain cold and dark with one or more stars inside it? Explain your answer.

20. If you placed your hand in boiling water (100°C) for even 1 second, you would get a very serious burn. If you placed your hand in a hot oven (200°C) for a second or two, you would hardly feel the heat. Explain that difference and how it relates to million-kelvin regions of the interstellar medium.

21. How do astronomers know that the Sun is located in a "local bubble" formed by a supernova?

22. Interstellar gas atoms typically cool by colliding with other gas atoms or grains of dust; during the collision, each gas atom loses energy and hence its temperature is lowered. How does that explain why very low-density gases are generally so hot, whereas dense gases tend to be so cold?

23. Explain how the 21-cm line discussed in the Process of Science Figure supports the cosmological principle (which states that the laws of physics must be the same everywhere).

24. Molecular hydrogen is very difficult to detect from the ground, but astronomers can easily detect carbon monoxide (CO) by observing its 2.6-cm microwave emission. Describe how observations of CO might help astronomers infer the amounts and distribution of molecular hydrogen within giant molecular clouds.

25. The Milky Way contains several thousand giant molecular clouds. Describe a giant molecular cloud and its role in star formation.

26. As a cloud collapses to form a protostar, the forces of gravity experienced by all parts of the cloud (which follow an inverse square law) become stronger and stronger. One might argue that under those conditions, the cloud should keep collapsing until it becomes a single massive object. Why doesn't that happen?

27. The internal structure of a protostar maintains hydrostatic equilibrium even as more material is falling onto it. Explain how that can be.

28. What are the similarities and differences between a brown dwarf and a giant planet such as Jupiter? Would you classify a brown dwarf as a supergiant planet? Explain your answer.

29. The H^- ion acts as a thermostat in controlling the surface temperature of a protostar. Explain that process.

30. How does the composition of the molecular cloud affect the type of planets and stars that form within it?

Applying the Concepts

31. In Chapter 13, you learned that astronomers can measure the temperature of a star by comparing its brightness in blue and yellow light. Does reddening by interstellar dust affect a star's temperature measurement? If so, how?

32. When a hydrogen atom is ionized, it splits into two components.
    a. Identify the two components.
    b. If both components have the same kinetic energy, which moves faster?

33. Estimate the typical density of dust grains (grains per cubic centimeter) in the interstellar medium. A typical grain has a mass of about 10^{-17} kilogram (kg). (Hint: You know the typical density of gas and the fraction of the interstellar medium's mass that is made of dust.)

34. Referring to Figure 15.3, estimate the effective blackbody temperature of the star as shown in part (b) (without dust) and part (c) (with dust). How significant are the effects of interstellar dust when observed data are used to determine the properties of a star?

35. A typical temperature of intercloud gas is 8000 K. Using Wien's law (see Working It Out 15.1 and Chapter 5), calculate the peak wavelength at which that gas would radiate.

36. Some parts of the Orion Nebula have a blackbody peak wavelength of 0.29 μm. What is the temperature of those parts of the nebula?

37. Stellar radiation can convert atomic hydrogen (H I) to ionized hydrogen (H II).
    a. Why does a B8 main-sequence star ionize far more interstellar hydrogen in its vicinity than does a K0 giant of the same luminosity?
    b. What properties of a star are important in determining whether it can ionize large amounts of nearby interstellar hydrogen?

38. A proton has 1,850 times the mass of an electron. If a proton and an electron have the same kinetic energy ($E_K = 1/2 \, mv^2$), how many times greater is the velocity of the electron than that of the proton?

39. If a typical hydrogen atom in a collapsing molecular-cloud core starts at a distance of 1.5×10^{12} km (10,000 AU) from the core's center and falls inward at an average velocity of 1.5 km/s, in how many years does it reach the newly forming protostar? Assume that a year is 3×10^7 seconds.

40. The ratio of hydrogen atoms (H) to carbon atoms (C) in the Sun's atmosphere is approximately 2,400:1 (see Table 13.1). We can reasonably assume that this ratio also applies to molecular clouds. If 2.6-cm radio observations indicate 100 M_{Sun} of carbon monoxide (CO) in a giant molecular cloud, what is the implied mass of molecular hydrogen (H_2) in the cloud? (Carbon represents 3/7 of the mass of a CO molecule.)

41. Neutral hydrogen emits radiation at a radio wavelength of 21 cm when an atom drops from a higher-energy spin state to a lower-energy spin state. On average, each atom remains in the higher-energy state for 11 million years (3.5×10^{14} seconds).
    a. What is the probability that any given atom will make the transition in 1 second?
    b. If 6×10^{59} atoms of neutral hydrogen are in a 500-M_{Sun} cloud, how many photons of 21-cm radiation will the cloud emit each second?
    c. How does that number compare with the 1.8×10^{45} photons emitted each second by a solar-type star?

42. The Sun took 30 million years to evolve from a collapsing cloud core to a star, with 10 million of those years spent on its Hayashi track. The Sun will spend 10 billion years on the main sequence. Suppose that the Sun's main-sequence lifetime were compressed into a single day.
    a. How long would the total collapse phase last?
    b. How long would the Sun spend on its Hayashi track?

43. A protostar with the mass of the Sun starts out with a temperature of about 3500 K and a luminosity about 200 times larger than the Sun's present value. Estimate that protostar's size and compare it with the size of the Sun today.

44. The star-forming region 30 Doradus is 160,000 light-years away in the nearby galaxy called the Large Magellanic Cloud and appears about one-sixth as bright as the faintest stars visible to the naked eye. If it were located at the distance of the Orion Nebula (1,300 light-years away), how much brighter than the faintest visible stars would it appear?

45. Assume that a brown dwarf has a surface temperature of 1000 K and approximately the same radius as Jupiter. What is its luminosity compared with that of the Sun? How many brown dwarfs like that one would be needed to produce the luminosity of a star like the Sun?

USING THE WEB

46. Go to the Astronomy Picture of the Day (APOD) website (https://apod.nasa.gov/apod), search for "molecular clouds," and pick out a few images. Were those pictures obtained from space or on the ground, and at what wavelengths? With which telescopes? What wavelengths do the colors in the images represent? Are they "real" or "false-color" images?

47. Go to NASA's Spitzer Space Telescope website (http://www.spitzer.caltech.edu). Click on "News" and find a recent story about star formation. What did Spitzer observe? What wavelengths do the colors in the picture represent? How does that "false color" help astronomers analyze the images? Why do astronomers study star formation in the infrared rather than in the visual part of the spectrum?

48. The Stratospheric Observatory for Infrared Astronomy (SOFIA) is a 2.5-meter telescope on a modified Boeing 747 aircraft. Go to the SOFIA website (http://sofia.usra.edu). Why have astronomers put an infrared telescope on an airplane? What has been detected with that telescope?

49. Citizen science: Go to the website for *Stardust* (http://stardustathome.ssl.berkeley.edu), a project in which volunteers use a virtual microscope to analyze digital scans of particles collected by the *Stardust* mission in 2006. The goal is to identify tiny interstellar dust grains. Follow the steps under "Get Started" (you need to create a log-in account) and help search for stardust. Click on "News." What has been learned from that project? Remember to save the images for your homework, if required.

50. Do a news search for a story about brown dwarfs. Is the story from an observatory? A NASA mission? A press release? What is new, and why is it interesting?

In this Exploration, you will see how the H⁻ thermostat works in the formation of stars. You will need about 20 coins (they do not have to be all the same type).

Place your coins on a sheet of paper and draw a circle around them—the smallest possible circle that will fit all the coins. Then divide the circle into three parts, as shown in **Figure 15.26.** This circle represents a star with a changing temperature. The coins represent H⁻ ions. Removing a coin from the circle means that the H⁻ ion has turned into a neutral hydrogen atom. Placing a coin in the circle means that the neutral hydrogen atom has become an H⁻ ion.

Place all the coins back on the circle.

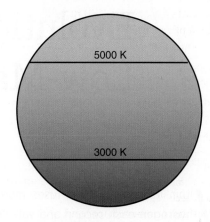

Figure 15.26 This circle represents a star with a changing temperature.

1 How many "H⁻ ions" are now in the star?

_____

_____

The "blanket" of H⁻ ions holds heat in the star, so the star begins to heat up until it reaches about 5000 K. At that surface temperature, the H⁻ ions begin to be destroyed. Now that the star is hot, begin removing coins one at a time, starting from the top of the circle and working downward. When you see the line marking 3000 K, stop removing coins.

2 How many "H⁻ ions" are now in the star?

_____

_____

3 What will happen to the surface temperature of the star now that it has fewer ions?

_____

_____

When the star cools off to about 3000 K, H⁻ ions begin to form. Place the coins back on the circle, starting from the bottom and working your way up to the line at 5000 K.

4 How many "H⁻ ions" are now in the star?

_____

_____

5 What will happen to the surface temperature of the star now that it has more ions?

_____

_____

Now that the star is hot, begin removing coins one at a time, starting from the top of the circle and working downward. When you see the line marking 3000 K, stop removing coins.

6 What should happen next?

_____

_____

7 Make a circular flowchart that includes the following steps in the proper order: the star heats up; the star cools down; H⁻ is formed; H⁻ is destroyed.

_____

_____

16

Evolution of Low-Mass Stars

Within its core, the Sun fuses more than 4 billion kilograms (kg) of hydrogen each second and will eventually run out of fuel. Although the Sun may seem immortal by human standards, roughly 5 billion years from now the Sun's time on the main sequence will come to an end. In this chapter, we look at how the mass of a star relates to its main-sequence lifetime. Then we examine what happens when a low-mass star like the Sun leaves the main sequence.

LEARNING GOALS

By the end of this chapter, you should be able to:

LG 1 Estimate the main-sequence lifetime of a star from its mass.

LG 2 Explain why low-mass stars grow larger and more luminous as they run out of fuel.

LG 3 Sketch post-main-sequence evolutionary tracks on a Hertzsprung-Russell (H-R) diagram and list the stages of evolution for low-mass stars.

LG 4 Describe how planetary nebulae and white dwarfs form.

LG 5 Explain how some close binary systems evolve differently from single stars.

When low-mass stars die, they leave behind an expanding nebula of gas and dust, as seen in this Hubble Space Telescope image of the Ring Nebula. ▶ ▶ ▶

Is this the
future of
our Sun?

439

Figure 16.5 As a star moves up the red giant branch in the H-R diagram, the luminosity of the star grows faster and faster. The fusion of hydrogen to helium in a shell surrounding a degenerate helium core feeds on itself, creating a cycle that speeds up as time goes on.

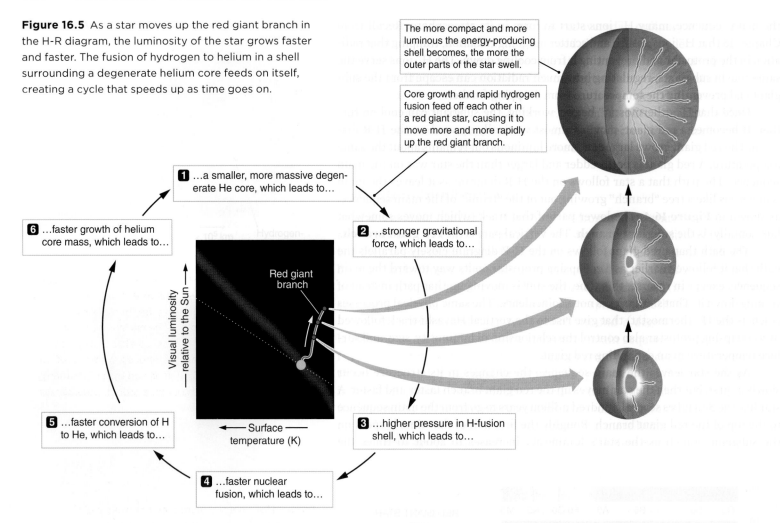

The more compact and more luminous the energy-producing shell becomes, the more the outer parts of the star swell.

Core growth and rapid hydrogen fusion feed off each other in a red giant star, causing it to move more and more rapidly up the red giant branch.

1 ...a smaller, more massive degenerate He core, which leads to...

6 ...faster growth of helium core mass, which leads to...

2 ...stronger gravitational force, which leads to...

Red giant branch

5 ...faster conversion of H to He, which leads to...

3 ...higher pressure in H-fusion shell, which leads to...

4 ...faster nuclear fusion, which leads to...

Visual luminosity —relative to the Sun →

← Surface — temperature (K)

luminosity of the Sun (L_{Sun}). (During the subgiant and red giant phases, the luminosity increases but the mass does not, so the main-sequence mass–luminosity relation no longer applies.) During the second half of that time, the hydrogen-fusing shell adds more helium to the degenerate helium core of the star. That helium core grows in mass but not in radius. The increasing mass of the ever-more-compact helium core increases the force of gravity in the heart of the star. That increased gravity once again increases the pressure and the temperature and thus the rate of nuclear fusion, this time in the hydrogen-fusing shell. Faster nuclear reactions in the shell convert hydrogen into helium more quickly, so the core grows more rapidly. The star has entered a feedback loop; increasing core mass leads to faster shell fusion, and faster shell fusion leads to faster core growth. As the core gains mass and the shell becomes more luminous, the outer layers swell. As a result, the star's luminosity climbs at an ever-higher rate from 10 to almost 1,000 L_{Sun}. The evolution of the star is illustrated in **Figure 16.5**.

CHECK YOUR UNDERSTANDING 16.2

The red giant branch is nearly vertical on the H-R diagram because: (a) the surface temperature rises, but the luminosity is nearly constant; (b) the luminosity rises, but the temperature is nearly constant; (c) the luminosity and the temperature rise significantly; (d) the surface temperature and the luminosity both remain nearly constant.

16.3 Helium Fuses in the Degenerate Core

On the main sequence, a star fuses hydrogen in the core. On the red giant branch, the core consists of nonfusing helium surrounded by a shell of fusing hydrogen. Once the hydrogen in the shell is used up, the core of the red giant contracts and heats. Eventually, the temperature and pressure rise enough to start the next stage of stellar evolution: helium fusion in the core.

Helium Fusion and the Triple-Alpha Process

As the red giant evolves, its helium core grows not only smaller and more massive but also hotter. That increase in temperature is due partly to the gravitational energy released as the core shrinks and partly to the energy released by hydrogen fusion in the surrounding shell. The thermal motions of the atomic nuclei in the core become more and more energetic. Eventually, at a temperature of about 100 million (10^8) K, the ^{4}He nuclei are slammed together hard enough for the strong nuclear force to act, and helium fusion begins.

Helium fuses in a two-stage process called the **triple-alpha process**, so named because it involves the fusion of three ^{4}He nuclei, which are sometimes called **alpha particles**. The process, illustrated in **Figure 16.6**, begins when two helium-4 (^{4}He) nuclei fuse to form a beryllium-8 (^{8}Be) nucleus consisting of four protons and four neutrons. The ^{8}Be nucleus is extremely unstable and quickly decays. But if, in that short time, it collides with another ^{4}He nucleus, the two nuclei will fuse into a stable nucleus of carbon-12 (^{12}C) consisting of six protons and six neutrons. The reaction rate depends on the temperature: higher temperatures increase the number of ^{8}Be nuclei that collide with a ^{4}He nucleus and therefore increase the reaction rate.

Recall that the core of the red giant is electron-degenerate, which means that as many electrons are packed into that space as possible. Those degenerate electrons prevent the core from collapsing. The atomic nuclei, however, behave like a normal gas, moving through the sea of degenerate electrons almost as though the electrons were not there. The atomic nuclei in the core move freely about, as shown in **Figure 16.7**, just as they do throughout the rest of the star. We

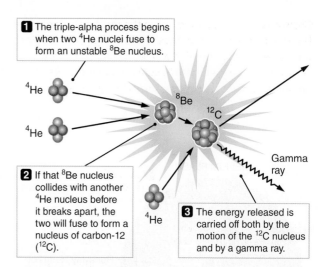

1 The triple-alpha process begins when two ^{4}He nuclei fuse to form an unstable ^{8}Be nucleus.

2 If that ^{8}Be nucleus collides with another ^{4}He nucleus before it breaks apart, the two will fuse to form a nucleus of carbon-12 (^{12}C).

3 The energy released is carried off both by the motion of the ^{12}C nucleus and by a gamma ray.

Gamma ray

Figure 16.6 The triple-alpha process produces a stable nucleus of carbon-12. Two helium-4 (^{4}He) nuclei fuse to form an unstable beryllium-8 (^{8}Be) nucleus. If that nucleus collides with another ^{4}He nucleus before breaking apart, the two will fuse to form a stable nucleus of carbon-12 (^{12}C). The energy produced is carried off both by the motion of the ^{12}C nucleus and by a high-energy gamma ray emitted in the second step of the process.

Figure 16.7 In a red giant star, the weight of the overlying layers is supported by electron degeneracy pressure in the core arising from electrons that are packed together as tightly as quantum mechanics allows. Atomic nuclei in the core can move freely about within the sea of degenerate electrons, so they behave as a normal gas.

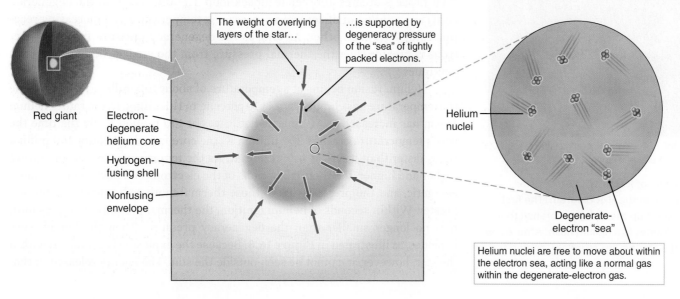

The weight of overlying layers of the star…

…is supported by degeneracy pressure of the "sea" of tightly packed electrons.

Red giant

Electron-degenerate helium core

Hydrogen-fusing shell

Nonfusing envelope

Helium nuclei

Degenerate-electron "sea"

Helium nuclei are free to move about within the electron sea, acting like a normal gas within the degenerate-electron gas.

can understand the fusion of helium by treating the nuclei as matter in a normal state. Once the pressure and temperature are high enough, those nuclei begin to fuse, as the hydrogen nuclei do in main-sequence stars. However, the energy released will not affect the degenerate-electron core in the same way that it affects the core in a main-sequence star.

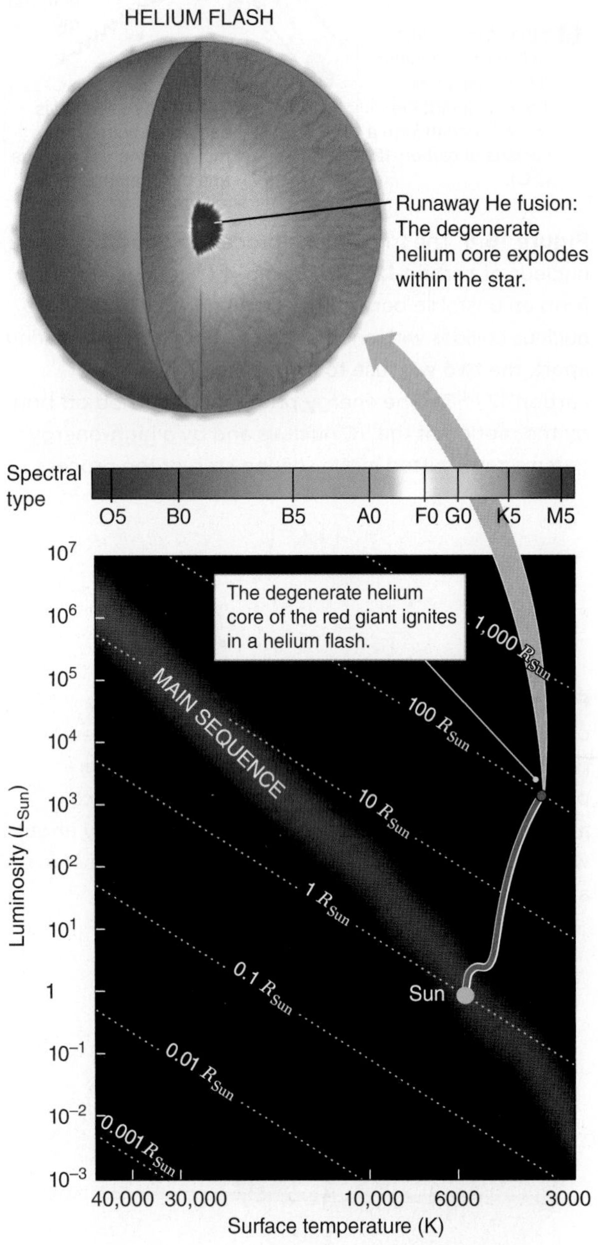

The Helium Flash

Degenerate material conducts thermal energy very well, so any differences in temperature within the core rapidly disperse. As a result, when helium fusion begins at the center of the core, the energy released quickly heats the entire core. Within a few minutes, the entire core is fusing helium into carbon by the triple-alpha process. Some of the carbon will fuse with an additional ^{4}He nucleus to form stable oxygen-16 (^{16}O).

In a normal gas, such as the air around you or the core of a main-sequence star, the pressure of the gas comes from the random thermal motions of the atoms, which are faster at higher temperatures. Increasing the temperature increases the pressure of the gas. In the core of a main-sequence star, heating it causes it to expand; the temperature, density, and pressure then decrease; nuclear reactions slow; and the star settles into a new balance between gravity and pressure. Those are exactly the sorts of changes steadily occurring within the core of a main-sequence star like the Sun as the structure of the star shifts in response to the changing composition in the star's core.

However, the degenerate core of a red giant is not a normal gas. The pressure in a red giant's degenerate core comes primarily from how tightly the electrons in the core are packed together. Heating the core does not change the number of electrons that can be packed into its volume, so the core's pressure does not respond to changes in temperature. Because the pressure does not increase, the core does not expand when heated, as a normal gas would.

Although the higher temperature does not change the pressure, it does cause the helium nuclei to collide more often and with greater force, so the nuclear reactions become more vigorous. More vigorous reactions increase the temperature, and higher temperature means even more vigorous reactions. That process creates another feedback loop. Helium fusion in the degenerate core runs wildly out of control as increasing temperature and increasing reaction rates feed each other. As long as the degeneracy pressure from the electrons is greater than the thermal pressure from the nuclei, the feedback loop continues.

Helium fusion begins at a temperature of about 100 million K. By the time the temperature has climbed by just 10 percent, to 110 million K, the rate of helium fusion has increased to 40 times what it was at 100 million K. By the time the core's temperature reaches 200 million K, the core is fusing helium 460 million times faster than it was at 100 million K. As the temperature in the core grows higher and higher, the thermal motions of the electrons and nuclei become more energetic, and the pressure due to those thermal motions becomes greater and greater. Within seconds of helium ignition, the thermal pressure increases until it is no longer smaller than the degeneracy pressure. Then the helium core explodes, as illustrated in **Figure 16.8**. Because the explosion is contained within the star, however, it cannot be seen outside the star. The energy released in that

Figure 16.8 At the end of its life, a low-mass star travels a complex path on the H-R diagram. The first part of that path takes it up the red giant branch to a point where helium ignites in a helium flash. After a few hours, the core of the star begins to inflate, ending the helium flash.

runaway thermonuclear explosion lifts the intermediate layers of the star, and as the core expands, the electrons can spread out. The drama is over within a few hours because the expanded helium-fusing core is no longer degenerate, and the star is on its way toward a new equilibrium. The explosion of helium fusion in the core at the top of the red giant branch is called a **helium flash**; however, it is not visible as a "flash" from outside the star.

Energy from helium fusion in the core does not emerge from the red giant as light, so the luminosity does not change. The tremendous energy released instead goes into fighting gravity and puffing up the core. Afterward, the core (which is no longer degenerate) is much larger, so the force of gravity within it and in the surrounding shell is much smaller. Weaker gravity means less weight pushing down on the core and the shell, which means lower pressure. Lower pressure, in turn, slows the nuclear reactions. After the helium flash, core helium fusion keeps the core of the star puffed up, and the star becomes less luminous than it was as a red giant.

The Horizontal Branch

The star takes 100,000 years or so to settle into stable helium fusion. It then spends about 100 million years fusing helium into carbon in a normal, nondegenerate core while hydrogen fuses to helium in a surrounding shell. The star is now about a hundred times less luminous than it was when the helium flash occurred. The lower luminosity means that the outer layers of the star are not as puffed up as they were when the star was a red giant. Gravity becomes stronger than the outward pressure of the escaping radiation and pulls the outer layers back in. The star shrinks, and its surface temperature climbs as gravitational energy is converted to thermal energy. The star moves horizontally across the H-R diagram, remaining at the same luminosity but increasing in surface temperature. That portion of a star's path on the H-R diagram is called the horizontal branch, shown in **Figure 16.9**. At that point in their evolution, low-mass stars with chemical compositions similar to that of the Sun lie on the H-R diagram just to the left of the red giant branch. Stars that contain much less iron than the Sun tend to distribute themselves away from the red giant branch along the horizontal branch.

The structure and behavior of a star on the horizontal branch are similar to those of a main-sequence star. However, instead of fusing hydrogen into helium in the core, the horizontal branch star fuses helium into carbon in the core, with a shell of hydrogen fusion around that core.

The star's time on the horizontal branch is much shorter than its time on the main sequence. The star is more luminous than it was on the main sequence, so it is consuming fuel more rapidly. As it fuses, helium releases less energy than hydrogen, so the star has to fuse even faster to maintain equilibrium. The horizontal branch star remains stable for about 100 million years.

The temperature at the center of a horizontal branch star is not high enough for carbon to fuse, so carbon builds up in the heart of the star. When the horizontal branch star has fused all the helium at its core, gravity once again begins to overwhelm the pressure of the escaping radiation. The nonfusing carbon core is crushed by the weight of the layers of the star above it until once again the electrons in the core are packed together as tightly as possible, given its pressure. The

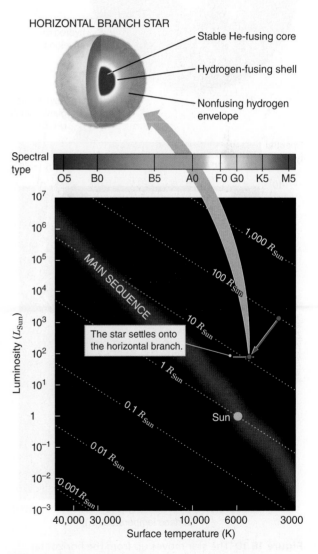

Figure 16.9 The star moves down from the red giant branch onto the horizontal branch. For about 100 million years or less, the star will remain on the horizontal branch and fuse helium in its core and hydrogen in a surrounding shell.

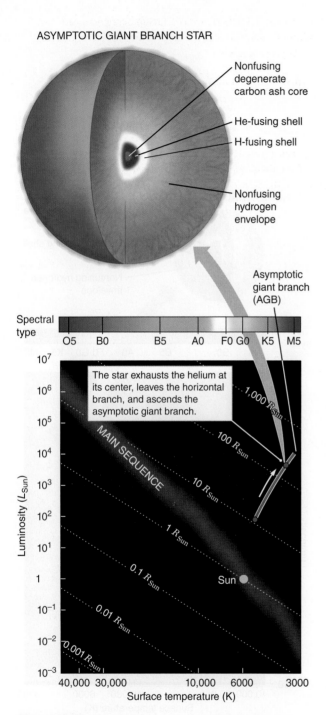

ASYMPTOTIC GIANT BRANCH STAR

Nonfusing degenerate carbon ash core

He-fusing shell

H-fusing shell

Nonfusing hydrogen envelope

Asymptotic giant branch (AGB)

The star exhausts the helium at its center, leaves the horizontal branch, and ascends the asymptotic giant branch.

MAIN SEQUENCE

1,000 R_{Sun}

100 R_{Sun}

10 R_{Sun}

1 R_{Sun}

0.1 R_{Sun}

Sun

0.01 R_{Sun}

0.001 R_{Sun}

Figure 16.10 The star moves up from the horizontal branch onto the asymptotic giant branch (AGB). An AGB star consists of a degenerate carbon core surrounded by helium-fusing and hydrogen-fusing shells. As the carbon core grows, the star brightens, accelerating up the AGB just as it earlier accelerated up the red giant branch while its degenerate helium core grew.

carbon core becomes electron-degenerate, with physical properties much like those of the degenerate helium core at the center of a red giant.

CHECK YOUR UNDERSTANDING 16.3

Stars begin fusing helium to carbon when the temperature rises in the core. That temperature increase is caused by (choose all that apply): (a) gravitational collapse; (b) fusion of hydrogen into helium in the core; (c) fusion of hydrogen into helium in a shell around the core; (d) electron degeneracy pressure.

16.4 Dying Stars Shed Their Outer Layers

After the horizontal branch, small changes in the properties of a star—mass, chemical composition, strength of the magnetic field, and even the rate of rotation—can lead to noticeable differences in how the star (especially its outer envelope) evolves. In this section, we follow a 1-M_{Sun} star with solar composition as it concludes its evolutionary stages, losing its outer layers and leaving behind a cooling carbon core.

Stellar-Mass Loss and the Asymptotic Giant Branch

After the horizontal branch, a small, dense, electron-degenerate carbon core remains. That core is very compact, causing the gravity in the inner parts of the star to be very high, which in turn drives up the pressure, which speeds up the nuclear reactions, which causes the degenerate core to grow more rapidly. Those internal changes should sound familiar—they are similar to the changes that took place at the end of the star's main-sequence lifetime, and the path the star follows as it leaves the horizontal branch parallels that earlier phase of evolution. Just as the star accelerated up the red giant branch as its degenerate helium core grew, the star now leaves the horizontal branch and once again begins to grow larger, redder, and more luminous as its degenerate carbon core grows. As shown in **Figure 16.10**, the path that the star follows, called the **asymptotic giant branch (AGB)** of the H-R diagram, approaches the red giant branch as the star grows more luminous. An AGB star fuses helium and hydrogen in nested concentric shells around a degenerate carbon core as the star moves once again up the H-R diagram.

AGB stars are huge objects. When the Sun becomes an AGB star, its outer layers will engulf the orbits of the inner planets, possibly including Earth and maybe even Mars. When a star expands to such a size, the gravitational force at its surface is only 1/10,000 as strong as the gravity at the surface of the present-day Sun. Pushing surface material away from the star thus takes little extra energy. Before the temperature in the carbon core becomes high enough for carbon to fuse, the radiation pressure pushing outward on the outer layers of the star becomes greater than the gravitational pull inward. Those outer layers begin to drift away into space in a process called **stellar-mass loss.** That mass loss actually begins when the star is still on the red giant branch: by the time a 1-M_{Sun} main-sequence star reaches the horizontal branch, it may have lost 10–20 percent of its total mass. As the star ascends the asymptotic giant branch, it loses another 20 percent or even more of its total mass. By the time it is well up on that branch, a star that began as a 1-M_{Sun} star may have lost more

16.2 Working It Out Escaping the Surface of an Evolved Star

Why are giant stars likely to lose mass? The escape velocity from the surface of a planet or star was given in Working It Out 4.2:

$$v_{esc} = \sqrt{\frac{2GM}{R}}$$

How does v_{esc} change when a star becomes a red giant? Let's look at the Sun as an example. When the Sun is on the main sequence, the escape velocity from its surface can be calculated by using $M_{Sun} = 1.99 \times 10^{30}$ kg, $R_{Sun} = 6.96 \times 10^5$ km, and $G = 6.67 \times 10^{-20}$ km³/(kg s²):

$$v_{esc} = \sqrt{\frac{2 \times [6.67 \times 10^{-20}\ \text{km}^3/(\text{kg s}^2)] \times (1.99 \times 10^{30}\ \text{kg})}{6.96 \times 10^5\ \text{km}}}$$

$$v_{esc} = \sqrt{3.81 \times 10^5\ \text{km}^2/\text{s}^2} = 618\ \text{km/s}$$

What will the escape velocity be when the Sun becomes a red giant, with a radius 50 times greater than the radius it has today and a mass 0.9 times its present mass?

$$v_{esc} = \sqrt{\frac{2 \times [6.67 \times 10^{-20}\ \text{km}^3/(\text{kg s}^2)] \times 0.9 \times (1.99 \times 10^{30}\ \text{kg})}{50 \times (6.96 \times 10^5\ \text{km})}}$$

$$v_{esc} = \sqrt{6.86 \times 10^3\ \text{km}^2/\text{s}^2} = 83\ \text{km/s}$$

The escape velocity from the surface of a red giant star is only 13 percent that of a main-sequence star. That is part of the reason that red giant and AGB stars lose mass. The Sun may eventually lose half its mass.

than half its original mass. Stellar-mass loss is further explored in **Working It Out 16.2**.

Mass loss on the AGB can be spurred on by the star's unstable interior. The extreme sensitivity of the triple-alpha process to temperature in the core can lead to episodes of fast fusion and rapid energy release, which can provide the extra kick needed to expel material from the star's outer layers. Even stars that are initially similar can behave very differently when they reach that stage in their evolution.

Planetary Nebula

Toward the end of an AGB star's life, mass loss itself becomes a runaway process. When a star loses a bit of mass from its outermost layers, the weight pushing down on the underlying layers of the star is reduced. Without that weight holding them down, the remaining outer layers of the star puff up further. The post-AGB star, which is now both less massive and larger, is even less tightly bound by gravity, so less energy is needed to push its outer layers away. Mass loss leads to weaker gravity, which leads to faster mass loss, which leads to weaker gravity, and so on. By the time the last layers are lost, much of the remaining mass of the star is ejected into space, typically at speeds of 20–30 kilometers per second (km/s).

After ejection of its outer layers, all that is left of the low-mass star is a tiny, very hot, electron-degenerate carbon core surrounded by a thin envelope in which hydrogen and helium are still fusing. The star is now less luminous than when it was at the top of the AGB, but it is still much more luminous than a horizontal branch star. The remaining hydrogen and helium in the star rapidly fuse to carbon, and as more and more of the mass of the star ends up in the carbon core, the star itself shrinks and becomes hotter and hotter. Over the course of only about 30,000 years after the beginning of runaway mass loss,

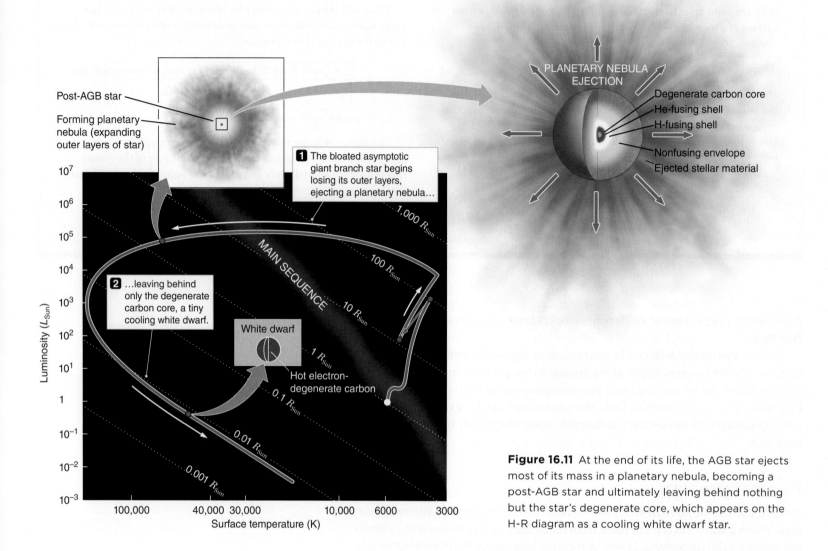

Post-AGB star

Forming planetary nebula (expanding outer layers of star)

1 The bloated asymptotic giant branch star begins losing its outer layers, ejecting a planetary nebula…

2 …leaving behind only the degenerate carbon core, a tiny cooling white dwarf.

MAIN SEQUENCE

White dwarf

Hot electron-degenerate carbon

Luminosity (L_{Sun})

Surface temperature (K)

$1,000\ R_{Sun}$
$100\ R_{Sun}$
$10\ R_{Sun}$
$1\ R_{Sun}$
$0.1\ R_{Sun}$
$0.01\ R_{Sun}$
$0.001\ R_{Sun}$

PLANETARY NEBULA EJECTION

Degenerate carbon core
He-fusing shell
H-fusing shell
Nonfusing envelope
Ejected stellar material

Figure 16.11 At the end of its life, the AGB star ejects most of its mass in a planetary nebula, becoming a post-AGB star and ultimately leaving behind nothing but the star's degenerate core, which appears on the H-R diagram as a cooling white dwarf star.

the star moves from right to left across the top of the H-R diagram, as shown in **Figure 16.11**.

The surface temperature of the star may eventually rise above 100,000 K. At such temperatures, the peak wavelength of the radiation is in the high-energy ultraviolet (UV) part of the spectrum, as determined by Wien's law. That intense UV light heats and ionizes the ejected, expanding shell of gas, causing it to glow in the same way that UV light from an O star causes an H II region to glow. When those glowing shells were first observed in small telescopes, they were named "planetary nebulae" because they appeared fuzzy like nebular clouds of dust and gas, but they were approximately round, like planets. Later imagery, such as the images in **Figure 16.12**, showed that those objects are not like planets at all. Rather, a **planetary nebula** is the remaining outer layers of a star, ejected into space at the end of the star's ascent of the AGB. A planetary nebula may be visible for 50,000 years or so before the gas ejected by the star disperses so far that the

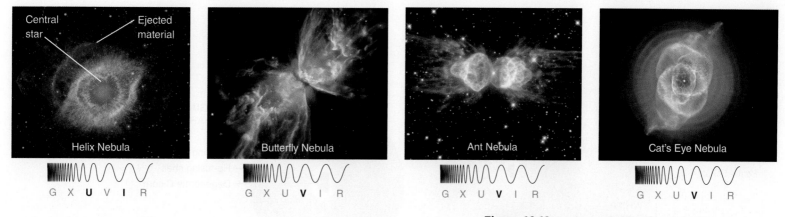

Central star · Ejected material

Helix Nebula

G X **U** V I R

Butterfly Nebula

G X **U** V I R

Ant Nebula

G X **U** V I R

Cat's Eye Nebula

G X **U** V I R

Figure 16.12 At the end of its life, a low-mass star ejects its outer layers and forms a planetary nebula consisting of an expanding shell of gas surrounding the white-hot remnant of the star. Planetary nebulae are not all simple spherical shells around their parent stars. These images of planetary nebulae from the Hubble Space Telescope and the Spitzer Space Telescope show the wealth of structures that result from the complex processes by which low-mass stars eject their outer layers.

nebula is too faint to be seen. Not all stars form planetary nebulae. Stars more massive than about 8 M_{Sun} pass through the post-AGB stage too quickly. Stars without enough mass take too long in the post-AGB stage, so their envelope evaporates before they can illuminate it. Astronomers do not know whether our own Sun will retain enough mass during its post-AGB phase to form a planetary nebula.

The detailed structure of a planetary nebula may contain concentric rings of varied density, indicating that the rate of stellar-mass loss varied—sometimes faster, sometimes slower. In the image of the Ring Nebula in the chapter-opening photograph, the colored rings are emission lines from ions located at different places in the nebula. When that nebula was forming, a lot of mass was lost nearly all at once. Then the mass loss ceased, resulting in a hollow shell around the central star. As the light makes its way out from the central star, higher-energy light gets "used up" in ionizing the inner layers. Outer layers have emission lines from atoms that require less energy to ionize. In other planetary nebulae, the material may be concentrated parallel to the equator or poles of the star, indicating that the stellar-mass loss was blocked in some directions. You can see that effect in the images of the Butterfly, Ant, and Cat's Eye nebulae in Figure 16.12.

The gas in a planetary nebula carries the chemical elements from the star's outer layers off into interstellar space. Planetary nebulae often show a greater percentage of elements such as carbon, nitrogen, and oxygen than the percentage of those elements present in the outer layers of the Sun. Those elements are by-products of nuclear fusion either from the star that produced the planetary nebula or from the stars of earlier generations. Once that chemically enriched material leaves the star, it mixes with interstellar gas, increasing the chemical diversity of the interstellar medium.

White Dwarfs

Within about 50,000 years, a post-AGB star fuses all the fuel remaining on its surface, leaving behind a nonfusing ball of carbon with a mass less than 70 percent of the mass of the original star. As that occurs, the post-AGB star becomes smaller and fainter but remains about the same temperature. Its position on the H-R diagram falls down the left side. Within a few thousand years, it shrinks to about the size of Earth, at which point it has become fully electron-degenerate and can shrink no further. That remnant of stellar evolution is called a **white dwarf**.

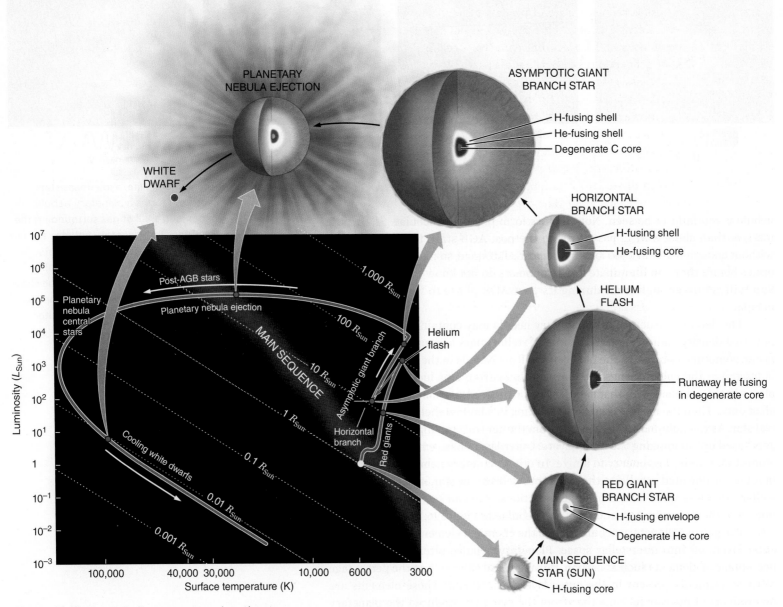

Figure 16.13 This H-R diagram summarizes the stages in the post-main-sequence evolution of a 1-M_{Sun} star.

The white dwarf, composed of nonfusing electron-degenerate carbon and maybe some oxygen, continues to radiate energy away into space. As it does so it cools, just like the heating coil on an electric stove once it is turned off. Because the white dwarf is electron-degenerate, its size remains nearly constant as it cools, so it follows a line of constant radius down and to the right on the H-R diagram. The white dwarf will remain very hot for awhile but then will slowly cool. Its tiny size means it is a thousand times less luminous than a main-sequence star like the Sun. Many white dwarfs are known, but none can be seen without a telescope. Sirius, the brightest star in Earth's sky, has a faint white dwarf as a binary companion.

Figure 16.13 summarizes the evolution of a solar-type, 1-M_{Sun} main-sequence star through its final existence as a ~0.6-M_{Sun} white dwarf. The star leaves the main sequence, climbs the red giant branch, falls to the horizontal branch, climbs back up the AGB, takes a left across the top of the diagram while ejecting a planetary nebula, and finally falls to its final resting place in the bottom left of the diagram. That process is representative of the fate of low-mass stars. Although every low-mass star forms a white dwarf at the end point of its evolution, the precise path a low-mass star follows from core hydrogen fusion on the main sequence to white dwarf depends on many details particular to the star.

Some stars less massive than the Sun may become white dwarfs composed largely of helium rather than carbon. Conversely, temperatures in the cores of evolved 2- to 3-M_{Sun} stars are high enough to allow additional nuclear reactions to occur, leading to the formation of more massive white dwarfs composed of materials such as oxygen, neon, and magnesium. Differences in chemical composition of a star can also lead to differences in its post-main-sequence evolution.

In our discussion of stellar evolution, we have focused on what happens after a star leaves the main sequence. The spectacle of a red giant or AGB star is ephemeral. Once it leaves the main sequence, the Sun will travel the path from red giant to white dwarf in less than one-tenth of the time it spent on the main sequence, steadily fusing hydrogen to helium in its core. Stars spend most of their luminous lifetimes on the main sequence, which is why most of the stars in the sky are main-sequence stars. The fainter white dwarfs constitute the final resting place for most of the stars that have been or ever will be formed.

The Fate of the Planets

What happens to the planets orbiting a low-mass star as it goes through those post-main-sequence stages? Because the first decade of extrasolar planet discoveries yielded primarily planets with orbits very close to their respective stars, astronomers assumed that any planets closer than 1–2 astronomical units (AU) would not survive the post-main-sequence expansion of the star. Therefore, they did not expect to find planets in orbit around evolved stars. However, many planets have been discovered in orbit around red giants, AGB stars, and horizontal branch stars.

Astronomers cannot be sure whether the extrasolar planets they observe have remained in the same orbital locations as when their stars were on the main sequence or whether the planets migrated to different orbits. The surface gravity of a red giant is low, and some of the mass in its outermost layers will blow away. That decrease in mass reduces the gravitational force between the star and its planets, which could lead to planetary orbits evolving outward away from the star. In some models of planetary migration, the tidal forces between planets or between a star and a planet are significant factors too, suggesting that some planetary orbits could evolve inward.

As a star loses mass during the evolutionary process, stellar or planetary companions may change their orbits. Rocky asteroids or smaller planets could migrate inward past the Roche limit (see Chapter 4) of the white dwarf and break up. Some of the material could remain in orbit around the white dwarf as a debris disk, similar to those seen in main-sequence stars with a planetary system. Those dusty debris disks have been observed around white dwarf stars with the Hubble Space Telescope and the Spitzer Space Telescope. Some of that dusty or rocky material may also fall onto the white dwarf, "polluting" its spectrum with heavy elements that were not produced in the stellar core.

What will happen to Earth? The age of the Sun as determined from radioactive dating of meteorites indicates that the Solar System formed 4.6 billion years ago. In Chapter 14, we estimated how long the Sun might last by calculating its rate of hydrogen fusion. The Sun's luminosity is about 30 percent higher now than it was early in the history of the Solar System, and it will continue to increase steadily over the rest of the Sun's main-sequence lifetime of another 5 billion years or so. The Sun's luminosity may increase enough—even while the Sun is a main-sequence star—that Earth will heat up to the point where the oceans evaporate, perhaps as soon as 1 billion to 2 billion years from now. By the time the Sun is a red giant, it might be impossible for any planet within the orbit of Jupiter to maintain liquid water on the surface.

Scientists are not certain whether the radius of the red giant Sun will extend past Earth to engulf the planet. The red giant Sun will have low surface gravity and will lose mass, which could cause Earth's orbit to expand, thereby enabling Earth to escape the encroaching solar surface. Alternatively, as the Sun expands in radius and its rotation rate slows (see Working It Out 7.1), tidal forces might pull Earth inward. The solar core will eventually become a white dwarf, perhaps with a dusty disk and a "polluted" atmosphere as the only remaining evidence of our rocky planet.

CHECK YOUR UNDERSTANDING 16.4

A planetary nebula forms from: (a) the ejection of mass from a low-mass star; (b) the collision of planets around a dying star; (c) the collapse of the magnetosphere of a high-mass star; (d) the remainders of the original star-forming nebula.

16.5 Binary Star Evolution

So far in this chapter, we have discussed the evolution of single stars in isolation. But many stars are members of binary systems. Those systems may form when a molecular-cloud core has too much angular momentum for a single star to form. Binary systems are common—as many as half of all stars may be in binary systems, and that fraction is largest among more massive stars. How is evolution different for stars in those systems? Sometimes, if the stars are close together and one star is more massive, their evolution after the main sequence may be linked. In this section, we will trace the steps from binary star system through the end point of their evolution: nova or supernova.

Mass Flows from an Evolving Star onto Its Companion

Think for a moment about what would happen if you were to travel in a spacecraft from Earth toward the Moon. When you are still near Earth, the force of Earth's gravity is far stronger than that of the Moon. As you move away from Earth and closer to the Moon, the gravitational attraction of Earth weakens, and the gravitational attraction of the Moon becomes stronger. You eventually reach an intermediate zone where neither body has the stronger pull. If you continue beyond that point, the lunar gravity begins to dominate until you find yourself firmly in the gravitational grip of the Moon. The regions surrounding the two objects—their gravitational domains—are called the **Roche lobes** of the system.

The same situation exists between two stars, as shown in **Figure 16.14**. Gas near each star clearly belongs to that star. When one star leaves the main sequence and swells up, its outer layers may cross that gravitational dividing line separating the star from its companion. Once a star expands past the boundary of its

Roche lobe, its material begins to fall onto the other star. That transfer of material from one star to the other is called **mass transfer**.

Evolution of a Close Binary System

The best way to understand how mass transfer affects the evolution of stars in a binary system is to apply what is known from studying the evolution of single low-mass stars. Figure 16.14a shows a binary system consisting of two low-mass stars: star 1 is more massive, and star 2 is less massive. That is an ordinary binary system, and each star is an ordinary main-sequence star for most of the system's lifetime.

More massive main-sequence stars evolve more rapidly than less massive main-sequence stars. Therefore, star 1 will be the first to use up the hydrogen at its center and begin to evolve off the main sequence, as shown in Figure 16.14b. If the two stars are close enough to each other, star 1 will eventually grow to overfill its Roche lobe, and material will transfer onto star 2, as shown in Figure 16.14c. The transfer of mass between the two stars results in a "drag" that causes the orbits of the two stars to shrink, bringing the stars closer together and further enhancing mass loss. In addition, as star 1 loses mass, its Roche lobe shrinks, further enhancing the mass transfer. The two stars can even reach the point at which they are effectively two cores sharing the same extended envelope of material, called a *common envelope*.

Despite those complexities, star 2 probably remains a basically normal main-sequence star throughout the process, fusing hydrogen in its core. However, over time, the mass of star 2 increases because of the accumulation of material from its companion. As it does so, the structure of star 2 must change to accommodate its new status as a higher-mass star. If we plotted star 2's position on the H-R diagram during that period, we would see it move up and to the left along the main sequence, becoming larger, hotter, and more luminous.

Star 1, because it is losing mass to star 2, never grows larger than its Roche lobe, so it does not become an isolated red giant or AGB star at the top of the H-R diagram. Yet star 1 continues to evolve, fusing helium in its core on the horizontal branch, proceeding through helium shell fusion, and finally losing its outer layers and leaving behind a white dwarf. Figure 16.14d shows the binary system after star 1 has completed its evolution. All that remains of star 1 is a white dwarf, orbiting about star 2, its bloated main-sequence companion.

Novae

As star 2 begins to evolve off the main sequence, it expands to fill its Roche lobe, as shown in Figure 16.14e. Like star 1 before it, star 2 grows to fill its Roche lobe: material from star 2 begins to pour through the "neck" connecting the Roche lobes of the two stars. However, this time the mass is not being added to a normal star but is drawn toward the tiny white dwarf left behind by star 1. Because the system is revolving and the white dwarf is so small, the infalling material generally misses the star, instead landing on an accretion disk around the white dwarf. That disk is similar to the accretion disk that forms around a protostar. As with star formation, the accretion disk accumulates material that has too much angular momentum to hit the white dwarf directly.

A white dwarf has a mass comparable to that of the Sun but a size comparable to that of Earth. Having a large mass and a small radius means strong

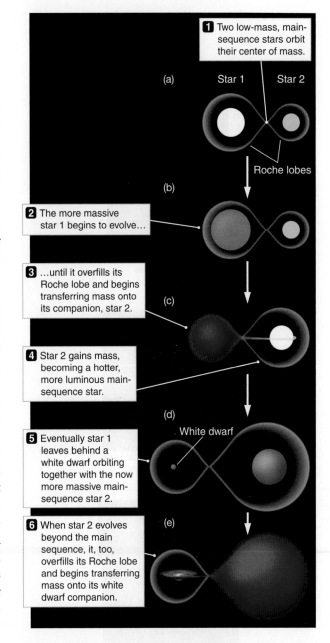

Figure 16.14 A compact binary system consisting of two low-mass stars that evolve through stages, transferring mass back and forth.

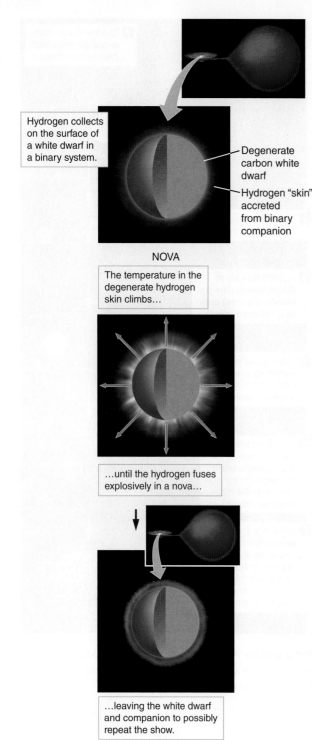

Hydrogen collects on the surface of a white dwarf in a binary system.

Degenerate carbon white dwarf

Hydrogen "skin" accreted from binary companion

NOVA

The temperature in the degenerate hydrogen skin climbs…

…until the hydrogen fuses explosively in a nova…

…leaving the white dwarf and companion to possibly repeat the show.

Figure 16.15 In a binary system in which mass is transferred onto a white dwarf, a layer of hydrogen builds up on the surface. That hydrogen may ignite on the surface of the white dwarf, producing a nova.

surface gravity. The material streaming toward the white dwarf in the binary system falls into an incredibly deep gravitational "well." The depth of that well affects the amount of energy with which matter impacts the white dwarf. A kilogram of material falling from space onto the surface of a white dwarf releases 100 times more energy than a kilogram of material falling from the outer Solar System onto the surface of the Sun. All that energy is turned into thermal energy. The spot where the stream of material from star 2 hits the accretion disk can be heated to millions of kelvins, where it glows in the far-UV and X-ray parts of the electromagnetic spectrum.

The infalling material accumulates on the surface of the white dwarf (**Figure 16.15**), where the enormous gravitational pull of the white dwarf compresses the material to a density close to that of the white dwarf itself. As more and more material builds up on the surface of the white dwarf, the white dwarf shrinks (just as the core of a red giant shrinks as it grows more massive). The density increases more and more while the release of gravitational energy drives the temperature of the white dwarf higher and higher. The infalling material comes from the outer, unfused layers of star 2, so it is composed mostly of hydrogen. That hydrogen is compressed to higher and higher densities and heated to higher and higher temperatures on the surface of the white dwarf.

Once the temperature at the base of the white dwarf's surface layer of hydrogen reaches about 10 million K, that hydrogen begins to fuse to helium. But that process is not the contained hydrogen fusion that takes place in the center of the Sun; it is explosive hydrogen fusion in a degenerate gas. Energy released by hydrogen fusion drives up the temperature. Because the surface is degenerate, that rising temperature does not cause an expansion, as in normal matter, but instead drives up the rate of hydrogen fusion. That runaway thermonuclear reaction is much like the runaway helium fusion that takes place during the helium flash, except now no overlying layers of a star are there to absorb the energy liberated by fusion. The result is a tremendous explosion that blows part of the layer covering the white dwarf out into space at speeds of thousands of kilometers per second, as shown in Figure 16.15. An exploding white dwarf of that kind is called a **nova**.

The explosion of a nova does not destroy the underlying white dwarf star. In fact, much of the material that had built up on the white dwarf may remain behind after the explosion. Afterward, the binary system is in much the same configuration as before: material from star 2 is still pouring onto the white dwarf. The nova can repeat many times as material builds up and ignites again and again on the surface of the white dwarf. If the underlying white dwarf is old and cooler, or the mass accumulates slowly on its surface, outbursts are separated by thousands of years, so most novae have been seen only once in historical times. But if the white dwarf is hotter, or mass is transferring quickly, such explosions can happen every few years or even every few months. Such recurring novae are called dwarf novae.

About 50 novae occur in our galaxy each year. About 10 of those can be detected with telescopes, but very few become visible to the eye. The rest are blocked from view by dust in the disk of our galaxy. Novae reach their peak brightness in only a few hours, and briefly they can be several hundred thousand times more luminous than the Sun. Although the brightness of a nova sharply drops in the weeks after the outburst, it can sometimes still be seen for years. During that time, the glow from the expanding cloud of ejected material is caused by the decay of radioactive isotopes created in the explosion. Large bursts of gamma rays have been observed to be associated with several novae.

Supernovae

A white dwarf with a mass greater than 1.4 M_{Sun} cannot remain stable. That value is referred to as the Chandrasekhar limit, named for Subrahmanyan Chandrasekhar (1910–1995), who derived it. Above that mass, even the pressure supplied by degenerate electrons is no longer enough to balance gravity, so the white dwarf will collapse. A white dwarf that is accumulating mass, however, likely does not reach the Chandrasekhar limit. As the star reaches about 1.38 M_{Sun}, core pressures and temperatures rise enough to ignite carbon and begin a simmering phase that holds off thermonuclear runaway for a while. Once the temperature reaches about 1.0×10^8 K, the runaway carbon fusion involves the entire white dwarf. Within about a second, the whole white dwarf is consumed in the resulting explosion, known as a **supernova**. In that instant, 100 times more energy is liberated than the Sun will give off over its 10-billion-year lifetime on the main sequence. Runaway fusion reactions convert a large fraction of the mass of the star into elements such as iron and nickel, and the explosion blasts the shards of the white dwarf into space at top speeds in excess of 20,000 km/s, enriching the interstellar medium with those heavier elements. The explosion is known as a **Type Ia supernova**.

How might a white dwarf gain mass and explode in that way? Three options exist, as shown in **Figure 16.16**.

First, in the binary system discussed above, star 2 eventually may simply go on to form a white dwarf, leaving behind a stable binary system consisting of two white dwarfs, as in Figure 16.16a. Those two white dwarfs may then eventually spiral together and merge. If the sum of their masses is greater than 1.4 M_{Sun}, the resulting merged star will explode. That explosion destroys both the original white dwarfs. The amount of mass involved in the explosion may range from 1.4 to 2.8 M_{Sun}. A majority of Type Ia supernovae may be of that type.

Second, star 2 may lose mass to the white dwarf while still a main-sequence star, as in Figure 16.16b. Through millions of years of mass transfer from star 2 onto the white dwarf, and possibly through countless nova outbursts, the white dwarf's mass slowly increases—until it approaches the Chandrasekhar limit and later explodes. Here, the white dwarf's mass is always almost exactly 1.4 M_{Sun}. Only one object of that type has so far been observed, from a pulse of blue light emitted when the white dwarf exploded and heated up star 2.

Third, star 2 may evolve off the main sequence to become a red giant, filling its Roche lobe. The material from that star flows onto the white dwarf over millions of years, as shown in Figure 16.16c. As in the above case, the white dwarf approaches the Chandrasekhar limit and then explodes. Here, the white dwarf's mass is almost exactly 1.4 M_{Sun} every time. A red giant companion has not yet been observed in a Type Ia supernova.

In a galaxy the size of the Milky Way, Type Ia supernovae occur about once a century. They can briefly shine with a luminosity billions of times that of our Sun, possibly outshining the galaxy itself. Those objects are particularly useful to astronomers because their luminosities can be approximately determined from a careful study of their light curves. Because that type of supernova happens during the explosion of an object with a mass between 1.4 and 2.8 M_{Sun}, the total energy involved will vary by at most a factor of 2. Therefore, the luminosity of Type Ia supernovae should vary by no more than a factor of 2. As we will see in the next few chapters, objects with known luminosity can be used to find distances to far away galaxies. Combining the luminosity of a Type Ia supernova with its

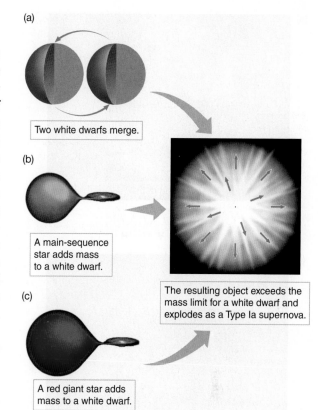

(a)

Two white dwarfs merge.

(b)

A main-sequence star adds mass to a white dwarf.

The resulting object exceeds the mass limit for a white dwarf and explodes as a Type Ia supernova.

(c)

A red giant star adds mass to a white dwarf.

Figure 16.16 A Type Ia supernova results when a white dwarf exceeds the mass limit. That can happen because (a) two white dwarfs merge; (b) mass from a main-sequence companion falls onto the white dwarf, increasing its mass to the limit; or (c) mass from an evolved companion falls onto the white dwarf, increasing its mass to the limit.

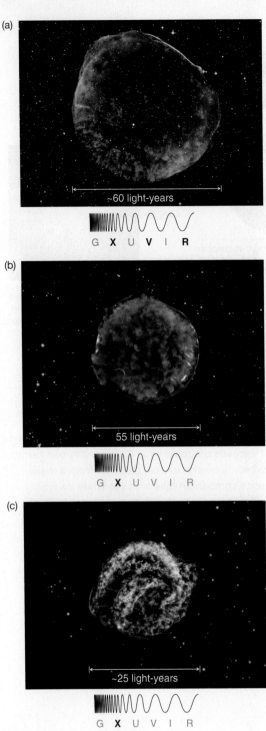

Figure 16.17 These images show the remnants of Type Ia supernovae. The material heated by the expanding blast wave from a supernova glows in the X-ray portion of the spectrum. (a) SN 1006 is the brightest recorded supernova and was observed in China, Japan, Europe, and the Middle East in 1006. X-ray data are shown in blue; radio data are shown in red. (b) Tycho's supernova was observed in 1572. Low-energy X-rays are red; high-energy X-rays are blue. (c) Kepler's supernova, observed in 1604, is shown in five X-ray wavelengths.

apparent brightness gives the distance to the host galaxy. Type Ia supernovae are still an active area of exploration in astronomy (see the **Process of Science Figure**).

Those explosions leave behind expanding shells of dust and gas called **supernova remnants**, as shown in **Figure 16.17**. The material is heated by the expanding blast wave from the supernova and glows in X-rays.

CHECK YOUR UNDERSTANDING 16.5

A white dwarf will become a supernova if: (a) the original star was more than 1.38 M_{Sun}; (b) it accretes an additional 1.38 M_{Sun} from a companion; (c) some mass falls onto it from a companion; (d) enough mass accretes from a companion to give the white dwarf a total mass of 1.38 M_{Sun}.

....

Origins Stellar Lifetimes and Biological Evolution

From fossil records and DNA analysis, scientists estimate that life took hold on Earth within 1 billion years after the Solar System and Earth formed 4.6 billion years ago. It took another 1.5 billion years for more complex cells to develop and another billion years to develop multicellular life. The first animals didn't appear on Earth until 600 million years ago, 4 billion years after the formation of the Sun and the Solar System.

The only known example of biology is the life on Earth. Extrapolating from one data point is always risky, and it is not known whether biology is widespread in the universe or whether Earth's biological timeline is typical. Still, reasoning from our one example is the only way to begin thinking about life in the universe. How does the preceding timeline of the evolution of life on Earth compare with the lifetimes of main-sequence stars? Table 16.1 indicates that the lifetime of an O5 star is less than half a million years; of a B5 star, about 120 million years; and of an A0 star, about 700 million years. Those stars would have run out of hydrogen in the core and started post-main-sequence evolution in less time than Earth took to settle down after its periods of heavy bombardment by debris early in the Solar System's history.

The 1-billion-year main-sequence lifetime of an A2 star corresponds to the time the simplest life-forms took to develop on Earth. The 3-billion-year lifetime for F0 stars corresponds to the time photosynthetic bacteria took to develop on Earth. Only stars cooler and less massive than F5 stars have main-sequence lifetimes longer than the 4 billion years life took to evolve into animals on Earth. Thus, searches for extrasolar planets survey stars that are F5 or cooler—because hotter and more massive stars probably don't live long enough on the main sequence for complex life to develop.

After a star leaves the main sequence, the helium-fusing red giant stage is estimated to last for about one-tenth of the main-sequence lifetime, so that doesn't help stars with short lifetimes last long enough for complex biology to evolve. Could life survive the transition of its star to a red giant? As noted earlier in the chapter, even if a planet is not destroyed, its orbit, temperature, and atmospheric conditions will drastically change, and any life might have to relocate if it is to survive.

Science Is Unfinished

Understanding Type Ia supernovae is at the foundation of measuring distances to the farthest galaxies and therefore for our conclusions about the universe as a whole. The observational evidence, though, remains incomplete.

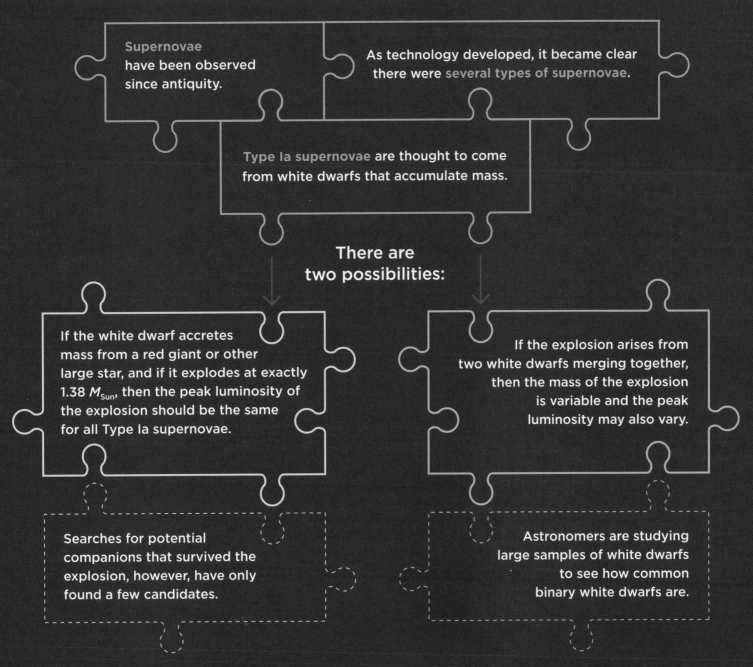

Supernovae have been observed since antiquity.

As technology developed, it became clear there were several types of supernovae.

Type Ia supernovae are thought to come from white dwarfs that accumulate mass.

There are two possibilities:

If the white dwarf accretes mass from a red giant or other large star, and if it explodes at exactly 1.38 M_{Sun}, then the peak luminosity of the explosion should be the same for all Type Ia supernovae.

If the explosion arises from two white dwarfs merging together, then the mass of the explosion is variable and the peak luminosity may also vary.

Searches for potential companions that survived the explosion, however, have only found a few candidates.

Astronomers are studying large samples of white dwarfs to see how common binary white dwarfs are.

When observational evidence is inconclusive, each scientist adds a piece to the puzzle. Some will conduct larger observational studies, whereas others will create new theoretical models. Eventually, these efforts bring clarity to a confusing situation.

Scientists report that the spectra of some dead stars show evidence of rocky material left over from planetary systems. ⊗

Scientists Solve Riddle of Celestial Archaeology

By University of Leicester Press Office

A decades old space mystery has been solved by an international team of astronomers led by Professor Martin Barstow of the University of Leicester and President-elect of the Royal Astronomical Society.

Scientists from the University of Leicester and University of Arizona investigated hot, young, white dwarfs—the super-dense remains of Sun-like stars that ran out of fuel and collapsed to about the size of the Earth. Their research is featured in MNRAS—the *Monthly Notices of the Royal Astronomical Society*, published by Oxford University Press.

It has been known that many hot white dwarf atmospheres, essentially of pure hydrogen or pure helium, are contaminated by other elements—like carbon, silicon, and iron. What was not known, however, was the origins of these elements, known in astronomical terms as metals.

"The precise origin of the metals has remained a mystery and extreme differences in their abundance between stars could not be explained," said Professor Barstow, a Pro-Vice-Chancellor at the University of Leicester whose research was assisted by his daughter Jo, a coauthor of the paper, during a summer work placement in Leicester. She has now gone on to be an astronomer working in Oxford—on extrasolar planets.

"It was believed that this material was 'levitated' by the intense radiation from deeper layers in the star," said Professor Barstow.

Now the researchers have discovered that many of the stars show signs of contamination by rocky material, the leftovers from a planetary system.

The researchers surveyed 89 white dwarfs, using the Far Ultraviolet Spectroscopic Explorer to obtain their spectra (dispersing the light by color) in which the "fingerprints" of carbon, silicon, phosphorus, and sulfur can be seen when these elements are present in the atmosphere.

"We found that in stars with polluted atmospheres the ratio of silicon to carbon matched that seen in rocky material, much higher than found in stars or interstellar gas.

"The new work indicates that around one-third of all hot white dwarfs are contaminated in this way, with the debris most likely in the form of rocky minor planet analogs. This implies that a similar proportion of stars like our Sun, as well as stars that are a little more massive like Vega and Fomalhaut, build systems containing terrestrial planets. This work is a form of celestial archaeology where we are studying the 'ruins' of rocky planets and/or their building blocks, following the demise of the main star.

"The mystery of the composition of these stars is a problem we have been trying to solve for more than 20 years. It is exciting to realize that they are swallowing up the leftovers from planetary systems, perhaps like our own, with the prospect that more detailed follow-up work will be able to tell us about the composition of rocky planets orbiting other stars," said Professor Barstow.

The study also points to the ultimate fate of the Earth billions of years from now—ending up as a contamination within the white dwarf Sun.

1. The article refers to the white dwarfs in the study as "hot, young, white dwarfs." What does "young" mean in that context?
2. The spectra described are compared to fingerprints. In what ways are white dwarf spectra like fingerprints?
3. Why were scientists surprised to find elements other than hydrogen and helium in the atmospheres of white dwarf stars? Where do they think those other elements originate?
4. Why does a telescope need to be in space to observe far-UV wavelengths?
5. How common are contaminated white dwarfs in the sample in the study? Compare that with the percentage of Sun-like stars with planets (22 percent). Does the finding in the article seem like a sensible number?

Summary

Stars with masses similar to that of the Sun form planetary nebulae like the one shown in the chapter-opening photograph. That is a likely fate for our own Sun at the end of its evolution. After leaving the main sequence, low-mass stars follow a convoluted path along the H-R diagram that includes the red giant branch, the horizontal branch, the asymptotic giant branch, and a path across the top and then down to the lower left of the diagram. Those stages of evolution are dominated by the balance between gravity and energy production from various fusion processes that "turn on and off" in the core of the star. That entire process takes much less time than the main-sequence lifetime of the star. Stars following that path have main-sequence lifetimes comparable to the evolutionary timescales of life on Earth. If life on Earth is typical, more massive stars with shorter lifetimes will not be stable long enough for complex life to evolve.

LG 1 **Estimate the main-sequence lifetime of a star from its mass.** All stars eventually exhaust their nuclear fuel as hydrogen fuses to helium in the cores of main-sequence stars. Less massive stars exhaust their fuel more slowly and have longer lifetimes than more massive stars.

LG 2 **Explain why low-mass stars grow larger and more luminous as they run out of fuel.** When a low-mass star uses up the hydrogen in its core, it begins to fuse hydrogen in a shell around the core, heating the gaseous interior. The star expands and becomes a red giant.

LG 3 **Sketch post-main-sequence evolutionary tracks on a Hertzsprung-Russell (H-R) diagram and list the stages of evolution for low-mass stars.** After exhausting its hydrogen, a low-mass star leaves the main sequence and swells to become a red giant, with a helium core made of electron-degenerate matter. The red giant fuses helium to carbon via the triple-alpha process, and quickly the core ignites in a helium flash. The star then moves onto the horizontal branch. A horizontal branch star accumulates carbon and sometimes oxygen in its core and then moves up the asymptotic giant branch. The star may become an AGB star and lose some of its mass.

LG 4 **Describe how planetary nebulae and white dwarfs form.** In their dying stages, some stars eject their outer layers to form planetary nebulae. All low-mass stars eventually become white dwarfs, which are very hot but very small.

LG 5 **Explain how some close binary systems evolve differently from single stars.** Transfer of mass within some binary systems can lead to a nuclear explosion. A nova occurs when hydrogen collects and ignites on the surface of a white dwarf in a binary system. If the mass of the white dwarf approaches $1.38\ M_{Sun}$, the entire star may become involved in a Type Ia supernova. That explosion can occur in several ways.

? Unanswered Questions

- Why do planetary nebulae have different shapes? Some are not simply chaotic but are well organized, with various types of symmetry. Some are spherically symmetric (like a ball), some have bipolar symmetry (like a long, hollow tube, pinched in the middle), and some are even point-symmetric (like the letter *S*). How can an essentially spherically symmetric object such as a star produce such beautifully organized outflows? Because stars are three-dimensional objects, and astronomers can view each object from only one direction, researchers can't easily determine how much of that variation is due to orientation and how much is due to actual differences in the shape of the object. For example, a bipolar nebula, viewed from one end, would appear spherically symmetric. That orientation effect, among other problems, complicates efforts to understand how those shapes are formed. No single explanation has yet satisfactorily covered all the object types.

- Could Earth be moved farther from the Sun to accommodate the Sun's inevitable changes in luminosity, temperature, and radius? One proposal suggests that Earth could capture energy from a passing asteroid and migrate outward, thus staying in the habitable zone while moving farther from the Sun as the Sun ages. Or a huge, thin "solar sail" could be constructed so that radiation pressure from the Sun would slowly push Earth into a larger orbit. Such feats of "astronomical engineering" are not feasible anytime soon, but perhaps they could be accomplished in millions of years (when they will be needed).

Questions and Problems

Test Your Understanding

1. Place the main-sequence lifetimes of the following stars in order from shortest to longest.
   a. the Sun: mass, 1 M_{Sun}; luminosity, 1 L_{Sun}
   b. Capella Aa: mass, 3 M_{Sun}; luminosity, 76 L_{Sun}
   c. Rigel: mass, 24 M_{Sun}; luminosity, 85,000 L_{Sun}
   d. Sirius A: mass, 2 M_{Sun}; luminosity, 25 L_{Sun}
   e. Canopus: mass, 8.5 M_{Sun}; luminosity, 13,600 L_{Sun}
   f. Achernar: mass, 7 M_{Sun}; luminosity, 3,150 L_{Sun}

2. Place the following steps in the evolution of a low-mass star in order.
   a. main-sequence star
   b. planetary nebula ejection
   c. horizontal branch
   d. helium flash
   e. red giant branch
   f. asymptotic giant branch
   g. white dwarf

3. If a star follows a horizontal path across the H-R diagram, the star
   a. maintains the same temperature.
   b. stays the same color.
   c. maintains the same luminosity.
   d. keeps the same spectral type.

4. Degenerate matter is different from normal matter because as the mass of degenerate material goes up,
   a. the radius goes down.
   b. the temperature goes down.
   c. the density goes down.
   d. the luminosity goes down.

5. The most massive stars have the shortest lifetimes because
   a. the temperature is higher in the core, so they fuse faster.
   b. they have less fuel in the core when the star forms.
   c. their fuel is located farther from the core.
   d. the temperatures are lower in the core, so they fuse their fuel more slowly.

6. If a main-sequence star suddenly started fusing hydrogen at a faster rate in its core, it would become
   a. larger, hotter, and more luminous.
   b. larger, cooler, and more luminous.
   c. smaller, hotter, and more luminous.
   d. smaller, cooler, and more luminous.

7. A low-mass main-sequence star's climb up the red giant branch is halted by
   a. the end of hydrogen shell burning.
   b. the beginning of helium fusion in the core.
   c. electron-degeneracy pressure in the core.
   d. instabilities in the star's expanding outer layers.

8. Post-main-sequence stars lose up to half their mass because
   a. jets from the poles release material at an increasing rate.
   b. the mass of the star drops because of mass loss from fusion.
   c. the magnetic field causes increasing numbers of coronal mass ejections.
   d. the star swells until the surface gravity is too weak to hold material.

9. A planetary nebula glows because
   a. it is hot enough to emit UV radiation.
   b. fusion is happening in the nebula.
   c. it is heating up the interstellar medium around it.
   d. light from the central star causes emission lines.

10. As an AGB star evolves into a white dwarf, it runs out of nuclear fuel, and one might guess that the star should cool off and move to the right on the H-R diagram. Why does the star move instead to the left?
    a. It becomes larger.
    b. More of the star is involved in fusion.
    c. As outer layers are lost, deeper layers are exposed.
    d. The temperature of the core rises.

11. When compressed, ordinary gas heats up but degenerate gas does not. Why, then, does a degenerate core heat up as the star continues shell fusion around it?
    a. It is heated by the radiation from fusion.
    b. It is heated by the gravitational collapse of the shell.
    c. It is heated by the weight of helium falling on it.
    d. It is insulated by the shell.

12. All Type Ia supernovae
    a. are at the same distance from Earth.
    b. always involve two stars of identical mass.
    c. are extremely luminous.
    d. always release the same amount of energy in fusion.

13. In Latin, *nova* means "new." That word is used for novae and supernovae because they are
    a. newly formed stars.
    b. newly dead stars.
    c. newly visible stars.
    d. new main-sequence stars.

14. When the Sun runs out of hydrogen in its core, it will become larger and more luminous because
    a. it will start fusing hydrogen in a shell around a helium core.
    b. it will start fusing helium in a shell and hydrogen in the core.
    c. infalling material will rebound off the core and puff up the star.
    d. energy balance will no longer hold, and the star will drift apart.

15. A white dwarf is located in the lower left of the H-R diagram. From that information alone, you can determine that the star is
    a. very massive.
    b. very dense.
    c. very hot.
    d. very bright.

Thinking about the Concepts

16. Can a star skip the main sequence and immediately begin fusing helium in its core? Explain your answer.

17. Suppose a main-sequence star suddenly started fusing hydrogen at a faster rate in its core. How would the star react? Discuss changes in size, temperature, and luminosity.

18. Describe some possible ways in which the temperature in the core of a star might increase while the density decreases.

19. Astronomers typically say that the mass of a newly formed star determines its destiny from birth to death. However, that statement is not true for one common environmental circumstance. Identify that circumstance and explain why the birth mass of a star might not fully account for the star's destiny.

20. Study the Process of Science Figure. Suppose that a new mechanism is found to explain Type Ia supernovae. In that mechanism, all Type Ia supernovae are more luminous than previously thought. Would the derived distances to galaxies be larger or smaller than we understand them to be now?

21. Do stars change structure while on the main sequence? Why or why not?

22. Suppose Jupiter is not a planet but a G5 main-sequence star with a mass of 0.8 M_{Sun}.
    a. How will life on Earth be affected, if at all?
    b. How will the Sun be affected as it comes to the end of its life?

23. Explain the similarity in the paths that a star follows along the H-R diagram as it forms from a protostar and as it leaves the main sequence to climb the red giant branch.

24. Why is a horizontal branch star (which fuses helium at a high temperature) less luminous than a red giant branch star (which fuses hydrogen at a lower temperature)?

25. Suppose the core temperature of a star is high enough for the star to begin fusing oxygen. Predict how the star will continue to evolve, including its path on the H-R diagram.

26. Why does a white dwarf move down and to the right along the H-R diagram?

27. Why does fusion in degenerate material always lead to a runaway reaction?

28. Suppose the more massive red giant star in a binary system engulfs its less massive main-sequence companion, and their nuclear cores combine. What structure will the new star have? Where will the star lie on the H-R diagram?

29. T Coronae Borealis is a well-known recurrent nova.
    a. Is it a single star or a binary system? Explain.
    b. What mechanism causes a nova to flare up?
    c. How can a nova flare-up happen more than once?

30. Why do astronomers prefer to search for planets around low-mass stars?

Applying the Concepts

31. Figure 16.1 contains the label "Straight-line (power law) approximation." What does that label tell you about the axes on the graph: are they linear or logarithmic? Explain why those data are plotted that way.

32. Use Figure 16.2 to estimate the percentage of the Sun's mass that is turned from hydrogen into helium over its lifetime.

33. Study Figure 16.13. How many times brighter is a star at the top of the giant branch than the same star (a) when it was on the main sequence and (b) when it was on the horizontal branch?

34. Study Figure 16.13. Make a graph of surface temperature versus time for the evolutionary track shown—from the time the star leaves the main sequence until it arrives at the dot showing that it is a white dwarf. Your time axis may be approximate, but it should show that the star spends different amounts of time in the different phases.

35. For most stars on the main sequence, luminosity scales with mass as $M^{3.5}$ (see Working It Out 16.1). What luminosity does that relationship predict for (a) 0.5-M_{Sun} stars, (b) 6-M_{Sun} stars, and (c) 60-M_{Sun} stars? Compare those numbers with the values in Table 16.1.

36. Calculate the main-sequence lifetimes for (a) 0.5-M_{Sun} stars, (b) 6-M_{Sun} stars, and (c) 60-M_{Sun} stars (see Working It Out 16.1). Compare those lifetimes with the values in Table 16.1.

37. What will the escape velocity be when the Sun becomes an AGB star with a radius 200 times greater and a mass only 0.7 times that of today? How will those changes in escape velocity affect mass loss from the surface of the Sun as an AGB star?

38. Each form of energy generation in stars depends on temperature.
    a. The rate of hydrogen fusion (proton-proton chain) near 10^7 K increases with temperature as T^4. If the temperature of the hydrogen-fusing core is raised by 10 percent, by how much does the hydrogen fusion energy increase?
    b. Helium fusion (the triple-alpha process) at 10^8 K increases with an increase in temperature at a rate of T^{40}. If the temperature of the helium-fusing core is raised by 10 percent, by how much does the helium fusion energy increase?

39. A planetary nebula has an expansion rate of 20 km/s and a lifetime of 50,000 years. Roughly how large will that planetary nebula grow before it disperses?

40. Suppose a companion star transferred mass onto a white dwarf at a rate of about 10^{-9} M_{Sun} per year. Roughly how long after mass transfer begins will the white dwarf explode as a Type Ia supernova? How does that length of time compare with the typical lifetime of a low-mass star? Assume that the white dwarf started with a mass of 0.6 M_{Sun}.

41. Use Kepler's third law to estimate how fast material in an accretion disk (size = 2×10^5 km) orbits around a 0.6 M_{Sun} white dwarf.

42. A white dwarf has a density of approximately 10^9 kilograms per cubic meter (kg/m^3). Earth has an average density of 5,500 kg/m^3 and a diameter of 12,700 km. If compressed to the same density as a white dwarf, what would Earth's radius be?

43. What is the density of degenerate material? Calculate how large the Sun would be if all its mass were degenerate.

44. Recall from Chapter 5 that the luminosity of a spherical object at surface temperature T is given by $L = 4\pi R^2 \sigma T^4$, where R is the object's radius. If the Sun became a white dwarf with a radius of 10^7 meters, what would its luminosity be at the following surface temperatures: (a) 10^8 K; (b) 10^6 K; (c) 10^4 K; (d) 10^2 K?

45. According to current astronomical evidence, our universe is approximately 13.8 billion years old. If that age is correct, what is the least mass that a star could possess to have already evolved into a red giant? What spectral type of star is that?

USING THE WEB

46. Go to the website for the Katzman Automatic Imaging Telescope at Lick Observatory (http://w.astro.berkeley.edu/bait/kait.html). What is that project? Why can a search for supernovae be automated? Pick a recent year. How many supernovae were discovered? Look at some of the images. How bright do the supernovae look in comparison with their galaxies?

47. Go to the website of the American Association of Variable Star Observers (AAVSO; https://aavso.org). What does that 100-year-old organization do? Read about the types of intrinsic variable stars. Click on "Getting Started." If you have access to dark skies, you can contribute to the study of variable stars. Go to the page for observers (http://aavso.org/observers) and click on each item in the "For New Observers" list, including the list of stars easy to observe. Assemble a group and observe a variable star from that list.

48. Go to the University of Washington's "Properties of Planetary Nebulae" Web page (https://sites.google.com/a/uw.edu/introductory-astronomy-clearinghouse/activities/stars/properties-of-planetary-nebulae) and download and complete the lab exercise.

49. Go to the Hubble Space Telescope's planetary nebula gallery (http://hubblesite.org/images/news/34-planetary-nebulae). For each of the three types of symmetry, find an example of a nebula that shows clearly the type of symmetry: spherical (being symmetric in every direction, like a circle), bipolar (having an axis about which they are symmetric, like a person's face), and point-symmetric (being symmetric about a point, like the letter S). Print each of the three images you chose, and label the type of symmetry each one represents. For all three nebulae, identify the location of the central star. For bipolar symmetry, draw a line that shows the axis about which the nebula is symmetric. For point symmetry, identify several features that are symmetric across the location of the central star.

50. In the Hubble telescope news archive, look up press releases on planetary nebulae (http://hubblesite.org/news/34-planetary-nebulae) and white dwarf stars (http://hubblesite.org/news/90-white-dwarfs). Pick a story for each. What observations were reported, and why were they important?

EXPLORATION

digital.wwnorton.com/astro6

The evolution of a low-mass star, as discussed in this chapter, corresponds to many twists and turns on the H-R diagram. In this Exploration, we return to the "H-R Diagram" simulation to investigate how those twists and turns affect the appearance of the star.

Visit the Digital Resources Page and on the Student Site open the "H-R Diagram" Interactive Simulation in Chapter 16. The box labeled "Size Comparison" shows an image of both the Sun and the test star. Initially, those two stars have identical properties: the same temperature, the same luminosity, and the same size.

Examine the box labeled "Cursor Properties." That box shows the temperature, luminosity, and radius of a test star located at the "X" in the H-R diagram. Before you change anything, answer the first three questions.

1   What is the temperature of the test star?

_____

2   What is the luminosity of the test star?

_____

3   What is the radius of the test star?

_____

As a star leaves the main sequence, it moves up and to the right on the H-R diagram. Move the cursor (the X on the diagram) up and to the right.

4   What changes about the image of the test star next to the Sun?

_____

5   What is the test star's temperature? What property of the image of the test star indicates that its temperature has changed?

_____

6   What is the test star's luminosity?

_____

7   What is the test star's radius?

_____

8   Ordinarily, the hotter an object is, the more luminous it is. Here, the temperature has gone down, but the luminosity has gone up. How can that be?

_____

The star then moves around a lot in that part of the H-R diagram. Look at Figure 16.13, and then use the cursor to approximate the motion of the star as it moves up the red giant branch, back down and onto the horizontal branch, and then back to the right and up the asymptotic giant branch.

9   Are the changes you observe in the image of the star as dramatic as the ones you observed for question 4?

_____

10 What is the most noticeable change in the star as it moves through that portion of its evolution?

_____

Next, the star begins moving across the H-R diagram to the left, maintaining almost the same luminosity. Drag the cursor across the top of the H-R diagram to the left, and study what happens to the image of the star in the "Size Comparison" box.

11 What changed about the star as you dragged it across the H-R diagram?

_____

12 How does the star's size now compare with that of the Sun?

_____

Finally, the star drops to the bottom of the H-R diagram and then begins moving down and to the right. Move the cursor toward the bottom of the H-R diagram, where the star becomes a white dwarf.

13 What changed about the star as you dragged it down the H-R diagram?

_____

_____

14 How does its size now compare with that of the Sun?

_____

_____

To thoroughly cement your understanding of stellar evolution, press the "Reset" button and then move the star from main sequence to white dwarf several times. The exercise will help you remember how that part of a star's life appears on the H-R diagram.

17

Evolution of High-Mass Stars

Most stars are smaller and less massive than the Sun and live a relatively long time. O and B stars, which are much more massive than the Sun, are rarer and live a much shorter time. Those massive stars die more quickly and explosively than stars like the Sun. In this chapter, we will explore the life and death of high-mass stars.

LEARNING GOALS

By the end of this chapter, you should be able to:

LG 1 Describe how the death of high-mass stars differs from that of low-mass stars.

LG 2 List the sequence of stages for evolving high-mass stars.

LG 3 Explain the origin of chemical elements up to and heavier than iron.

LG 4 Identify how Hertzsprung-Russell (H-R) diagrams of clusters enable astronomers to measure the ages of stars and test theories of stellar evolution.

The Crab Nebula is the remains of a massive star whose explosion was observed in 1054. The center is an X-ray image of the pulsar. ▶ ▶ ▶

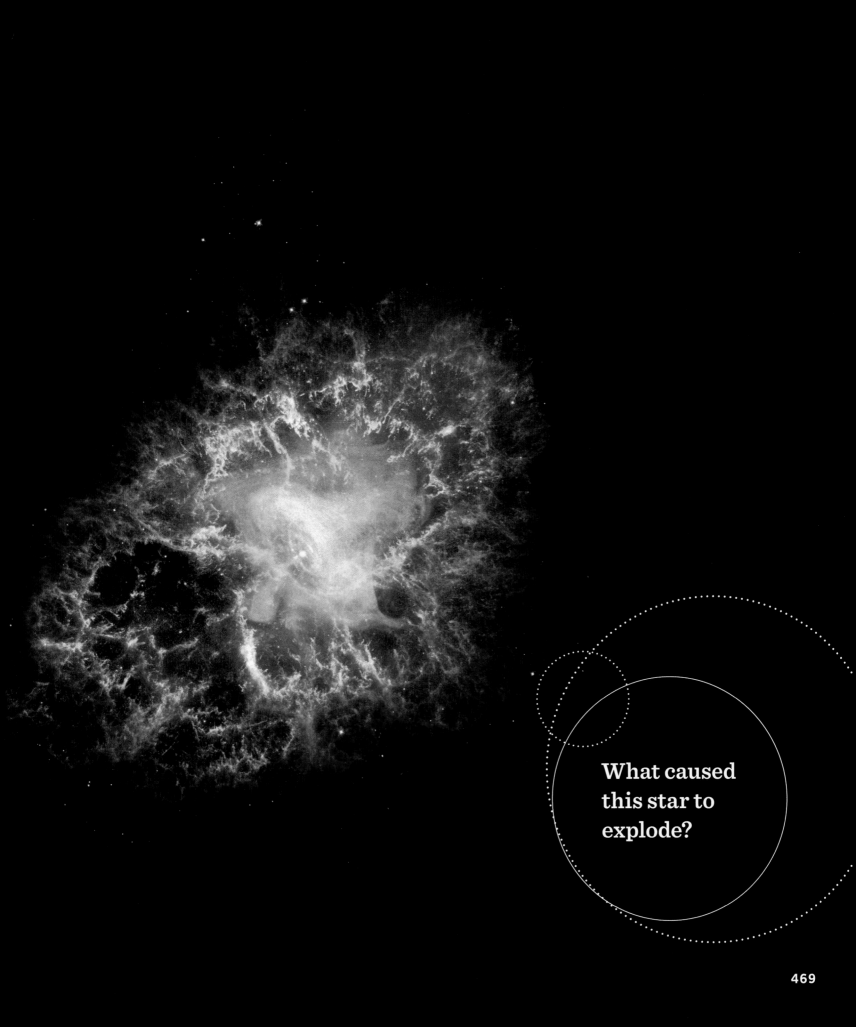

What caused
this star to
explode?

17.1 High-Mass Stars Follow Their Own Path

High-mass stars are those with masses greater than about 8 M_{Sun}. Those stars not only have several times more mass than the Sun, but they also have luminosities thousands or even millions of times greater than the Sun's. Even though they have more fuel to begin with, high-mass stars use it up much faster than low-mass stars and therefore have shorter lives. Whereas low-mass stars live for billions of years, high-mass stars live only hundreds of thousands to millions of years. In this section, we will discuss the stages in the evolution of high-mass (greater than 8 M_{Sun}) and medium-mass (3–8 M_{Sun}; also called intermediate mass) stars.

The CNO Cycle

A high-mass star has greater gravitational force pressing down on the interior than a low-mass star does. That greater force leads to higher temperature and pressure. Those two factors increase the rate of nuclear fusion, and therefore the stars are more luminous. At those higher temperatures and pressures at the center of a high-mass star, additional nuclear reactions become possible beyond the fusion of hydrogen into helium. Recall from Chapter 14 that the hydrogen nucleus has only one proton—a single positive charge—so hydrogen fuses at lower temperatures (a few million kelvins) than any other atomic nucleus. However, the probability that any two hydrogen atoms will fuse is low. The low probability of that first step in the proton-proton chain limits how rapidly the entire process can move forward.

If the star contains elements from a previous generation of stars that produced more massive elements, carbon and other elements will be mixed with the hydrogen. In the core of a massive star, the temperature and pressure are high enough that the **carbon-nitrogen-oxygen (CNO) cycle**, a nuclear fusion process that converts hydrogen to helium in the presence of carbon, occurs. That process is illustrated in **Figure 17.1a**, which shows each step. First, a hydrogen nucleus fuses with a carbon-12 (^{12}C) nucleus to form nitrogen-13 (^{13}N). Second, a proton in that ^{13}N nucleus decays, so that the atom is once again carbon. However, that carbon nucleus has an "extra" neutron, so it is now carbon-13 (^{13}C), not carbon-12. Third and fourth, two more hydrogen nuclei then fuse with that ^{13}C nucleus, creating nitrogen-14 (^{14}N) and then oxygen-15 (^{15}O). Fifth, a proton in the oxygen nucleus decays, creating nitrogen-15 (^{15}N). One more proton enters the nucleus, causing the ejection of a helium-4 (^{4}He) nucleus and leaving behind a ^{12}C nucleus, which can participate in the cycle again.

Notice that only one of two reactions occurs at most steps in that process: either a hydrogen nucleus fuses with another nucleus to create a new element with higher atomic number or a proton spontaneously decays to a neutron to create a new element with lower atomic number. Along the way, several by-products are formed: a positron and a neutrino are ejected each time a proton decays to form a neutron. Each positron then annihilates with an electron to produce a gamma ray. Additional gamma rays are released each time fusion occurs. **Figure 17.1b** shows the net reaction: a carbon-12 nucleus and four hydrogen nuclei combine with two electrons to produce a carbon-12 nucleus, a helium-4 nucleus, two neutrinos, and seven gamma rays. Getting a hydrogen nucleus past the

Simulation: CNO Cycle

Figure 17.1 (a) In high-mass stars, carbon serves as a catalyst for the fusion of hydrogen to helium. That process is the carbon-nitrogen-oxygen (CNO) cycle. (b) The CNO cycle takes carbon-12, hydrogen, and electrons as inputs and produces carbon-12, helium, neutrinos, and gamma rays.

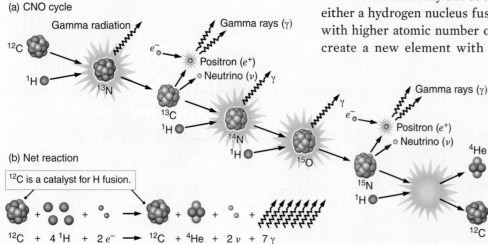

(a) CNO cycle

(b) Net reaction

^{12}C is a catalyst for H fusion.

^{12}C + 4 ^{1}H + 2 e^- ⟶ ^{12}C + ^{4}He + 2 ν + 7 γ

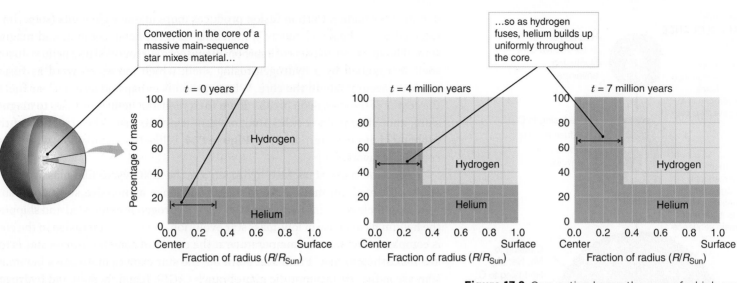

Figure 17.2 Convection keeps the core of a high-mass main-sequence star well mixed, so the composition remains uniform throughout the core as it evolves from zero age in the top graph to age 7 million years in the bottom graph. (Evolution times are for a 25-M_{Sun} star.)

electric barrier set up by a carbon nucleus, with its six protons, takes a lot of energy. But when that barrier can be overcome at high pressure and temperature, fusion is much more probable than it is in the interaction between two hydrogen nuclei. The CNO cycle is far more efficient than the proton-proton chain in stars more massive than about 1.3–1.5 M_{Sun}.

In a high-mass star, the temperature drops so quickly from the center of the star to the outside of the core that convection occurs within the core, "stirring" the core like the water in a boiling pot. As shown in **Figure 17.2**, helium residue spreads uniformly throughout the core of a high-mass star as the star consumes its hydrogen. That is unlike the case for low-mass stars, in which the helium residue builds up from the center outward (compare this figure with Figure 16.2). That difference is a result of the different processes of helium fusion in high-mass and low-mass stars.

The High-Mass Star Leaves the Main Sequence

As a high-mass star runs out of hydrogen in its core, the weight of the overlying star compresses the core, just as in a low-mass star. Yet in a high-mass star, the pressure and temperature in the core become high enough for helium fusion to begin (10^8 K) before the high-mass star forms an electron-degenerate core. The star makes a fairly smooth transition from hydrogen fusion to helium fusion as it leaves the main sequence.

Recall that when a low-mass star leaves the main sequence, the path it follows on the Hertzsprung-Russell (H-R) diagram is nearly vertical (see Figure 16.4) as the star becomes more luminous at roughly constant temperature. In contrast, as a high-mass star leaves the main sequence, the radius grows while its surface temperature falls, so its evolutionary track is nearly horizontal on the H-R diagram, as shown in **Figure 17.3**. The massive star fuses helium in its core and hydrogen in a surrounding shell. Recall that this is the same structure as a low-mass horizontal branch star. Stars more massive than 10 M_{Sun} become red supergiants during their helium-fusing phase. They have very cool surface temperatures (about 4000 K) and radii as much as 1,000 times that of the Sun.

The next stage in the evolution of a high-mass star has no analog in low-mass stars. When the high-mass star exhausts the helium in its core, the core again begins to collapse. Carbon fusion begins when the core reaches temperatures of

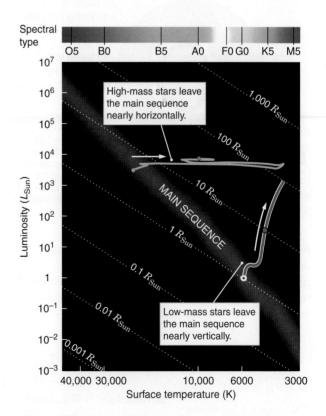

Figure 17.3 When high-mass stars leave the main sequence, they move horizontally across the H-R diagram, unlike low-mass stars, which move nearly vertically.

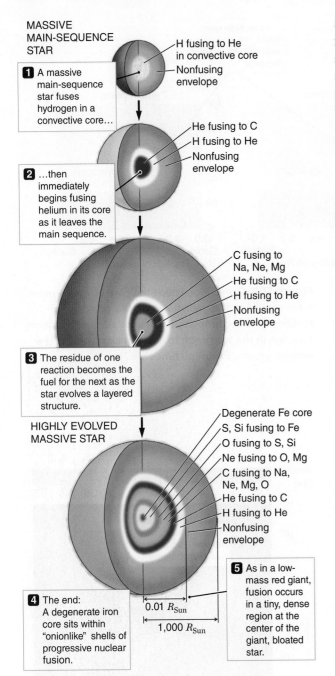

MASSIVE MAIN-SEQUENCE STAR

H fusing to He in convective core
Nonfusing envelope

1 A massive main-sequence star fuses hydrogen in a convective core...

He fusing to C
H fusing to He
Nonfusing envelope

2 ...then immediately begins fusing helium in its core as it leaves the main sequence.

C fusing to Na, Ne, Mg
He fusing to C
H fusing to He
Nonfusing envelope

3 The residue of one reaction becomes the fuel for the next as the star evolves a layered structure.

HIGHLY EVOLVED MASSIVE STAR

Degenerate Fe core
S, Si fusing to Fe
O fusing to S, Si
Ne fusing to O, Mg
C fusing to Na, Ne, Mg, O
He fusing to C
H fusing to He
Nonfusing envelope

5 As in a low-mass red giant, fusion occurs in a tiny, dense region at the center of the giant, bloated star.

0.01 R_{Sun}
1,000 R_{Sun}

4 The end: A degenerate iron core sits within "onionlike" shells of progressive nuclear fusion.

Figure 17.4 As a high-mass star evolves, it builds up a layered structure like that of an onion, with progressively more advanced stages of nuclear fusion found deeper and deeper within the star. The bottom image has been reduced to fit on the page.

Simulation: H-R Explorer

8×10^8 K or higher. Carbon fusion produces more massive elements (sometimes referred to as "heavy elements"), including oxygen, neon, sodium, and magnesium. The star then consists of a carbon-fusing core surrounded by a helium-fusing shell surrounded by a hydrogen-fusing shell, which moves outward as fusion products accumulate in the core. When carbon is exhausted as a nuclear fuel at the center of the star, neon breaks down to oxygen and helium or fuses to magnesium; and when neon is exhausted, oxygen begins to fuse. The structure of the evolving high-mass star, shown in **Figure 17.4**, is like that of an onion, with many concentric layers.

Medium-mass stars fuse hydrogen via the CNO cycle like massive stars. Stars with medium mass leave the main sequence as massive stars do, fusing helium in their cores immediately after their hydrogen is exhausted and skipping the helium flash phase of low-mass star evolution. When helium fusion in the core is complete, however, the temperature at the center of a medium-mass star is too low for carbon to fuse. From that point on, the star evolves more like a low-mass star, ascending the asymptotic giant branch (AGB), fusing helium and hydrogen in shells around a degenerate core, and then ejecting its outer layers and leaving behind a white dwarf.

Stars on the Instability Strip

While undergoing post-main-sequence evolution, a star may make one or more passes through a region of the H-R diagram known as the **instability strip**, as shown by the dashed-line region in **Figure 17.5**. **Variable stars** are stars whose brightness changes. In that region of the H-R diagram, stars grow and shrink repeatedly, so they appear to pulsate, regularly growing bright and faint and then bright again. The time for one pulsation is called the **period**. Such stars are called **pulsating variable stars**. A pulsating variable star does not achieve a steady balance between pressure and gravity; instead, it repeatedly overshoots the equilibrium radius, shrinking too far before being pushed back out by pressure or expanding too far before being pulled back in by gravity.

The most luminous pulsating variable stars are **Cepheid variables**, named after the prototype star Delta Cephei. Type I, or Classical, Cepheids are massive and luminous yellow supergiants. A Cepheid variable takes between 1 and 100 days to complete one pulsation period. The period correlates with the luminosity, creating a **period-luminosity relationship** in which stars with longer periods are more luminous, as shown in **Figure 17.6**. Henrietta Leavitt (1868–1921) experimentally determined that period-luminosity relationship for Cepheid variables in 1912. That relationship is important because it allows astronomers to use Cepheid variables as distance indicators. That process has three steps: First, astronomers observe the period. Then they use that period, combined with the period-luminosity relationship, to find the luminosity. Finally, they combine that luminosity with the observed brightness in the sky to find the distance to the star.

Thermal energy powers the pulsations of stars like Cepheid variables. An ionized gas is nearly opaque because ions can interact with light of all wavelengths, scattering many of the photons back toward the interior. In contrast, a neutral gas can absorb only light with energies that match its electron transitions. As the helium atoms in the atmosphere of the star alternate between ionized and neutral states, the atmosphere of the star alternates between opaque and transparent. That in turn alternately traps and releases light from the star. The star's atmosphere is much like a lid on a pot of boiling water. The pot builds up pressure enough to pop open the lid and let steam escape. Then gravity pulls the lid back

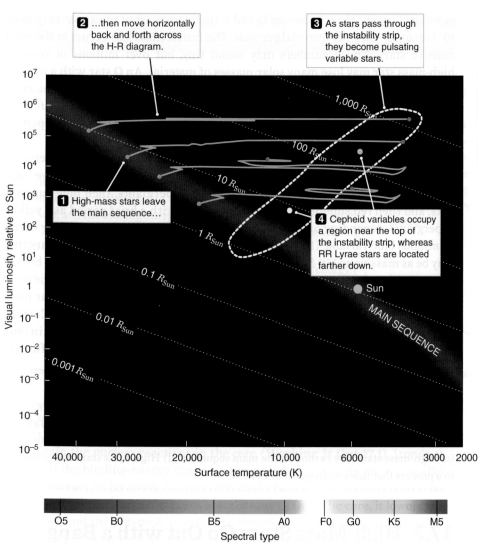

2 ...then move horizontally back and forth across the H-R diagram.

3 As stars pass through the instability strip, they become pulsating variable stars.

1 High-mass stars leave the main sequence...

4 Cepheid variables occupy a region near the top of the instability strip, whereas RR Lyrae stars are located farther down.

Figure 17.5 The paths of high-mass stars along the H-R diagram take them through a region known as the instability strip, shown by the region surrounded by the white dashed line. Pulsating variable stars such as Cepheid variables and RR Lyrae stars are found in that region of the H-R diagram.

down, and pressure builds again to repeat the process. Those pulsations do not affect the nuclear fusion in the star's interior; they affect just the light escaping from the star. From Chapter 13, recall that the luminosity, temperature, and radius of stars are all related, so both the luminosity and the temperature of the star change as the star expands and shrinks. The star is at its brightest and hottest (and therefore bluest) while it expands through its equilibrium size and is at its faintest and coolest (and therefore reddest) while it falls back inward.

High-mass Type I Cepheid variables are not the only type of variable star. The low-mass horizontal branch also passes through the instability strip on the H-R diagram. Low-mass (~0.8-M_{Sun}) stars can be Type II Cepheid variables or **RR Lyrae variables**, which have periods of less than a day. RR Lyrae stars pulsate by the same mechanism as Cepheid variables but are typically hundreds of times less luminous. They, too, follow a period-luminosity relationship, shown in Figure 17.6b.

High-mass stars expel a significant percentage of their mass back into space throughout their lifetimes, changing their composition. Even while on the main sequence, massive O and B stars have low-density winds with velocities as high as 3,000 kilometers per second (km/s). The pressure of the intense radiation generated in the core overcomes the star's gravity and drives away material from its

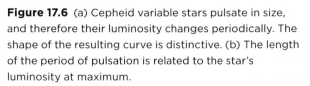

(a) Luminosity vs. time

(b) Period-luminosity relationship

Type I (Classical) Cepheids

Type II Cepheids

RR Lyrae

Figure 17.6 (a) Cepheid variable stars pulsate in size, and therefore their luminosity changes periodically. The shape of the resulting curve is distinctive. (b) The length of the period of pulsation is related to the star's luminosity at maximum.

18

Relativity and Black Holes

S ome stars leave behind a black hole at the end of their lives. Black holes have conditions so extreme that the laws of Newtonian physics can't describe them. To discuss black holes, we must understand how Albert Einstein changed the way physicists thought about the nature of space and time in the early 20th century. Einstein's special theory of relativity shows that matter behaves differently when it is traveling near the speed of light. Einstein's general theory of relativity shows that space itself is warped near very massive objects. That warping of space is so extreme at a black hole, it is as though a hole exists in space. Here in Chapter 18, we move beyond Newtonian ideas of space and time to understand black holes.

....

LEARNING GOALS

By the end of this chapter, you should be able to:

LG 1 Describe how the motion of the observer affects the observed velocity of objects.

LG 2 Discuss the observable consequences of the relationship between space and time.

LG 3 Explain how gravity is a consequence of the way mass distorts the very shape of spacetime.

LG 4 Explain the formation of black holes from the most massive stars, and describe the key properties and observational consequences of those stellar black holes.

This artist's rendition of the black hole Cygnus X-1 shows the accretion disk and the massive star that feeds it. ▶ ▶ ▶

How are space
and time changed
near a black hole?

(a)

Car stationary

Car moving

(b)

Car stationary

Car moving

Figure 18.1 On a windless day, the direction in which rain falls depends on the frame of reference in which it is viewed. (a) From outside the car, the rain is seen to fall vertically downward whether the car is stationary or moving. (b) From inside the car, the rain is seen to fall vertically downward if the car is stationary; but if the car is moving, the rain is seen to fall at an angle determined by the speed and direction of the car's motion.

Figure 18.2 When we observe stars, their apparent positions are deflected slightly toward the direction in which Earth is moving. As Earth orbits the Sun, stars appear to trace out small loops in the sky. That effect is called *aberration of starlight*.

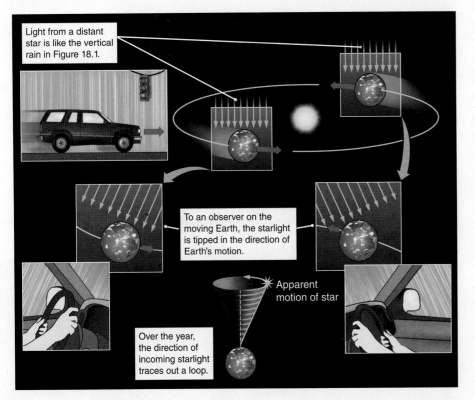

Light from a distant star is like the vertical rain in Figure 18.1.

To an observer on the moving Earth, the starlight is tipped in the direction of Earth's motion.

Apparent motion of star

Over the year, the direction of incoming starlight traces out a loop.

18.1 Relative Motion Affects Measured Velocities

All observers, whatever their motion, measure the same speed of light. That observed fact has profound implications for relative motion, space, and time. In this section, we will lay the groundwork for our discussion of relativity by considering how relative motion affects measurements and how those effects change at very high speeds.

Aberration of Starlight

The first direct measurement of the effect of Earth's motion around the Sun was made in the 18th century. Imagine that you are sitting in a car in a windless rainstorm (**Figure 18.1**). If the car is stationary and the rain is falling vertically, you see raindrops falling straight down when you look out your side window. When the car is moving forward, however, the situation is different. Between the time a raindrop appears at the top of your window and the time it disappears beneath the bottom of your window, the car has moved forward. The raindrop disappears beneath the window behind the point at which it appeared at the top of the window, which means the raindrop looks as though it is falling at an angle, even though in reality it is falling straight down. As you go faster, the apparent front-to-back slant of the raindrops increases, and their apparent paths become more slanted. An observer by the side of the road would say the raindrops are coming from directly overhead, but to you in the moving car they are coming from a direction in front of the car. You are observing that apparent motion of the raindrops from within your own unique frame of reference.

The light from a distant star arrives at Earth from the direction of the star (**Figure 18.2**). However, just as the raindrops appeared to be coming from in front of the moving car in Figure 18.1b, an observer on the moving Earth sees the starlight coming from a slightly different direction. Because the direction of Earth's motion around the Sun continuously changes during the year, the apparent position of a star in the sky moves in a small loop, a phenomenon known as the **aberration of starlight**. That shift in apparent position was first detected in the 1720s by two astronomers—Samuel Molyneux and James Bradley. Measurement of the aberration of starlight shows that Earth moves on its path about the Sun with an average speed of just under 30 kilometers per second (km/s). Because distance equals speed multiplied by time, astronomers used that measurement to determine the circumference, and therefore the radius, of Earth's orbit. The speed of Earth (29.8 km/s) multiplied by the number of seconds in 1 year (3.16×10^7 seconds) gives a circumference of 9.42×10^8 km. Astronomers used that circumference to estimate the radius of Earth's orbit (1.5×10^8 km). The aberration of starlight is an astronomical example of how relative motion affects a measurement.

Relative Speeds Close to the Speed of Light

Every observer inhabits a reference frame, a set of coordinates in which the observer measures distances and speeds. A nonaccelerating reference frame, in which no net force is acting, is known as an **inertial reference frame**. Newton's laws of motion predict how the motion of an object will be measured by observers in different inertial reference frames. For example, the observed motion of a ball thrown from a moving car depends on the reference frame of the observer (recall the analogy for the Coriolis effect in Figure 2.11). If light behaved like other objects, the speed of light should differ from one observer to the next as a result of the observer's motion, just like the speed of a ball thrown from a moving car. However, the results of laboratory experiments with light in the late 19th century showed something very different. Scientists found instead that *all* observers measure the same value for the speed of light, regardless of the observer's frame of reference.

Imagine that you are in a red car traveling at 100 kilometers per hour (km/h) on a highway (**Figure 18.3a**), and you throw a ball at a speed of 50 km/h out the window at an oncoming green car, also traveling at 100 km/h. An observer standing by the side of the road watches the entire event. In *your* reference frame (Figure 18.3a, top panel), the red car is stationary; that is, you and the car are moving together, so the car does not move relative to you. In your reference frame, then, you measure the ball traveling at 50 km/h. To the observer standing by the road (Figure 18.3a, middle panel), however, the ball is moving at 150 km/h—50 km/h from throwing it plus 100 km/h from the motion of the car. To passengers in the oncoming green car moving at 100 km/h (Figure 18.3a, bottom panel), the speed of the ball is 250 km/h because the ball approaches them at 150 km/h, and they approach the ball at 100 km/h. That example shows that the speed of the ball depends on how the observer, the cars, and the ball are moving relative to one another. The velocities add together to yield the velocity of one object relative to another. That is Galilean relativity, which you use in everyday life.

Figure 18.3b shows how light differs from the ball in the previous example. Imagine that you are riding in the yellow spaceship at half the speed of light (0.5c) and you shine a beam of laser light forward (Figure 18.3b, top panel). You measure the speed of the beam of light to be c, or 3×10^8 meters per second (m/s)— as expected because you are holding the source of the light. But the observer on the planet also measures the speed of the passing beam of light to be c, not 1.5c, which would be the sum of the speed of light plus the speed of the spaceship (Figure 18.3b, middle panel). Even a

Figure 18.3 (a) The Newtonian rules of motion apply in daily life; however, those rules break down when speeds approach the speed of light. (b) Light itself always travels at the same speed for any observer, and that fact is the basis of special relativity, which implies that velocities don't simply add together, as they do in the Newtonian world.

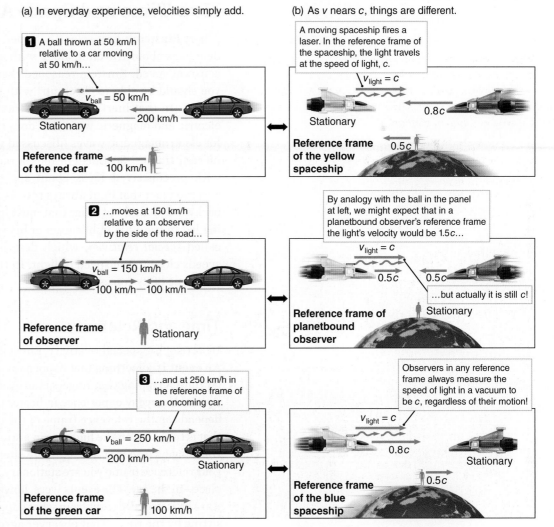

(a) In everyday experience, velocities simply add.

1 A ball thrown at 50 km/h relative to a car moving at 50 km/h…

v_{ball} = 50 km/h
200 km/h
Stationary

Reference frame of the red car 100 km/h

2 …moves at 150 km/h relative to an observer by the side of the road…

v_{ball} = 150 km/h
100 km/h — 100 km/h

Reference frame of observer Stationary

3 …and at 250 km/h in the reference frame of an oncoming car.

v_{ball} = 250 km/h
200 km/h
Stationary

Reference frame of the green car 100 km/h

(b) As v nears c, things are different.

A moving spaceship fires a laser. In the reference frame of the spaceship, the light travels at the speed of light, c.

v_{light} = c
0.8c
Stationary

Reference frame of the yellow spaceship 0.5c

By analogy with the ball in the panel at left, we might expect that in a planetbound observer's reference frame the light's velocity would be 1.5c…

v_{light} = c
0.5c 0.5c
…but actually it is still c!

Reference frame of planetbound observer Stationary

Observers in any reference frame always measure the speed of light in a vacuum to be c, regardless of their motion!

v_{light} = c
0.8c
Stationary

Reference frame of the blue spaceship 0.5c

passenger in an oncoming blue spacecraft traveling at 0.5*c* (Figure 18.3b, bottom panel) finds that the beam from your light is traveling at exactly *c* in her own reference frame, and not at 2.0*c*, which would be the sum of the speeds of the spacecraft and the laser light. At speeds close to the speed of light, speeds do not add simply. That statement is true not only for light but also for all ordinary objects moving at nearly the speed of light. According to Einstein's relativistic formulas, the relative speed between the two spacecraft in the top and bottom panels of Figure 18.3b (0.5*c* + 0.5*c*) adds to 0.8*c*, not 1.0*c*! At **relativistic speeds** (speeds close to the speed of light), everyday experience no longer holds true. Every observer always finds that light in a vacuum travels at the same speed, *c*, regardless of his or her own motion or the motion of the light source. That observation directly conflicts with Newtonian theory and Galilean relativity.

CHECK YOUR UNDERSTANDING 18.1

Which beam of light is moving faster? (a) the one from the headlight of a parked car; (b) the one from the headlight of a moving car; (c) the one from the headlight of a moving spaceship; (d) all the beams are moving at *c*.

18.2 Special Relativity Explains How Time and Space Are Related

Albert Einstein's first scientific paper, written when he was a 16-year-old student, was about traveling along with a light wave, moving in a straight line at constant speed. Einstein reasoned that according to Newton's laws of motion, you should be able to "keep up" with light so that you are moving right along with it. In that inertial reference frame, the light is stationary: an oscillating electric and magnetic wave that does not move. However, Maxwell's equations for electromagnetic waves (discussed in Chapter 5) are actually the same in all inertial frames. So Einstein took a radical new approach. Instead of starting with preconceived ideas about space and time, Einstein started with the observed fact that light always travels at the same speed, and then he reasoned backward to find out what that must imply about space and time. That reasoning led to the 1905 publication of his **special theory of relativity**, sometimes called *special relativity*, which describes the effects of traveling at constant speeds close to the speed of light. In this section, we explore special relativity and some of its consequences.

Time and Relativity

In developing special relativity, Einstein focused his thinking on pairs of *events*. An **event** is something that happens at a particular location in space at a particular time. Snapping your fingers is an event because that action has both a time and a place. Everyday experience indicates that the distance between any two events depends on the reference frame of the observer. Imagine you are sitting in a car that is traveling on the highway in a straight line at a constant 60 km/h. You snap your fingers (event 1), and a minute later you snap your fingers again (event 2). In your reference frame you are stationary, and the two events happened at the same place—in the car. The events were, however, separated by a minute in time. What you saw is very different from what happens in the reference frame of an observer sitting by the road. That observer agrees that the second snap of your fingers (event 2) occurred a minute after the first snap of your fingers (event 1), but to that observer the two events were separated from each other in space by a kilometer, the distance your car traveled in the minute between snaps. In that "Newtonian" view, the distance between two events depends on the motion of the observer, but the *time* between the two events does not. Special relativity says instead that both the distance *and* the time between events vary depending on the motion of the observer.

The notion that different observers will measure time differently is a *very* counterintuitive idea, but it is central to special relativity and therefore to our scientific understanding of the universe as well. To see how Einstein arrived at the concept of relative time, consider his thought experiment known as the boxcar experiment. In that experiment, observer 1 is in a boxcar of a train moving to the right. Observer 1 has a lamp, a mirror (mounted on the roof of the boxcar), and a clock. Observer 2 is standing on the ground outside. The clock is based on a value that everyone can agree on—such as the speed of light.

Figure 18.4a shows the experimental setup as seen by observer 1, who is stationary with respect to the clock. At time t_1, event 1 happens: the lamp gives off a pulse of light. The light bounces off a mirror at a distance l meters away and then heads back toward its source. At time t_2, event 2 happens: the light arrives at the clock and is recorded by a photon detector. (Note that the light beam is shown leaving and arriving at two locations. The artist did that so that you can see both events in the figure. Both events occur at the same location.) The time between events 1 and 2 is just the distance the light travels ($2l$ meters) divided by the speed of light: $t_2 - t_1 = 2l/c$.

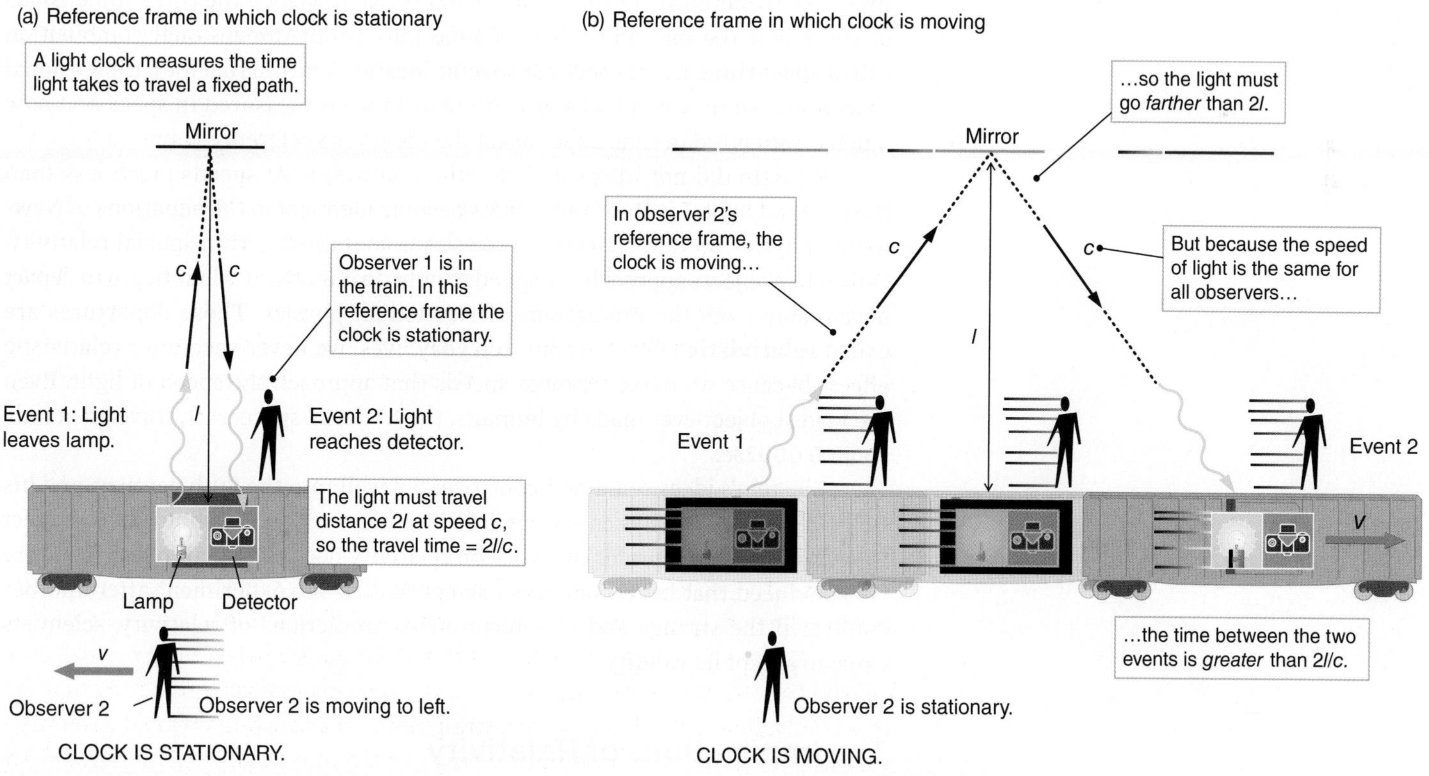

(a) Reference frame in which clock is stationary

A light clock measures the time light takes to travel a fixed path.

Mirror

c c

Observer 1 is in the train. In this reference frame the clock is stationary.

Event 1: Light leaves lamp.

l

Event 2: Light reaches detector.

The light must travel distance $2l$ at speed c, so the travel time = $2l/c$.

Lamp   Detector

v

Observer 2   Observer 2 is moving to left.

CLOCK IS STATIONARY.

(b) Reference frame in which clock is moving

...so the light must go *farther* than $2l$.

Mirror

In observer 2's reference frame, the clock is moving...

c c

But because the speed of light is the same for all observers...

Event 1

l

Event 2

v

...the time between the two events is *greater* than $2l/c$.

Observer 2 is stationary.

CLOCK IS MOVING.

Figure 18.4 The "tick" of a light clock is different when seen in two reference frames: (a) stationary, as in the reference frame of observer 1 on the boxcar, and (b) moving, as in the reference frame of observer 2 on the tracks. As Einstein's thought experiment demonstrates, if the speed of light is the same for every observer, moving clocks *must* run more slowly than stationary clocks.

Figure 18.4b shows the experiment as seen by observer 2, who is stationary on the ground outside the train, which is moving at speed *v*. In observer 2's reference frame, the clock moves to the right between the two events, so the light has farther to go because of the horizontal distance. The time between the two events is still the distance traveled divided by the speed of light, but now that distance is *longer* than 2*l* meters. Because the speed of light is the same for all observers, the time between the two events must be longer as well.

The two events are the same two events, regardless of the reference frame from which they are observed. Because the speed of light is the same for all observers, more time must pass between the two events when viewed from a reference frame in which the clock is moving (observer 2). The seconds of a moving clock are stretched. That is, a moving clock takes more time than a stationary clock to complete one "tick." Therefore, the passage of time must depend on an observer's frame of reference. Because both frames of reference are equally good places to do physics, both time measurements are valid in their own frames, even though they differ from each other.

In that experiment, light travels farther between events in a moving boxcar than between events in a stationary boxcar and consequently takes longer to travel between the events in the moving boxcar. Einstein realized that the *only* way the speed of light can be the same for all observers is *if the passage of time is different from one observer to the next*. For moving observers, the time is stretched out, so that the time interval between events is longer, a phenomenon known as **time dilation**.

The Newtonian view of the world describes a three-dimensional space through which time marches steadily onward. Einstein discovered, however, that time flows differently for different observers. He reshaped the three dimensions of space and the one dimension of time into a four-dimensional combination called **spacetime**. Events occur at specific locations within that four-dimensional spacetime, but how much of a spacetime distance is measured in space and how much is measured in time depends on the observer's reference frame.

Einstein did not "disprove" Newtonian physics. At speeds much less than the speed of light, Einstein's equations become identical to the equations of Newtonian physics, so that Newtonian physics is contained within special relativity. Only when objects approach the speed of light do our observations begin to depart measurably from the predictions of Newtonian physics. Those departures are called **relativistic** effects. In our everyday lives, we never encounter relativistic effects because we never travel at speeds that approach the speed of light. Even the fastest object ever made by humans, the *Helios II* spacecraft, traveled at only about 0.00023*c*.

Einstein's ideas remained controversial well into the 20th century, and his 1921 Nobel Prize in Physics was awarded for his work on the photoelectric effect (see Chapter 5) and not for his work on relativity because not all physicists were yet convinced that his theory was correct. But as one experiment after another confirmed the strange and counterintuitive predictions of relativity, scientists came to accept its validity.

The Implications of Relativity

Today, special relativity shapes our thinking about the motions of both the tiniest subatomic particles and the most distant galaxies. In this subsection, we discuss only a few of the essential insights that come from Einstein's work.

MASS AND ENERGY   What we think of as "mass" and what we think of as "energy" are actually closely related. The energy of an object depends on its speed: the faster it moves, the more energy it has. But Einstein's famous equation, $E = mc^2$, says that even a stationary object has an intrinsic "rest" energy that equals the mass (m) of the object multiplied by the speed of light (c) squared. The speed of light is a very large number, so a small mass has a very large rest energy. A single tablespoon of water, for example, has a rest energy equal to the energy released in the explosion of more than 300,000 tons of TNT. We used that relationship between mass and energy in Chapter 14 when discussing the nuclear fusion that makes stars shine.

In Chapter 3, we connected the *mass* of an object to its inertia—its resistance to changes in motion. At relativistic speeds, adding to the energy of motion of an object increases its inertia. For example, a proton in a high-energy particle accelerator may travel so close to the speed of light that its total energy is 1,000 times greater than its rest energy. Such an energetic proton is harder to "push around" (in other words, it has more inertia) than a proton at rest. It also more strongly attracts other masses through gravity.

THE ULTIMATE SPEED LIMIT   We already discussed Einstein's insight that if traveling at the speed of light were possible, then in that reference frame light would cease to be a traveling wave, and the laws of electromagnetism wouldn't be valid. You also can think about that limit in terms of the equivalence of mass and energy just discussed. As the speed of an object gets closer and closer to the speed of light, the energy of that object, and therefore its mass, becomes greater and greater, so it becomes increasingly resistant to further changes in its motion. Only a photon or other massless particle can travel at the speed of light. We rely on the fact that light travels at a constant speed in a vacuum whenever we use the travel time of light to describe astronomical distances.

Adding energy to an object will cause its velocity to get closer and closer to the speed of light, but the velocity will never actually reach the speed of light. To accelerate an object with a nonzero rest mass to the speed of light would take an *infinite* amount of energy. Not enough energy exists in the entire universe to accelerate even one electron to the speed of light. The electron can get arbitrarily close to that number—0.9999999999999999999999 . . . × c is possible (at least in principle)—but not enough energy is available to accelerate the electron beyond that to the speed of light. **Figure 18.5** shows how a rocket ship, which experiences a constant acceleration equal to that of gravity on Earth (so that its occupants will feel "normal" gravity), moves faster and faster but never reaches the speed of light. In the Newtonian view shown by the purple curve, that limit is not present, and the speed of the spaceship would continuously increase at the same rate.

TIME   Time passes more slowly in a moving reference frame: for moving objects, the seconds are stretched out by time dilation. No inertial reference frame is special. If you compared clocks with an observer moving at nine-tenths the speed of light ($0.9c$) relative to you, you would find that the other observer's clock was running 2.29 times slower than your clock. The other observer would find instead that *your* clock was running 2.29 times slower. To you, the other observer may be moving at $0.9c$, but to the other observer, *you* are the one who is moving. Both frames of reference are equally valid, so you would each find the other's clock to be slower than your own. That time dilation effect

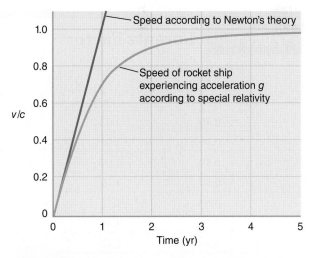

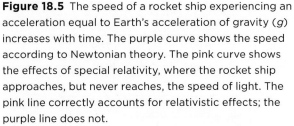

Figure 18.5 The speed of a rocket ship experiencing an acceleration equal to Earth's acceleration of gravity (g) increases with time. The purple curve shows the speed according to Newtonian theory. The pink curve shows the effects of special relativity, where the rocket ship approaches, but never reaches, the speed of light. The pink line correctly accounts for relativistic effects; the purple line does not.

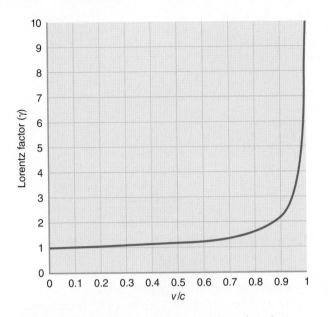

Figure 18.6 The Lorentz factor, γ, is plotted against v/c. That factor doesn't become significant until velocities are about 50 percent of the speed of light.

TABLE 18.1	Lorentz Factor

v/c	γ
0.10	1.005
0.20	1.02
0.30	1.05
0.40	1.09
0.50	1.15
0.60	1.25
0.70	1.40
0.80	1.67
0.90	2.29
0.95	3.20
0.99	7.09
0.995	10.01
0.999	22.37
0.9999	70.71

increases with speed, and that symmetry holds as long as neither frame accelerates. **Figure 18.6** and **Table 18.1** show that how much the time is stretched depends on the object's speed. The factor of 2.29 by which time is stretched in the preceding example is called the Lorentz factor and is usually denoted by the symbol γ.

A scientific observation demonstrates time dilation in nature. Fast particles called cosmic-ray muons (**Figure 18.7**) are produced 15 km up in Earth's atmosphere when high-energy **cosmic rays**—elementary particles moving at nearly the speed of light—strike atmospheric atoms or molecules. Muons at rest decay very rapidly into other particles. That decay happens so quickly that even if they could move at the speed of light, virtually all muons would have decayed long before traveling the 15 km to reach Earth's surface. Time dilation slows the muons' clocks, however, so the particles live longer and can travel farther and reach the ground. As muons move faster and faster, their clocks run slower and slower, and more of them can reach the ground. The same general principle is observed in particle accelerators, in which particles traveling at speeds near the speed of light live longer before decaying. **Working It Out 18.1** shows some examples of time dilation.

LENGTH An object appears shorter in motion than it is at rest. Moving objects are compressed in the direction of their motion by a factor of $1/\gamma$, where γ is the same Lorentz factor introduced in the discussion of time dilation. That phenomenon is called length contraction. A meterstick moving at $0.9c$ appears to be 0.44 meters long. That finding also explains our muon experiment from the perspective of the muons themselves. In the reference frame of the fast-moving muon produced at a height of 15 km, Earth's atmosphere is moving fast and appears to be much shorter than 15 km; indeed, it is so compressed from the muon's perspective that the muon may be able to reach the ground before decaying. That length contraction effect also increases with speed.

TWIN PARADOX Suppose you take a trip into space and leave your identical twin back on Earth. You accelerate to nearly the speed of light as you leave Earth. After you arrive at your destination, you return to Earth, again traveling at a speed close to c. To your twin on Earth, you were the one in a moving reference frame, so your twin measures your time as running much slower (see Working It Out 18.1), in which case you should return younger than your twin. However, from your perspective, the spaceship didn't move. Instead, Earth receded from you, stopped, and returned. Your twin is the one who moved at just under the speed of light, and your twin's time ran more slowly than yours, so your twin has aged much less. Both you and your twin cannot be correct, and that is the nature of the paradox.

One key difference between you and your twin resolves the paradox. You experienced acceleration during your trip, whereas your twin did not. Accelerated motion is not uniform motion. As a result, you changed reference frames during your trip. You changed reference frames when you left Earth, changed again when you stopped at your destination, changed a third time when you left your destination to return home, and changed reference frames one final time when you arrived back at Earth. Your twin, however, remained in Earth's reference frame. Upon your return, you would find that more time has passed for your twin on Earth than for you and that your twin has aged more than you have.

SPACE TRAVEL   What is the actual reality of human travel in space? Some U.S. astronauts have been to the Moon and back, and several robotic spacecraft have been sent to explore objects throughout the Solar System. (To date, only a few robotic spacecraft are traveling at a high enough velocity to leave the Solar System.) But humans could not easily visit and explore other planetary systems in the Milky Way Galaxy because of the constraints of energy and the ultimate speed limit of light. With current technology, engineers can construct rockets that can travel at speeds of up to 20,000 m/s. At such a speed, a one-way trip to Earth's nearest neighbor star, Proxima Centauri, at a distance of 4.2 light-years, would take well over 50,000 years. Travel to more distant stars would take even longer.

In principle, travel just under the speed limit *c* is possible, so one could take advantage of relativistic time dilation to make such adventures well within the lifetime of a space traveler. In the example in Working It Out 18.1, an astronaut experiences a round-trip travel time to planets 20 light-years away in just 5.6 years at a speed of 0.99*c*. Or an astronaut could travel to the center of the Milky Way Galaxy and back in just 2 years by traveling at 0.9999999992*c*.

Although theoretically possible, travel at those speeds in practice would require an impractical amount of energy. If *M* is the mass of the astronauts, rocket ship, and fuel, just accelerating the rocket ship to such a high speed would take γMc^2 of energy, or $10Mc^2$ in the Proxima Centauri example and $25,000Mc^2$ in the

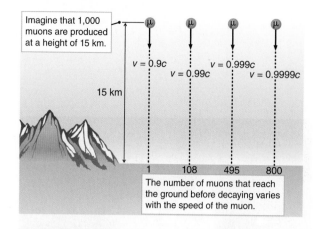

Figure 18.7 Muons created by cosmic rays high in Earth's atmosphere decay long before reaching the ground if they are not traveling at nearly the speed of light. Here, we show what happens to 1,000 muons produced at an altitude of 15 km for a variety of speeds. Faster muons have slower clocks, so more of them survive long enough to reach the ground—many more than would be expected simply because of the faster speed.

18.1 Working It Out Time Dilation

Physicist Hendrik Lorentz (1853–1928) derived the equation for how much time is dilated and how much space is contracted when something is traveling at velocities near the speed of light. That *Lorentz factor* (abbreviated γ) is given by

$$\gamma = \frac{1}{\sqrt{1 - \dfrac{v^2}{c^2}}}$$

Figure 18.6 shows the Lorentz factor plotted against velocity, and Table 18.1 gives the calculated value of γ for different values of velocities *v/c*. For something moving at half the speed of light, the Lorentz factor is 1.15. For something moving at 90 percent of *c*, however, the factor is 2.29, and it goes up quickly from there, becoming arbitrarily large as the velocity approaches, but never quite reaches, the speed of light.

Suppose you take a trip to a star 20 light-years away to study the "super-Earth" planets in orbit around it. You travel at 0.99*c*, whereas your twin stays behind on Earth. Our common Newtonian experience on Earth would tell you how long the trip takes, using the familiar equation that time equals distance divided by speed:

$$\text{Time passed} = \frac{20\,\text{light-years}}{0.99c} = 20.2\,\text{years}$$

The round trip would take 40.4 years. When you return, you will find that 40.4 years passed on Earth.

To you on the spaceship moving at 0.99*c*, special relativity says time will pass more slowly, by the Lorentz factor of 7.09. Your trip to the star would take 1/7 × 20.2 years = 2.8 years and another 2.8 years for the return trip. So as you traveled to that star and back, 5.6 years would have passed for you in the spaceship, but *40.4 years* would have passed on Earth. Your twin on Earth would be almost 35 years older than you are!

For a more practical experiment that can be performed on Earth, we've noted that time dilation can be seen with subatomic particles. Suppose one type of particle, called a pion, can "live" for 20 nanoseconds (ns) before it decays into different particles. If pions are produced in a particle accelerator at a speed of 0.999*c*, how long will physicists see the pions before they decay? Here *v* = 0.999*c*, so from Table 18.1, γ = 22.37. In the reference frame of the moving pions, they are still living for 20 ns: they are like the twin who traveled into space at high speed. In the reference frame of the physicists, though, special relativity predicts that the particle will last longer, 20 ns × 22.37, or 447 ns. Indeed, physicists observe that pions moving at nearly the speed of light "live" longer and travel farther before decay (like the muons in Figure 18.7).

The same factor γ applies for mass and length. If physicists measured a ruler moving at 0.999*c*, the ruler would be 22.37 times shorter than its length when at rest. In the previous example, the pions traveling at 0.999*c* will behave as though their mass were 22.37 times larger: if the high-speed pions collided with other particles, the energy of the collision would be as though they had the higher mass.

example of the trip to the center of the Milky Way. For the second example, the energy to accelerate just the astronaut (not even including the spaceship) to such energies is more than that contained in 10 billion nuclear weapons. So, although not theoretically impossible, visits to other stars in our galaxy will not take place anytime soon.

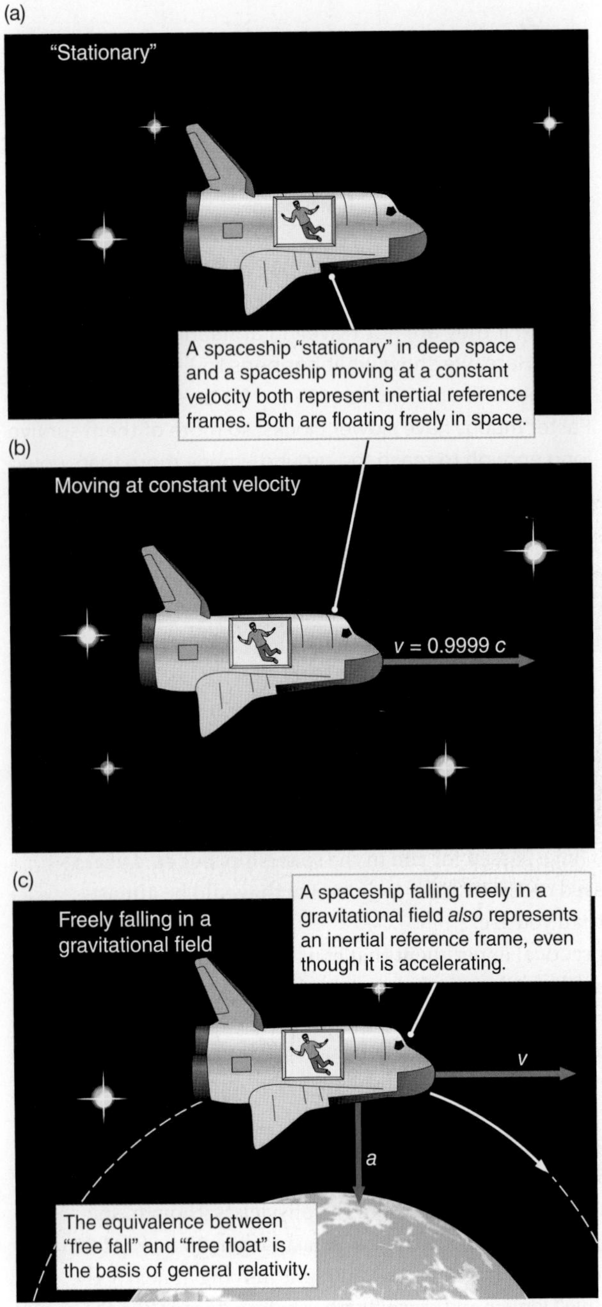

CHECK YOUR UNDERSTANDING 18.2

Suppose that your friend flies past you in a spaceship she measures to be 20 m long, and both of you measure the time the spaceship takes to pass your location. Which of the following is true? (a) The time you measure is *greater* than the time your friend measures. (b) The time you measure is *less* than the time your friend measures. (c) You both measure the same amount of time.

18.3 Gravity Is a Distortion of Spacetime

Our exploration of special relativity began with the observation that the speed of light is always the same regardless of the motion of an observer or the motion of the source of the light. We have seen that three-dimensional space and time are actually just the result of a particular, limited perspective on a four-dimensional spacetime that is different for each observer. That four-dimensional spacetime is itself warped and distorted by the masses it contains. As we discuss the properties of black holes—indeed, of all massive objects in the universe—the concepts of space and time will diverge even further from the absolutes of Newtonian physics. In this section, we explore the general theory of relativity, which describes how mass affects space and time.

The Equivalence Principle

We have already discussed one fundamental connection between gravity and spacetime in Chapter 4 and showed that the *inertial mass* of an object—the mass appearing in Newton's equation $F = ma$—is *exactly* the same as the object's gravitational mass. In addition, any two objects at the same location and moving with the same velocity will follow the same path through spacetime, regardless of their masses. The astronaut in an orbiting spaceship falls around Earth, moving in lockstep with the spaceship itself. A feather dropped by an *Apollo* astronaut standing on the Moon falls toward the surface of the Moon at the same rate as a dropped hammer does. Instead of thinking of gravity as a force that acts on objects, it is more accurate to think of gravity as a consequence of the warping of spacetime in the presence of a mass. *Gravitation is the result of the shape of spacetime that objects move through.* That is one of the key insights of the **general theory of relativity**, Einstein's theory of gravity.

Special relativity tells us that any inertial reference frame is as good as any other. No experiment can distinguish between sitting in an enclosed spaceship floating "stationary" in deep space (**Figure 18.8a**) and sitting in an enclosed spaceship traveling at $0.9999c$ (**Figure 18.8b**). Those two situations feel the same because neither observer feels an acceleration. Both are equally valid inertial reference frames. As long as nothing accelerates either spaceship, neither observer can distinguish between the two spacecraft.

But what if an acceleration occurs? An acceleration can change either the speed or the direction of an object. Consider an astronaut inside a spaceship

Figure 18.8 According to special relativity, no difference exists between (a) a reference frame floating stationary in space and (b) one moving through the galaxy at constant velocity. General relativity adds that no difference exists between those inertial reference frames and (c) an inertial reference frame falling freely in a gravitational field. Free fall is the same as free float, as far as the laws of physics are concerned.

Labels within figure:

(a) "Stationary"

A spaceship "stationary" in deep space and a spaceship moving at a constant velocity both represent inertial reference frames. Both are floating freely in space.

(b) Moving at constant velocity

$v = 0.9999\,c$

(c) Freely falling in a gravitational field

A spaceship falling freely in a gravitational field *also* represents an inertial reference frame, even though it is accelerating.

The equivalence between "free fall" and "free float" is the basis of general relativity.

orbiting Earth, as shown in **Figure 18.8c**. That spaceship is accelerating because the direction of its velocity is constantly changing as it orbits Earth. The astronaut also is accelerating. Because he feels weightless, the astronaut has no way to tell the difference between being inside the spaceship as it falls around Earth and being inside a spaceship floating through interstellar space. Even though the ship's velocity is constantly changing as it falls, the inside of a spaceship orbiting Earth is an inertial frame of reference just as an object drifting along a straight line through interstellar space is an inertial frame of reference.

The idea that a freely falling reference frame is equivalent to a freely floating reference frame is called the **equivalence principle**. If you close your eyes and jump off a diving board, for the brief time that you are falling freely through Earth's gravitational field, the sensation you feel is the same as the sensation you would feel floating in interstellar space.

The natural path that an object will follow through spacetime in the absence of other forces is called the object's **geodesic**. In the absence of a gravitational field, the geodesic of an object is a straight line, in accordance with Newton's first law: an object will move at a constant speed in a constant direction unless acted on by a net external force. However, the shape of spacetime becomes distorted in the presence of mass, so an object's geodesic becomes curved.

Figure 18.9a shows two examples of inertial frames: an astronaut in a spaceship coasting through space and a person in a box falling toward the ground with an acceleration g. Both people are following their geodesics, so those two inertial reference frames are equivalent, and neither observer can distinguish between them. The equivalence principle also applies in cases of accelerations that result in a change in speed. Imagine an observer sitting in a closed box on the surface of Earth (**Figure 18.9b**). The floor of the box pushes on him to keep him from following his geodesic, and he feels that force. Now imagine the box is inside a spaceship accelerating through deep space at a rate of 9.8 meters per second per second (m/s^2) in the direction of the arrow shown in Figure 18.9b. The floor of the box pushes on the observer to overcome his inertia and cause him to accelerate at 9.8 m/s^2, so he feels as though he is being pushed into the floor of the box. In both cases, a force acts on the observer so that he does not follow the same free-falling geodesic. The observer feels as though he is being pushed into the floor of the box, so he feels the acceleration, and his frame of reference is not inertial. According to the equivalence principle, sitting in an armchair in a spaceship traveling with an acceleration of 9.8 m/s^2 is equivalent to sitting in an armchair on the surface of Earth reading this book. In each case, it is the same mass—the mass that gives an object inertia—that resists the change. Gravitational mass and inertial mass are the same thing.

The equivalence principle has an important caveat. In an accelerated reference frame such as an accelerating spaceship, the same acceleration (both magnitude and direction) is experienced everywhere.

Figure 18.9 According to the equivalence principle, (a) an object falling freely in a gravitational field is in an inertial reference frame, and (b) an object at rest in a gravitational field is in an accelerated reference frame. As a result, sitting in a spaceship accelerating at 9.8 m/s^2 feels the same as sitting still on Earth.

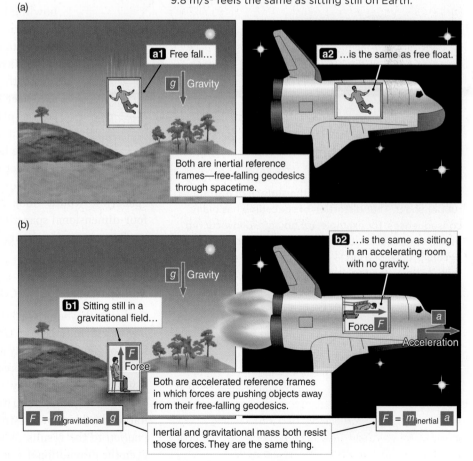

(a)

a1 Free fall…

g Gravity

a2 …is the same as free float.

Both are inertial reference frames—free-falling geodesics through spacetime.

(b)

g Gravity

b1 Sitting still in a gravitational field…

b2 …is the same as sitting in an accelerating room with no gravity.

Force F

a

Acceleration

Both are accelerated reference frames in which forces are pushing objects away from their free-falling geodesics.

$F = m_{gravitational}\, g$

$F = m_{inertial}\, a$

Inertial and gravitational mass both resist those forces. They are the same thing.

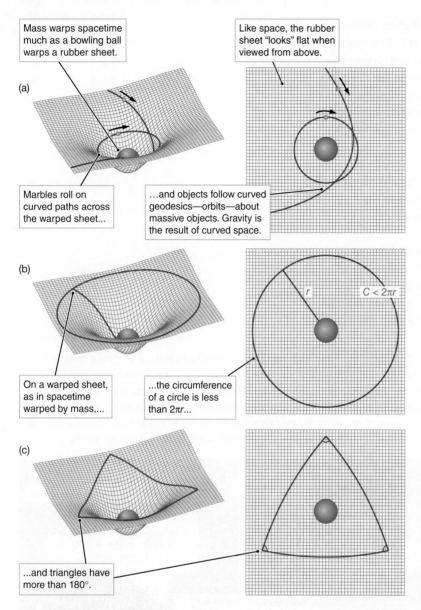

Mass warps spacetime much as a bowling ball warps a rubber sheet.

Like space, the rubber sheet "looks" flat when viewed from above.

(a)

Marbles roll on curved paths across the warped sheet...

...and objects follow curved geodesics—orbits—about massive objects. Gravity is the result of curved space.

(b)

On a warped sheet, as in spacetime warped by mass,...

...the circumference of a circle is less than $2\pi r$...

r $C < 2\pi r$

(c)

...and triangles have more than 180°.

Figure 18.10 Mass warps the geometry of spacetime much as a bowling ball warps the surface of a stretched rubber sheet. That distortion of spacetime has many consequences; for example, (a) objects follow curved paths, or geodesics, through curved spacetime; (b) the circumference of a circle around a massive object is less than 2π times the radius of the circle; and (c) angles in triangles need not add to exactly 180 degrees.

In contrast, the curvature of space by a massive object is weaker farther from the object. The effects of gravity and acceleration are equivalent only locally; that is, the equivalence principle is valid only as long as attention is restricted to small enough volumes of space so that differences in gravity within that volume can be ignored.

Mass Distorts Spacetime

The general theory of relativity describes how mass distorts the geometry of spacetime. Imagine the surface of a tightly stretched, flat rubber sheet. A marble will roll in a straight line across the sheet. Euclidean geometry—the geometry of everyday life—applies on the surface of the sheet: If you draw a circle, its circumference is equal to 2π times its radius, r. If you draw a triangle, the angles add up to 180 degrees. Lines that are parallel anywhere are parallel everywhere.

Now place a bowling ball in the middle of the rubber sheet, creating a deep depression, or "well," as in **Figure 18.10**. The surface of the sheet is no longer flat, and Euclidean geometry no longer applies. If you roll a marble across the sheet (Figure 18.10a), its path dips and curves. You can roll the marble so that it moves around and around the bowling ball, like a planet orbiting about the Sun. If you draw a circle around the bowling ball (Figure 18.10b), the circumference of the circle is less than $2\pi r$. If you draw a triangle (Figure 18.10c), the angles add up to more than 180 degrees.

Mass affects the fabric of spacetime like the bowling ball affects the fabric of the rubber sheet. The bowling ball stretches the sheet, changing the distances between any two points on the surface of the sheet. Similarly, mass distorts spacetime, changing the distance between any two locations or events. Larger masses produce larger distortions in spacetime. We can see how a rubber sheet with a bowling ball on it is stretched through a third spatial dimension, but most people cannot picture what a curved four-dimensional spacetime would "look like." Yet experiments verify that the geometry of four-dimensional spacetime is distorted much like the rubber sheet, whether or not it can be easily pictured.

When One Physical Law Supplants Another

Earlier in this book, we described gravity as a force that obeys Newton's universal law of gravitation: $F = Gm_1m_2/r^2$. Here in Chapter 18, we have introduced the ideas of general relativity and asked you to change the way you think about gravity. If general relativity is correct, does that mean Newton's formulation of gravity is wrong? If so, then why does Newton's law continue to be used?

Those questions go to the heart of how science progresses. As long as a gravitational field is not too strong, Newton's law of gravitation is a very close *approximation* to the results of a calculation using general relativity. In that context, even the gravitational field near the core of a massive main-sequence star would be considered "weak." An astronomer will obtain the same results if she uses a general relativistic formulation of gravity rather than Newton's laws to calculate the structure of a main-sequence star. Similarly, even though spacetime is curved by the presence of mass, that curvature near Earth is so slight that over small regions it can be ignored entirely, and flat (Euclidean) geometry can be used. That is the same kind of approximation people use when they navigate with a flat road map, despite the curvature of Earth.

Similarly, Newton's laws of motion are *approximations* of special relativity and quantum mechanics. In fact, Newton's laws can be mathematically derived from special relativity and quantum mechanics by using the assumptions that speeds of objects are much less than the speed of light and that objects are much larger than the particles from which atoms are made. Newton's laws of motion and gravitation are used most of the time because they are far easier to apply than general relativity and because any inaccuracies introduced by using Newtonian approximations are usually far too tiny to measure. That was illustrated in Figure 18.5, in which the behavior of an accelerated object moving at a velocity much less than c is the same whether or not relativity is used. Calculations based on general relativity are required only when conditions are very different from those of everyday life (for example, the behavior of the gravitational field of a black hole) or in special cases when extremely high accuracy is needed (for example, the precise timing used by the global positioning system [GPS] satellite network).

If a new theory is to replace an earlier, highly successful scientific theory, the new theory must hold the old theory within it—that is, the new theory must be able to reproduce the successes of the earlier theory. Special relativity contains Newton's laws of motion, and general relativity holds within it the successful Newtonian description of gravity that we have relied on throughout this book (see the **Process of Science Figure**).

The Observable Consequences of General Relativity

The curved spacetime of general relativity does have observable consequences. Indeed, in our own Solar System, observations can be made that distinguish general relativity from Newtonian physics. In Newton's theory, orbits are elliptical and fixed in space. In contrast, general relativity predicts that the long axis of an elliptical orbit slowly rotates, or precesses. **Figure 18.11** illustrates the difference between an orbit in Newton's theory, which remains stationary (left panel), and one in general relativity, which precesses (right panel). In our Solar System, even after one accounts for the effects of other planets on Mercury, a very small shift in its axis remains, equal to 43 arcseconds (arcsec) per century—which cannot be explained by Newton's laws alone. General relativity predicts exactly that precession for Mercury.

General relativity has other unique implications. A beam of light moving through empty space travels in a straight line, but a beam of light moving through the distorted spacetime around a massive object is bent by gravity, just as the lines in Figure 18.10 are bent by the curvature of the sheet. That bending of the light path by curved spacetime is called **gravitational lensing** because optical lenses also bend light paths. The first measurement of gravitational lensing came during the total solar eclipse of 1919. Several months before the eclipse, astrophysicist Sir Arthur Stanley Eddington (1882–1944) measured the positions of several stars in the direction of the sky where the eclipse would occur. He then

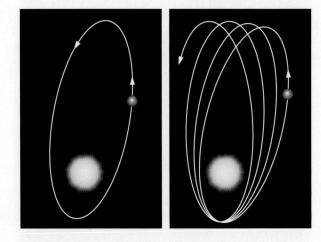

Figure 18.11 The left panel denotes an elliptical orbit about the Sun. In the Newtonian view, that elliptical orbit remains stationary. In the right panel, Mercury's orbit precesses as a result of the warped spacetime near the Sun.

Questions and Problems

Test Your Understanding

1. Rank the following (from least massive to most massive) in terms of the mass of the star that produces each.
   a. neutron star
   b. black hole
   c. white dwarf

2. A car approaches you at 50 km/h. A fly inside the car is flying toward the back of the car at 7 km/h. From your point of view by the side of the road, the fly is moving at _____ km/h.
   a. 7
   b. 28.5
   c. 43
   d. 57

3. A car approaches you at 50 km/h. The driver turns on the headlights. From your point of view, the light from the headlights is moving at
   a. $c + 50$ km/h.
   b. $c - 50$ km/h.
   c. $(c + 50$ km/h$)/2$.
   d. c.

4. Imagine that you are on a spaceship. A second spaceship rockets past yours at $0.5c$. You start a stopwatch and stop it 10 seconds later. For an astronaut in the other spaceship, the number of seconds that have ticked by during the 10 seconds on your stopwatch is
   a. more than 10 seconds.
   b. equal to 10 seconds.
   c. less than 10 seconds.

5. The International Space Station flies overhead. Using a telescope, you take a picture and measure its length to be _____ than its length as it would be measured if it were sitting on the ground.
   a. much greater
   b. slightly greater
   c. slightly less
   d. much less

6. Astronauts in the International Space Station
   a. have no mass.
   b. have no energy.
   c. are outside Earth's gravitational field.
   d. are in free fall.

7. Einstein's formulation of gravity
   a. is approximately equal to Newton's universal law of gravitation for small gravitational fields.
   b. is always used to calculate gravitational effects in modern times.
   c. explained why Newton's universal law of gravitation describes the motions of masses.
   d. both a and c

8. As the mass of a black hole increases, its Schwarzschild radius
   a. increases as the square of the mass.
   b. increases proportionately.
   c. stays the same.
   d. decreases proportionately.
   e. decreases as the square of the mass.

9. If a neutron star is more than ~2.5 times as massive as the Sun, it collapses because
   a. the force of electron degeneracy is stronger than gravity.
   b. gravity overpowers the force of electron degeneracy.
   c. gravity overpowers the force of neutron degeneracy.
   d. the force of neutron degeneracy is stronger than gravity.

10. Relative motion between two objects is apparent
    a. even at everyday speeds, such as 10 km/h.
    b. only at very large speeds, such as $0.8c$.
    c. only near very large masses.
    d. only when both objects are in the same reference frame.

11. If a spaceship approaches you at $0.5c$, and a light on the spaceship is turned on pointing in your direction, how fast will the light be traveling when it reaches you?
    a. $1.5c$
    b. between $1.0c$ and $1.5c$
    c. exactly c
    d. between $0.5c$ and $1.0c$

12. Imagine two protons traveling past each other at a distance d, with relative speed $0.9c$. Compared with two *stationary* protons a distance d apart, the gravitational force between those two protons when at their closest will be
    a. smaller because they interact for less time.
    b. smaller because the moving proton acts as though it has less mass.
    c. the same because the particles have the same mass.
    d. larger because the moving proton acts as though it has more mass.

13. If two spaceships approach each other, each traveling at $0.5c$ relative to an outside observer, spaceship 1 will measure spaceship 2 to be traveling
    a. much faster than c.
    b. slightly faster than c.
    c. at c.
    d. more slowly than c.

14. Some black holes are rapidly flickering X-ray sources. That observation indicates that
    a. X-rays are not light, because even light cannot escape from a black hole.
    b. the X-rays are coming from a very small source—so small that it could be a black hole.
    c. black holes are (at least sometimes) surrounded by hot gas.
    d. black holes have very high temperature.

15. The current model of long-duration GRBs includes jets from the collapsed star. You have seen jets like that before, when studying (choose all that apply)
    a. giant planets.
    b. white dwarfs.
    c. pulsars.
    d. star formation.
    e. supernovae.
    f. planetary nebulae.

Thinking about the Concepts

16. An astronomer sees a redshift in the spectrum of an object. Without any other information, can she determine whether it is an extremely dense object (exhibiting gravitational redshift) or one that is receding from her (exhibiting Doppler redshift)? Explain your answer.

17. Imagine you are traveling in a spacecraft at 0.9999999c. You point your laser pointer out the back window of the spacecraft. At what speed does the light from the laser pointer travel away from the spacecraft? What speed would be observed by someone on a planet traveling at 0.000001c?

18. Einstein's special theory of relativity says that no object can travel faster than, or even at, the speed of light. Recall that light is both an electromagnetic wave and a particle called a photon. If light acts as a particle, how can a photon travel at the speed of light?

19. Twin A takes a long trip in a spacecraft and returns younger than twin B, who stayed behind. Could twin A ever return before twin B was born? Explain.

20. In one frame of reference, event A occurs before event B. In another frame of reference, could the two events be reversed, so that B occurs before A? Explain.

21. Imagine you are watching someone whizzing by at very high speed in a spacecraft. Will that person's pulse rate appear to be extremely fast or extremely slow?

22. Suppose you had a density meter that could instantly measure the density of an object. You point the meter at a person in a spacecraft zipping by at very high speed. Is that person's density value larger or smaller than an average person's? Explain.

23. Imagine a future astronaut traveling in a spaceship at 0.866 times the speed of light. Special relativity says that along the direction of flight, the spaceship is only half as long as it was when at rest on Earth. The astronaut checks that prediction with a meter stick that he brought with him. Will his measurement confirm the contracted length of his spaceship? Explain your answer.

24. Within a speedy spacecraft, astronauts are playing soccer with a spherical ball. How would the shape of the ball appear to an observer watching the spacecraft speed by?

25. You observe a meter stick traveling past you at 0.9999c. You measure the meter stick to be 1 meter long. How is the meter stick oriented relative to you?

26. Suppose astronomers discover a 3-M_{Sun} black hole located a few light-years from Earth. Should they be concerned that its tremendous gravitational pull will lead to Earth's untimely demise?

27. If you could watch a star falling into a black hole, how would the color of the star change as it approached the event horizon?

28. Why don't people detect the effects of special and general relativity in their everyday lives here on Earth?

29. Many movies and television programs (such as *Star Wars*, *Star Trek*, and *Battlestar Galactica*) are premised on faster-than-light travel. How likely is it that such technology will be developed in the near future?

30. How could a gamma-ray burst in our galaxy potentially affect life on Earth?

Applying the Concepts

31. Does the angle of the rain falling outside the car in Figure 18.1b depend on the speed of the car? Knowing only that angle and the information on your speedometer, how could you determine the speed of the falling rain?

32. Compare Figure 18.1 with Figure 18.2. If you knew the speed of Earth in its orbit from a prior experiment, how could you determine the speed of light from the angle of the aberration of starlight?

33. According to Einstein, mass and energy are equivalent. So, which weighs more on Earth: a cup of hot coffee or a cup of iced coffee? Why? Do you think the difference is measurable?

34. As explained in the Process of Science Figure, a new theory should contain the old theory within it. Study Figure 18.5, which compares two imaginary rocket ships experiencing the same acceleration, *g*.
    a. Approximately how fast (as a fraction of the speed of light) are the two spaceships going when the effects of relativity begin to be significant?
    b. Convert that speed to kilometers per hour. How does that speed compare with the speeds at which you usually travel? Why do you not usually see relativistic effects in your life?

35. Figure 18.6 shows how the Lorentz factor depends on speed. At about what speed (in terms of *c*) does the Lorentz factor begin to differ noticeably from 1? What happens to the Lorentz factor as the speed of an object approaches the speed of light?

36. Imagine that the perihelion of Mercury advances 2 degrees per century. How many arcseconds does the perihelion advance in a year? (Recall that 60 arcseconds are in an arcminute and 60 arcminutes in a degree.) Can Mercury's position be measured well enough to measure the advance of perihelion in 1 year?

37. If the Sun were twice as massive in Figure 18.12, would the distance between the apparent positions of stars 1 and 2 increase or decrease during an eclipse?

38. Follow Working It Out 18.1 to find out how much younger than your twin you would be if you made the journey described there at 0.5c.

39. Use Working It Out 18.1 to predict how much younger than your twin you would be if you traveled at 0.999c instead of 0.5c, as in question 38. Then calculate the difference in ages and compare the calculated result with your prediction.

40. What is the Schwarzschild radius of a black hole that has a mass equal to the average mass of a person (~70 kilograms)?

41. What is the mass of a black hole with a Schwarzschild radius of 1.5 km?

42. The Moon has a mass equal to $3.7 \times 10^{-8} M_{Sun}$. Suppose the Moon suddenly collapsed into a black hole.
    a. What would be the Schwarzschild radius of the black-hole Moon?
    b. How would that collapse affect tides raised by the Moon on Earth? Explain.
    c. Would that event generate gravitational waves? Explain.

43. If a spaceship approaching Earth at 0.9c shines a laser beam at Earth, how fast will the photons in the beam be moving when they arrive at Earth?

44. Suppose you discover signals from an alien civilization coming from a star 25 light-years away, and you go to visit it by using the spaceship described in the discussion of the twin paradox in Working It Out 18.1.
    a. How long will you take to reach that planet, according to your clock? According to a clock on Earth? According to the aliens on the other planet?
    b. How likely is it that someone you know will be here to greet you when you return to Earth?

45. Working It Out 18.2 relates the mass of a binary pair to the period and the size of the orbit. Suppose that a spaceship orbited a black hole at a distance of 1 AU, with a period of 0.5 year. What assumptions could you make that would allow you to calculate the mass of the black hole from that information? Make those assumptions, and calculate the mass.

USING THE WEB

46. Go to NASA's Swift Gamma-Ray Burst Mission website (https://www.nasa.gov/mission_pages/swift/main). Locate a recent result related to supernovae, gamma-ray bursts, or stellar black holes. Why would two merging neutron stars be likely to form a black hole?

47. NASA's Fermi Gamma-ray Space Telescope (https://www.nasa.gov/content/fermi-gamma-ray-space-telescope and https://fermi.gsfc.nasa.gov/) is exploring the gamma-ray universe. What objectives of that mission relate to the study of black holes? What is a recent news story related to black holes?

48. Go to the LIGO website (https://ligo.org/science.php) and read about gravitational waves. Click on "Sources of Gravitational Waves" and listen to the example. What are the differences among the four listed sources of gravitational waves? Click on "Advanced LIGO flyer." What's new with the project?

49. NuSTAR (Nuclear Spectroscopic Telescopic Array) is a NASA mission to study black holes, gamma-ray bursts, and neutron stars. Go to the NuSTAR website (https://www.nustar.caltech.edu/news). What wavelengths and energies does that telescope observe? What has been observed? What new science has been learned?

50. Go to the "Inside Black Holes" website (https://jila.colorado.edu/~ajsh/insidebh/), enter, and click on "Schwarzschild." Work your way down the page, watching the videos. What does it look like when you go into a black hole? Why does gravitational lensing occur when Earth is in orbit? What happens when you fall through the event horizon: is everything black? Click on "Reissner-Nordström" to see an electrically charged black hole. What is a wormhole? Why is a warning at the top of the page? Click on "4D perspective." What does it look like if you move toward the Sun at the speed of light?

Because grabbing a black hole and bringing it into the lab is not possible, and because Earth has never actually been close to one, astronomers can conduct only thought experiments to explore the properties of black holes. Following are a few thought experiments to help you think about what's happening near a black hole.

Imagine a big rubber sheet. It is very stiff and not easily stretched, but it does have some "give" to it. At the moment, it is perfectly flat. Imagine rolling some golf balls across it.

1 Describe the path of the golf balls across the sheet.

_____

_____

Now imagine putting a bowling ball (very much heavier than a golf ball) in the middle of the sheet, so that it makes a big, slope-sided pit. Roll some more golf balls.

2 What happens to the path of the golf balls when they are very far from the bowling ball?

_____

_____

3 What happens to the path of the golf balls when they come just inside the edge of the dip?

_____

_____

_____

4 What happens to the path of the golf balls when they go directly toward the bowling ball?

_____

_____

_____

5 How do each of the three cases in questions 2–4 change if the golf balls are moving very, very fast? What if they are moving very slowly?

_____

_____

6 What happens to the depth and width of the pit as the golf balls fall into the center near the bowling ball? (Imagine putting *lots* of golf balls in.)

_____

_____

All the preceding thought experiments relate to ordinary stuff. Stars, people, planets—everything interacts in that way because of gravity. With black holes, things are different. Here, it is more accurate to think of the bowling ball as a hole in the sheet that pulls it down rather than as an object that sits on it. But the bowling ball still affects the sheet in the same way. The hole is a good analogy for the event horizon of a black hole. Objects outside the event horizon will know that the black hole is there because the sheet is sloping, but they won't be captured unless they come within the event horizon. Think about light for a moment as though it were, say, grains of sand rolling across the sheet.

7 What happens to the light as it passes far from the pit? What happens if it reaches the hole?

_____

_____

Now suppose you roll another bowling ball across the sheet.

8 What happens to the sheet when the second bowling ball falls in after the first? Would that change affect your golf balls and grains of sand? How? What happens to the hole? What happens to the size of the pit?

_____

_____

_____

_____

None of those thought experiments take into account relativistic effects (length contraction and time dilation). Imagine for a moment that you are traveling close to the black hole.

9 Look out into the galaxy and describe what you see. Consider the lifetimes of stars, the distances between them, their motions in your sky, and how they die. Add anything else that occurs to you.

_____

_____

_____

_____

19

Galaxies

L ess than a century has passed since astronomers realized that the universe is filled with huge collections of stars, gas, and dust called galaxies. Just as stars vary in their mass or their stage of evolution, galaxies come in many forms. Chapter 19 begins our discussion of galaxies with a survey of the types of galaxies and their basic properties to understand better the differences among them. Chapter 20 then looks in detail at our galaxy, the Milky Way, and in later chapters we describe the evolution of galaxies and of the universe itself.

....

LEARNING GOALS

By the end of this chapter, you should be able to:

LG 1 Determine a galaxy's type from its appearance and describe the motions of its stars.

LG 2 Explain the distance ladder and how distances to galaxies are measured.

LG 3 Describe the evidence suggesting that galaxies are composed mostly of dark matter.

LG 4 Discuss the evidence indicating that most—perhaps all—large galaxies have supermassive black holes at their centers.

The large, barred spiral galaxy NGC 1300 is about 20 million parsecs away. ▶ ▶ ▶

How do
galaxies get
their shapes?

19.2 Astronomers Use Several Methods to Find Distances to Galaxies

To determine the distances to galaxies, astronomers started with the closest objects and looked for patterns that would help them find distances to the farthest objects—just as they did for stars. Distances are measured in a series of different methods called the **distance ladder**, which relates distances on a variety of overlapping scales, each method building on the previous one. In this section, we discuss those distance methods and how they led to the surprising discovery that the galaxies are moving away from us.

The Distance Ladder

The distance ladder is summarized in **Figure 19.10**. In the 18th and 19th centuries, astronomers used several methods to estimate the astronomical unit (AU; the average distance from Earth to the Sun) and thus the distances to the planets. (Since the 1960s, distances within the Solar System have been found by using radar and signals from space probes.) Once the value of the AU was known, astronomers used trigonometric parallax (see Chapter 13) to measure distances to nearby stars and thereby to build up the Hertzsprung-Russell (H-R) diagram. For more distant stars, astronomers use the spectral and luminosity classification of a star to determine its position on the H-R diagram. That position provides a star's luminosity, enabling astronomers to then estimate its distance by comparing its apparent brightness with its luminosity through using spectroscopic parallax (see Chapter 13 and Appendix 7).

Astronomers measure the distance to relatively nearby galaxies by using *standard candles*, a term borrowed from an old unit of light intensity that was based on actual candles. **Standard candles** are objects that have a known luminosity, usually because they have been observed within the Milky Way.

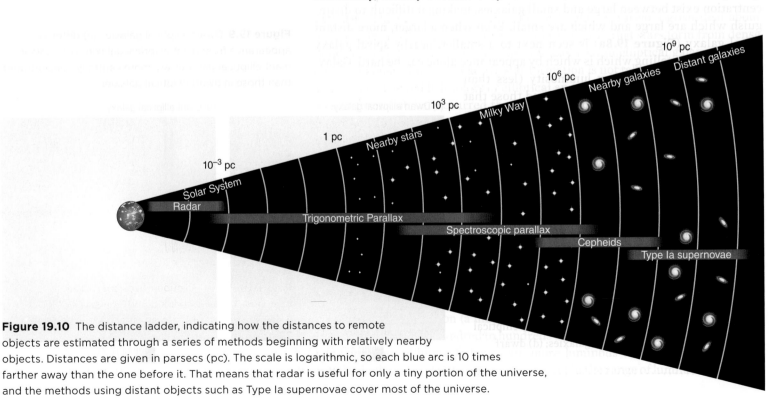

Figure 19.10 The distance ladder, indicating how the distances to remote objects are estimated through a series of methods beginning with relatively nearby objects. Distances are given in parsecs (pc). The scale is logarithmic, so each blue arc is 10 times farther away than the one before it. That means that radar is useful for only a tiny portion of the universe, and the methods using distant objects such as Type Ia supernovae cover most of the universe.

The objects must also be bright enough to be recognizable in the distant galaxy. Astronomers assume that the luminosity of each object is the same as that for a similar object of that type in the Milky Way, and then they compare the luminosity and apparent brightness of the standard candle to find its distance.

Objects that can be used as standard candles include main-sequence O stars, globular clusters, planetary nebulae, novae, Cepheid variable stars, and supernovae. For example, Hubble Space Telescope (HST) observations of Cepheid variables enable astronomers to measure distances accurately to galaxies as far away as 30 million parsecs (also called 30 megaparsecs [Mpc]). Even more luminous than the Cepheids, and thus detectable at greater distances, are Type Ia supernovae.

Recall from Chapter 16 that Type Ia supernovae can occur when gas flows from an evolved star onto its white dwarf companion, pushing the white dwarf up toward the Chandrasekhar limit for the mass of an electron-degenerate object: 1.4 M_{Sun}. When that happens, the white dwarf burns carbon, collapses, and then explodes. At first, astronomers thought that those Type Ia supernovae all occurred in white dwarfs of just below 1.4 M_{Sun}. In that situation, all such explosions would occur at the same mass and have similar luminosity, with some calibration adjustment for the rate at which the brightness declines after it peaks. But now astronomers estimate that 80 percent of Type Ia supernovae come from double-degenerate systems—for example, two white dwarfs that merge and then explode. The double white dwarf supernova could have up to twice the mass of a single white dwarf supernova. In addition, Type Ia supernovae at large distances may have different colors or compositions than nearby ones, since they formed when galaxies were younger. To test whether they all have about the same luminosity, astronomers observe nearby Type Ia supernovae in galaxies with distances determined from other methods, such as by Cepheid variables, and see that those do have similar luminosities. With a peak luminosity that can outshine a billion Suns (**Figure 19.11**), Type Ia supernovae can be seen and measured with modern telescopes at very large distances (**Working It Out 19.1**).

The motions of the stars in galaxies give rise to two "secondary" methods for estimating the distance to a galaxy. In rotating spiral galaxies, some of the light is approaching Earth and is thus blueshifted, whereas some light is moving away from Earth and is thus redshifted. Those redshifts and blueshifts together broaden a spectral line in the galaxy, and the amount of broadening indicates how fast the galaxy is rotating. Astronomers can measure the broadening from the Doppler shifts of the 21-centimeter (21-cm) radio emission line of hydrogen, which tells them the speed of rotation, which then relates to the galaxy's mass by Newton's version of Kepler's third law. The more massive galaxies have more stars and are therefore more luminous. That empirical relation between the measured width of the 21-cm line and the luminosity of the spiral galaxy is called the Tully-Fisher relation. Once the luminosity of the galaxy is known, it can be compared with the galaxy's observed apparent brightness to estimate its distance. That method is thought to work out to about 100 Mpc.

Elliptical galaxies and the bulges of S0 galaxies do not rotate, so instead astronomers look at the distribution of the surface brightness of a galaxy. Closer galaxies show more variations in the surface brightness because the distribution of stars throughout the galaxy isn't perfectly uniform. For more distant galaxies, those variations are less noticeable, and the surface brightness appears more uniform across the galaxy. That method yields less accurate results than the Tully-Fisher method for spirals, but generally it also is thought to work out to about 100 Mpc.

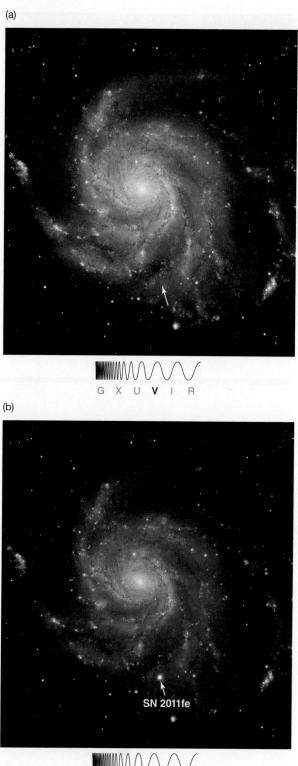

Figure 19.11 Type Ia supernovae are extremely luminous standard candles. The Pinwheel Galaxy is 6.4 Mpc away, shown before (a) and after (b) Supernova 2011fe.

How do astronomers use a standard candle to estimate distance? Figure 19.11 shows the Pinwheel Galaxy (M101), with a supernova that was observed in 2011. That supernova was detected before its brightness reached its peak. Astronomers can then compare the peak observed brightness of this supernova with the peak luminosity for that type of supernova to compute the distance.

In Section 13.1, we gave the equation relating brightness, luminosity, and distance:

$$\text{Brightness} = \frac{\text{Luminosity}}{4\pi d^2}$$

Rearranging to solve for distance gives:

$$d = \sqrt{\frac{\text{Luminosity}}{4\pi \times \text{Brightness}}}$$

The maximum observed brightness of the supernova was 7.5×10^{-12} watts per square meter (W/m^2). The graph in Figure 17.11 shows that the typical maximum luminosity L of a Type Ia supernova is 9.5×10^9 times the luminosity of the Sun:

$$L = 9.5 \times 10^9 \times L_{\text{Sun}}$$

$$L = 9.5 \times 10^9 \times (3.9 \times 10^{26}\,\text{W}) = 3.7 \times 10^{36}\,\text{W}$$

Thus, we can solve the equation:

$$d = \sqrt{\frac{3.7 \times 10^{36}\,\text{W}}{4\pi \times 7.5 \times 10^{-12}\,\text{W/m}^2}} = 2.0 \times 10^{23}\,\text{m}$$

To put that value into megaparsecs:

$$d = \frac{2.0 \times 10^{23}\,\text{m}}{3.1 \times 10^{22}\,\text{m/Mpc}} = 6.4\,\text{Mpc}$$

The distance is 6.4 Mpc. Because the Pinwheel Galaxy is relatively close, other standard candles can be observed in that galaxy to help calibrate the distance. Also note that 6.4 Mpc = 21 million light-years, so that supernova explosion took place 21 million years ago.

The Discovery of Hubble's Law

In the 1920s, Hubble and his coworkers were studying the properties of a large collection of galaxies. Another astronomer, Vesto Slipher (1875–1969), was obtaining spectra of those galaxies at Lowell Observatory in Flagstaff, Arizona. Slipher's galaxy spectra looked like the spectra of ensembles of stars with a bit of glowing interstellar gas mixed in. But he was surprised to find that the emission and absorption lines in the spectra of those galaxies were rarely seen at the same wavelengths as in the spectra of stars observed in the Milky Way Galaxy. The lines were almost always shifted to longer wavelengths, as seen in **Figure 19.12**.

Slipher characterized most of the observed shifts in galaxy spectra as redshifts because the light from those galaxies is shifted to longer (redder) wavelengths. Hubble interpreted Slipher's redshifts as Doppler shifts, concluding that almost all the galaxies in the universe are moving away from the Milky Way. Recall from Chapter 5 that objects with larger Doppler redshifts are moving away more quickly than those with smaller redshifts. When Hubble combined the measurements of galaxy velocities with his own estimates of the distances to those galaxies, he found that distant galaxies are moving away from Earth more rapidly than are nearby galaxies. Specifically, *the velocity at which a galaxy is moving away from an observer is proportional to the distance of that galaxy.* That relationship between distance and recession velocity is known as **Hubble's law**.

AstroTour: Hubble's Law

Figure 19.13 plots the measured recession velocities of galaxies against their measured distances. Because the velocity and distance are proportional to each other, the points lie along a line on the graph with a slope equal to the proportionality constant H_0, called the **Hubble constant**. Notice how well the data line up along the line. That strong correlation indicates that the nearby universe follows Hubble's law; for example, a galaxy 30 Mpc from Earth moves away twice as fast as a galaxy 15 Mpc distant. The original value for the Hubble constant was

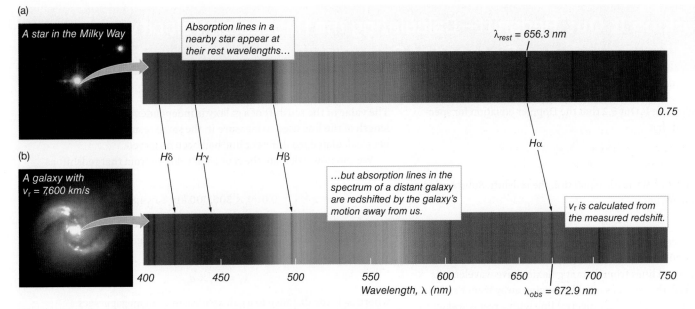

Figure 19.12 (a) The spectrum of a star in our galaxy shows absorption lines, which here lie at the rest wavelength. (b) A distant galaxy, shown with its spectrum at the same scale as that of the star, has lines redshifted to longer wavelengths. v_r is recession velocity, or radial velocity.

8 times too large, which led to inconsistencies, but the problem was resolved when astronomers realized that two types of Cepheids exist with slightly different period-luminosity relationships. Today, astronomers have measured the Hubble constant to an accuracy of a few percent by several methods, using data from several space observatories. The value is likely to be further refined in the years to come. In this text, we use a value of 70 km/s/Mpc as an approximation to the best current measured values of 67 to 74 km/s/Mpc, with uncertainties of 1–3 km/s/Mpc, depending on the measurement. The unit means that for every additional megaparsec of distance, the galaxy's motion includes an additional 70 km/s.

Using that constant, Hubble's law is written as $v_r = H_0 \times d_G$, where d_G is the distance to the galaxy, and v_r is the galaxy's recession velocity (**Working It Out 19.2**). Hubble's law gives astronomers a practical tool for measuring distances to remote galaxies. Once astronomers know the value of H_0, they can use a straightforward measurement of the redshift of a galaxy to find its distance. In other words, once H_0 is known, Hubble's law makes the once-difficult task of measuring distances in the universe relatively easy, giving astronomers a tool to map the structure of the observable universe. We will return to Hubble's law and its implications for understanding the universe as a whole in Chapter 21.

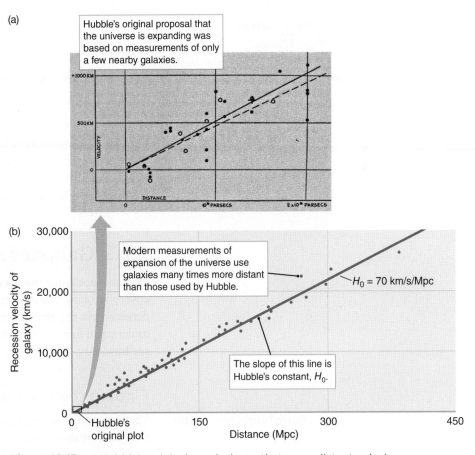

Figure 19.13 (a) Hubble's original graph shows that more distant galaxies are receding faster than less distant galaxies. (b) Modern data on galaxies many times farther away than those studied by Hubble show that recession velocity is proportional to distance.

19.2 Working It Out Redshift—Calculating the Recession Velocity and Distance of Galaxies

Recall from Working It Out 5.2 that the Doppler equation for spectral lines showed that

$$v_r = \frac{\lambda_{obs} - \lambda_{rest}}{\lambda_{rest}} \times c$$

The fraction in front of the c is equal to z, the redshift. Substituting z for the fraction, we get

$$v_r = z \times c$$

(Note: That correspondence between velocity and redshift requires a correction as velocities approach the speed of light.)

Because spectral lines from distant galaxies have wavelengths shifted to the red, the galaxies must be moving away from Earth. Suppose astronomers observe a spectral line with a rest wavelength of 373 nanometers (nm) in the spectrum of a distant galaxy. If the observed wavelength of the spectral line is 379 nm, its redshift (z) is

$$z = \frac{\lambda_{obs} - \lambda_{rest}}{\lambda_{rest}}$$

$$z = \frac{379\,nm - 373\,nm}{373\,nm} = 0.0161$$

The value of the redshift of a galaxy is independent of the wavelength of the line used to measure it: the same result would have been calculated if a different line had been observed.

We can now calculate the recession velocity from that redshift as follows:

$$v_r = z \times c = 0.0161 \times 300,000\ km/s = 4,830\ km/s$$

How far away is that distant galaxy? Here is where *Hubble's law* and the *Hubble constant* apply. Hubble's law relates a galaxy's recession velocity to its distance as

$$v_r = H_0 \times d_G$$

where d_G is the distance to a galaxy measured in megaparsecs. Dividing through by $H_0 = 70$ km/s/Mpc yields

$$d_G = \frac{v_r}{H_0} = \frac{4,830\,km/s}{70\,km/s/Mpc} = 69\ Mpc$$

From a measurement of the wavelength of a spectral line, we see that the distant galaxy is approximately 69 Mpc away.

CHECK YOUR UNDERSTANDING 19.2

Hubble's law was discovered by using measurements of two properties of a galaxy: _____ and _____. (a) size; mass (b) distance; rotation speed (c) distance; recession velocity (d) size; recession velocity

19.3 Galaxies Are Mostly Dark Matter

Efforts to measure the masses of galaxies during the 20th century led to the discovery of dark matter—mass that does not interact with light and cannot be detected via the light they emit. To understand that discovery, we first need to understand how astronomers go about measuring the mass of a galaxy and then see how they concluded that much of the mass in a galaxy is dark matter.

Finding the Mass of a Galaxy

AstroTour: Dark Matter

To measure the mass of a galaxy, astronomers add up the mass of the stars, dust, and gas that they observe. Because a galaxy's spectrum is composed primarily of starlight, once astronomers know what types of stars are in the galaxy, they can use what is known about stellar evolution to estimate the total stellar mass from the galaxy's luminosity. Astronomers then estimate the mass of the dust and gas by using the physics of radiation from interstellar gas at X-ray, infrared, and radio wavelengths. Together, the stars, gas, and dust in a galaxy are called **luminous matter** (or simply **normal matter**) because that matter emits or scatters electromagnetic radiation.

For determining mass, an independent method is available that does not involve luminosity—namely, the effect of gravity on an object's motion. Stars in disks follow orbits that are much like the orbits of planets around their parent stars and binary stars around each other (see Working It Out 13.4). To measure the mass of a spiral galaxy, astronomers apply Kepler's laws, just as they do for those other systems.

Dark Matter

Astronomers originally hypothesized that the mass and the light in a galaxy are distributed in the same way; that is, they assumed that all the mass in a galaxy is luminous matter. They observed that the light of all galaxies, including spiral galaxies, is highly concentrated toward the center (**Figure 19.14a**). On the basis of the observed location of light, astronomers predicted that nearly all mass in a spiral galaxy is concentrated toward its center (**Figure 19.14b**). The situation is much like that in the Solar System, where nearly all the mass is in the Sun—at the center of the Solar System. Therefore, they predicted faster orbital velocities near the center of the spiral galaxy and slower orbital velocities farther out (**Figure 19.14c**).

To test that prediction, astronomers used the Doppler effect to measure orbital motions of stars, gas, and dust at various distances from a galaxy's center. The velocities of stars are obtained from observations of absorption lines in their spectra. The velocities of interstellar gas are obtained using emission lines such as those produced by hydrogen alpha (Hα) emission or 21-cm emission from neutral hydrogen (see Chapter 15). Once the velocities have been found, astronomers create a graph—called a **rotation curve**—that shows how orbital velocity in a galaxy varies with distance from the galaxy's center. The rotation curve of a spiral galaxy enables astronomers to determine directly how the mass in that galaxy is distributed.

Vera Rubin (1928–2016) pioneered work on galaxy rotation rates in the 1970s. She discovered that, contrary to earlier prediction (Figure 19.14c), the rotation velocities of spiral galaxies remain about the same out to the most distant measured parts of the galaxies (**Figure 19.14d**). Observations of 21-cm radiation from neutral hydrogen show that the rotation curves appear level, or "flat," in their outer parts even well outside the extent of the visible disks. Those observations indicated that mass and light are distributed differently.

What mass distribution would cause that unexpected rotation curve? Recall from Chapter 4 that only the mass inside a given radius contributes to the net gravitational force experienced by an orbiting object. From the rotation velocity, you can calculate the mass within the orbit of the object (**Figure 19.15**). The black line shows the measured speed of rotation of that galaxy at a particular radius. The pink line shows how much luminous mass is seen inside that radius. To produce a rotation curve like the one shown in black, that galaxy must have a second component consisting of matter that does not show up in the census of stars, gas, and dust. That material, which does not interact with light, and therefore reveals itself only by the influence of its gravity, is called **dark matter**. The purple line

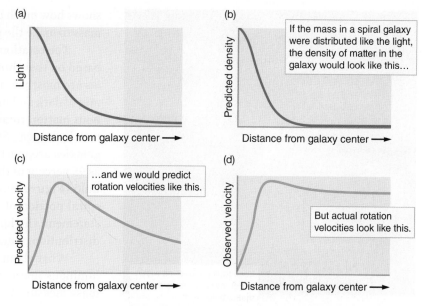

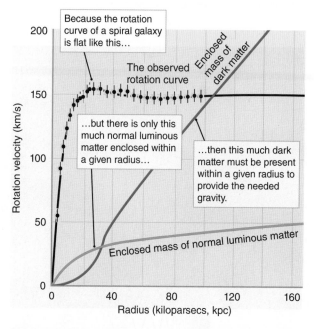

Figure 19.14 (a) The profile of visible light in a typical spiral galaxy drops off with distance from the center. (b) The predicted mass density of stars and gas located at a given distance from the galaxy's center follows the light profile. If stars and gas accounted for all of the mass of the galaxy, then the galaxy's rotation curve would be as shown in (c). However, observed galaxy rotation curves look more like the curve shown in (d).

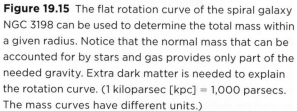

Figure 19.15 The flat rotation curve of the spiral galaxy NGC 3198 can be used to determine the total mass within a given radius. Notice that the normal mass that can be accounted for by stars and gas provides only part of the needed gravity. Extra dark matter is needed to explain the rotation curve. (1 kiloparsec [kpc] = 1,000 parsecs. The mass curves have different units.)

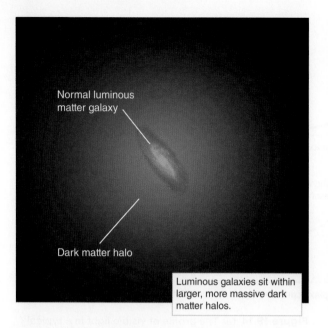

Normal luminous
matter galaxy

Dark matter halo

Luminous galaxies sit within
larger, more massive dark
matter halos.

Figure 19.16 In addition to the matter that is visible, galaxies are surrounded by halos containing a large amount of dark matter.

G X U V I R

Figure 19.17 In this combined visible-light and X-ray image of elliptical galaxy NGC 1132, the false-color blue and purple halo is X-ray emission from hot gas surrounding the galaxy. The hot gas extends well beyond the visible light from stars.

shows how much dark matter must be inside a particular radius to supply enough mass to make the galaxy rotate as it does.

The rotation curves of the inner parts of spiral galaxies match predictions based on their luminous matter, indicating that the inner parts of spiral galaxies are mostly luminous matter. Within the entire *visual* image of a galaxy, the mix of dark and luminous matter is about half and half. However, past the luminous matter, rotation curves can be measured with 21-cm radiation from neutral hydrogen. Such measurements indicate that the outer parts of spiral galaxies are mostly dark matter. Astronomers currently estimate that as much as 95 percent of the total mass in some spiral galaxies consists of a **dark matter halo** (**Figure 19.16**), which can extend up to 10 times farther than the visible spiral portion of the galaxy located at the galaxy's center. That is a startling statement: the luminous part of a spiral galaxy is only a part of a much larger distribution of mass dominated by some type of invisible dark matter.

What about elliptical galaxies? Again, astronomers need to compare the luminous mass measured from the light they can see with the gravitational mass measured from the effects of gravity. Because elliptical galaxies do not rotate, astronomers cannot use Kepler's laws to measure the gravitational mass from the rotation of a disk. Instead, astronomers noticed that an elliptical galaxy's ability to hold on to its hot, X-ray–emitting gas depends on its mass. If the galaxy is not massive enough, the hot atoms and molecules will escape into intergalactic space. To find the mass of an elliptical galaxy, astronomers first infer the total amount of gas from X-ray images, such as the (false color) blue and purple halo seen in **Figure 19.17**. Then they calculate the total mass needed to hold on to the gas and compare that gravitational mass with the luminous mass. The amount of dark matter is the difference between what is needed to hold on to the inferred amount of gas and the observed amount of luminous matter.

Some elliptical galaxies contain up to 20 times as much mass as can be accounted for by their stars and gas alone, so they must be dominated by dark matter, just like spiral galaxies. As with spirals, the luminous matter in ellipticals is more centrally concentrated than the dark matter. The transition from the inner parts of galaxies (where luminous matter dominates) to the outer parts (where dark matter dominates) is remarkably smooth. Some galaxies may contain less dark matter than others, but about 90–95 percent of the total mass in a typical galaxy is in the form of dark matter.

The high percentage of dark matter distinguishes smaller dwarf galaxies from globular clusters, which do not have dark matter. That difference is an important observation that will need to be explained in the context of the evolution of galaxies.

The Composition of Dark Matter

What is the dark matter that makes up most of a galaxy? Several candidates have been investigated, including objects such as large planets, compact stars, black holes, and exotic unknown elementary particles.

At first, astronomers considered small main-sequence M stars, Jupiter-sized planets, white dwarfs, neutron stars, or black holes as dark matter candidates. Those are collectively referred to as **MaCHOs**, which stands for *massive compact halo objects*. If the dark matter in a galaxy's halo consists of MaCHOs, a lot of those objects must be present, and they must each exert gravitational force but not emit much light. If astronomers were observing a distant star and a MaCHO passed between Earth and the star, the star's light would be deflected and, if the geometry were just right, focused by the intervening MaCHO as it passed across their line of sight. Astronomers monitored tens of millions of stars in two small companion galaxies of the Milky Way for several years, but they did not see enough of those gravitational lensing events to account for the amount of dark matter in the halo of our galaxy. Thus, MaCHOs with mass less than 100 M_{Sun} are ruled out, implying that dark matter in the Milky Way (and therefore other galaxies) must be composed of something other than MaCHOs.

Other types of candidates for dark matter are unknown elementary particles, called **WIMPs**, which stands for *weakly interacting massive particles*. WIMPs are predicted to be similar to neutrinos in that they would barely interact with ordinary matter, yet they would be more massive and would move more slowly. Many current experiments are seeking to directly or indirectly detect WIMPs. There are experiments at the Large Hadron Collider to produce and detect a WIMP. There are also multiple experiments in laboratories and in deep mines to detect halo dark matter particles as they pass through Earth. Other observations are attempting to detect the (very weak) interactions of dark matter particles in the halos of galaxies. As of this writing, there have been no confirmed direct or indirect detections of dark matter particles.

Thus, physicists and astronomers are constructing larger detectors to better catch WIMPs. Researchers are considering other types of exotic elementary particles as potential dark matter candidates. Some are exploring theoretical alternatives to dark matter, in which Newtonian dynamics or Newtonian gravity behave differently under conditions like those found in galaxies. Some of those alternative models can produce the observed rotation curves in spiral galaxies, but they cannot replace dark matter as an explanation for other aspects of galaxy evolution, as we will see in later chapters.

CHECK YOUR UNDERSTANDING 19.3

Astronomers detect dark matter: (a) by comparing luminous mass with gravitational mass; (b) because it blocks background light; (c) because more distant galaxies move away faster; (d) because it emits lots of X-rays.

19.4 Most Galaxies Have a Supermassive Black Hole at the Center

Studying the centers of galaxies is difficult because so many stars and so much dust and gas are in the way that astronomers cannot get a clear picture of the center, even for nearby galaxies. Instead, observations of the most distant objects in the universe yielded the clues to understanding what lies in the centers of massive galaxies.

The Discovery of Quasars

In the late 1950s, radio surveys detected several bright, compact objects that at first seemed to have no optical counterparts. Improved radio positions revealed that the radio sources coincided with faint, very blue, starlike objects. Unaware of

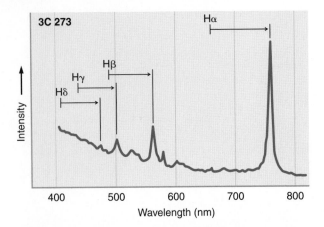

Figure 19.18 This graph shows the spectrum of quasar 3C 273, one of the closest and most luminous known quasars. The emission lines are redshifted by $z = 0.16$ from marked rest wavelengths, indicating that the quasar is at a distance of about 750 Mpc.

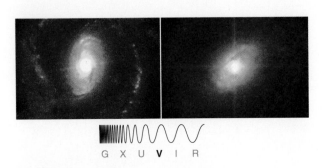

Figure 19.19 These HST images show quasars embedded in the centers of galaxies.

Figure 19.20 Radio galaxy Centaurus A is the closest AGN to Earth, at a distance of 3.4 Mpc. In this composite image, the visible-light image shows the galaxy, the X-ray image (pink) shows the hot gas and an energetic jet blasting from the AGN, and the radio image (purple) shows the jets and lobes.

the true nature of those objects, astronomers called them "radio stars." Obtaining spectra of the first two radio stars was a laborious task, requiring 10-hour exposures. Astronomers were greatly puzzled by the results because those spectra did not display the expected absorption lines characteristic of blue stars. Instead, the spectra showed only a single pair of emission lines that were broad—indicating very rapid motions within the objects—and that did not seem to correspond to the lines of any known substances.

For several years, astronomers believed they had discovered a new type of star until astronomer Maarten Schmidt realized that those broad spectral lines, shown in **Figure 19.18**, were the highly redshifted lines of ordinary hydrogen. The implications were surprising: those "stars" were not stars. They were extraordinarily luminous objects at enormous distances. Those "quasi-stellar radio sources" were named **quasars**. Other quasars were soon found by the same techniques, and astronomers began cataloging them.

Quasars are phenomenally powerful, shining with the luminosity of a trillion to a thousand trillion (10^{12} to 10^{15}) Suns. They also are very distant from Earth: hundreds or thousands of megaparsecs. Billions of galaxies are closer to Earth than the nearest quasar is. Recall that the distance to an object also indicates how much time has passed since the light from that object left its source. The fact that quasars are seen only at great distances implies that they are rare in the universe now but were once much more common. The discovery that quasars existed in the distant and therefore earlier universe gave astronomers one of the first pieces of evidence demonstrating that the universe has evolved.

Quasars are now recognized as the result of the most extreme form of activity that can occur in the nuclei of galaxies (**Figure 19.19**), often resulting from interactions with other galaxies. Quasars are a type of **active galactic nucleus** (or **AGN**, which can stand for either *nucleus* or *nuclei*). The distinct types of active nuclei are identified from the spectrum of the galaxy. A "normal" galaxy has an absorption spectrum that is a composite of the light from its billions of stars. A galaxy with an AGN exhibits emission lines in addition to the stellar absorption spectrum. Thus, AGN are identified by the emission lines in their spectra, which distinguishes them from normal galaxies.

AGN come in several types and can occur in spiral or elliptical galaxies. **Seyfert galaxies**, named after Carl Seyfert (1911–1960), who discovered them in 1943, are spiral galaxies whose centers contain AGN. The luminosity of a typical Seyfert nucleus can be 10 billion to 100 billion L_{Sun}, comparable to the luminosity of the rest of the host galaxy as a whole. Similarly, **radio galaxies** are elliptical galaxies whose centers contain AGN; their emission is usually most prominent in radio wavelengths. Radio galaxies and the more distant and luminous quasars are often the sources of slender jets that extend outward millions of light-years from the galaxy, powering twin lobes of radio emission (**Figure 19.20**).

Much of the light from AGN is synchrotron radiation—the same type that comes from extreme environments such as the Crab Nebula supernova remnant. Synchrotron radiation comes from relativistic charged particles spiraling around the direction of a magnetic field. The fact that AGN accelerate large amounts of material to nearly the speed of light indicates that they are very violent objects. In addition to the continuous spectrum of synchrotron emission, the spectra of many quasars and Seyfert nuclei also show emission lines that are smeared out by the Doppler effect across a wide range of wavelengths. That observation implies that gas in AGN is swirling around the centers of those galaxies at speeds of thousands or even tens of thousands of kilometers per second.

AGN Are the Size of the Solar System

The enormous radiated power and mechanical energy of AGN are made even more spectacular by the fact that all that power emerges from a region that can be no larger than a light-day or so across—comparable in size to the Solar System. Although the HST and large, ground-based telescopes show faint fuzz—light from the surrounding galaxy—around the images of some quasars and other AGN, the objects themselves remain as unresolved points of light.

To understand why astronomers think that AGN are compact objects, think about the halftime show at a football game. **Figure 19.21** illustrates a problem faced by every director of a marching band. When a band is all together in a tight formation at the center of the field, the notes you hear in the stands are clear and crisp; the band plays together beautifully. But, as the band spreads out across the field, its sound begins to get mushy. That is not because the marchers are poor musicians. Instead, it is because sound travels at a finite speed. Sound travels at a speed of about 330 meters per second (m/s). At that speed, sound takes approximately 1/3 of a second to travel from one end of the football field to the other. Even if every musician on the field plays a note at exactly the same instant in response to the director's cue, in the stands you hear the instruments

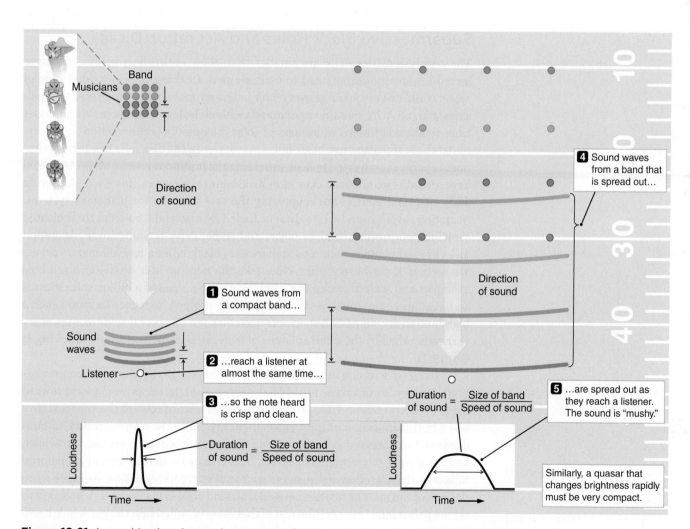

Figure 19.21 A marching band spread out across a field cannot play a clean note. Similarly, AGN must be very compact to explain their rapid variability.

AstroTour: Active Galactic Nuclei

Astronomy in Action: Size of Active Galactic Nuclei

close to you first but have to wait longer for the sound from the far end of the field to arrive.

If the band is spread from one end of the field to the other, the beginning of a note will be smeared out over about 1/3 of a second, the time sound takes to travel from one end of the field to the other. If the band were spread out over two football fields, the sound from the most distant musician would take about 2/3 of a second to arrive at your ear. If the marching band were spread out over a kilometer, it would be roughly 3 seconds—the time sound takes to travel a kilometer—before you heard a crisply played note start and stop. Even with your eyes closed, you could easily tell whether the band was in a tight group or spread out across the field.

The same principle applies to the light observed from AGN. Quasars and other AGN change their brightness dramatically over only a day or two—and sometimes as quickly as in a few hours. That rapid variability sets an upper limit on the size of the AGN, just as hearing clear music from a marching band indicates that the musicians are close together. The AGN powerhouse must therefore be no more than a light-day or so across because if it were larger, the light astronomers see could not possibly change in a day or two. An AGN has the light of up to 10,000 galaxies pouring out of a region of space that would come close to fitting within the orbit of Neptune.

Supermassive Black Holes and Accretion Disks

The conclusion that AGN are smaller than their entire host galaxy and have incredible energy outputs had to be explained. In thinking about what type of object could be very small yet very energetic, astronomers hypothesized that galaxies with an AGN contain **supermassive black holes**—black holes with masses from thousands to tens of billions of solar masses. Violent accretion disks surround those supermassive black holes. As matter in the disk falls inward, gravitational energy is converted to heat, providing the luminous energy of the AGN. You have already encountered accretion disks several times in this book. Accretion disks surround young stars, supplying the raw material for planetary systems. Accretion disks around white dwarfs, fueled by material torn from their bloated evolving companions, lead to novae and some Type Ia supernovae. Accretion disks around neutron stars and stellar-mass black holes a few kilometers across are seen as X-ray binary stars. Now take the neutron star or stellar black hole examples and scale them up to a black hole with a mass of a billion solar masses and a radius comparable in size to that of the orbit of Neptune. To attain such a high luminosity, an AGN has an accretion disk fed by several solar masses every year rather than by the small amounts of material siphoned off a star (**Working It Out 19.3**).

In our discussion of star formation in Chapter 15, we showed that gravitational energy is converted to thermal energy as material moves inward toward the growing protostar. Here, as material moves inward toward the supermassive black hole, conversion of gravitational energy heats the accretion disk to hundreds of thousands of kelvins, causing it to glow brightly in visible, ultraviolet, and X-ray light. Conversion of gravitational energy to thermal energy as material falls onto the accretion disk also is a source of energetic emission. As much as 20 percent of the mass of infalling material around a supermassive black hole is converted to luminous energy. The rest of that mass is pulled into the black hole itself, causing it to grow even more massive.

19.3 Working It Out The Size, Density, and Power of a Supermassive Black Hole

Size. What are the sizes of supermassive black holes? Recall from Chapter 18 that the Schwarzschild radii of stellar-mass black holes are kilometers in size. The formula for the Schwarzschild radius is given by

$$R_S = \frac{2GM_{BH}}{c^2}$$

where G is the gravitational constant and c is the speed of light.

The largest supermassive black holes observed have about 10 billion solar masses (M_{Sun}). For example, the black hole at the center of the galaxy M87 is 6.6 billion M_{Sun}. To compute its size, recall that $M_{Sun} = 1.99 \times 10^{30}$ kilograms (kg), $c = 3 \times 10^5$ km/s, and $G = 6.67 \times 10^{-20}$ km³/(kg s²). Then, a 6.6 billion-M_{Sun} black hole has a Schwarzschild radius of

$$R_S = \frac{2 \times [6.67 \times 10^{-20} \, \text{km}^3/(\text{kg s}^2)] \times (6.6 \times 10^9 \times 1.99 \times 10^{30} \, \text{kg})}{(3 \times 10^5 \, \text{km/s})^2}$$

$$R_S = 2.0 \times 10^{10} \, \text{km}$$

We can convert that value into astronomical units (AU), with 1 AU = 1.5×10^8 km. Therefore, that supermassive black hole has a radius of 130 AU—about 4 times the radius of Neptune's orbit. We know that light takes 8.3 minutes to reach Earth from the Sun at a distance of 1 AU, so 130 AU corresponds to a distance of 1,080 light-minutes, or 18 light-hours.

Density. What is the average density of that object inside of the event horizon? The mass of the black hole divided by the volume within the Schwarzschild radius is

$$\text{Density} = \frac{\text{Mass}}{\text{Volume}} = \frac{(6.69 \times 10^9) \times (1.99 \times 10^{30} \, \text{kg})}{4/3 \times \pi \times (2.0 \times 10^{10} \, \text{km})^3}$$

$$\text{Density} = 3.9 \times 10^8 \, \text{kg/km}^3 = 0.39 \, \text{kg/m}^3$$

That is about 1/3 of the density of air at sea level, or 1/2,500 the density of water. Supermassive black holes do not have the extremely high densities of stellar-mass black holes.

Feeding an AGN. Power for an AGN is produced when matter falls onto the accretion disk around the central supermassive black hole. Some of that high-velocity mass is radiated away according to Einstein's mass-energy equation: $E = mc^2$. About how much material has to be accreted to produce the observed luminosities? Astronomers estimate the efficiency of the accretion to be about 10–20 percent. Here, we'll assume that 15 percent of the infalling matter is radiated away as energy, or

$$E = 0.15 \, mc^2$$

Astronomers can measure how much energy is produced by the infalling material and radiated to space. For a relatively weak AGN like that of the galaxy M87, $L = 5 \times 10^{35}$ joules per second (J/s), or 5×10^{35} kg m²/s² each second. Recall that $c = 3 \times 10^8$ m/s. Dividing both sides of Einstein's equation by 0.15 c^2 gives us the mass consumed each second:

$$m = \frac{E}{0.15 \, c^2} = \frac{5 \times 10^{35} \, \text{kg m}^2/\text{s}^2}{0.15 \times (3 \times 10^8 \, \text{m/s})^2}$$

$$m = 3.7 \times 10^{19} \, \text{kg}$$

Multiplying that result (the mass consumed each second) by 3.2×10^7 seconds per year shows that the AGN accretes 10^{27} kg, or about half the mass of Jupiter, each year, which is then radiated away as energy.

If we consider a quasar with a luminosity (L) of 10^{39} J/s = 10^{39} kg m²/s² each second (= 2.5 trillion L_{Sun}), the mass accreted each second is given by

$$m = \frac{10^{39} \, \text{kg m}^2/\text{s}^2}{0.15 \times (3 \times 10^8 \, \text{m/s})^2}$$

$$m = 7.4 \times 10^{22} \, \text{kg}$$

Multiplying by 3.2×10^7 seconds per year yields a mass of 2.4×10^{30} kg per year. The mass of the Sun is 1.99×10^{30} kg, so to radiate that much energy, the quasar supermassive black hole is accreting about 1.2 M_{Sun} each year. A quasar with 10 times that luminosity would be accreting 10 times the mass.

The interaction of the accretion disk with the black hole creates powerful radio jets that emerge perpendicular to the disk (as in the jet in the upper left of Figure 19.20). Throughout, twisted magnetic fields accelerate charged particles such as electrons and protons to relativistic speeds, producing synchrotron emission. Gas in the accretion disk or in nearby clouds orbiting the central black hole at high speeds produces emission lines smeared out by the Doppler effect into the broad lines seen in AGN spectra. That accretion disk surrounding a supermassive black hole is the "central engine" that powers AGN.

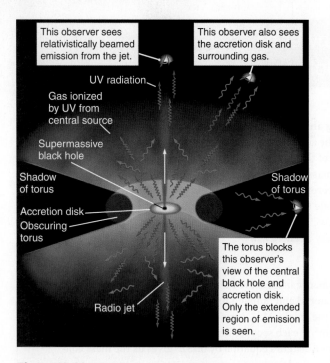

This observer sees relativistically beamed emission from the jet.

This observer also sees the accretion disk and surrounding gas.

UV radiation

Gas ionized by UV from central source

Supermassive black hole

Shadow of torus

Shadow of torus

Accretion disk

Obscuring torus

The torus blocks this observer's view of the central black hole and accretion disk. Only the extended region of emission is seen.

Radio jet

Figure 19.22 The basic model of an active galactic nucleus, with a supermassive black hole surrounded by an accretion disk at the center. A larger, dusty torus sometimes blocks the view of the black hole. The mass of the central black hole, the rate at which it is being fed, and the viewing angle determine the observational properties of an AGN.

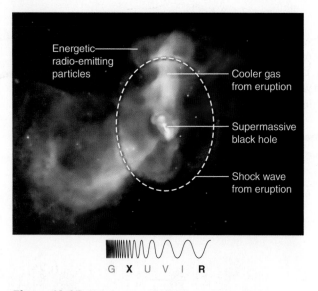

Energetic radio-emitting particles

Cooler gas from eruption

Supermassive black hole

Shock wave from eruption

G X U V I R

Figure 19.23 This image of M87 in radio and X-rays shows the location of the supermassive black hole.

The Unified Model of AGN

Astronomers have developed the basic picture of a supermassive black hole surrounded by an accretion disk into a more complete AGN model. The **unified model of AGN** attempts to explain all types of AGN—quasars, Seyfert galaxies, and radio galaxies. **Figure 19.22** shows the various components of that AGN model, in which an accretion disk surrounds a supermassive black hole. Much farther out from the accretion disk lies a large torus, or "doughnut" of gas and dust consisting of material feeding the central engine. Located far from the inner turmoil of the accretion disk, and far larger than the central engine, some of that torus is ionized by UV light from the AGN (see the **Process of Science Figure**).

In the unified model of AGN, the various AGN observed from Earth are partly explained by astronomers' view of the central engine. The torus of gas and dust obscures that view in different ways, depending on the viewing angle. Variation in that angle, in the mass of the black hole, and in the rate at which it is being fed accounts for a wide range of observed AGN properties. When the AGN is viewed edge on, astronomers see emission lines from the surrounding torus and other surrounding gas. Astronomers also can sometimes see the torus in absorption against the background of the galaxy. From that nearly edge-on orientation, they cannot see the accretion disk itself, so they do not see the Doppler-smeared lines that originate closer to the supermassive black hole. If jets are present in the AGN, though, they should be visible emerging from the center of the galaxy.

If astronomers observe the accretion disk more face on, they can see over the edge of the torus and thus get a more direct look at the accretion disk and the location of the black hole. Then they see more of the synchrotron emission from the region around the black hole and the Doppler-broadened lines produced in and around the accretion disk. **Figure 19.23** shows an image of one such object, the galaxy M87, at an intermediate inclination. M87 is a source of powerful jets that continue outward for 100,000 light-years but originate in the tiny engine at the heart of the galaxy. Spectra of the disk at the center of that galaxy show the rapid rotation of material around a central black hole that has a mass of 3 billion (3×10^9) M_{Sun}.

The material in an AGN jet travels very close to the speed of light, so what astronomers see is strongly influenced by relativistic effects. One of those effects is called **relativistic beaming**: matter traveling at close to the speed of light concentrates any radiation it emits into a tight beam pointed in the direction in which it is moving. So, astronomers often observe only one side of the jets from AGN, even though the radio lobes of radio galaxies are usually two-sided. The jet moving away is just too faint to observe.

In rare instances when the accretion disk in a quasar or radio galaxy is viewed almost directly face on, relativistic beaming dominates the observations. In those *blazars*, emission lines and other light coming from hot gas in the accretion disk are overwhelmed by the bright glare of the jet emission beamed directly at Earth.

Normal Galaxies and AGN

The essential elements of an AGN are a central engine (an accretion disk surrounding a supermassive black hole) and a source of fuel (gas and stars flowing onto the accretion disk). Without a source of matter falling onto the black hole, an AGN would no longer be an active nucleus. Astronomers looking at such an object would observe a normal galaxy with a supermassive black hole sitting in its center.

Finding the Common Thread

When first discovered, active galactic nuclei (AGNs) seemed to come in many different types, with dramatically different spectra. Later, it was realized that this could be an orientation effect: the many different types of objects could all be explained with one type of object, viewed from different angles.

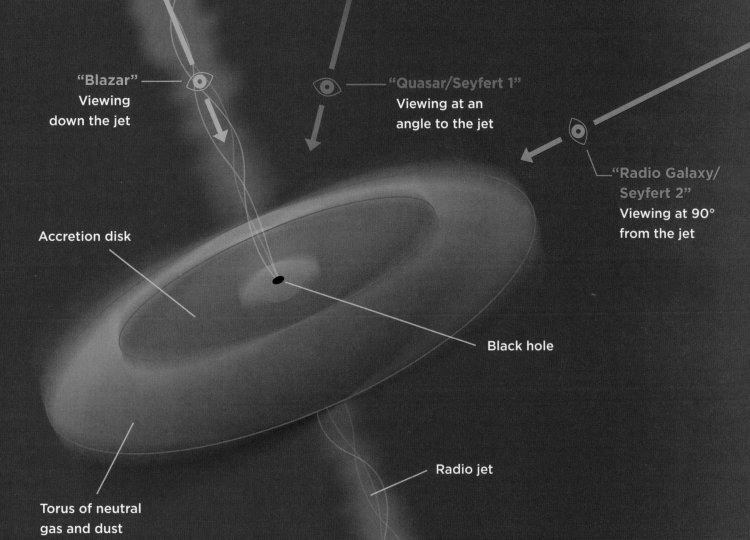

"Blazar"
Viewing
down the jet

"Quasar/Seyfert 1"
Viewing at an
angle to the jet

"Radio Galaxy/
Seyfert 2"
Viewing at 90°
from the jet

Accretion disk

Black hole

Radio jet

Torus of neutral
gas and dust

Scientists seek underlying principles that explain more than one phenomenon. In the case of AGNs, scientists developed one model to explain many types of objects. This model encompasses the idea that our classification can be affected by our viewing angle, which is true for galaxy shapes generally. A unified model is much simpler than having a different model for every type of object.

Only a few percent of present-day galaxies contain AGN as luminous as the host galaxy. But when astronomers look at more distant galaxies (and therefore look further back in time), the percentage of galaxies with AGN is much larger. These observations show that when the universe was younger, many more AGN existed than are there today. If astronomers' understanding of AGN is correct, all the supermassive black holes that powered those dead AGN should still be around. If astronomers combine what they know of the number of AGN in the past with ideas about how long a given galaxy remains in an active AGN phase, they are led to predict that many massive galaxies today contain supermassive black holes.

If supermassive black holes are present in the centers of normal galaxies, those black holes should reveal themselves in several ways. For one thing, such a concentration of mass at the center of a galaxy should draw surrounding stars close to it. The central region of such a galaxy should be much brighter than could be explained if stars alone were responsible for the gravitational field in the inner part of the galaxy. Stars experiencing the gravitational pull of a supermassive black hole in the center of a galaxy should also orbit at very high velocities and therefore show large Doppler shifts. Astronomers have found those large Doppler shifts in every normal galaxy with a substantial bulge in which a careful search has been conducted. The masses inferred for those black holes range from 10,000 to 20 billion M_{Sun}. The mass of the supermassive black hole seems to be related to the mass of the bulge in which it is found. Most large galaxies, whether elliptical or spiral, probably contain supermassive black holes. Those observations reveal something remarkable about the structure and history of normal galaxies.

Apparently, the only difference between a normal galaxy and an active galaxy is whether the supermassive black hole at its center is being fed when we see that galaxy. The rarity of present-day galaxies with very luminous AGN does not indicate which galaxies have the potential for AGN activity. Rather, it indicates which galaxy centers are being lit up at the moment. If a large amount of gas and dust were dropped directly into the center of any large galaxy, that material would fall inward toward the central black hole, forming an accretion disk and a surrounding torus. That process would change the nucleus of that galaxy into an AGN.

In Chapter 23, we will discuss galaxy evolution and note that many of the observed properties of galaxies discussed here in Chapter 19—including the type of galaxy that forms, spiral structure, star formation, and AGN—depend on the interactions and mergers between galaxies. To account for the many large galaxies visible today, interactions and mergers must have been much more prevalent in the past when the universe was younger; that is one explanation for the larger number of AGN that existed in the past. Computer models show that galaxy-galaxy interactions can cause gas located thousands of parsecs from the center of a galaxy to fall inward toward the galaxy's center, where the gas can provide fuel for an AGN. During mergers, a significant fraction of a galaxy might wind up being cannibalized. HST images of quasars often show that quasar host galaxies are tidally distorted or are surrounded by other visible matter that is probably still falling into the galaxies. Galaxies that show evidence of recent interactions with other galaxies are more likely to house AGN in their centers (**Figure 19.24**). Any large galaxy might be only an encounter away from becoming an AGN.

Figure 19.24 The Swift Gamma-Ray observatory has detected active black holes (circles) in these merging galaxies.

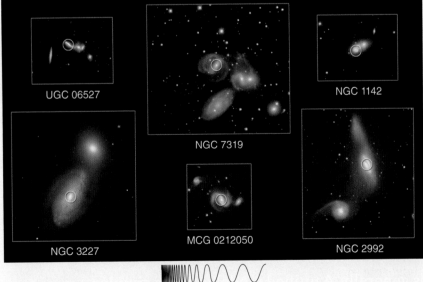

UGC 06527

NGC 7319

NGC 1142

NGC 3227

MCG 0212050

NGC 2992

G X U V I R

Supermassive black holes: (a) are extremely rare—only a handful exist in the universe; (b) are completely hypothetical; (c) occur in most, perhaps all, large galaxies; (d) occur only in the space between galaxies.

Origins Habitability in Galaxies

Now that we know something about the types of galaxies that have been observed, what can we say about their potential for life? No solid information is available because those galaxies are too far away for astronomers to have detected any planets around their stars. Instead, all we can do is speculate about the habitability of other galaxies. Two key requirements are the presence of heavy elements to form planets (and life) and an environment without too much radiation that might be damaging to life.

A study of the host stars of exoplanet candidates discovered by the Kepler telescope suggests that stars with a higher percentage of heavier elements may be more likely to have planets. (That finding fits with the core accretion models of planet formation discussed in Chapter 7.) The first generation of stars made from the hydrogen and helium of Big Bang nucleosynthesis do not have heavy elements. Recall from the discussion of stellar evolution that elements heavier than helium are created in the cores of dying stars and then are scattered into the galactic environment through planetary nebulae, stellar winds, and supernova explosions. So the amount of heavier elements in a star depends on the cosmic history of the material from which the star formed. Therefore, astronomers must consider the galactic environment of the star, which varies among types of galaxies and locations within the galaxies.

Spiral galaxies have had more continual star formation in their disks throughout their history. They contain more stars born from recycled material and therefore more stars with a higher fraction of heavy elements. Elliptical (and S0) galaxies have older, redder populations of stars and very little current star formation. Old, massive ellipticals have a larger percentage of lower-mass stars than that of smaller ellipticals or spirals. Astronomers had previously thought that difference meant that large elliptical galaxies would not be good environments for planet formation. But the Kepler telescope has found many planets around small, red, main-sequence stars like the ones that populate elliptical galaxies. One study of two elliptical galaxies showed that both had some fraction of stars with a heavy-element fraction similar to that of the stars hosting Kepler exoplanets in the Milky Way, so ellipticals are not ruled out.

Another issue is the presence of radiation that might be hazardous to life. That radiation would most likely come from the center of the galaxy. Galaxies in an active AGN state might have too much radiation in regions close to their centers to be conducive to life. Stars whose orbits cross spiral arms many times also might be exposed to higher-than-average levels of radiation, but that is not the case for most stars in a galaxy.

The conditions in those galaxies may also change as the galaxies evolve. Galaxy mergers can shake up stellar orbits and move stars and their planets to different locations. Mergers also may affect the growth and activity level of supermassive black holes and thus the presence of radiation. Some galactic environments just may not remain habitable for the length of time—billions of years— that life took to evolve from bacteria to intelligence on Earth.

20.1 Astronomers Have Measured the Size and Structure of the Milky Way

The universe is full of galaxies of many sizes and types (see Chapter 19). Because Earth is embedded within the Milky Way, the details of the shape and structure are actually harder to determine for the Milky Way than for other galaxies. Comparing observations of the Milky Way with observations of more distant galaxies improves our understanding of the Milky Way. In this section, we explore how astronomers use observations from our position in the galaxy to infer the size and structure of the Milky Way

Spiral Structure in the Milky Way

Figure 20.1a shows the Milky Way Galaxy in Earth's night sky. From a dark location at night, you can see dark bands of interstellar gas and dust that obscure much of the central plane of the Milky Way. The view of the Milky Way from inside offers a different and much closer perspective of a galaxy than can be obtained by viewing external galaxies. Compare **Figure 20.1b**, which shows an edge-on spiral galaxy. The similarities between those images suggest that the Milky Way is a spiral galaxy and that we are viewing it edge on, from inside the disk.

Finding further information about the size and shape of the Milky Way requires more extensive observations in the visible, infrared, and radio regions of the electromagnetic spectrum. Recall from Chapter 15 that neutral hydrogen emits radiation at a wavelength of 21 centimeters (cm), in the radio region of the spectrum. Maps of that radiation show spiral structure in other galaxies and

(a)

(b)

G X U **V** I R

Figure 20.1 (a) We see the Milky Way as a luminous band across the night sky. Prominent dark lanes caused by interstellar dust obscure the light from more distant stars. (b) The edge-on spiral galaxy NGC 891, whose disk greatly resembles that of the Milky Way.

suggest spiral structure in the Milky Way. In addition, observations of ionized hydrogen gas in visible light show two spiral arms in the Milky Way, with concentrations of young, hot O and B stars. Those observations confirm that the Milky Way is a spiral galaxy.

In 2005, Spitzer Space Telescope observations of the distribution and motions of stars in the inner part of the galaxy confirmed that the Milky Way has a substantial bar with a modest bulge at its center. **Figure 20.2** shows an artist's rendering of the major features of the Milky Way. Two major spiral arms—Scutum-Centaurus and Perseus—connect to the ends of the central bar and sweep through the galaxy's disk, just like the arms observed in external spiral galaxies. Our galaxy has several smaller arm segments, including the Orion Spur, which contains the Sun and Solar System. Astronomers conclude that the Milky Way is a giant barred spiral that is more luminous than an average spiral. Viewed from the outside, the Milky Way would look much like the barred spiral galaxy M109 (**Figure 20.3**).

Spiral Arms and Star Formation

In pictures of other spiral galaxies, the arms are often the most prominent feature, as in the Andromeda Galaxy (**Figure 20.4**). The spiral arms are prominent in the ultraviolet image (Figure 20.4a), and although they are less prominent in visible light (Figure 20.4b), they are still clearly defined. You might then conclude that most stars in the disk of a spiral galaxy are concentrated in the spiral arms. That turns out *not* to be the case: although stars are slightly concentrated in spiral arms, the concentration is not strong enough to account for their prominence. Structures associated with star formation, however, such as molecular clouds and associations of luminous O and B stars, are all concentrated in spiral arms. Spiral arms are prominent because star formation is occurring there, so the arms contain significant concentrations of young, massive, hot, and therefore luminous stars which emit strongly in blue light.

Stars form when dense interstellar clouds become so dense that they begin to collapse under the force of their own gravity (see Chapter 15). Because stars

Figure 20.2 Infrared and radio observations contribute to an artist's model of the Milky Way Galaxy. The galaxy's two major arms (Scutum-Centaurus and Perseus) are seen attached to the ends of a thick central bar.

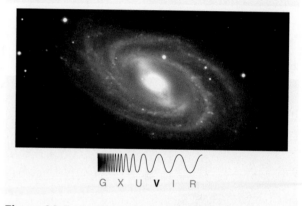

G X U V I R

Figure 20.3 From the outside, the Milky Way would look much like this barred spiral galaxy, M109.

(a) Ultraviolet light   (b) Visible light

G X U V I R   G X U V I R

Figure 20.4 The Andromeda Galaxy in ultraviolet light (a) and visible light (b). The spiral arms, dominated by hot, young stars, are most prominent in ultraviolet light. The spiral arms are less prominent in visible light.

(a)

Dust lane in spiral arm

G X U **V I** R

(b)

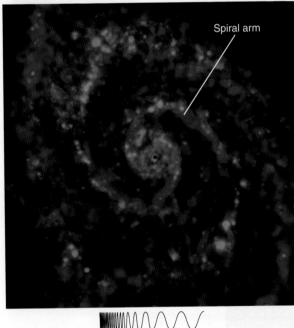

Spiral arm

G X U **V I** R

Figure 20.5 These two images of a face-on spiral galaxy show the spiral arms. (a) This visible-light image also shows dust absorption. (b) This image shows the distribution of neutral interstellar hydrogen (green), carbon monoxide (CO) emission from cold molecular clouds (blue), and hydrogen alpha (Hα) emission from ionized gas (red).

form in spiral arms, spiral arms must be places where clouds of interstellar gas and dust pile up and are compressed. Those clouds can be observed where the dust and gas block starlight (**Figure 20.5a**) or where gases such as neutral hydrogen or carbon monoxide emit at various wavelengths (**Figure 20.5b**).

Any disturbance in the disk of a spiral galaxy will cause a spiral pattern because the disk rotates. Disks of galaxies do not rotate like a solid body. Instead, material close to the center takes less time to travel around the galaxy than material farther out, and so the inner part of the disk gets ahead of the outer part. **Figure 20.6** illustrates the point: In the second frame, you can see that the outer part of the line is trailing behind the inner part. As the galaxy rotates, a straight line through the center becomes a spiral. In the time that objects in the inner part of the galaxy take to complete several rotations, objects in the outer part of the galaxy may not have completed even a single rotation.

A spiral galaxy can be disturbed, for example, by gravitational interactions with other galaxies or by a burst of star formation. However, a single disturbance will not produce a *stable* spiral-arm pattern. Spiral arms produced from a single disturbance will wind themselves up completely in two or three rotations of the disk and then disappear. Disturbances that are repetitive, however, can sustain spiral structure indefinitely. When the bulge in the center of a spiral galaxy is elongated (as seems to be the case for most spiral galaxies), the bulge gravitationally disturbs the disk. As the disk rotates through that disturbance, repeated episodes of star formation occur, and stable spiral arms form.

Many galaxies show clear evidence of a relationship between the shapes of their bulges and the structure of their spiral arms. Barred spirals, for example, have a characteristic two-armed spiral pattern that is connected to the elongated bulge, as seen in Figure 20.3. Even the bulges of galaxies that are not obviously barred may be elongated enough to contribute to the formation of a two-armed spiral structure. Smaller galaxies in orbit about larger galaxies also can give rise to a periodic gravitational disturbance, triggering the same sort of two-armed structure.

Star formation itself also can create spiral structure. Regions of star formation release considerable energy into their surroundings through UV radiation, stellar winds, and supernova explosions. That energy compresses clouds of gas and triggers more star formation. Typically, many massive stars form in the same region at about the same time, and their combined mass outflows and supernova explosions after short lifetimes occur one after another in the same region of space over only a few million years. The result can be large, expanding bubbles of hot gas that sweep out cavities in the interstellar medium and concentrate the swept-up gas into dense, star-forming clouds, much like the snow that piles up in front of a snowplow. In that way, star formation can propagate through the disk of a galaxy. Rotation bends the resulting strings of star-forming regions into spiral structures.

Stars move in and out of arms as they orbit the center of a galaxy. Consequently, the stars in an arm today are not the same stars that were in the arm 20 million years ago. The conditions are roughly analogous to a traffic jam on a busy highway. The cars in the jam are changing all the time, yet the traffic jam persists as a place of higher density—a place with more cars than usual. The traffic jam itself moves slowly backward, even as the cars move forward and pass through it. Just like the traffic jam, the disturbance of the spiral arm also moves at a different speed from that of the individual stars. Those disturbances

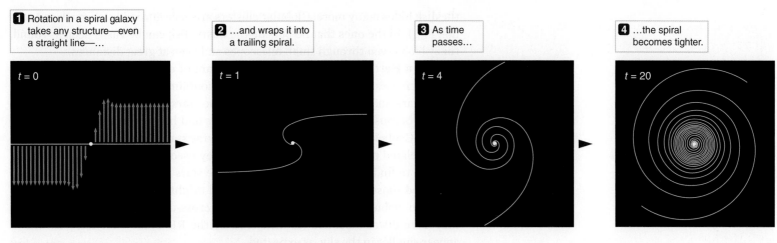

1 Rotation in a spiral galaxy takes any structure—even a straight line—...

2 ...and wraps it into a trailing spiral.

3 As time passes...

4 ...the spiral becomes tighter.

$t = 0$

$t = 1$

$t = 4$

$t = 20$

Figure 20.6 The rotation of a spiral galaxy will naturally take even an originally linear structure and wrap it into a progressively tighter spiral as time (t) goes by.

in the disks of spiral galaxies are called **spiral density waves** because they are waves of greater mass density and increased pressure in the galaxy's interstellar medium. Those waves move around a disk in the pattern of a two-armed spiral that does not rotate at the same rate as the stars, gas, or dust. As material in the disk orbits the center of the galaxy, it passes through those spiral density waves.

A spiral density wave has very little effect on the motions of stars as they pass through it, but it does compress the gas that flows through it. Gas flows into the spiral density wave and piles up. Stars form in the resulting compressed gas. Massive stars are concentrated in the arms because they have such short lives (typically 10 million years or so) that they never have the chance to drift far from the spiral arms where they were born. Less massive stars, however, have plenty of time to move away from their places of birth, so they form a smooth underlying disk.

The Size of the Milky Way Galaxy

If you go out on a dark night, away from any streetlights, and look toward the center of the Milky Way—located in the constellation Sagittarius—you will see the dark lane of dusty clouds shown in Figure 20.1a. Because the Solar System is inside the dusty disk of the Milky Way, the visible-light view of the galaxy itself is badly obscured. To probe the structure of the Milky Way, modern astronomers use long-wavelength infrared and radio radiation that can penetrate the dust in the disk. The most powerful tool for that work is the same 21-cm line from neutral interstellar hydrogen, which (as we described in Chapter 19) is used to measure the rotation of other galaxies. Even in those long-wavelength parts of the spectrum, the distance to the center of the galaxy cannot be measured directly.

In the 1920s, Harlow Shapley made a three-dimensional map of globular clusters in the Milky Way, which led to the first determination of the size of the Milky Way and that the Sun is offset from the center. Recall from Chapter 17 that globular clusters are large, spheroidal groups of stars held together by gravity. The Milky Way contains more than 150 cataloged globular clusters, and dust in the disk hides many more. Globular clusters are very luminous (as much as 1 million L_{Sun}), so the ones that lie outside the dusty disk can be easily seen as round, fuzzy blobs even through small telescopes and even at great distances.

In a Hertzsprung-Russell (H-R) diagram of an old cluster, the horizontal branch crosses the instability strip, which contains pulsating stars such as RR Lyrae stars and Cepheid variables. RR Lyrae stars are easy to spot in globular clusters because they are relatively luminous and have a distinctive light curve. As with Cepheid variables, the time an RR Lyrae star takes to undergo one pulsation is related to the star's luminosity. Shapley used that period-luminosity relationship to find the luminosities of RR Lyrae stars in globular clusters. He then combined those luminosities with measured brightnesses to determine the distances to globular clusters. Finally, Shapley cross-checked his results by noting that more distant clusters (as measured by the RR Lyrae stars) also tended to appear smaller in the sky, as expected.

Those globular clusters trace out the luminous part of the galactic halo of the Milky Way Galaxy, a large spherical volume of space surrounding the disk and bulge. The center of the distribution of globular clusters coincides with the gravitational center of the galaxy. Shapley realized that because he could determine the distance to the center of that distribution, he had actually determined the Sun's distance from the center of the Milky Way, as well as the size of the galaxy itself. His map showed that globular clusters occupy a roughly spherical region of space with a diameter of about 90 kiloparsecs (kpc), or 90,000 parsecs (pc). Shapley did not know about gas and dust, however, so he overestimated the distance to the globular clusters (see the **Process of Science Figure**). A modern determination indicates that the Sun is located about 8,300 pc (27,000 light-years) from the center of the galaxy, or roughly halfway out toward the edge of the disk.

CHECK YOUR UNDERSTANDING 20.1

List three pieces of evidence supporting the theory that the Milky Way is a spiral galaxy.

20.2 The Components of the Milky Way Reveal Its Evolution

Figure 20.7 illustrates the Sun's position and relationship to the rest of the Milky Way. The Sun is a middle-aged disk star located among other middle-aged stars that orbit around the galaxy within the galactic disk, as do the gas and dust in the disk. The stars in the halo move in random orbits similar to those of stars in elliptical galaxies, sometimes at high velocities. Some of those stars can be observed near the Sun as their orbits carry them swiftly through the disk. Most of those stars are much older than the Sun. The bar in the galactic bulge of the Milky Way is shaped primarily by stars and gas moving not only in highly elongated orbits up and down the long axis of the bar but also in short orbits aligned perpendicular to the bar. All those stellar orbits determine the shapes of the various parts of the galaxy, and stellar orbits are easier to measure in the Milky Way than in other galaxies. Using the ages, chemical abundances, and motions of nearby stars, astronomers can differentiate between disk and halo stars to learn more about the galaxy's structure. In this section, we describe how astronomers study the constituents of the Milky Way to find direct clues about how spiral galaxies form.

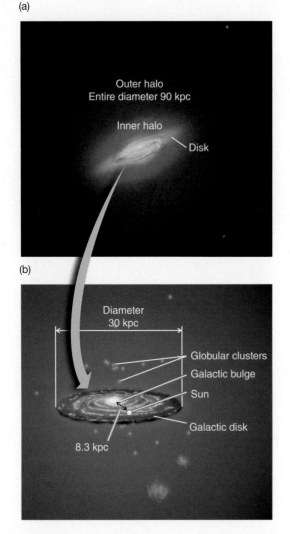

(a)

Outer halo
Entire diameter 90 kpc

Inner halo

Disk

(b)

Diameter
30 kpc

Globular clusters
Galactic bulge
Sun
Galactic disk

8.3 kpc

Figure 20.7 Parts of the Milky Way include the disk and the inner and outer halos (a) and the galactic bulge and disk (b). Globular clusters are located in the halo, and the Sun is located in the disk, about 8.3 kpc from the center. (1 kpc = 1,000 pc.)

Unknown Unknowns

Initial efforts to measure the Milky Way did not include dark matter, because astronomers did not know about it.

Fritz Zwicky
1898–1974

Fritz Zwicky was the first to predict that dark matter existed, based on observations of clusters of galaxies.

Vera Rubin
1928–2016

Vera Rubin subsequently discovered dark matter in the Andromeda galaxy, which rotated faster than could be accounted for by the observed mass. Following up with observations of other galaxies, she confirmed that dark matter was an important component in nearly every galaxy.

Today, astronomers routinely account for dark matter, which represents about 90% of the mass of the Milky Way.

Often, scientists don't know what is unknown until a set of observations fails to make sense. Once this "unknown unknown" is discovered, prior results must be modified to incorporate the new knowledge.

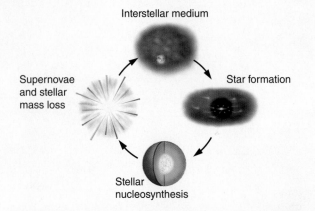

Figure 20.8 Matter moves from the interstellar medium into stars and back again in a progressive cycle that has enriched today's universe with massive elements.

Figure 20.9 As subsequent generations of stars form, live, and die, they enrich the interstellar medium with massive elements—the products of stellar nucleosynthesis. The chemical evolution of the Milky Way and other galaxies can be traced in many ways, including by the strength of interstellar emission lines and stellar absorption lines.

Age and Chemical Compositions of Stars

Over time, the chemical content of a galaxy changes as stars are born, live and die, and progressively enrich the interstellar medium (**Figure 20.8**). The interstellar medium therefore reflects all the stellar evolution that has taken place up to the present time. Gas rich in massive elements has gone through a great deal of stellar processing. That evolution is similarly reflected in the composition of stars within the galaxy. As illustrated in **Figure 20.9**, the chemical composition of a star's atmosphere reflects the cumulative amount of star formation that occurred before that star formed. Although the details are complex, several clear and important lessons can be learned from the observed patterns in the amounts of massive elements in the galaxy.

Stars in globular clusters, being among the earliest stars to form, should contain only very small amounts of massive elements. Some globular-cluster stars contain only 0.5 percent as much of those massive elements as the Sun. That relationship between age and abundances of massive elements is evident throughout much of the galaxy. Within the disk, younger stars typically have higher abundances of massive elements than those of older stars. Similarly, older stars in the outer parts of the galaxy's bulge have lower massive-element abundances than those of young stars in the disk. All the stars in the galaxy's halo have smaller amounts of heavy elements. That includes globular-cluster stars, which constitute only a minority of stars in the galactic halo.

Globular clusters orbit in the halo of the Milky Way; open clusters orbit in the disk of the Milky Way. As with globular clusters, the stars in an open cluster all formed at about the same time, so astronomers can use the H-R diagram to find the cluster's age. Some open clusters are very young, containing the very youngest stars known, whereas other open clusters contain stars older than the Sun; but all open clusters are younger than the youngest globular cluster. Because the globular clusters are old and located in the halo, astronomers infer that stars in the halo formed before stars in the disk. That epoch of halo star formation did

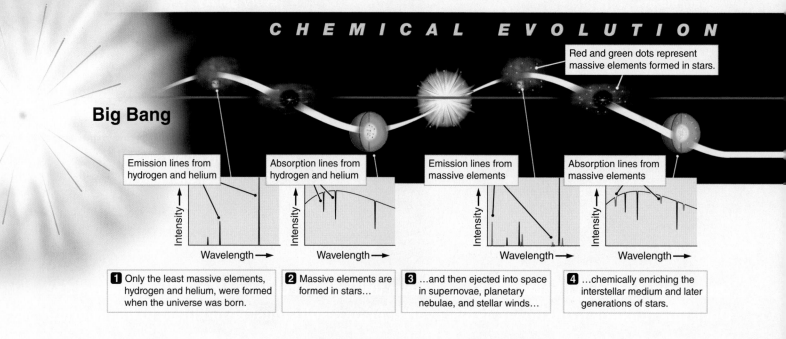

CHEMICAL EVOLUTION

Red and green dots represent massive elements formed in stars.

Big Bang

Emission lines from hydrogen and helium

Absorption lines from hydrogen and helium

Emission lines from massive elements

Absorption lines from massive elements

Intensity → / Wavelength →

1 Only the least massive elements, hydrogen and helium, were formed when the universe was born.

2 Massive elements are formed in stars...

3 ...and then ejected into space in supernovae, planetary nebulae, and stellar winds...

4 ...chemically enriching the interstellar medium and later generations of stars.

not last long. Star formation in the disk started later and has been continuing ever since.

Within the galaxy's disk, astronomers observe differences in abundances of massive elements from place to place, related to the rate of star formation in different regions. Star formation is generally more active in the denser, inner part of the Milky Way than in the outer parts. Observations of chemical abundances in the interstellar medium, based both on interstellar absorption lines in the spectra of stars and on emission lines in glowing clouds of neutral hydrogen gas known as H II regions, confirm that prediction by showing a smooth decline in abundances of massive elements from the inner to the outer parts of the disk. Astronomers have observed similar trends in other galaxies. Within a galactic disk, relatively old stars near the center of a galaxy often have greater massive-element abundances than those of young stars in the outer parts of the disk.

The basic idea that higher massive-element abundances should follow from the more prodigious star formation in the inner galaxy seems correct, but the full picture is not that simple. New material falling into the galaxy might affect the amounts of heavy elements in the interstellar medium. Chemical elements produced in the inner disk might be blasted into the halo in great "fountains" powered by the energy of massive stars, only to fall back onto the disk elsewhere. Past interactions with other galaxies might have stirred the Milky Way's interstellar medium, mixing gas from those other galaxies with gas already there. The variations of chemical abundances within the Milky Way and other galaxies—and what those variations tell us about the history of star formation and the formation of elements—remain active topics of research.

Even the very oldest globular-cluster stars contain some chemical elements fused in previous generations of more massive stars. That observation implies that globular-cluster stars and other halo stars were not the first stars in the Milky Way to form. At least one generation of massive stars lived and died, ejecting newly synthesized massive elements into space, before even the oldest globular clusters

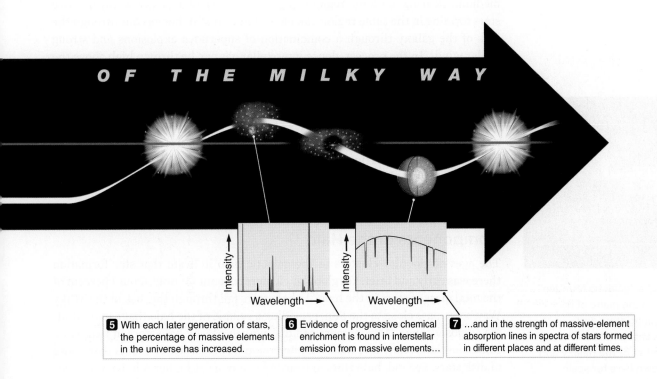

5 With each later generation of stars, the percentage of massive elements in the universe has increased.

6 Evidence of progressive chemical enrichment is found in interstellar emission from massive elements…

7 …and in the strength of massive-element absorption lines in spectra of stars formed in different places and at different times.

(a)

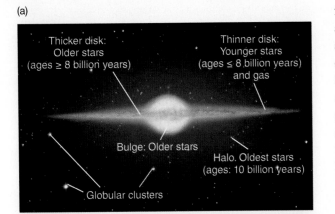

(b)

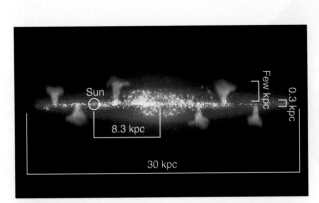

Figure 20.10 (a) This illustration shows the disks, bulge, and inner halo of the Milky Way and the location of globular clusters. (b) An all-sky view of the interstellar dust that fills the Milky Way Galaxy, from observations by ESA's *Gaia* satellite.

Figure 20.11 Artist's concept of a "galactic fountain," in which gas is pushed away from the plane of the galaxy by energy released by young stars and supernovae and then falls back onto the disk. The distance of the Sun from the center of the galaxy and the diameter of the disk are shown here for scale.

formed. (We return to these first stars in Chapter 23.) Every star less massive than about 0.8 M_{Sun} that ever formed is still around as a main-sequence star today. Even so, astronomers find no disk stars with exceptionally low massive-element abundances. The gas that wound up in the plane of the Milky Way must have seen a significant amount of star formation before it settled into the disk of the galaxy and made stars. Still, even a chemically "rich" star like the Sun, which is made of gas processed through approximately 9 billion years of previous generations of stars, is composed of less than 2 percent massive elements. Luminous matter in the universe is still dominated by hydrogen and helium formed just after the Big Bang, long before the first stars.

Components of the Disk

The disk of the Milky Way has a thinner component and a thicker component (**Figure 20.10a**). The youngest stars in the galaxy are most strongly concentrated in the galactic plane, defining a thinner disk about 300 pc (1,000 light-years) thick but more than 30,000 pc (100,000 light-years) across. That ratio of the diameter to the thickness of the disk is similar to that of a DVD. The older population of disk stars, distinguishable by lower abundances of massive elements, occupy the thicker disk, about 3,700 pc (12,000 light-years) thick. The youngest stars are concentrated closest to the plane of the galaxy because that is where the molecular clouds and gas are. Those stars show a decrease in heavy-element content as the distance from the galactic center increases. Older stars make up the thicker parts of the disk, with the oldest stars closer to the galactic center. Astronomers are still debating how distinct those two disks are and how they originated. Dust is also concentrated in the plane of the galaxy, as seen in a recent image from the *Gaia* mission (**Figure 20.10b**).

The interstellar medium is a dynamic place—energy from star-forming regions can shape it into impressively large structures. As mentioned, energy from regions of star formation can form interesting structures in the interstellar medium, clearing out large regions of gas in the disk of a galaxy. Many massive stars forming in the same region can blow "fountains" of hot gas out through the disk of the galaxy through a combination of supernova explosions and strong stellar winds. In the process, dense interstellar gas can be thrown high above the plane of the galaxy (see the artist's depiction in **Figure 20.11**). Once the gas is a few kiloparsecs above or below the disk, it radiates and cools, falling back to the disk. Maps of the 21-cm emission from neutral hydrogen in the galaxy, X-ray observations, and visible-light images of hydrogen emission from some edge-on external galaxies show many vertical structures in the interstellar medium of disk galaxies. Those vertical structures are often interpreted as the "walls" of the fountains. If enough massive stars are formed together, enough energy may be deposited to blast holes all the way through the plane of the galaxy.

Components of the Halo

The ages of globular clusters in the galactic halo indicate that star formation there was early and brief. Yet globular clusters account for only about 1 percent of the total mass of stars in the halo. As halo stars fall through the disk of the Milky Way, some pass close to the Sun, providing a sample of the halo that can be studied at closer range. Astronomers can distinguish nearby halo stars in two ways. First, most halo stars have much smaller amounts of massive elements than those of disk stars. Second, halo stars appear to be moving at higher relative velocities than disk stars. Most stars near the Sun are disk stars, so they orbit the center of the galaxy at nearly the same speed, in roughly the same direction as the Sun. In contrast, halo stars orbit the center of the galaxy in random directions, so the relative velocity between the halo stars and the Sun tends to be high. Those stars are known as high-velocity stars.

By studying the orbits of high-velocity stars, astronomers have determined that the halo has two separate components: an inner halo that includes stars up to about 15 kpc (50,000 light-years) from the center and an outer halo that extends far beyond that (see Figure 20.7a). The stars in the outer halo have lower fractions of heavier elements, suggesting that those stars formed very early. Many are moving in a direction opposite to the rotation of the galaxy, suggesting that the outer halo may have its origins in a merger with a small dwarf galaxy long ago. The orbits of halo stars suggest that those stars fill a volume of space similar to that occupied by the globular clusters in the halo.

X-ray observations indicate that a halo of hot gas surrounds the Milky Way (see the artist's concept in **Figure 20.12**). That gas halo may extend for about 100–200 kpc from the galactic center, encompassing two nearby small galaxies and containing as much mass as that of all the stars in our galaxy. The halo's temperature is estimated to be about 2 million kelvins (K), so the gas particles are ionized and moving very quickly. The gas is extremely diffuse, however, so the particles are not colliding with one another and transferring energy. The gas wouldn't "feel" hot, much like the solar corona (see Chapter 14).

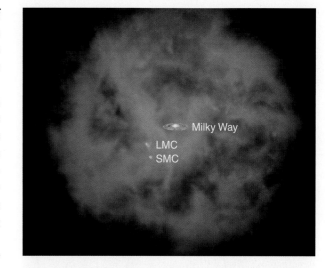

Figure 20.12 This artist's concept shows the Milky Way surrounded by a halo of hot gas, which may contain as much mass as that of all the stars in the galaxy combined. The Large Magellanic Cloud (LMC) and Small Magellanic Cloud (SMC) are two of the many nearby dwarf galaxies.

Magnetic Fields and Cosmic Rays Fill the Galaxy

The interstellar medium of the Milky Way is laced with magnetic fields that are wound up and compressed by the rotation of the galaxy's disk. The total interstellar magnetic field, however, is about a hundred thousand times weaker than Earth's magnetic field. Charged particles and magnetic fields interact strongly; the particles spiral around magnetic fields, moving along the field rather than across it. Conversely, magnetic fields cannot freely escape from a cloud of gas containing even a small number of charged particles. The dense clouds of interstellar gas in the midplane of the Milky Way (**Figure 20.13**) anchor the galaxy's magnetic field to the disk, in turn anchoring high-energy charged particles, known as cosmic rays, to the galaxy.

Cosmic rays are charged particles that originate mainly outside the Solar System and travel close to the speed of light. Despite their name, cosmic rays are not a form of electromagnetic radiation: they were named before their true nature was known. Most cosmic-ray particles are protons, but some are nuclei of helium, carbon, and other elements produced by nucleosynthesis. A few are high-energy electrons and other subatomic particles.

Cosmic rays span an enormous range in particle energy. Astronomers can observe the lowest-energy cosmic rays with interplanetary spacecraft. Those cosmic rays have energies as low as about 10^{-11} joule (J), which corresponds to the energy of a proton moving at a velocity of 1/3 the speed of light. In contrast, the most energetic cosmic rays are 10 trillion (10^{13}) times as energetic as the lowest-energy cosmic rays, and they move very close to the speed of light, $0.999999c$. Those high-energy cosmic rays are constantly hitting Earth and are detected from the showers of elementary particles that they cause when crashing through Earth's atmosphere. Those

Figure 20.13 The mass of interstellar clouds anchors the magnetic field of the Milky Way to the disk of the galaxy. The galaxy's magnetic field in turn traps cosmic rays.

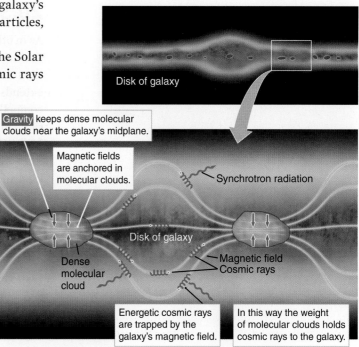

Gravity keeps dense molecular clouds near the galaxy's midplane.

Magnetic fields are anchored in molecular clouds.

Synchrotron radiation

Disk of galaxy

Dense molecular cloud

Magnetic field
Cosmic rays

Energetic cosmic rays are trapped by the galaxy's magnetic field.

In this way the weight of molecular clouds holds cosmic rays to the galaxy.

(a)

(b)

Figure 20.14 (a) Artist's sketch of the Pierre Auger Observatory in Argentina, an array of stations designed to catch the particles that shower from collisions of cosmic rays with the upper atmosphere. (b) Each station in the array is equipped with its own particle collectors, carefully protected from the elements.

particle showers are observed by special telescopes such as the High Energy Stereoscopic System (HESS) imaging telescopes in Namibia, or the Pierre Auger Observatory in Argentina (**Figure 20.14**).

Astronomers hypothesize that most of those cosmic rays are accelerated to those incredible energies by the shock waves produced in supernova explosions. The very highest-energy cosmic rays are as much as a hundred million times more energetic than any particle ever produced in a particle accelerator on Earth. Those extremely high energies make those cosmic rays much more difficult to explain than those with lower energies.

The disk of the galaxy glows from synchrotron radiation (see Chapter 17) produced by cosmic rays, mostly electrons, spiraling around the galaxy's magnetic field. Such synchrotron emission is seen in the disks of other spiral galaxies as well, indicating that they, too, have magnetic fields and populations of energetic cosmic rays. The very highest-energy cosmic rays are moving much too fast to be confined by the gravitational force of their originating galaxy. Any such cosmic rays that formed in the Milky Way would soon stream away from the galaxy into intergalactic space. Thus, some of the energetic cosmic rays reaching Earth probably originated in energetic events outside the Milky Way Galaxy.

The total energy of all the cosmic rays in the galactic disk can be estimated from the energy of the cosmic rays reaching Earth. The strength of the interstellar magnetic field can be measured by observing how the field affects the properties of radio waves passing through the interstellar medium. Those measurements indicate that in the Milky Way Galaxy, the magnetic-field energy and the cosmic-ray energy are about equal. Both are comparable to the energy present in other energetic components of the galaxy, including the motions of interstellar gas and the total energy of electromagnetic radiation within the galaxy.

CHECK YOUR UNDERSTANDING 20.2

What parts of the Milky Way contain old stars, and what parts contain young stars?

20.3 Most of the Milky Way Is Unseen

As in other galaxies, the most interesting parts of the Milky Way may be the parts that can't be seen directly but are detected only by their gravitational influence on the stars around them. Dark matter accounts for most of the mass in a galaxy and extends far beyond a galaxy's visible boundary. As in all other spiral galaxies, compelling evidence indicates that dark matter dominates the Milky Way. From radio and infrared observations, astronomers can figure out how the disk of the Milky Way moves, and from that motion, in turn, they can determine its mass.

The supermassive black hole at the center of the Milky Way poses a different kind of observation problem. That object also is detected by its gravitational effects on the stars nearby but cannot be seen directly. In this section, we explore what can be inferred about those two components of the Milky Way from how their gravity affects other objects.

Dark Matter in the Milky Way

The rotation of the disk of the Milky Way can be determined from observations of the relative velocities of interstellar hydrogen measured from 21-cm radiation. **Figure 20.15** shows how those velocities vary with viewing direction from the

Sun. Looking toward the center of the galaxy, we see that the gas is stationary relative to the Sun. When we look in the direction of the Sun's motion around the galactic center, hydrogen clouds appear to be moving toward Earth, whereas in the opposite direction, clouds are moving away from Earth. In other directions, the measured velocities are complicated by Earth's moving vantage point within the disk and so are more difficult to interpret at a glance. That is a pattern of the rotation velocity of gas in a disk like those you learned about in Chapter 19. The only difference is that instead of looking at it from outside, we see the Milky Way Galaxy's rotation curve from a vantage point located within—and rotating with—the galaxy. Even so, observed velocities of neutral hydrogen enable astronomers to measure the Milky Way Galaxy's rotation curve and even determine the structure present throughout its disk.

Recall (Chapter 19) that observations of rotation curves led astronomers to conclude that the masses of spiral galaxies consist mostly of dark matter. **Figure 20.16** shows the rotation curve of the Milky Way as inferred primarily from 21-cm observations. The orbital motion of the nearby dwarf galaxy called the Large Magellanic Cloud provides data for the outermost point in the rotation curve, at a distance of roughly 50,000 pc (160,000 light-years) from the center of the galaxy. Like other spiral galaxies, the Milky Way has a fairly flat rotation curve, which indicates the presence of large amounts of dark matter. The mass outside the Sun's orbit does not greatly affect the Sun's orbit.

The total mass of the Milky Way Galaxy is currently estimated to be about 1.0 trillion to 1.5 trillion times the mass of the Sun. The luminous mass, however, estimated by adding the masses of stars, dust, and gas, is only about one-tenth as much. Like other spiral galaxies, the Milky Way's mass consists mainly of dark

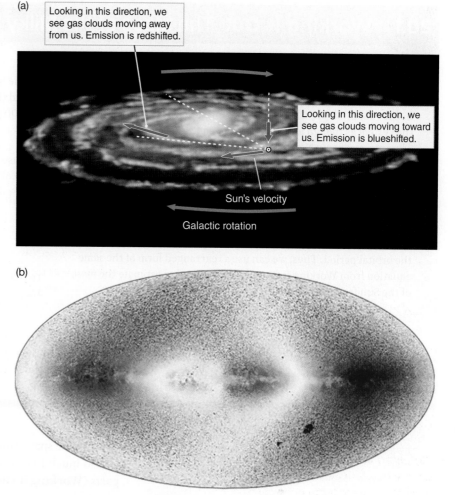

Figure 20.15 (a) From Earth's perspective within the Solar System, the signature of a rotating disk is clear from views of either side of the galactic center. (b) Those velocities vary between redshift and blueshift as an observer looks around in the plane of the Milky Way.

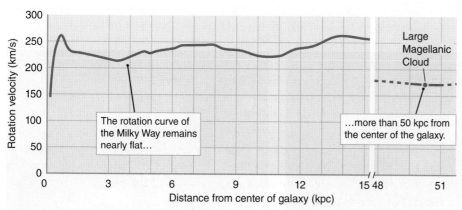

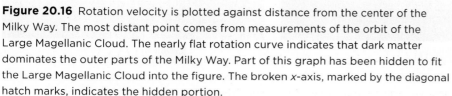

Figure 20.16 Rotation velocity is plotted against distance from the center of the Milky Way. The most distant point comes from measurements of the orbit of the Large Magellanic Cloud. The nearly flat rotation curve indicates that dark matter dominates the outer parts of the Milky Way. Part of this graph has been hidden to fit the Large Magellanic Cloud into the figure. The broken x-axis, marked by the diagonal hatch marks, indicates the hidden portion.

20.1 Working It Out The Mass of the Milky Way inside the Sun's Orbit

The Sun orbits about the center of the Milky Way Galaxy. Even though the gravitational pull on the Sun comes from all the material inside that orbit (see Chapter 4), Newton showed that we could treat the system as though all the mass were concentrated at the center. Thus, we can apply Newton's and Kepler's laws to calculate the mass of the Milky Way inside the Sun's orbit. Newton's version of Kepler's third law relates the period of the orbit to the orbital radius and the masses of the objects. But here, the mass of the galaxy is much larger than the mass of the Sun, so the Sun's mass is negligible by comparison. In addition, what astronomers can *measure* is the orbital speed of the Sun or other stars about the galactic center, rather than the orbital period. Thus, we can use a rearranged form of the same equation from Working It Out 4.2 that we used to estimate the mass of the Sun from the orbit of Earth:

$$M = \frac{r v_{circ}^2}{G}$$

The Sun orbits the center of the galaxy at 240 kilometers per second (km/s). The distance of the Sun from the center of the galaxy is 8,300 pc, which converts to kilometers as follows:

$$8{,}300 \text{ pc} \times (3.09 \times 10^{13} \text{ km/pc}) = 2.56 \times 10^{17} \text{ km}$$

Because we know that the gravitational constant, $G = 6.67 \times 10^{-20} \text{ km}^3/(\text{kg s}^2)$, we can calculate the mass of the portion of the Milky Way inside the Sun's orbit:

$$M = \frac{(2.56 \times 10^{17} \text{ km}) \times (240 \text{ km/s})^2}{6.67 \times 10^{-20} \text{ km}^3/(\text{kg s}^2)}$$

$$M = 2.21 \times 10^{41} \text{ kg}$$

To put that result in units of the Sun's mass, we divide the answer by $M_{Sun} = 1.99 \times 10^{30}$ kg, yielding

$$M = \frac{2.21 \times 10^{41} \text{ kg}}{1.99 \times 10^{30} \text{ kg}/M_{Sun}} = 1.11 \times 10^{11} \, M_{Sun}$$

The mass of the Milky Way inside the Sun's orbit is about 111 billion times the mass of the Sun.

matter. The spatial distribution of dark and normal matter within the Milky Way is also much like that of other galaxies, with dark matter dominating its outer parts (**Working it Out 20.1**).

The Supermassive Black Hole

Figure 20.17 shows images of the Milky Way's center taken with the Chandra X-ray Observatory and the Spitzer Space Telescope. The X-ray view (Figure 20.17a) shows the location of a strong radio source called Sagittarius A* (abbreviated

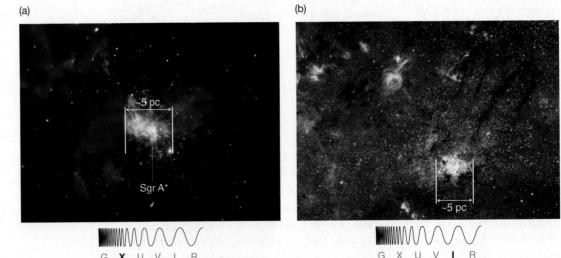

(a)

(b)

Figure 20.17 (a) This X-ray view of the Milky Way's central region shows the active source, Sagittarius A* (Sgr A*), as the brightest spot at the middle of the image. Lobes of superheated gas (shown in red) are evidence of recent, violent explosions happening near Sgr A*. (b) This infrared view of the central core of the Milky Way shows hundreds of thousands of stars. The bright white spot at the lower right marks Sgr A*, the location of the supermassive black hole.

~5 pc

Sgr A*

~5 pc

G **X** U V I R

G **X** U V I R

Sgr A*), which lies at the center of the Milky Way. The infrared image (Figure 20.17b) cuts through the dust to reveal the galaxy's crowded, dense core containing hundreds of thousands of stars.

Studies of the motions of stars closest to the Sgr A* source suggest a central mass very much greater than that of the few hundred stars orbiting there. Furthermore, observations of the galaxy's rotation curve show rapid rotation velocities very close to the galactic center. Stars closer than 0.1 light-year from the galactic center follow Kepler's laws, indicating that their motion is dominated by mass within their orbit. The closest stars studied are only about 0.01 light-year from the center of the galaxy—so close that their orbital periods are only about a dozen years. The positions of those stars change noticeably over time, and astronomers can see them speed up as they whip around what can only be a supermassive black hole at the focus of their elliptical orbits (**Figure 20.18**). Using Newton's version of Kepler's third law, we can then estimate that the black hole at the center of the Milky Way Galaxy is a relative lightweight, having a mass of "only" 4 million times the mass of the Sun (**Working It Out 20.2**).

Clouds of interstellar gas at the galaxy's center are heated to millions of degrees by shock waves from supernova explosions and colliding stellar winds blown outward by young, massive stars. Superheated gas produces X-rays, and the Chandra X-ray Observatory has detected more than 9,000 X-ray sources within the central region of the galaxy. Those sources include frequent, short-lived X-ray flares near Sgr A* (see Figure 20.17a), which provide direct evidence that matter falling toward the supermassive black hole fuels the energetic activity at the galaxy's center.

The Fermi Gamma-ray Space Telescope has observed gamma-ray-emitting bubbles that extend 8 kpc (25,000 light-years) above and below the galactic plane. The bubbles may have formed after a burst of star formation a few million years ago produced massive star clusters near the center of the galaxy. If some of the gas formed stars and about 2,000 $M_{\rm Sun}$ of material fell into the supermassive black hole, enough energy could have been released to power the bubbles. More recently, faint gamma-ray signals were observed that look like jets coming from the center,

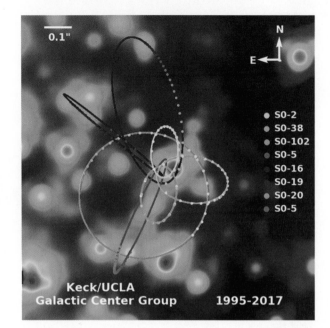

Figure 20.18 The orbits of eight stars within 0.03 pc (0.1 light-year, or about 6,000 astronomical units [AU]) of the Milky Way's center. The Keplerian motions of those stars reveal the presence of a 4-million-$M_{\rm Sun}$ supermassive black hole at the galaxy's center. Colored dots show the measured positions of each star over many years: the dots progress from lighter in 1995 to darker in 2016.

20.2 Working It Out The Mass of the Milky Way's Central Black Hole

Figure 20.18 illustrates data points for the stars in the central region orbiting closely to the central black hole of the Milky Way Galaxy. Those stars have highly elliptical orbits with changing speeds, but the orbital periods are short enough that they can be observed and measured. Star S0-2 in the figure has a measured orbital period of 15.8 years. The semimajor axis of its orbit is estimated to be 1.5×10^{11} km = 1,000 AU. With that information, we can use Newton's version of Kepler's third law to estimate the mass inside S0-2's orbit. Setting up the equation as we did in Working It Out 13.4:

$$\frac{m_{\rm BH}}{M_{\rm Sun}} + \frac{m_{\rm S0\text{-}2}}{M_{\rm Sun}} = \frac{A_{\rm AU}^3}{P_{\rm years}^2}$$

The mass of star S0-2 is much less than the mass of the black hole, so the sum of the two is very close to the mass of the black hole. Therefore, we can write

$$\frac{m_{\rm BH}}{M_{\rm Sun}} = \frac{A_{\rm AU}^3}{P_{\rm years}^2} = \frac{1,000^3}{15.8^2} = 4.0 \times 10^6$$

$$m_{\rm BH} = 4.0 \times 10^6 \, M_{\rm Sun}$$

The supermassive black hole at the center of the Milky Way has a mass 4 million times that of the Sun. That value is considerably less than the billion-solar-mass black holes in some active galactic nuclei discussed in Chapter 19.

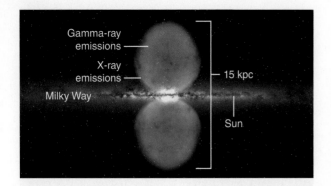

Figure 20.19 The Fermi Gamma-ray Space Telescope observed gamma-ray bubbles (purple) extending 8 kpc above and below the galactic plane. Hints of the edges of the bubbles were first observed in X-rays (blue) in the 1990s. In this artist's conceptual view from outside the galaxy, the gamma-ray jets are in magenta.

Figure 20.20 This graphical map shows some of the members of the Local Group of galaxies. Most are dwarf galaxies. Spiral galaxies are shown in yellow. The closest galaxies to the Milky Way (such as the Large Magellanic Cloud and Small Magellanic Cloud) are not seen on this scale. (1 Mpc = 1 megaparsec = 1 million parsecs = 3.26 million light-years.)

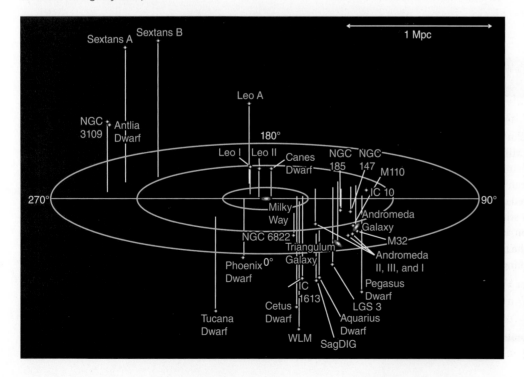

within the bubbles (see the artist's depiction in **Figure 20.19**). Using the Hubble Space Telescope, astronomers have estimated that those jets originated from material falling into the supermassive black hole about 6 million to 9 million years ago. Some astronomers predict that gas clouds are heading toward the center and will soon be accreted by the black hole. Currently, the observed activity is not as intense as that seen in active galactic nuclei with central, supermassive black holes. The inner Milky Way is a reminder that it was almost certainly "active" in the past and could become active once again.

CHECK YOUR UNDERSTANDING 20.3

Which property is detectable for both dark matter and the supermassive black hole at the center of the Milky Way? (a) luminosity; (b) temperature; (c) gravity; (d) composition

20.4 The History and Future of the Milky Way

A fundamental goal of stellar astronomy is to understand the life cycle of stars, including how they form from clouds of interstellar gas. In Chapter 15, we described that process, at least as it occurs today, and tied that story strongly to observations of Earth's galactic neighborhood. Galactic astronomy has a similar basic goal. Astronomers would like to have a complete and well-tested theory of how the Milky Way formed and to be able to make predictions about its future. The distribution of stars of different ages with different amounts of heavy elements is one clue. Additional clues come from studying other galaxies at different distances (and therefore of different ages), their supermassive black holes, and their merger history. In this section, we explore the history and the future of the Milky Way.

The Local Group

Galaxies do not exist in isolation. Most galaxies are parts of gravitationally bound collections of galaxies, the smallest and most common of which are called **galaxy groups**. A galaxy group contains as many as several dozen galaxies, most of them dwarf galaxies. The Milky Way is a member of the **Local Group** (see Chapter 1), first identified by Edwin Hubble in 1936. Hubble labeled 12 galaxies as part of the Local Group, but now astronomers count at least 50. The Local Group (**Figure 20.20**) includes the two giant barred spirals—the Milky Way Galaxy and the Andromeda Galaxy—along with a few ellipticals and irregulars and at least 30 smaller dwarf galaxies in a volume of space about 3 million pc (10 million light-years) in diameter. Almost 98 percent of all the galaxy mass in the Local Group resides in just those two giant galaxies. The third-largest galaxy, Triangulum, is an unbarred spiral with a few percent the mass of the Milky Way or Andromeda. Most, but not all, of the dwarf elliptical and dwarf spheroidal galaxies in the group are satellites of the Milky Way or Andromeda. The Local Group interacts with a few nearby groups, which is discussed further in Chapter 23.

More than 50 of those satellite dwarf galaxies exist, although it's not certain that all are gravitationally bound to the Milky Way. Some of the fainter dwarf galaxies were discovered only very recently because of their low luminosity. The dwarf galaxies are the lowest-mass galaxies observed, and they are dominated by an even greater percentage of invisible dark matter than are other known galaxies. They also contain stars very low in elements more massive than helium. Those old ultrafaint dwarf galaxies offer clues to the formation of the Local Group. In addition, observations of the motions and speeds of the dwarf galaxies about the Milky Way may lead to new estimates of the dark matter mass within the Milky Way itself.

The Formation of the Milky Way

We have seen that globular clusters and high-velocity stars must have been among the first stars formed in the Milky Way that still exist. The fact that they are not concentrated in the disk or bulge of the galaxy indicates that they formed from clouds of gas well before those clouds settled into the galaxy's disk. That hypothesis is supported by observations that globular clusters are very old and that the youngest globular cluster is older than the oldest disk stars. The presence of extremely small amounts of massive elements in the atmospheres of halo stars also indicates that at least one generation of stars must have lived and died *before* the formation of the halo stars visible today. That line of reasoning implies that the Milky Way formed from the merger of several smaller clumps of matter, which included both stars and clouds of dust and gas.

Combining that conclusion with the presence of the central, supermassive black hole and the number of nearby dwarf galaxies, astronomers conclude that the Milky Way must have formed when the gas within a huge "clump" of dark matter collapsed into many small protogalaxies. Some of those protogalaxies then merged to form the large, barred spiral galaxies in the Local Group, but some of those smaller protogalaxies are still around today in the form of the small, satellite dwarf galaxies near the Milky Way. The largest among them are the Large Magellanic Cloud and the Small Magellanic Cloud (**Figure 20.21**), which are easily seen by the naked eye in the Southern Hemisphere. The Magellanic Clouds were named for Ferdinand Magellan (1480–1521), who headed a European expedition that ventured far enough into the Southern Hemisphere to see them.

The Future of the Milky Way

Mergers and collisions of Local Group galaxies continue today. Among the closest companions to the Milky Way is the Sagittarius Dwarf Galaxy, which is plowing through the disk of the Milky Way on the other side of the bulge. Astronomers have observed streams of stars, as sketched in **Figure 20.22**, from Sagittarius Dwarf and some of the other dwarf galaxies that are being tidally disrupted by the Milky Way. Those dwarf galaxies will be incorporated into the Milky Way—an indication that the galaxy is still growing. Computer simulations suggest that the spiral-arm structure could be the result of such mergers.

Large Magellanic Cloud

Small Magellanic Cloud

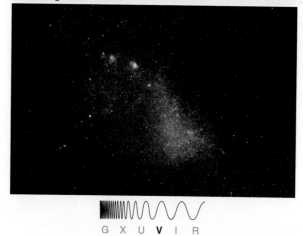

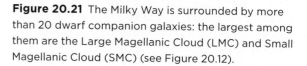

G X U V I R

Figure 20.21 The Milky Way is surrounded by more than 20 dwarf companion galaxies: the largest among them are the Large Magellanic Cloud (LMC) and Small Magellanic Cloud (SMC) (see Figure 20.12).

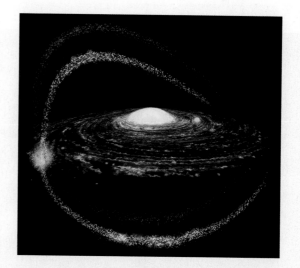

Figure 20.22 This artist's impression shows tidal tails of stars from the Sagittarius Dwarf elliptical galaxy (reddish-orange). Those stars have been stripped from the dwarf galaxy by the much more massive Milky Way, and in billions of years the two galaxies will eventually merge.

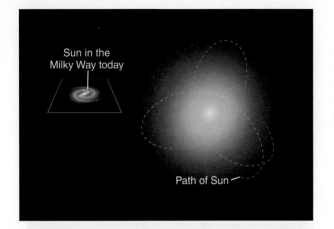

Figure 20.23 A computer simulation of the new orbit of the Sun within the Milky Way–Andromeda merger remnant 10 billion years from now.

The Andromeda Galaxy appears to violate Hubble's law because its spectrum shows blueshifts, indicating that the galaxy is moving toward, not away from, the Milky Way at 110 km/s (400,000 km/h). Andromeda and the Milky Way are the two largest galaxies in the Local Group, about 770 kpc (2.5 million light-years) apart. (If each galaxy were the size of a quarter, they would be about an arm's length apart.) They are so massive that their mutual gravitational attraction is strong and dominates over cosmological expansion, so they are moving toward each other and we see blueshifts.

Astronomers used the Hubble Space Telescope to measure the motions of Andromeda and determine whether a collision will be head-on, partial, or a total miss. They concluded that the two galaxies will "collide" head-on in about 4 billion years. Most of a galaxy is space between the stars, however, so actual collisions between the stars themselves are unlikely: although diffuse gas in the interstellar medium will collide and cause hot shocks, most of the material of one galaxy will pass through the other. The two galaxies will take another 2 billion years to merge completely and form one giant elliptical galaxy. The third spiral in the Local Group, the Triangulum Galaxy, may merge first with the Milky Way, or with Andromeda, or be ejected from the Local Group.

Note the timing here. This first close encounter with Andromeda in 4 billion years will occur *before* the Sun runs out of hydrogen in its core, although the Sun will probably have increased its luminosity enough by then that Earth's habitability will have been affected. In 6 billion years, the Sun and its planets could end up near the center of the merged galaxy, or more probably they will have a new location farther from the center with a different orbit and in a different stellar neighborhood (**Figure 20.23**). If by chance the encounter leads to a star passing close to the Sun, the orbits of the Solar System planets could be disrupted, causing them to be at different distances from the Sun. **Figure 20.24** shows how that merged galaxy might look like in Earth's sky—if anyone is still around to see it!

CHECK YOUR UNDERSTANDING 20.4

Why is Andromeda now moving toward us?

Figure 20.24 This computer simulation depicts the view in the sky on Earth when the Andromeda and Milky Way galaxies collide. The spiral Andromeda appears larger in the sky as it gets closer, and then the sky becomes brighter as the two collide.

Origins The Galactic Habitable Zone

In Chapter 19, we discussed the concept of galactic habitable zones in general; in this Origins, we focus more specifically on ideas about the habitable zone of the Milky Way. Stars situated too far from the galactic center may have protoplanetary disks with insufficient quantities of heavy elements—such as oxygen, silicon, iron, and nickel (to make up rocky planets like Earth), or carbon, nitrogen, and oxygen (to make up the molecules of life). Stars too close to the galactic center may have planets too strongly affected by its high-energy radiation environment (X-rays and gamma rays from the supermassive black hole) and by supernova explosions and gamma-ray bursts. The bulge has a higher density of stars, creating a strong radiation field, and the halo and thick disk have stars with smaller amounts of heavy elements, so perhaps only stars in the thin disk of the galaxy are candidates for residence in a galactic habitable zone.

Astronomers also must compare stellar lifetimes against the 4 billion years after the formation of Earth that life took to evolve into land animals—so only stars with masses low enough that they will live at least 4 billion years on the main sequence are considered potential hosts of more complex life. In that simple model based on the evolution times for life on Earth, the galactic habitable zone could be a doughnut-shaped region around the galactic center. In one version of the model, that zone is estimated to contain stars born 4 billion to 8 billion years ago located 7–9 kpc from the galactic center, with the Sun exactly in the middle of the doughnut. The doughnut would grow larger as heavier-element formation spread outward from the galactic center.

Some researchers have conducted additional studies and proposed more complex models. For example, one group searched for the molecule formaldehyde (H_2CO)—a key "prebiotic" molecule—by observing molecular clouds in the outer parts of the Milky Way, 12–23.5 kpc from the galactic center. Formaldehyde was detected in two-thirds of the group's sample of 69 molecular clouds, suggesting that at least one important prebiotic molecule is available far from the galactic center.

Another computer model took into account details of the evolution of individual stars within the Milky Way, including birth rates, locations, distribution within the galaxy, abundances of heavy elements, stellar masses, main-sequence lifetimes, and the likelihood that stars became or will become supernovae. The model also assumed that the development of complex life would take 4 billion years. One result of that model is that stars in the inner part of the Milky Way are more likely to be affected by supernova explosions, but those stars are even more likely to have the heavier elements for the formation of planets. In that model, then, the inner part of the galaxy, about 2.5–4 kpc from the center, in and near the midplane of the thin disk, is the most likely place for habitable planets. Here, the Sun is *not* in the middle of the most probable zone for habitable planets. As more extrasolar planets are discovered, astronomers will have a better idea of their distribution throughout the Milky Way.

Mergers with other galaxies could cause stars to migrate into or out of the galactic habitable zone. The uncertainties increase as the assumptions in those models move from the astronomical to the biological. For example, maybe life evolved faster on other planets, so stars of higher mass and shorter lifetimes should be included. Or maybe life evolved slower elsewhere, in which case the older stars would be the best candidates. Scientists do not know whether intense radiation from a supernova or a gamma-ray burst would permanently sterilize a planet or only affect evolution for a while. For example, if Earth's ozone layer were temporarily destroyed, life on land might die out, but life in the oceans would continue. Habitability in the Milky Way Galaxy is complicated, with many unanswered questions.

You Are Here: Scientists Unveil Precise Map of More Than a Billion Stars

By Nell Greenfieldboyce, National Public Radio

Wednesday was the day astronomers said goodbye to the old Milky Way they had known and loved and hello to a new view of our home galaxy.

A European Space Agency mission called *Gaia* just released a long-awaited treasure trove of data: precise measurements of 1.7 billion stars (**Figure 20.25**).

It's unprecedented for scientists to know the exact brightness, distances, motions, and colors of more than a billion stars. The information will yield the best three-dimensional map of our galaxy ever.

"This is a very big deal. I've been working on trying to understand the Milky Way and the formation of the Milky Way for a large fraction of my scientific career, and the amount of information this is revealing in some sense is thousands or even hundreds of thousands of times larger than any amount of information we've had previously," said David Hogg, an astrophysicist at New York University and the Flatiron Institute. "We're really talking about an immense change to our knowledge about the Milky Way."

The *Gaia* spacecraft launched in 2013 and is orbiting our sun, about a million miles away from Earth. Although it has surveyed a huge number of stars, *Gaia* is charting only about 1 percent of what is out there. The Milky Way contains around 100 billion stars.

For 7 million stars, *Gaia* even has measurements showing their velocity as they move toward or away from the spacecraft. "This is just a fantastic new addition," said Anthony Brown of the Netherlands' Leiden University, with the *Gaia* Data Processing and Analysis Consortium.

In 2016, *Gaia* released some initial results, but that was just a tease compared with Wednesday's massive data dump. Astronomers around the world have been planning how to attack it for months, knowing that discoveries are just waiting in there for the taking.

"Enjoy it," Antonella Vallenari, another member of the consortium from the Astronomical Observatory of Padua in Italy, said at a news conference as she announced that the data was becoming available.

The European team presented new images of our galaxy based on some initial analyses of the data, along with little movies that simulated what it would be like to fly through the stars. "It's not fake, it's real measurements," Brown said. "We know exactly where the stars are."

Scientists around the world who were watching the event online struggled to take it all in. "This is like massive scientific progress and we were blinked an image," said Jackie Faherty, an astronomer at the American Museum of Natural History, one of more than a dozen astronomers who woke up before dawn and gathered together at the Flatiron Institute in New York City to watch.

Then it was time to play, as the scientists bent their heads over their laptops and tried to download data. Researchers are expecting a slew of discoveries in the coming hours and days, but analyses will go on for years and even decades.

"This is the data we're going to be working on for the rest of my career. Probably no data set will rival this," Faherty said. "It's the excitement of the day that we see it. It's why we were up at 5 a.m. to get here. It's exciting to be around each other and trying to get the data all at once. It's a day we're going to remember."

Figure 20.25 *Gaia*'s all-sky view of our Milky Way Galaxy and neighboring galaxies, from measurements of 1.7 billion stars. The map shows the total brightness and color of stars observed by the ESA satellite in each portion of the sky between July 2014 and May 2016.

1. What observations are needed to make a 3-dimensional map of the Milky Way?
2. How do astronomers know which direction stars are moving?
3. Go to the "360° View of Gaia's Sky" at http://sci.esa.int/gaia/60219-360-view-of-gaia-s-sky/. As you scan the image, how do you know when you are looking toward the center of the Milky Way and when you are looking away from it?
4. Do a search for "Gaia Milky Way." What has been discovered from those data?

SUMMARY

Astronomers compare images and spectra of other galaxies with observations of the Milky Way. From that, they determine that the Milky Way is a barred spiral of type SBbc. The Milky Way formed from a collection of smaller protogalaxies that collapsed out of a halo of dark matter. The idea of a galactic habitable zone is that certain parts of the Milky Way may be more suitable for the existence of habitable planets. That zone would have enough heavy elements for the formation of rocky planets and organic molecules, but not so much radiation that it would damage any life.

LG 1 **Explain how astronomers discovered the size and spiral structure of the Milky Way.** The distances to globular clusters can be found from the luminosity of variable stars within them. Because those globular clusters are symmetrically distributed around the center of the Milky Way, the center of the distribution is located at the center of the galaxy. The Sun is located about 8,300 pc (27,000 light-years) from the Milky Way's center, and the Milky Way is 30,000 pc (100,000 light-years) across. Radio observations of neutral hydrogen gas and observations of star-forming nebulae provide evidence for the Milky Way's spiral structure.

LG 2 **List the clues of galaxy formation that can be found from the components of the Milky Way.** The chemical composition of the Milky Way has evolved as material has cycled between stars and the interstellar medium. A generation of stars must have existed before the formation of the oldest halo and globular-cluster stars we see today. The Milky Way has a disk consisting of two parts, the thick disk of old stars and the thin disk of young stars, implying that gas and dust from merging galaxies settled onto the disk while the stars passed through. The galactic halo consists of an inner halo and outer halo of stars and globular clusters, as well as a large, hot gas halo. The abundance of heavy elements in the Milky Way has increased as each generation of stars has produced more of those elements during the final phases of the stars' lives.

LG 3 **Explain the evidence for the dark matter halo and for the supermassive black hole at the center of the Milky Way.** The Doppler velocities of radio spectral lines show that the rotation curve of the Milky Way is flat, like those of other galaxies. But the inferred mass cannot be accounted for by the mass that is observed directly. That finding indicates that the Milky Way's mass is mostly in the form of dark matter. Evidence for the black hole at the center of the galaxy includes rapid orbital velocities of nearby stars and symmetric X-ray and gamma-ray outflows of material.

LG 4 **Describe the Local Group of galaxies and how it offers clues about the evolution of the Milky Way.** The Milky Way is part of the Local Group of galaxies, which consists of two large, barred spirals and several dozen smaller galaxies. Collisions and mergers between those galaxies probably happened in the past, and a merger with the Andromeda Galaxy may be part of the Milky Way's future. The dwarf satellites and other neighbors in the Local Group are evidence that the Milky Way is growing through accretion.

? Unanswered Questions

- Do many more ultrafaint dwarf galaxies exist than have so far been detected? Those types of dwarf galaxies are so faint that they are hard to detect even when close to the Milky Way. As we discuss in Chapter 23, the giant spirals such as the Milky Way were built up by mergers of those small, faint galaxies, so models predict that hundreds or even thousands of them should exist. They could be so dominated by dark matter that they are not at all visible, or too small ever to have formed stars, or they might have merged with the Milky Way or other Local Group members long ago.

- Will the merger of the Milky Way and Andromeda form a quasar at the center of the new elliptical galaxy? If the dust and gas at the center of the galaxy did not block the view from Earth, such a quasar could be brighter than the full Moon. The supermassive black hole at the center of Andromeda is thought to be 25–50 times larger than the one in the Milky Way, but both black holes combined still would be on the lower end of the masses of the black holes in active galactic nuclei discussed in Chapter 19.

Questions and Problems

Test Your Understanding

1. The size of the Milky Way is determined from studying _____ stars in globular clusters.
   a. Cepheid variable
   b. blue supergiant
   c. RR Lyrae
   d. Sun-like

2. Detailed observations of the structure of the Milky Way are difficult because
   a. the Solar System is embedded in the dust and gas of the disk.
   b. the Milky Way is mostly dark matter.
   c. too many stars are in the way.
   d. the galaxy is rotating too fast (about 200 km/s).

3. In general, as time passes, the Milky Way
   a. has the same chemical composition.
   b. has more abundant hydrogen.
   c. has more abundant heavy elements.
   d. has less abundant heavy elements.

4. The magnetic field of the Milky Way has been detected by
   a. synchrotron radiation from cosmic rays.
   b. direct observation of the field.
   c. the field's interaction with Earth's magnetic field.
   d. studying molecular clouds.

5. Evidence of a supermassive black hole at the center of the Milky Way comes from
   a. direct observations of stars that orbit it.
   b. visible light from material that is falling in.
   c. strong radio emission from the black hole itself.
   d. streams of cosmic rays from the center of the galaxy.

6. The Large Magellanic Cloud and Small Magellanic Cloud will probably
   a. become part of the Milky Way.
   b. remain orbiting forever.
   c. become attached to another passing galaxy.
   d. escape from the gravity of the Milky Way.

7. Globular clusters are important to understanding the Milky Way because
   a. they are so young that they provide information about current star formation.
   b. they provide information about dwarf ellipticals from which the Milky Way formed.
   c. they reveal the size of the Milky Way and Earth's location in it.
   d. the stars in them are highly enhanced in metals.

8. A globular cluster with no variable stars would have been left out of Shapley's study because
   a. the cluster would have been too far away.
   b. the distance to the cluster could not have been determined.
   c. the cluster would have been too faint to see.
   d. the cluster would have been too young to determine its evolutionary state.

9. The best evidence for the presence of dark matter in the Milky Way comes from the observation that the rotation curve
   a. is flat at great distances from the center.
   b. rises swiftly in the interior.
   c. falls off and then rises again.
   d. has a peak at about 2,000 light-years from the center.

10. Cosmic rays are
    a. a form of electromagnetic radiation.
    b. high-energy particles.
    c. high-energy dark matter.
    d. high-energy photons.

11. What kind of galaxy is the Milky Way?
    a. elliptical
    b. spiral
    c. barred spiral
    d. irregular

12. Where are the youngest stars in the Milky Way Galaxy?
    a. in the core
    b. in the bulge
    c. in the disk
    d. in the halo

13. Halo stars are found near the Sun. What observational evidence does *not* distinguish them from disk stars?
    a. their direction of motion
    b. their speed
    c. their composition
    d. their temperature

14. Why are most of the Milky Way's satellite galaxies so difficult to detect?
    a. They are very small.
    b. They are very far away.
    c. The halo of the Milky Way blocks the view.
    d. They are very faint.

15. The concept of a galactic habitable zone does not consider
    a. the radiation field.
    b. the ages of stars.
    c. the amount of heavy elements.
    d. the distance of a planet from its central star.

Thinking about the Concepts

16. Scientists can never know everything, especially at the beginning of a set of research programs. For example, Shapley did not know about dust in the Milky Way. Yet, scientists must often set aside that problem of the unknowns they don't know and do the best they can with the knowledge they have. Explain why that is a necessary step toward scientific understanding.

17. Describe the distribution of globular clusters within the Milky Way, and explain what that distribution implies about the size of the galaxy and our distance from its center.

18. In which parts of the Milky Way do astronomers find open clusters? In which parts do they find globular clusters?

19. Old stars in the inner disk of the Milky Way have higher abundances of massive elements than those of young stars in the outer disk. Explain how that difference might have developed.

20. How do 21-cm radio observations reveal the rotation of the Milky Way Galaxy?

21. What does the rotation curve of the Milky Way indicate about the presence of dark matter in the galaxy?

22. Halo stars are found near the Sun. What observational evidence distinguishes them from disk stars?

23. What is one source of synchrotron radiation in the Milky Way, and where is it found?

24. Why must astronomers use X-ray, infrared, and 21-cm radio observations to probe the center of the galaxy?

25. What is Sgr A*, and how was it detected?

26. Explain the evidence for a supermassive black hole at the center of the Milky Way. How does the mass of the supermassive black hole at the center of our galaxy compare with that found in most other spiral galaxies?

27. To observers in Earth's Southern Hemisphere, the Large Magellanic Cloud and Small Magellanic Cloud look like detached pieces of the Milky Way. What are those "clouds," and why is it not surprising that they look so much like pieces of the Milky Way?

28. What is the origin of the Milky Way's satellite galaxies? What has been the fate of most of the Milky Way's satellite galaxies? Why are most of the Milky Way's satellite galaxies so difficult to detect?

29. Use your imagination to describe how Earth's skies might appear if the Sun and Solar System were located (a) near the center of the galaxy; (b) near the center of a large globular cluster; (c) near the center of a large, dense molecular cloud.

30. What factors do astronomers consider when thinking about a galactic habitable zone in the Milky Way?

Applying the Concepts

31. From Figure 20.7, estimate the ratio between the radius of the Milky Way's outer halo and the radius of the disk.

32. According to Figure 20.16, what is the rotation velocity of a disk star located 6,000 pc from the center of the Milky Way? If you assume a circular orbit, how long does that star take to orbit once?

33. From the data in Figure 20.16, estimate the time the Large Magellanic Cloud would take to orbit the Milky Way if the Large Magellanic Cloud were on a circular orbit.

34. The Sun completes one trip around the center of the galaxy in approximately 230 million years. How many times has the Solar System made the circuit since its formation 4.6 billion years ago?

35. The Sun is located about 8,300 pc from the center of the galaxy, and the galaxy's disk probably extends another 9,000 pc farther out from the center. Assume that the Sun's orbit takes 230 million years to complete.
    a. With a truly flat rotation curve, how long would a globular cluster located near the edge of the disk take to complete one trip around the center of the galaxy?
    b. How many times has that globular cluster made the circuit since its formation about 13 billion years ago?

36. Parallax measurements of the variable star RR Lyrae indicate that it is located 230 pc from the Sun. A similar star observed in a globular cluster located far above the galactic plane appears 160,000 times fainter than RR Lyrae.
    a. How far from the Sun is that globular cluster?
    b. What does your answer to part (a) tell you about the size of the galaxy's halo in comparison with the size of its disk?

37. How do astronomers know that the center of the Milky Way contains a black hole and not just dark matter?

38. Given what you have learned about the distribution of massive elements in the Milky Way and what you know about the terrestrial planets, where do you think such planets are most likely and least likely to form?

39. A cosmic-ray proton is traveling at nearly the speed of light $(3 \times 10^8 \, \text{m/s})$.
    a. Using Einstein's relationship between mass and energy, $E = mc^2$, show how much energy (in joules) the cosmic-ray proton would have if m were based only on the proton's rest mass $(1.7 \times 10^{-27} \, \text{kg})$.
    b. The actual measured energy of the cosmic-ray proton is 100 J. What, then, is the relativistic mass of the cosmic-ray proton?
    c. How much greater is the relativistic mass of that cosmic-ray proton than the mass of a proton at rest?

40. One of the fastest cosmic rays ever observed had a speed of $(1.0 - [1.0 \times 10^{-24}]) \times c$ (that is, very, very close to c). Assume that the cosmic ray and a photon left a source at the same instant. To a stationary observer, how far behind the photon would the cosmic ray be after traveling for 100 million years?

41. Consider a black hole with a mass of 5 million M_{Sun}. A star's orbit about the black hole has a semimajor axis of 0.02 light-year (1.9×10^{14} meters). Calculate the star's orbital period. (Hint: You may want to refer to Chapter 4.)

42. A star in a circular orbit about the black hole at the center of the Milky Way (whose mass $M_{BH} = 8 \times 10^{36}$ kg) has an orbital radius of 0.0131 light-year (1.24×10^{14} meters). What is the average speed of that star in its orbit? (Hint: You may want to refer to Chapter 4.)

43. What is the Schwarzschild radius of the black hole at the center of the Milky Way? What is its density? How does that value compare with the density of a stellar black hole?

44. A star is observed in a circular orbit about a black hole with an orbital radius of 1.5×10^{11} km and an average speed of 2,000 km/s. What is the mass of that black hole in solar masses?

45. One model of the galactic habitable zone contains stars in a doughnut-shaped region between 7 and 9 kpc from the center of the galaxy, in the disk. If you assume that this doughnut is as thick as the disk itself, what fraction of the disk of the Milky Way lies in that habitable zone?

USING THE WEB

46. Go to the "Night Sky" Web page of the National Park Service (NPS; https://www.nps.gov/subjects/nightskies/). Click on and read "Light Pollution" under "Science of Light" and "Measuring Lightscapes" under "Managing Lightscapes." Why is it becoming more rare for people to see the Milky Way? Why does the NPS consider viewing the Milky Way an important part of the experience for people visiting the park?

47. a. Go to the Astronomy Picture of the Day (APOD) app or website for July 2, 2012, and watch the video clip "Zooming into the Center of the Milky Way" (https://apod.nasa.gov/apod/ap120702.html). Why do the selected pictures show a shift in wavelengths? On APOD, run a search for "Milky Way" to look at some of the best photographs of it. Where were the pictures taken from? Can you see the Milky Way from *your* location on a clear night?

    b. Why does the galactic center have to be observed in infrared and X-ray wavelengths? Go to the "Milky Way Galaxy" page of the Chandra X-ray telescope website (http://chandra.harvard.edu/photo/category/milkyway.html) and to the Spitzer infrared telescope website (http://www.spitzer.caltech.edu/). Are new images of the galactic center available? What has been learned?

48. Go to the websites of the two main groups studying the supermassive black hole at the galactic center: the UCLA Galactic Center Group (http://astro.ucla.edu/~ghezgroup/gc) and the Galactic Center Research group at the Max Planck Institute for Extraterrestrial Physics (http://mpe.mpg.de/ir/GC). Watch some of the time-lapse animations of the stars orbiting something unseen. Why is it assumed that the unseen object is a black hole? What new results are those groups reporting?

49. Citizen science: Go to the Milky Way Project website (https://www.zooniverse.org/projects/povich/milky-way-project). That project examines Spitzer and WISE telescope observations of the dusty material in the galaxy. Log in with your Zooniverse account name, and read the information under "About." Participants in the project have already discovered some of these bubbles (http://spitzer.caltech.edu/images/4938-sig12-002-Finding-Bubbles-in-the-Milky-Way). Classify some images.

50. Go to the Hubble Space Telescope website and watch the video simulation of the possible collision of the Milky Way and Andromeda (http://hubblesite.org/news_release/news/2012-20; click on "Release videos". What is different when Triangulum is included (scroll down)? Why will Andromeda "eat" the Milky Way instead of the other way around?

EXPLORATION

Astronomers once thought that the Sun was at the center of the Milky Way. In this Exploration, you will repeat Shapley's globular-cluster experiment that led to a more accurate picture of the size and shape of the Milky Way.

Imagine that the disk of the Milky Way is a flat, round plane, like a pizza. Globular clusters are arranged in a rough sphere around that plane. To map globular clusters on **Figure 20.26**, imagine that a line is drawn straight "down" from a globular cluster to the plane of the Milky Way. The "projected distance" in kiloparsecs is the distance from the Sun to the place where the line hits the plane. The galactic longitude indicates the direction toward that point; it is marked around the outside of the graph, along with the several constellations.

Make a dot at the location of each globular cluster by finding the galactic longitude indicated outside the circle and then coming in toward the center to the projected distance. The two globular clusters in boldface in **Table 20.1** have been plotted for you as examples. After plotting all the globular clusters, estimate the center of their distribution and mark it with an X. That is the center of the Milky Way.

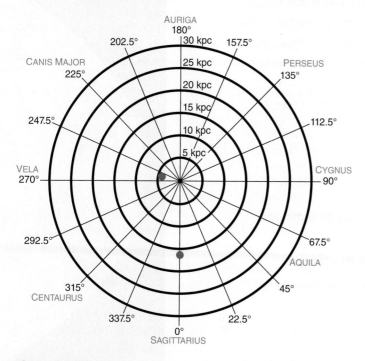

Figure 20.26 This polar graph can be used to plot distance and direction.

1 What is the approximate distance from the Sun to the center of the Milky Way?

_____

2 What is the galactic longitude of the center of the Milky Way?

_____

3 How do astronomers know that the Sun is not at the center of the Milky Way?

_____

| TABLE 20.1 | Globular Cluster Data | | | | |

Cluster	Galactic Longitude	Projected Distance (kpc)	Cluster	Galactic Longitude	Projected Distance (kpc)
104	306	3.5	6273	357	7
362	302	6.6	**6287**	**0**	**16.6**
2808	283	8.9	6333	5	12.6
4147	**251**	**4.2**	6356	7	18.8
5024	333	3.4	6397	339	2.8
5139	309	5	6535	27	15.3
5634	342	17.6	6712	27	5.7
Pal 5	1	24.8	6723	0	7
5904	4	5.5	6760	36	8.4
6121	351	4.1	Pal 10	53	8.3
O 1276	22	25	Pal 11	32	27.2
6638	8	15.1	6864	20	31.5
6171	3	15.7	6981	35	17.7
6218	15	6.7	7089	54	9.9
6235	359	18.9	Pal 12	31	25.4
6266	353	11.6	288	147	0.3
6284	358	16.1	1904	228	14.4
6293	357	9.7	Pal 4	202	30.9
6341	68	6.5	4590	299	11.2
6366	18	16.7	5053	335	3.1
6402	21	14.1	5272	42	2.2
6656	9	3	5694	331	27.4
6717	13	14.4	5897	343	12.6
6752	337	4.8	6093	353	11.9
6779	62	10.4	6541	349	3.9
6809	9	5.5	6626	7	4.8
6838	56	2.6	6144	352	16.3
6934	52	17.3	6205	59	4.8
7078	65	9.4	6229	73	18.9
7099	27	9.1	6254	15	5.7

21

The Expanding Universe

The cosmological principle has been at the center of astronomers' conceptual understanding of the universe. The cosmological principle states that the conclusions we reach about our universe should be more or less the same, regardless of whether we live in the Milky Way or in a galaxy billions of light-years away. Here in Chapter 21, we look at the evidence for the cosmological principle and for the Big Bang.

....

LEARNING GOALS

By the end of this chapter, you should be able to:

LG 1 Explain in detail the cosmological principle.

LG 2 Describe how the Hubble constant can be used to estimate the age of the universe.

LG 3 Describe the observational evidence for the Big Bang.

LG 4 Explain which chemical elements were created in the early hot universe.

This image shows the oldest light in the universe, the cosmic microwave background radiation, detected by the Planck space observatory. ▶ ▶ ▶

What is the
evidence for
the Big Bang?

21.1 The Cosmological Principle

In Chapter 19, we described observations of galaxy motions and distances. From those observations, Edwin Hubble created a graph that showed the galaxies moving apart, with the more distant ones moving faster—a result known as Hubble's law. In this section, we further explore the implications of Hubble's law.

The Homogeneous and Isotropic Universe

Cosmology is the study of space and time and the dynamics of the universe as a whole. In the 1920s, around the same time that astronomers were first measuring distances to galaxies, theoretical physicists were applying Einstein's general theory of relativity to cosmology. The cosmologist Alexander Friedmann (1888–1925) produced mathematical models of the universe that assumed the **cosmological principle**, which requires that *the physical laws and the properties of the universe are the same everywhere and in all directions*. The cosmological principle forms the basis of our study of distant galaxies and of the universe itself and is now a fundamental tenet of modern cosmology. The principle implies that we occupy no special place in the universe. Many observations have since validated that assumption.

We have used that principle throughout this book when applying laws of physics and chemistry to objects near and far in the universe. For example, according to the cosmological principle, gravity works the same way in distant galaxies as it does here on Earth. The cosmological principle is a testable scientific theory. An important prediction of the principle is that the conclusions that scientists reach about the universe should be the same whether we observe the universe from the Milky Way or from a galaxy billions of parsecs away. In other words, if the cosmological principle is correct, the universe is **homogeneous**, having the same composition and properties at all places.

In an absolute sense, the universe is *not* homogeneous because the conditions on Earth are very different from those in deep space or in the center of the Sun. In cosmology, homogeneity of the universe means that stars and galaxies in Earth's part of the universe are much the same, and behave in the same manner, as stars and galaxies in remote corners of the universe. Homogeneity also means that stars and galaxies everywhere are distributed in space in much the same way that they are distributed in Earth's cosmic neighborhood and that observers in those galaxies would see the same properties for the universe that astronomers see from here. However, homogeneity does not mean that the Universe is unchanging over time; it was different in the past and will be different in the future.

Directly verifying the prediction of homogeneity is not easy. Scientists cannot travel from the Milky Way to a galaxy in the remote universe to see whether conditions are the same. They can, however, compare light arriving from closer and farther locations in the distant universe and see the ways in which features look the same or different. For example, astronomers can look at how galaxies are distributed in distant space and see whether that distribution is similar (that is, homogeneous) to the distribution nearby.

In addition to predicting that the universe is homogeneous, the cosmological principle requires that all observers measure the same properties of the universe, regardless of the direction in which they are looking. If something is the same in all directions, it is **isotropic**. That prediction of the cosmological principle is much easier to test directly than is homogeneity. If galaxies were lined up in rows, for example, astronomers would measure very different properties of the universe, depending on the direction they looked. The universe would still be homogeneous, but not isotropic—so it would not satisfy the cosmological principle.

Usually, isotropy goes together with homogeneity, and the cosmological principle requires both. **Figure 21.1** shows examples of how the universe could have violated the cosmological principle by not being homogeneous or isotropic, as well as examples of how the universe might satisfy the cosmological principle. All observations show that the properties of the universe are basically the same, regardless of the direction in which observers are looking. When averaged over very large scales (thousands of millions of parsecs), the universe appears homogeneous as well.

The Hubble Expansion

According to Hubble's law (Chapter 19), the distance of a galaxy (d_G) is proportional to its recession velocity (v_r): $d_G = v_r/H_0$. Hubble's law helps astronomers investigate whether the universe is homogeneous and isotropic. They can confirm

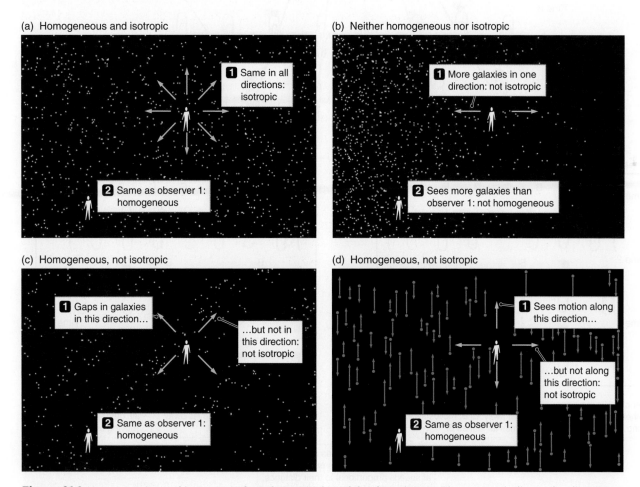

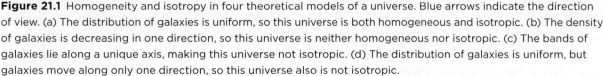

Figure 21.1 Homogeneity and isotropy in four theoretical models of a universe. Blue arrows indicate the direction of view. (a) The distribution of galaxies is uniform, so this universe is both homogeneous and isotropic. (b) The density of galaxies is decreasing in one direction, so this universe is neither homogeneous nor isotropic. (c) The bands of galaxies lie along a unique axis, making this universe not isotropic. (d) The distribution of galaxies is uniform, but galaxies move along only one direction, so this universe also is not isotropic.

its isotropy by observing that galaxies in one direction in the sky obey the same Hubble law as galaxies in other directions in the sky. Hubble's law says that Earth is located in an expanding universe and that the expansion looks the same regardless of the location of the observer. To help you visualize that uniformity, we now turn to a useful model that you can build for yourself with materials you probably have in your desk.

Figure 21.2 shows a long rubber band with paper clips attached along its length. If you stretch the rubber band, the paper clips—which represent galaxies in an expanding universe—get farther and farther apart. To get a sense of what that expansion would look like up close, imagine that you're an ant riding on paper clip A. As the rubber band is stretched, you notice that all the paper

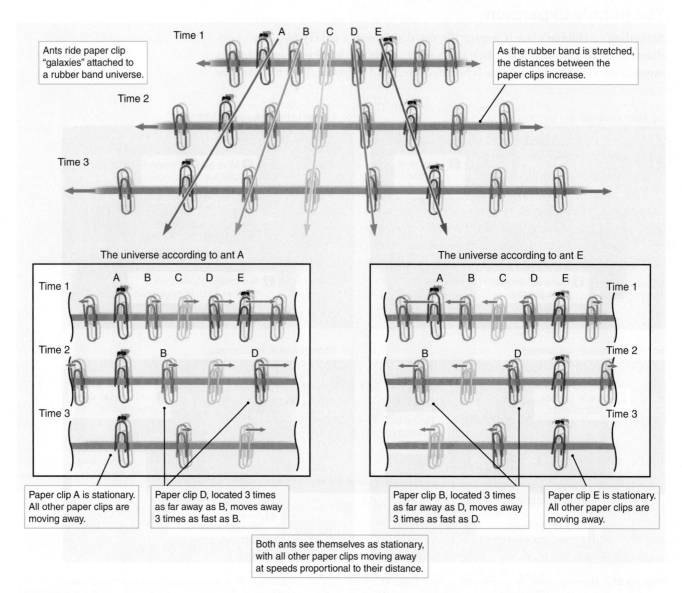

Ants ride paper clip "galaxies" attached to a rubber band universe.

As the rubber band is stretched, the distances between the paper clips increase.

The universe according to ant A

The universe according to ant E

Paper clip A is stationary. All other paper clips are moving away.

Paper clip D, located 3 times as far away as B, moves away 3 times as fast as B.

Paper clip B, located 3 times as far away as D, moves away 3 times as fast as D.

Paper clip E is stationary. All other paper clips are moving away.

Both ants see themselves as stationary, with all other paper clips moving away at speeds proportional to their distance.

Figure 21.2 In this one-dimensional analogy of Hubble's law, a rubber band with paper clips evenly spaced along its length is stretched. As the rubber band stretches, an ant riding on clip A observes clip C moving away twice as fast as clip B. Similarly, an ant riding on clip E sees clip C moving away twice as fast as clip D. Any ant sees itself as stationary, regardless of which paper clip it is riding, and it sees the other clips moving away with speed proportional to distance.

clips are moving away from you. Clip B, the closest one, is moving away slowly. Clip C, located twice as far from you as clip B, is moving away twice as fast as clip B. Clip E, located 4 times as far away as clip B, is moving away 4 times as fast as clip B. From your perspective as an ant riding on clip A, all the other paper clips on the rubber band are moving away with a velocity proportional to their distance. The paper clips located along the rubber band obey a Hubble-like law.

The key insight to the analogy comes from realizing that nothing is special about the perspective of paper clip A. If, instead, the ant were riding on clip E, clip D would be the one moving away slowly and clip A would be moving away 4 times as fast. Repeat the experiment for any paper clip along the rubber band, and you will arrive at the same result. For an ant on any paper clip along the rubber band, the speed at which other clips are moving away from the ant is proportional to their distance. The stretching rubber band, like the universe, is "homogeneous." The same Hubble-like law applies, regardless of where the observer ant is located.

The observation that nearby paper clips move away slowly and distant paper clips move away more rapidly does not mean that the paper clip selected as a vantage point is at the center of anything. Instead, it means that the rubber band is being stretched uniformly along its length. Similarly, Hubble's law for galaxies means that nearby galaxies are carried away slowly by expanding space, and distant galaxies are carried away more rapidly (see the **Process of Science Figure**).

The expanding universe has no center; the universe is expanding uniformly. Any observer in any galaxy sees nearby galaxies moving away slowly and more distant galaxies moving away more rapidly. The same Hubble law applies from their vantage point as applies from our vantage point on Earth in our galaxy. The expansion of the universe is homogeneous. The motion of galaxies as a result of the expansion of the universe is called the **Hubble flow**.

The only exception to that rule is for galaxies that are close together, in which case gravitational attraction dominates over the expansion of space. As you saw in Chapter 20, the Andromeda Galaxy and the Milky Way are being pulled together by their gravity. The Andromeda Galaxy is approaching the Milky Way at about 110 kilometers per second (km/s), so the light from the Andromeda Galaxy is blueshifted, not redshifted. That velocity is caused by the gravitational interaction between the two nearby galaxies rather than the expansion of space. The fact that gravitational or electromagnetic forces can overwhelm the expansion of space also explains why the Solar System is not expanding, and neither are you.

The Universe in Space and Time

Hubble's law gives astronomers a practical tool for measuring distances to remote objects. Once astronomers know the value of the Hubble constant, H_0, they can use a straightforward measurement of the redshift of a galaxy to find its distance. In other words, once H_0 is known, Hubble's law makes the once-difficult task of measuring distances in the universe relatively easy, giving astronomers a tool to map the structure of the observable universe. Astronomers find H_0 from nearby galaxies by using standard candles to get the distance and Doppler shifts of spectral lines to get v. Then they use that value of H_0 to find the distances to different, more distant galaxies.

21.1 Working It Out Expansion and the Age of the Universe

We can use Hubble's law to estimate the age of the universe. Consider two galaxies located 30 Mpc ($d_G = 9.3 \times 10^{20}$ km) away from each other (**Figure 21.3**). If those two galaxies are moving apart, then at some time in the past they must have been together in the same place at the same time. According to Hubble's law, and on the assumption that $H_0 = 70$ km/s/Mpc, the distance (d_G) between those two galaxies is increasing at the following rate:

$$v_r = H_0 \times d_G$$
$$v_r = 70 \text{ km/s/Mpc} \times 30 \text{ Mpc}$$
$$v_r = 2,100 \text{ km/s}$$

Knowing the velocity (v_r) at which they are traveling, we can calculate the time the two galaxies took to become separated by 30 Mpc:

$$\text{Time} = \frac{\text{Distance}}{\text{Velocity}} = \frac{9.3 \times 10^{20} \text{ km}}{2,100 \text{ km/s}} = 4.4 \times 10^{17} \text{ s}$$

Dividing by the number of seconds in a year (about 3.16×10^7 s/yr) gives

$$\text{Time} = 1.4 \times 10^{10} \text{ yr} = 14 \text{ billion yr}$$

In other words, *if* expansion of the universe has been constant, two galaxies that today are 30 Mpc apart started out at the same place about 14 billion years ago.

Now let's do the same calculation with two galaxies 60 Mpc (18.6×10^{20} km) apart. Those two galaxies are twice as far apart, but the distance between them is increasing twice as rapidly:

$$v_r = H_0 \times d_G = 70 \text{ km/s/Mpc} \times 60 \text{ Mpc} = 4,200 \text{ km/s}$$

Therefore,

$$\text{Time} = \frac{18.6 \times 10^{20} \text{ km}}{4,200 \text{ km/s}} = 4.4 \times 10^{17} \text{ s} = 1.4 \times 10^{10} \text{ yr}$$

Again, we calculate time as distance divided by velocity (twice the distance divided by twice the velocity) to find that those galaxies also took about 14 billion years to reach their current locations. We can do that calculation again and again for any pair of galaxies in the universe today. The farther apart the two galaxies are, the faster they are moving. But all galaxies took the *same* amount of time to get to where they are today. (Small differences in the intermediate steps of those calculations are the result of rounding and are not significant to the argument.) The most precise measurements give a result of 13.8 billion years.

Working out the example with words instead of numbers makes it clear why the answer is always the same. Because the velocity we are calculating comes from Hubble's law, velocity equals the Hubble constant multiplied by distance. Writing that out as an equation, we get

$$\text{Time} = \frac{\text{Distance}}{\text{Velocity}} = \frac{\text{Distance}}{H_0 \times \text{Distance}}$$

Distance divides out to give

$$\text{Time} = \frac{1}{H_0}$$

where $1/H_0$ is the Hubble time. That approach is one way to estimate the age of the universe.

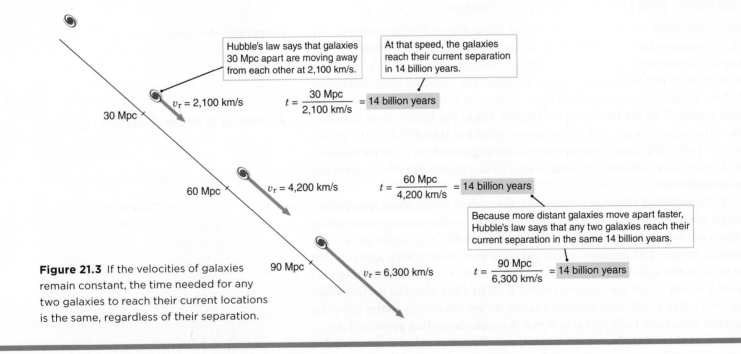

Hubble's law says that galaxies 30 Mpc apart are moving away from each other at 2,100 km/s.

At that speed, the galaxies reach their current separation in 14 billion years.

$v_r = 2,100$ km/s

30 Mpc

$t = \dfrac{30 \text{ Mpc}}{2,100 \text{ km/s}} = 14 \text{ billion years}$

$v_r = 4,200$ km/s

60 Mpc

$t = \dfrac{60 \text{ Mpc}}{4,200 \text{ km/s}} = 14 \text{ billion years}$

Because more distant galaxies move apart faster, Hubble's law says that any two galaxies reach their current separation in the same 14 billion years.

90 Mpc

$v_r = 6,300$ km/s

$t = \dfrac{90 \text{ Mpc}}{6,300 \text{ km/s}} = 14 \text{ billion years}$

Figure 21.3 If the velocities of galaxies remain constant, the time needed for any two galaxies to reach their current locations is the same, regardless of their separation.

state. That hot, dense beginning, 13.8 billion years ago, is called the **Big Bang** (**Figure 21.4**).

Georges Lemaître (1894–1966) was the first to propose the theory of the Big Bang. The idea greatly troubled many astronomers in the early and middle years of the 20th century. Several suggestions were put forward to explain the observed fact of Hubble expansion without resorting to the idea that the universe came into existence in an extraordinarily dense "fireball" billions of years ago. As more and more distant galaxies have been observed, however, and as more discoveries about the structure of the universe have been made, the Big Bang theory has grown stronger. The major predictions of the Big Bang theory have proven to be correct.

The implications of Hubble's law forever changed the scientific concepts of the origin, history, and possible future of the universe. At the same time, Hubble's law has pointed to many new questions about the universe. To address them, we next need to consider exactly what the term *expanding universe* means.

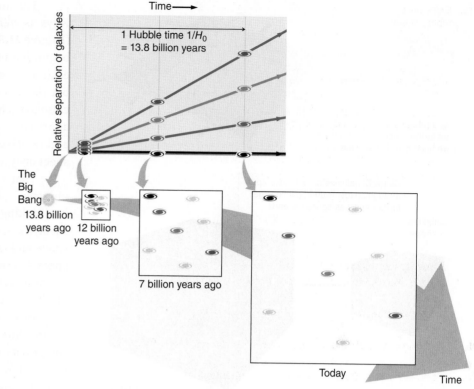

Figure 21.4 Looking backward in time, the distance between any two galaxies is smaller and smaller, until all matter in the universe is concentrated at the same point: the Big Bang.

Galaxies and the Expansion

At this point in our discussion, you may be picturing the expanding universe as a cloud of debris from an explosion flying outward through surrounding space. That is a common depiction of the Big Bang in movies and television shows, which portray a tiny bright spot that explodes to fill the screen. However, the Big Bang is not an explosion in the usual sense of the word, and in fact no surrounding space exists into which the universe is expanding.

We must draw a distinction between the universe and the observable universe. The observable universe is the part of the universe that we can see. The observable universe extends 13.8 billion light-years in every direction. That limit exists because it is the length of time the universe has been around. The light from more distant regions has not yet had time to travel to us, and so we cannot see it yet.

Where did the Big Bang take place? The Big Bang took place *everywhere*. Wherever anything is in the universe today, it is at the site of the Big Bang. The reason is that galaxies are not flying apart through space at all. Rather, space itself is expanding, carrying the stars and galaxies that populate the universe along with it.

We have already dealt with the basic ideas that explain the expansion of space. In our discussion of black holes (Chapter 18), you encountered Einstein's general theory of relativity. General relativity says that space is distorted by the presence of mass and that the consequence of that distortion is gravity. For example, the mass of the Sun, like any other object, distorts the geometry of spacetime around it; so Earth, coasting along in its inertial frame of reference, follows a curved path around the Sun. We illustrated that phenomenon in Figure 18.10 with the analogy of a ball placed on a stretched rubber sheet, showing how the ball distorted the surface of the sheet.

Astronomy in Action: Observable versus Actual Universe

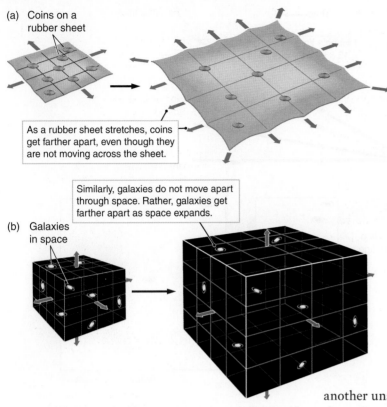

(a) Coins on a rubber sheet

As a rubber sheet stretches, coins get farther apart, even though they are not moving across the sheet.

Similarly, galaxies do not move apart through space. Rather, galaxies get farther apart as space expands.

(b) Galaxies in space

Figure 21.5 (a) As a rubber sheet is stretched, coins on its surface move farther apart, even though they are not moving with respect to the sheet itself. Any coin on the surface of the sheet observes a Hubble-like law in every direction. (b) Galaxies in an expanding universe are not flying apart through space. Rather, space itself is stretching.

The surface of a rubber sheet can be distorted in other ways as well. Imagine several coins placed on a rubber sheet (**Figure 21.5**). Then imagine grabbing the edges of the sheet and beginning to pull them outward. As the rubber sheet stretches, each coin remains at the same location on the surface of the sheet, but the distances between the coins increase. Two coins sitting close to each other move apart only slowly, whereas coins farther apart move away from each other more rapidly. The distances and relative motions of the coins on the surface of the rubber sheet obey a Hubble-like relationship as the sheet is stretched.

That movement is analogous to what is happening in the universe, with galaxies taking the place of the coins and space itself taking the place of the rubber sheet. For coins on a rubber sheet, the sheet can be stretched only so far before it breaks. With space and the real universe, no such limit exists. The fabric of space can, in principle, go on expanding forever. Hubble's law is the observational consequence of the fact that the space making up the universe is expanding.

How will that expansion of the universe behave in the future? Most of the mass in galaxies consists of dark matter, and the gravity of the dark matter slows down the expansion of the universe. In Chapter 22, you will also learn that the universe has another unseen constituent, called *dark energy*, which causes the expansion of the universe to *accelerate*. At the current stage in the expansion of the universe, the accelerating effect of dark energy dominates over the slowing effect of dark matter, suggesting that the universe will continue to expand at an ever-faster rate.

CHECK YOUR UNDERSTANDING 21.2

Where in the universe did the Big Bang take place? (a) near the Milky Way Galaxy; (b) near the center of the universe; (c) near some unknown location on the other side of the universe; (d) everywhere in the universe

---

21.3 Expansion Is Described with a Scale Factor

As the universe expands, the distance between any two objects increases because of the stretching of space. Astronomers find it useful to discuss that expansion in terms of the *scale factor* of the universe.

Scale Factor

Let's return to the analogy of the rubber sheet. Suppose you place a ruler on the rubber sheet and draw a tick mark every centimeter (**Figure 21.6a**). To measure the distance between two points on the sheet, you can count the marks between the two points and multiply by 1 centimeter (cm) per tick mark.

As the sheet is stretched, however, the distance between the tick marks does not remain 1 cm. When the sheet is stretched to 150 percent of the size it had when the tick marks were drawn, each tick mark is separated from its neighbors by 1 1/2 times the original distance, or 1.5 cm. The distance between two points can still come from counting the marks, but you need to scale up the distance between tick marks by 1.5 to find the distance in centimeters. If the sheet were twice the size it was when the tick marks were drawn (**Figure 21.6b**), each mark would correspond to 2 cm of actual distance. Astronomers use the term **scale factor** to indicate the size of the sheet relative to its size when the tick marks were drawn. The scale factor also indicates how much the distance between points on the sheet has changed. In the first example, the scale factor of the sheet is 1.5; in the second, the scale factor is 2.

Suppose astronomers choose today to lay out a "cosmic ruler" on the fabric of space, placing an imaginary tick mark every 10 Mpc. The scale factor of the universe at this time is defined to be 1. In the past, when the universe was smaller, distances between the points in space marked by the cosmic ruler would have been less than 10 Mpc. The scale factor of that younger, smaller universe would have been less than today's scale factor and therefore less than 1. In the future, as the universe continues to expand, the distances between the tick marks on the cosmic ruler will grow to more than 10 Mpc, and the scale factor of the universe will be greater than 1. Astronomers use the scale factor, usually written as R_U, to keep track of the changing scale of the universe.

The laws of physics are themselves unchanged by the expansion of the universe, just as stretching a rubber sheet does not change the properties of the coins on its surface. At noncosmological scales, the nuclear and electromagnetic forces within and between atoms, as well as the gravitational forces between relatively close objects, dominate over the expansion. As the universe expands, the sizes and other physical properties of atoms, stars, and galaxies also remain unchanged.

As we look back in time, the scale factor of the universe gets smaller and smaller, approaching zero as it comes closer and closer to the Big Bang. The fabric of space that today spans billions of parsecs spanned much smaller distances when the universe was young. When the universe was only a day old, all the space visible today amounted to a region only a few times the size of the Solar System. When the universe was 1/50 of a second old, the vast expanse of space that makes up today's observable universe (and all the matter in it) occupied a volume only the size of today's Earth. As we go backward in time approaching the Big Bang itself, the space that makes up today's observable universe becomes smaller and smaller—the size of a grapefruit, a marble, an atom, a proton. Every point in the fabric of space that makes up today's universe was right there at the beginning, a part of that unimaginably tiny, dense universe that emerged from the Big Bang.

The Big Bang did not occur at a specific point in space because space itself came into existence with the Big Bang. The Big Bang happened everywhere; no particular point in today's universe marked the site of the Big Bang. A Big Bang universe is homogeneous and isotropic, consistent with the cosmological principle.

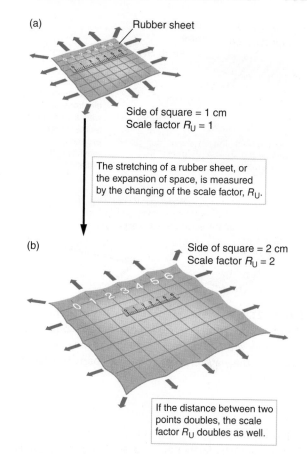

Figure 21.6 (a) On a rubber sheet, tick marks are drawn 1 cm apart. As the sheet is stretched, the tick marks move farther apart. (b) When the spacing between the tick marks is 2 cm, or twice the original value, the scale factor of the sheet, R_U, is said to have doubled. A similar scale factor, R_U, is used to describe the expansion of the universe.

Labels in figure:
(a) Rubber sheet
Side of square = 1 cm
Scale factor $R_U = 1$

The stretching of a rubber sheet, or the expansion of space, is measured by the changing of the scale factor, R_U.

(b) Side of square = 2 cm
Scale factor $R_U = 2$

If the distance between two points doubles, the scale factor R_U doubles as well.

CHECK YOUR UNDERSTANDING 21.3A

The scale factor keeps track of: (a) the movement of galaxies through space; (b) the current distances between many galaxies; (c) the changing distance between any two galaxies; (d) the location of the center of the universe.

Redshift Is Due to the Changing Scale Factor of the Universe

The ideas of general relativity discussed in Chapter 18 are powerful tools for interpreting Hubble's great discovery of the relationship between the velocity and distance of galaxies and of the expanding universe. Those velocities were determined from the redshift of the galaxies by using the Doppler effect. Although it is true that the distance between galaxies is increasing as a result of the expansion of the universe, and that we can use the equation for Doppler shifts to measure the

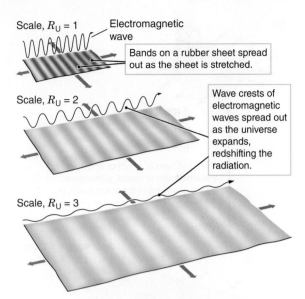

Scale, $R_U = 1$ — Electromagnetic wave

Bands on a rubber sheet spread out as the sheet is stretched.

Scale, $R_U = 2$

Wave crests of electromagnetic waves spread out as the universe expands, redshifting the radiation.

Scale, $R_U = 3$

Figure 21.7 Bands drawn on a rubber sheet represent the positions of the crests of a light wave in space. As the rubber sheet is stretched—that is, as the universe expands—the wave crests get farther apart. The light is redshifted.

redshifts of galaxies, those redshifts are not due to Doppler shifts in the same way that we described for a moving star. Light that comes from very distant objects was emitted when the universe was younger and therefore smaller. As that light comes toward Earth from distant galaxies, the scale factor of the space through which the light travels is constantly increasing, and as it does, the distance between adjacent light-wave crests increases as well. The light is "stretched out" as the space it travels through expands.

Recall the rubber sheet analogy that we used when discussing black holes in Chapter 18. Then, we imagined space as a two-dimensional rubber sheet. **Figure 21.7** uses that rubber-sheet analogy to explain why the increasing distance between galaxies causes the light to be redshifted. If you draw a series of bands on the rubber sheet to represent the crests of an electromagnetic wave, you can watch what happens to the wave as the sheet is stretched out. By the time the sheet is stretched to twice its original size—that is, by the time the scale factor of the sheet is 2—the distance between wave crests has doubled. When the sheet has been stretched to 3 times its original size (a scale factor of 3), the wavelength of the wave will be 3 times what it was originally.

When light left a distant galaxy, the scale factor of the universe was smaller than it is today. As light comes toward us from distant galaxies, the space through which the light travels is stretching, and the light is also "stretched out" as the space through which it travels expands. The universe expanded while the light was in transit, and as it did so, the wavelength of the light grew longer in proportion to the increasing scale factor of the universe. The redshift of light from distant galaxies is therefore a direct measure of how much the universe has expanded since the radiation left its source. Redshift measures how much the scale factor of the universe, R_U, has changed since the light was emitted. That redshift can be greater than 1 if the galaxy is far enough away that the recession velocity is relativistic (**Working It Out 21.2**).

CHECK YOUR UNDERSTANDING 21.3B

What is the interpretation of a redshift larger than 1? (a) The object is moving faster than the speed of light. (b) The universe has more than doubled in size since the light from that object was emitted. (c) The light was shifted to longer wavelengths from gravitational radiation. (d) The rate of expansion of the universe is increasing.

21.4 Astronomers Observe Cosmic Microwave Background Radiation

What is the evidence that the Big Bang actually took place? One piece of evidence comes from observations of the early universe across the entire sky. This section introduces the cosmic microwave background radiation, one of the major confirming observations of the theory that the universe had a beginning.

Radiation from the Big Bang

In the late 1940s, cosmologists Ralph Alpher (1921–2007), Robert Herman (1914–1997), and George Gamow (1904–1968) reasoned that because a compressed gas cools as it expands, the universe also should be cooling as it expands. When the universe was very young and small, it must have consisted of an extraordinarily hot, dense gas. As with any hot, dense gas, that early universe would have been awash in blackbody radiation that exhibits a Planck blackbody spectrum (see Chapter 5).

21.2 Working It Out: When Redshift Exceeds 1

From our discussion of the Doppler shift (Chapter 5), recall that $(\lambda_{obs} - \lambda_{rest})/\lambda_{rest}$ is equal to the velocity of an object moving away (v_r) divided by the speed of light (c). Edwin Hubble used that result to interpret the observed redshifts of galaxies as evidence that galaxies throughout the universe are moving away from the Milky Way. Einstein's special theory of relativity says that nothing can move faster than the speed of light. Hubble initially assumed that redshifts are due to the Doppler effect. The resulting relation, $z = v_r/c$, would then seem to imply that no object can have a redshift (z) greater than 1. Yet that is not the case. Astronomers routinely observe redshifts significantly in excess of 1. As of this writing, the most distant objects known have redshifts as large as 9–11. How can redshifts exceed 1?

To arrive at the expression for the Doppler effect, $v_r/c = (\lambda_{obs} - \lambda_{rest})/\lambda_{rest}$, we have to *assume* that v_r is much less than c. If v_r was close to c, we would have to consider more than just the fact that the waves from an object are stretched out by the object's motion away. We also would have to consider relativistic effects, including the fact that moving clocks run slowly (see Working It Out 18.1). When combining those effects, we would find that as the speed of an object approaches the speed of light, its redshift approaches infinity, as shown in **Figure 21.8**.

Doppler's original formula is essentially correct—for objects close enough to us that their measured velocities are far less than the speed of light. When astronomers look at the motions of orbiting binary stars or the peculiar velocities of galaxies relative to the Hubble flow, that equation works just fine. But for any redshift of 0.4 or greater, relativity must be taken into account.

Another source of redshift is the gravitational redshift discussed in Chapter 18. As light escapes from deep within a gravitational well, it loses energy, so photons are shifted to longer and longer wavelengths. If the gravitational well is deep enough, the observed redshift of that radiation can be boundlessly large. In fact, the event horizon of a black hole—that is, the surface around the black hole from which not even light can escape—is where the gravitational redshift becomes infinite.

Cosmological redshift, which is most relevant to this chapter, results from the amount of "stretching" that space has undergone while the light from its original source has been en route to Earth. The amount of stretching is given by the factor $1 + z$. When astronomers observe light from a distant galaxy whose redshift of $z = 1$, the wavelength of that light is twice as long as when it left the galaxy. When the light left its source, the universe was half the size that it is today. When they see light from a galaxy with $z = 2$, the wavelength

of the radiation is 3 times its original wavelength, and they are seeing the universe when it was one-third its current size. That direct relationship enables astronomers to use the observed redshift of the galaxy to calculate the size of the universe at the look-back time to that galaxy. Nearby, that means that distance and look-back time are proportional to z. As astronomers look back closer and closer to the Big Bang, however, redshift climbs more and more rapidly, approaching infinity as the look-back time approaches the age of the universe.

Written as an equation, the scale factor of the universe that astronomers see when looking at a distant galaxy is equal to 1 divided by 1 plus the redshift of the galaxy:

$$R_U = \frac{1}{1 + z}$$

For example, when astronomers report they have observed a galaxy with a redshift of 9, the scale factor when the light was emitted was

$$R_U = 1/(1 + 9) = 1/10.$$

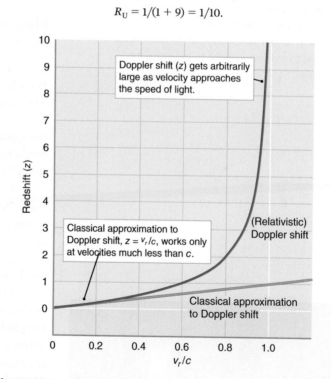

Figure 21.8 This graph shows the plot of the redshift (z) of an object versus its recession velocity (v_r) as a fraction of the speed of light. According to special relativity, the redshift becomes large without limit as v_r approaches c.

Gamow and Alpher took that idea a step further, noting that as the universe expanded, that radiation would have been redshifted to longer and longer wavelengths. According to Wien's law (Chapter 5), the temperature associated with Planck blackbody radiation is inversely proportional to the peak wavelength: $T = (2,900,000 \text{ nm K})/\lambda_{peak}$. Shifting the Planck radiation to longer and

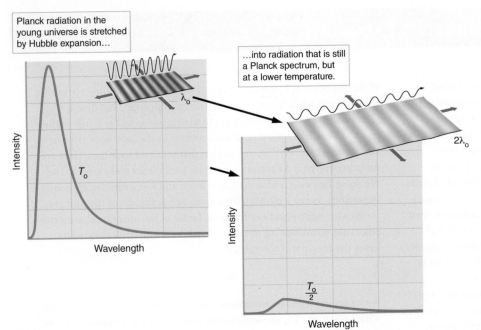

Planck radiation in the young universe is stretched by Hubble expansion...

...into radiation that is still a Planck spectrum, but at a lower temperature.

λ_0

$2\lambda_0$

Intensity

Wavelength

T_0

Intensity

Wavelength

$\frac{T_0}{2}$

Figure 21.9 As the universe expanded, Planck radiation left over from the hot young universe was redshifted to longer wavelengths. Redshifting a Planck spectrum is equivalent to lowering its temperature.

longer wavelengths is therefore the equivalent of shifting the temperature of the radiation to lower and lower values. As illustrated in **Figure 21.9**, doubling the wavelength of the photons in a Planck blackbody spectrum by stretching space and doubling the scale factor of the universe is equivalent to cutting the temperature of the Planck spectrum in half. As a result, that radiation should still be detectable today and should have a Planck blackbody spectrum with a temperature of about 5–50 kelvins (K). Alpher searched for the signal, but the technology of the late 1940s and early 1950s was not advanced enough. A decade later, physicist Robert Dicke (1916–1997) and his colleagues at Princeton University also predicted a hot early universe, arriving independently at the same basic conclusions that Alpher and Gamow had reached earlier.

Measuring the Temperature of the Cosmic Microwave Background Radiation

In the early 1960s, two physicists at Bell Laboratories—Arno Penzias and Robert Wilson (**Figure 21.10**)—detected a faint microwave signal in all parts of the sky (see Chapter 6). Astronomers interpreted the signal as the radiation left behind by the hot early universe. The strength of the detected signal was consistent with the glow from a blackbody with a temperature of about 3 K, very close to the predicted value. Their results, published in 1965, reported the discovery of the "glow" left behind by the Big Bang.

That radiation left over from the early universe is called the **cosmic microwave background radiation** (**CMB**). When the universe was young, it was hot enough that all atoms were ionized, so that the electrons were separate from the atomic nuclei. Recall from our discussion of the structure of the Sun and stars (Chapter 14) that free electrons in such conditions interact strongly with radiation, blocking its progress, so radiation does not travel well through ionized plasma.

The conditions within the early universe were much like the conditions within a star: hot, dense, and opaque (**Figure 21.11a**). As the universe expanded, the gas filling it cooled. By the time the universe was about 380,000 years old, and about a thousandth of its current size, the temperature had dropped to a few thousand kelvins. Hydrogen and helium nuclei combined with electrons to form neutral atoms for the first time—an event called the **recombination** of the universe (**Figure 21.11b**).

Hydrogen atoms block radiation much less effectively than do free electrons, so when recombination occurred, the universe suddenly became transparent to radiation. Since that time, the radiation left behind from the Big Bang has traveled largely unimpeded throughout the universe. At the time of recombination, when the temperature of the universe was about 3000 K, the wavelength of that radiation peaked at about 1 micron (μm), according to Wien's law (see Chapter 5). As the universe expanded, that radiation was redshifted to longer and longer wavelengths—and therefore cooler temperatures, as in Figure 21.9. Today, the scale of the universe has increased a thousandfold since recombination, and the peak wavelength of the cosmic background radiation has increased by a

Figure 21.10 Arno Penzias (left) and Robert Wilson next to the Bell Labs radio telescope antenna with which they discovered the cosmic background radiation. That antenna is now a U.S. National Historic Landmark.

In the ionized early universe, light was trapped by free electrons. Radiation had a blackbody spectrum.

(a)

At that time, it was as though the universe was filled with a thick fog.

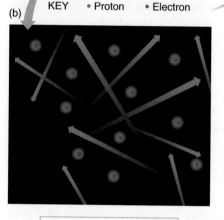

KEY • Proton • Electron   Path of photon

(b)

At recombination, the universe became transparent, and the blackbody radiation traveled freely through the universe.

Recombination was like the fog suddenly clearing.

Figure 21.11 The cosmic microwave background radiation we see originated at the moment the universe became transparent. (a) Before recombination, the universe was like a foggy day, except that the "fog" was a sea of electrons and protons. Radiation interacted strongly with free electrons and so could not travel far. The trapped radiation had a Planck blackbody spectrum. The tree in the picture is analogous to a source emitting just before recombination. (b) When the constituents of the universe recombined to form neutral hydrogen atoms, the fog cleared and that radiation was free to travel unimpeded.

thousandfold as well, to a value close to 1 millimeter (mm). The spectrum of the CMB still has the shape of a Planck blackbody spectrum, but with a characteristic temperature a thousandth what it was at the time of recombination.

Variations in the CMB

The presence of cosmic background radiation with a Planck blackbody spectrum is a strong prediction of the Big Bang theory. Penzias and Wilson had confirmed that a signal with the correct strength was there, but they could not say for certain whether the signal they saw had the spectral shape of a Planck blackbody spectrum. From the late 1960s to the 1990s, experiments at different wavelengths supported those same conclusions. The Cosmic Background Explorer (COBE) satellite made extremely precise measurements of the CMB at many wavelengths, from a few microns out to 1 cm. The COBE measurements showed the CMB to be a Planck blackbody spectrum with a temperature of 2.73 K (**Figure 21.12**). The observed spectrum so perfectly matches the one predicted by Big Bang cosmology that no real doubt can exist that this is the residual radiation left behind from the primordial fireball of the early universe. Astrophysicists John Mather and George Smoot won the Nobel Prize in Physics in 2006 for their work on COBE.

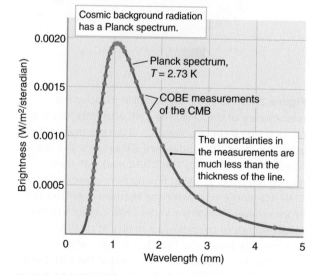

Cosmic background radiation has a Planck spectrum.

Planck spectrum, T = 2.73 K

COBE measurements of the CMB

The uncertainties in the measurements are much less than the thickness of the line.

Figure 21.12 The spectrum of the CMB, as measured by the Cosmic Background Explorer (COBE) satellite, is shown by the red dots. The uncertainty in the measurement at each wavelength is much less than the size of a dot. The line running through the data is a blackbody spectrum with a temperature of 2.73 K.

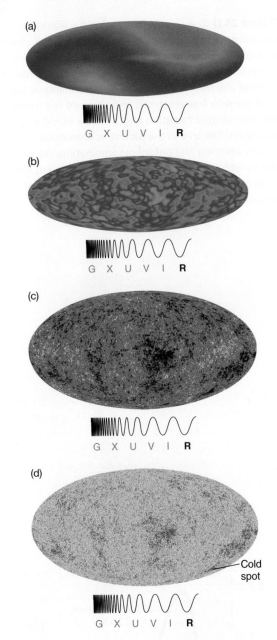

Figure 21.13 (a) The COBE satellite mapped the temperature of the CMB. The CMB is slightly hotter (by about 0.003 K) in one direction in the sky than in the other direction. That difference is due to Earth's motion relative to the CMB. (b) The COBE map with Earth's motion removed, showing tiny ripples remaining in the CMB. (c) *WMAP* (the Wilkinson Microwave Anisotropy Probe) showed variations in the CMB, confirming the fundamentals of cosmological theory at small and intermediate scales. (d) The Planck space observatory has yielded the highest resolution yet of the variations of the CMB and has detected some surprises, such as the "cold spot." The radiation seen here was emitted less than 400,000 years after the Big Bang. Blue spots are cooler, whereas red spots are warmer.

COBE data included much more than a measurement of the spectrum of the cosmic background radiation. **Figure 21.13a** shows a map obtained by COBE of the CMB from the entire sky. The different colors in the map correspond to variations of about 0.1 percent in the temperature of the CMB. Most of that range of temperature is present because one side of the sky looks slightly warmer than the opposite side of the sky. That difference has nothing to do with the large-scale structure of the universe itself but rather is the result of the Earth's motion relative to the gas that last interacted with and created the CMB.

Because the laws of physics are the same in any inertial reference frame, no frame of reference is better than any other. Yet in a certain sense a preferred frame of reference exists at every point in the universe. That is the frame of reference that is at rest with respect to the expansion of the universe and in which the CMB is isotropic, or the same in all directions. The COBE map shows that one side of the sky is slightly hotter than the other because Earth and the Sun are moving at a velocity of 370 km/s in the direction of the constellation Crater relative to that cosmic reference frame. Radiation coming from the direction in which Earth is moving is slightly blueshifted (and thus shifted to a higher temperature) by that motion, whereas radiation coming from the opposite direction is Doppler-shifted toward the red (or cooler temperatures). Earth's motion is due to a combination of factors, including the motion of the Sun around the center of the Milky Way Galaxy and the motion of the Milky Way relative to the CMB.

When that asymmetry in the CMB caused by the motion of Earth is subtracted from the COBE map, only slight variations, about 0.001 percent, in the CMB remain (**Figure 21.13b**). Those slight variations might not seem like much, but they are actually crucial in the history of the universe. Recall from Chapter 18 that gravity itself can create a redshift. Those tiny fluctuations in the cosmic background radiation are the result of gravitational redshifts caused by concentrations of mass that existed in the early universe. Those concentrations later gave rise to galaxies and the rest of the structure that is evident in the universe today.

From 2001 to 2010, NASA's *WMAP* (Wilkinson Microwave Anisotropy Probe) satellite made more precise measurements of the variations of the CMB. **Figure 21.13c** shows the ripples measured by *WMAP* with much higher resolution than could be detected by COBE. The European Space Agency's Planck space observatory collected data of even higher resolution from 2009 to 2013 (**Figure 21.13d**, and chapter opening figure). The much higher-resolution maps obtained by *WMAP* and Planck enable astronomers to refine their ideas about the development of structure in the early universe, which we discuss in Chapters 22 and 23.

CHECK YOUR UNDERSTANDING 21.4

The existence of the cosmic microwave background radiation tells us that the early universe was: (a) much hotter than it is today; (b) much colder than it is today; (c) about the same temperature as today but was much more dense.

Origins Big Bang Nucleosynthesis

The expansion of the universe and the cosmic microwave background radiation are two of the key pieces of observational evidence supporting the Big Bang theory. The third major piece of supporting evidence comes from observations of the number and types of chemical elements in the universe. For a short time after the Big Bang, the temperature and density of the universe were high enough for nuclear reactions to take place. Collisions between protons and neutrons in the early universe built up low-mass nuclei, including deuterium (heavy hydrogen) and isotopes of helium, lithium, and beryllium. That process of element creation, called **Big Bang nucleosynthesis**, determined the final chemical composition of the matter that emerged from the hot phase of the Big Bang. Because the universe was rapidly expanding and cooling, the density and temperature of the universe fell too low for fusion to heavier elements such as carbon to occur. Therefore, all elements more massive than beryllium, including most of the atoms that make up Earth and its life, must have formed in later generations of stars.

Figure 21.14 shows the observed and calculated predictions of the amounts of deuterium, helium, and lithium from Big Bang nucleosynthesis, plotted as a function of the observed present-day density of normal (luminous) matter in the universe. Observations of current abundances are shown as horizontal bands. Theoretical predictions, which depend on the density of the universe, are shown as darker, thick lines. Big Bang nucleosynthesis predicts that about 24 percent of the mass of the normal matter formed in the early universe should have ended up in the form of the very stable isotope ^{4}He, regardless of the total density of matter in the universe. Indeed, that is what is observed. That agreement between theoretical predictions and observation serves as powerful evidence that the universe began in a Big Bang.

Unlike with helium, most other isotope abundances depend on the density of normal matter in the universe, so comparing current abundances with models of isotope formation in the Big Bang helps pin down the density of the early universe. Beginning with the abundances of isotopes such as ^{2}H (deuterium) and ^{3}He found in the universe today (shown as the horizontal, light-colored bands in Figure 21.14) and comparing them with predictions in different models of how abundant isotopes *should* be when formed at different densities (dark-colored curves in Figure 21.14), cosmologists can find the density of normal matter (the vertical yellow band in Figure 21.14). The best current measurements give a value of about 3.9×10^{-28} kg/m^3 for the average density of normal matter in the universe today (at sea level, the density of air is about 1.2 kg/m^3). That value lies well within the range predicted by the observations shown in Figure 21.14. The agreement is remarkable, and it holds for many isotopes.

Turning that process around, cosmologists can begin with an observation of the amount of normal matter in and around galaxies and then compare that with calculations of what the chemical composition emerging from the Big Bang should have been. The observations agree remarkably well with the amounts of those elements actually found in nature. The idea that light elements originated in the Big Bang is resoundingly confirmed.

That agreement also creates a powerful constraint on the nature of dark matter, which dominates the mass in the universe. Dark matter cannot consist of normal matter made up of neutrons and protons; if it did, the density of neutrons and protons in the early universe would have been much higher, and the resulting abundances of light elements in the universe would have been much different from what is actually observed.

AstroTour: Big Bang Nucleosynthesis

Figure 21.14 Observed and calculated abundances of the products of Big Bang nucleosynthesis, plotted against the density of normal matter in today's universe. Big Bang nucleosynthesis correctly predicts the amounts of those isotopes found in the universe today. (Note the two scale breaks on the *y*-axis.)

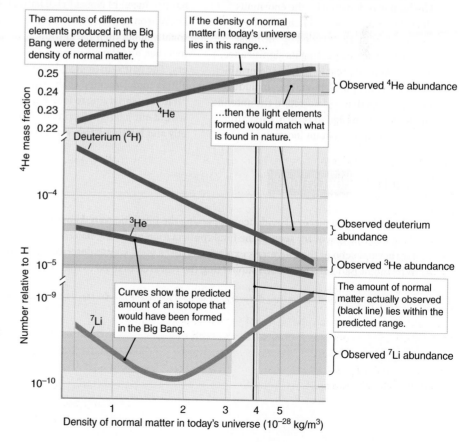

50th Anniversary of the Big Bang Discovery

By **JOANNE COLELLA**, *The Journal* (NJ)

A unique gathering of some of the most brilliant minds in the world was held last month on Holmdel's Crawfords Hill on a beautiful spring day, celebrating the 50th anniversary of a truly stellar discovery: the detection of cosmic microwave background radiation (CMB), the thermal echo of the universe's explosive birth, and the evidence that proved the famed Big Bang theory, which would have taken place about 13.8 billion years ago. On May 20, 1964, American radio astronomers Robert Wilson and Arno Penzias confirmed that discovery, admittedly by accident but only after an exhaustive amount of investigative research to rule out every possible explanation for an odd buzzing sound that came from all parts of the sky at all times of day and night. The hum was detected by the enormous Horn Antenna at the Bell Labs site, now a national landmark. Puzzled by the noise, but initially not suspecting its significance, the pair went to great lengths to determine, or rule out, any possible source—including some pigeons that had nested in the antenna and were determined to return, even after being shipped to a distant location.

In 1978, Dr. Wilson and Dr. Penzias won the esteemed Nobel Prize in Physics for their work. Now 78 and 81 years old, respectively, the two came together again at the Horn Antenna site to celebrate the momentous anniversary with current and former Bell Labs colleagues. The event was headed up by Bell Labs President and Corporate CTO Marcus Weldon, who lauded their achievement and spoke about the company initiative to return "back to the future"— back to the classic model of Bell Labs, the research arm of Alcatel-Lucent, working to invent the future. Regarding the Big Bang theory, Mr. Weldon brought chuckles by stating, "In the beginning there was nothing and then it exploded . . . and then there was Arno and Bob." Subtle humor was a recurring theme by some speakers, who even poked fun at myths about Bell Labs, which has produced 12 Nobel Prize laureates . . . . Dr. Wilson and Dr. Penzias each spoke personally about their backgrounds, their work together, and the meaning of their discovery. The pair's distinct styles and personalities complemented each other perfectly. "It is very satisfying to look back and see we did our job right," said Dr. Wilson quietly and humbly. "This was pretty heady stuff," stated the talkative Dr. Penzias. "This is as close to being religious as I can be."

Throughout the presentations and celebratory luncheon, held under oversized tents on the expansive grounds, the air of excitement, pride, and mutual admiration among the attendees was palpable, as they looked back—and looked ahead—to the incredible legacy and pool of talent shared by Bell Labs personnel over the years. Robert Wilson and his wife still reside in Holmdel, as do many other Bell Labs employees, and the company has been an integral component of Holmdel and surrounding communities.

During the event, details were also announced about the establishment of the Bell Labs Prize, an annual competition to give scientists around the globe the chance to introduce their ideas in the fields of information and communications technology. The challenge offers a grand prize of $100,000, second prize of $50,000, and third prize of $25,000. Winners may also get the chance to develop their ideas at Bell Labs. The program is intended to inspire world-changing discoveries and innovations by young researchers.

1. What other main pieces of evidence supporting the Big Bang were known before that discovery?
2. Why was that discovery seen as the final confirmation of the Big Bang?
3. Is "thermal echo of the universe's explosive birth" a good way to describe the CMB?
4. Why might Bell Labs have been supporting the research that led to that discovery in 1964?

Source: "50th anniversary of the Big Bang discovery," by Joanne Colella. *The Journal*, May 29, 2014. Reprinted by permission of *The Holmdel Journal*, Shrewsbury, NJ.

Summary

The universe has been expanding since the Big Bang, which occurred nearly 13.8 billion years ago. The Big Bang happened everywhere: it is not an explosion spreading out from a single point. Three major pieces of evidence exist for the Big Bang. Hubble's law states that more distant galaxies have higher redshifts, and those observed redshifts result from the increasing space between them. Observations of the cosmic microwave background radiation independently confirm that the universe had a hot, dense beginning. Finally, the amounts of helium and trace amounts of other light elements measured today agree with what would be expected from nuclear reactions of normal matter in the hot early universe.

LG 1 **Explain in detail the cosmological principle.** Observations suggest that on the largest scales, the universe is homogeneous (looks the same to all observers) and isotropic (looks the same in all directions), in agreement with the cosmological principle.

LG 2 **Describe how the Hubble constant can be used to estimate the age of the universe.** Hubble found that all galaxies are moving away, with a recession velocity proportional to distance, indicating that the universe is expanding. The expansion does not affect local physics and the structure of objects. Running the Hubble expansion backward leads to the Hubble time, and the age of the universe, found from the inverse of the Hubble constant.

LG 3 **Describe the observational evidence for the Big Bang.** Hubble's law states that light from distant galaxies is redshifted; that occurs because space itself is expanding. Hubble's law suggests that the universe was once very hot and very dense, a beginning known as the Big Bang. Big Bang theory predicts that we should be able to observe the radiation from a few hundred thousand years after the Big Bang. That radiation, called the cosmic microwave background radiation, has the same spectrum in every direction.

LG 4 **Explain which chemical elements were created in the early hot universe.** During the first few minutes after the Big Bang, the universe was hot enough for nucleosynthesis to take place through fusion. Deuterium, helium, lithium, and beryllium, but not the heavier elements, were created before the universe cooled too quickly for that fusion to take place further.

? Unanswered Questions

- What existed before the Big Bang? The usual (and less than satisfying) answer is that the Big Bang was the beginning of space and *time*, so time could not exist before it happened. A more recent answer is that this universe may be just one of many universes, and the Big Bang was the beginning of this universe only. We return to that topic in Chapter 22.

- If evidence existed that the cosmological principle were not true, then could we deduce anything about the evolution of the universe?

Questions and Problems

Test Your Understanding

1. What do astronomers mean when they say that the universe is homogeneous?
   a. The universe looks the same from every perspective.
   b. Galaxies are generally distributed evenly throughout the universe.
   c. All stars in all galaxies have planetary systems just like ours.
   d. The universe has looked the same at all times in its history.

2. What do astronomers mean when they say that the universe is isotropic?
   a. More distant parts of the universe look just like nearby parts.
   b. Intergalactic gas has the same density everywhere in the universe.
   c. The laws of physics apply everywhere in the universe.
   d. The universe looks the same in every direction.

3. Cosmological redshifts are calculated from observations of spectral lines from
   a. individual stars in distant galaxies.
   b. clouds of dust and gas in distant galaxies.
   c. spectra of entire galaxies.
   d. rotations of the disks of distant galaxies.

4. When they look into the universe, astronomers observe that nearly all galaxies are moving away from the Milky Way. That observation suggests that
   a. the Milky Way is at the center of the universe.
   b. the Milky Way must be at the center of the expansion.
   c. the Big Bang occurred at the current location of the Milky Way.
   d. an observer in a distant galaxy would make the same observation.

5. Some galaxies have redshifts z that if equated to v_r/c correspond to velocities greater than the speed of light. Special relativity is not violated then
   a. because of relativistic beaming.
   b. because of superluminal motion.
   c. because redshifts carry no information.
   d. because those velocities do not measure motion through space.

6. The Big Bang theory predicted (select all that apply)
   a. the Hubble law.
   b. the cosmic microwave background radiation.
   c. the cosmological principle.
   d. the abundance of helium.
   e. the period-luminosity relationship of Cepheid variables.

7. The simplest way to estimate the age of the universe is from
   a. using the slope of Hubble's law.
   b. the age of Moon rocks.
   c. models of stellar evolution.
   d. measurements of the abundances of elements.

8. The CMB includes information about (select all that apply)
   a. the age of the universe.
   b. the temperature of the early universe.
   c. the density of the early universe.
   d. density fluctuations in the early universe.
   e. the motion of Earth around the center of the Milky Way.

9. Repeated measurements showing that the current helium abundance is much less than the value predicted by the Big Bang would imply that
   a. some part of the Big Bang theory is incorrect or incomplete.
   b. the current helium abundance is wrong.
   c. scientists don't know how to measure helium abundances.

10. The cosmological principle says that
    a. the universe is expanding.
    b. the universe began in the Big Bang.
    c. the rules that govern the universe are the same everywhere.
    d. the early universe was 1,000 times hotter than the characteristic temperature of the CMB.

11. Why is the Milky Way Galaxy not expanding together with the rest of the universe?
    a. Because it is at the center of the expansion.
    b. It is expanding, but the expansion is too small to measure.
    c. The Milky Way is a special location in the universe.
    d. Local gravity dominates over the expansion of the universe.

12. The scale factor keeps track of
    a. the movement of galaxies through space.
    b. the current distances between many galaxies.
    c. the changing distance between any two galaxies.
    d. the location of the center of the universe.

13. The Big Bang is
    a. the giant supernova explosion that triggered the formation of the Solar System.
    b. the explosion of a supermassive black hole.
    c. the eventual demise of the Sun.
    d. the beginning of space and time.

14. The CMB is essentially uniform in all directions in the sky. That fact is an example of
    a. anisotropy.
    b. isotropy.
    c. thermal fluctuations.
    d. Wien's law.

15. Which of the following was *not* created as a result of Big Bang nucleosynthesis?
    a. helium   b. lithium
    c. hydrogen   d. deuterium
    e. carbon

Thinking about the Concepts

16. Imagine that you are standing in the middle of a dense fog.
    a. Would you describe your environment as isotropic? Why or why not?
    b. Would you describe it as homogeneous? Why or why not?

17. As the universe expands from the Big Bang, galaxies are not actually flying apart from one another. What is really happening?

18. We see the universe around us expanding, which gives distant galaxies an apparent velocity of 70 km/s/Mpc. If you were an astronomer living today in a galaxy located 1 billion light-years away from us, at what rate would you see the galaxies moving away from you?

19. Does Hubble's law imply that our galaxy is sitting at the center of the universe? Explain.

20. What does the value of R_U, the scale factor of the universe, tell us?

21. Does the expansion of the universe make the Sun bigger? What about the Milky Way? Why or why not?

22. Science's greatest strength is the self-correcting nature of scientific inquiry: minds can be changed in the face of new data, as described in the Process of Science Figure. Consider a modern paradigm of science such as the Big Bang, climate change, or the power source of stars. In a short paragraph, describe your current viewpoint, and give a piece of evidence that would make you change your mind if it were true. For example, Einstein believed the universe was static. Measurements of the movement of galaxies caused him to change his mind.

23. Name two predictions of the standard Big Bang theory that have been verified by observations.

24. Knowing that you are studying astronomy, a curious friend asks where the center of the universe is located. You answer, "Right here and everywhere." Explain in detail why you give this answer.

25. The general relationship between recession velocity (v_r) and redshift (z) is $v_r = cz$. That simple relationship fails, however, for very distant galaxies with large redshifts. Explain why.

26. Why is it significant that the CMB displays a Planck black-body spectrum?

27. What is the significance of the tiny brightness variations observed in the CMB?

28. What important characteristics of the early universe are revealed by today's observed abundances of various isotopes, such as ^2H and ^3He?

29. Why were only a few of the chemical elements created in the Big Bang?

30. How do astronomers know that some of the observed helium is left over from the Big Bang?

Applying the Concepts

31. In Figure 21.6, a rubber sheet is shown as an analogy to help you think about the scale factor. Between the moments shown in parts (a) and (b), each square doubles in size on every edge. How does the area of a square change? Imagine that the sheet is now a block of rubber, expanding in three dimensions instead of two. How would the volume of a cube change between the moment shown in part (a) and the moment shown in part (b)?

32. Error bars have not been plotted in Figure 21.12. Why not? Was that a very precise measurement or a very imprecise measurement? How does the precision of the measurement affect your confidence in the conclusions drawn from it?

33. Figure 21.14 includes both predictions and observations.
    a. What do the vertical yellow bar and the slanted lines and curves represent: theory or observation?
    b. What do the pastel horizontal lines and the vertical black line represent: predictions or observations?
    c. Do the predictions and observations match? Choose one example, and explain how you know.

34. The Hubble time ($1/H_0$) represents the age of a universe that has been expanding at a constant rate since the Big Bang. Calculate the age of the universe in years if H_0 equals 80 km/s/Mpc. (Note: 1 year = 3.16×10^7 seconds, and 1 Mpc = 3.09×10^{19} km.)

35. Throughout the latter half of the 20th century, estimates of H_0 ranged from 50 to 100 km/s/Mpc. Calculate the age of the universe in years for each of those estimated values of H_0. Do any answers contradict the ages obtained from stellar evolution or geology on Earth?

36. Suppose a galaxy is observed with a redshift of $z = 2$. How much has the universe expanded since that light was emitted from those galaxies?

37. How much has the universe expanded since light was emitted from a galaxy with a redshift of $z = 8$?

38. You observe a distant quasar in which a spectral line of hydrogen with rest wavelength $\lambda_{rest} = 121.6$ nm is found at a wavelength of 547.2 nm. What is its redshift? When the light from that quasar was emitted, how large was the universe in comparison with its current size?

39. A distant galaxy has a redshift of $z = 5.82$ and a recession velocity $v_r = 287,000$ km/s (about 96 percent of the speed of light).
    a. If $H_0 = 70$ km/s/Mpc and if Hubble's law remains valid out to such a large distance, how far away is that galaxy?
    b. Assuming a Hubble time of 13.8 billion years, how old was the universe at the look-back time of that galaxy?
    c. What was the scale factor of the universe at that time?

40. The spectrum of the CMB is shown as the red dots in Figure 21.12, along with a blackbody spectrum for a blackbody at a temperature of 2.73 K. From the graph, determine the peak wavelength of the CMB spectrum. Use Wien's law to find the temperature of the CMB. How does that rough measurement that you just made compare with the accepted temperature of the CMB?

41. COBE observations show that the Solar System is moving in the direction of the constellation Crater at a speed of 368 km/s relative to the cosmic reference frame. What is the blueshift (negative value of z) associated with that motion?

42. The average density of normal matter in the universe is 4×10^{-28} kg/m^3. The mass of a hydrogen atom is 1.66×10^{-27} kg. On average, how many hydrogen atoms are present in each cubic meter in the universe?

43. To get a feeling for the emptiness of the universe, compare its density (4×10^{-28} kg/m^3) with that of Earth's atmosphere at sea level (1.2 kg/m^3). How much denser is Earth's atmosphere? Write that ratio in standard notation.

44. Assume that the most distant galaxies have a redshift of $z = 10$. The average density of normal matter in the universe today is 4×10^{-28} kg/m^3. What was its density when light was leaving those distant galaxies? (Hint: Keep in mind that volume is proportional to the *cube* of the scale factor.)

45. What was the size of the universe (compared with the present) when the CMB was emitted, at $z = 1{,}000$?

USING THE WEB

46. For more details on the history of the discovery of the expanding universe, go to the American Institute of Physics' "Cosmic Journey: A History of Scientific Cosmology" website (https://history.aip.org/exhibits/cosmology/index.htm). Read through the sections titled "Island Universes," "The Expanding Universe," and "Big Bang or Steady State?" Why was Albert Einstein "irritated" by the idea of an expanding universe? What was the contribution of Belgian astrophysicist (and Catholic priest) Georges Lemaître? What is the steady-state theory, and what was the main piece of evidence against it?

47. Go to the American Institute of Physics page on the age of the universe, https://history.aip.org/exhibits/curie/age-of-earth.htm. How does the age of the universe from Hubble's law compare with the age of the Solar System? How does that age compare with the age of the oldest globular clusters? Do a search for "age of the universe"—has the value been updated?

48. A series of public lectures at Harvard commemorating the 50th anniversary of the discovery of the CMB can be found here: https://astronomy.fas.harvard.edu/news/public-talk-50-year-anniversary-discovery-cosmic-micrrowave-background. Robert Wilson's talk starts at the 27-minute mark. Why does he say this is the most important thing he ever measured? How did he identify that radiation?

49. Updated results from observations of the CMB from the Planck space observatory are reported on the European Space Agency's Planck website (http://www.esa.int/Our_Activities/Space_Science/Planck). What has been learned from that mission? What is their value of the Hubble constant?

50. Go to the University of Washington Astronomy Department's "Hubble's Law: An Introductory Astronomy Lab" Web page (https://sites.google.com/a/uw.edu/introductory-astronomy-clearinghouse/activities/galaxies-and-cosmology/hubbles-law) and download the lab exercise, which uses real data from galaxies to calculate Hubble's constant. Your instructor will indicate whether you should use the "current" or the "shortened" version.

EXPLORATION

digital.wwnorton.com/astro6

The expansion of the universe is extremely difficult to visualize, even for professional astronomers. In this Exploration, you will use the surface of a balloon to get a feel for how an "expansion" changes distances between objects. Throughout this Exploration, remember to think of the surface of the balloon as a two-dimensional object, much as the surface of Earth is a two-dimensional object for most people. The average person can move east or west, or north or south, but moving into Earth and out to space are not options. For this Exploration you will need a balloon, 11 small stickers, a piece of string, and a ruler. A partner is helpful as well. **Figure 21.15** shows some of the steps involved.

Blow up the balloon partially and hold it closed, but *do not tie it shut*. Stick the 11 stickers on the balloon (those represent galaxies) and number them. Galaxy 1 is the reference galaxy.

Measure the distance between the reference galaxy and each of the galaxies numbered 2–10. The easiest way to do that is to use your piece of string. Lay it along the balloon between the two galaxies and then measure the length of the string. Record those data in the "Distance 1" column of a table like the one shown later.

Simulate the expansion of your balloon universe by *slowly* blowing up the balloon the rest of the way. Have your partner count the seconds you take to do that, and record that number in the "Time Elapsed" column of the table (each row has the same time elapsed because the expansion occurred for the same amount of time for each galaxy). Tie the balloon shut. Measure the distance between the reference galaxy and each numbered galaxy again. Record those data under "Distance 2."

Subtract the first measurement from the second. Record the difference in the table.

Divide that difference, which represents the distance traveled by the galaxy, by the time that blowing up the balloon took. Distance divided by time gives an average speed.

Make a graph with velocity on the *y*-axis and distance 2 on the *x*-axis to get "Hubble's law for balloons." You may wish to roughly fit a line to those data to clarify the trend.

1 Describe your data. If you fit a line to them, is it horizontal or does it trend upward or downward?

_____

2 Is anything special about your reference galaxy? Is it different in any way from the others?

_____

3 If you had picked a different reference galaxy, would the trend of your line be different? If you are not sure of the answer, get another balloon and try it.

_____

Figure 21.15 Measure the distance around a curved balloon by using a string.

Galaxy Number	Distance 1	Distance 2	Difference	Time Elapsed	Velocity
1 (reference)	0	0	0		0
2					
3					
4					
5					
6					
7					
8					
9					
10					
11					

4 The expansion of the universe behaves similarly to the movement of the galaxies on the balloon. We don't want to carry the analogy too far, but let's think about one more thing. In your balloon, some areas probably expanded less than others because the material was thicker; more "balloon stuff" was holding it together. How is that analogy similar to some places in the actual universe?

_____

_____

_____

22

Cosmology

Cosmology is the study of the large-scale universe, including its nature, origin, evolution, and ultimate destiny. Here in Chapter 22, we take a closer look at the nature of the universe, how it has evolved, and its ultimate fate. We also discuss the physics of the smallest particles, which is necessary for describing the earliest moments of the universe.

LEARNING GOALS

By the end of this chapter, you should be able to:

LG 1 Explain how mass within the universe and the gravitational force it produces affect the history, shape, and fate of the universe.

LG 2 Describe the evidence for the accelerating expansion of the universe.

LG 3 Describe the early period of rapid expansion of the universe known as inflation.

LG 4 Explain how the events that occurred in the earliest moments of the universe are related to the forces that operate in the modern universe.

The arrows point to a Type Ia supernova observed in the galaxy NGC 1365. Such supernovae provide evidence about the fate of the universe. ▶ ▶ ▶

What will be
the fate of
the universe?

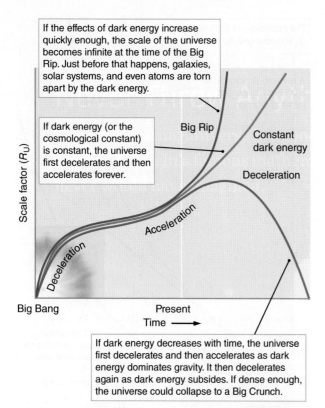

If the effects of dark energy increase quickly enough, the scale of the universe becomes infinite at the time of the Big Rip. Just before that happens, galaxies, solar systems, and even atoms are torn apart by the dark energy.

If dark energy (or the cosmological constant) is constant, the universe first decelerates and then accelerates forever.

Big Rip

Constant dark energy

Deceleration

Acceleration

Deceleration

Deceleration

Big Bang

Present

Time ⟶

If dark energy decreases with time, the universe first decelerates and then accelerates as dark energy dominates gravity. It then decelerates again as dark energy subsides. If dense enough, the universe could collapse to a Big Crunch.

Figure 22.4 The scale factor R_U of the universe varies depending on how dark energy changes.

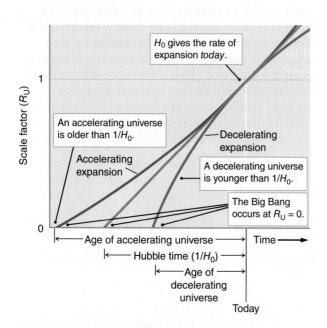

H_0 gives the rate of expansion *today*.

An accelerating universe is older than $1/H_0$.

Accelerating expansion

Decelerating expansion

A decelerating universe is younger than $1/H_0$.

The Big Bang occurs at $R_U = 0$.

Age of accelerating universe

Hubble time ($1/H_0$)

Age of decelerating universe

Time ⟶

Today

Figure 22.5 This graph shows plots of the scale factor R_U versus time for three possible universes. If the universe has expanded at a constant rate, its age is equal to the Hubble time, $1/H_0$. If the expansion of the universe has slowed, the universe is younger than the Hubble time. If the expansion has sped up, the universe is older than $1/H_0$.

about 0.3 and 0.7, respectively. Thus, the expansion of the universe is apparently accelerating under the dominant effect of dark energy and has been doing so for 5 billion to 6 billion years.

Because scientists do not yet understand the origin of dark energy, it is possible that it is not really a constant of nature and instead could be either increasing or decreasing. A changing cosmological constant would significantly change the future of the universe (**Figure 22.4**). For example, if dark energy were to decrease rapidly enough with time, the accelerating expansion of the universe that is observed now would change to a *deceleration* as the mass once again dominates over dark energy. In fact, if the universe were much denser than measured values indicate, the expansion could reverse, and the universe could collapse to what astronomers call the **Big Crunch**. If the effect of dark energy were to increase with time, by contrast, the universe would accelerate its expansion at an ever-increasing rate. Ultimately, expansion could be so rapid that the scale factor would become infinite within a finite period—a phenomenon called the **Big Rip**. In the Big Rip, the repulsive force of dark energy would become so dominant that the entire universe would come apart. First, gravity would no longer keep groups of galaxies together; then, gravity would no longer be able to hold individual galaxies together; and so on. Just before the end, the Solar System would come apart, and even atoms would be ripped into their constituent components. Don't worry too much about the Big Crunch or the Big Rip, though: the best observational data seem consistent with constant dark energy. In that case, the universe continues expanding, and after 100 trillion years or so, the universe is cold and dark, filled with black holes, dead stars, and dead planets (sometimes called the *Big Chill*).

The Age of the Universe

The values for Ω_m and Ω_Λ not only affect predictions for the future of the universe but also influence how astronomers interpret the past. **Figure 22.5** shows plots of the scale factor of the universe versus time. Measurement of the Hubble constant (H_0) indicates how fast the universe is expanding *today*. That is, the Hubble constant indicates the slope of the curves in Figure 22.5 at the current time. If the expansion of the universe has not changed, the plot of the scale factor versus time is the straight red line in Figure 22.5. The age of the universe in that case is equal to the Hubble time: $1/H_0$. If the expansion of the universe has been slowing down (green line in Figure 22.5), the universe is actually younger than the Hubble time. If the expansion of the universe has been speeding up (blue line), the true age of the universe is greater than the Hubble time. That scenario is also illustrated in **Figure 22.6**, where looking down the blue time line from the present to the past shows that the different models take different amounts of time to go from the Big Bang to the present.

Recall that a Hubble constant of $H_0 = 70$ km/s/Mpc corresponds to a Hubble time ($1/H_0$) of about 13.8 billion years. If the expansion of the universe has slowed, the universe is actually younger than 13.8 billion years. Having a younger universe is a problem if the measured ages of globular clusters—13 billion years—is correct because globular clusters cannot be older than the universe that contains them. But if the expansion of the universe has sped up, as suggested by the observations of Type Ia supernovae and of the 2.7-K cosmic background radiation, the universe is at least 13.8 billion years old—comfortably older than globular clusters. The effects from gravity and dark energy have nearly canceled out now.

The Shape of the Universe

We have already discussed such properties of the universe as density, dark energy, and age. The universe also has another key property: its shape in spacetime. Recall the concept of spacetime as described by general relativity. Space is a "rubber sheet" that has stretched outward from the Big Bang. In Chapter 18, you saw that the rubber sheet of space is also curved by the presence of mass. You saw how the shape of space around a massive object is detected through changes in geometric relationships, such as the ratio of the circumference of a circle to its radius or the sum of the angles in a triangle. Recall that the mass of a star, planet, or black hole causes a distortion in the shape of space; similarly, the mass of everything in the universe—including galaxies, dark matter, and dark energy—distorts the shape of the universe *as a whole*.

Three basic shapes are possible for the universe (**Figure 22.7**). Which shape actually describes the universe is determined by the total amount of mass and energy—in other words, the sum of Ω_m and Ω_Λ. As we continue with the rubber-sheet analogy, the first possibility is a **flat universe** (Figure 22.7a), corresponding to $\Omega_m + \Omega_\Lambda = 1$. A flat universe is described overall by the rules of the basic Euclidean geometry. That is, circles in a flat universe have a circumference of 2π times their radius ($2\pi r$), and triangles contain angles whose sum is 180 degrees. A flat universe stretches on forever.

The second possibility, an **open universe** (Figure 22.7b), corresponds to $\Omega_m + \Omega_\Lambda < 1$ and is shaped something like a saddle. In an open universe, the circumference of a circle is greater than $2\pi r$, and triangles contain less than 180 degrees. An open universe, like a flat universe, is infinite.

The third possibility, a **closed universe** (Figure 22.7c) corresponds to $\Omega_m + \Omega_\Lambda > 1$ and is like the surface of a sphere. The geometric relationships on a sphere are similar to those near a massive object (see Chapter 18). Thus the circumference of a circle on a sphere is less than $2\pi r$, and triangles contain more than 180 degrees. Space in a closed universe is finite and closes back on itself. Imagine a universe with two spatial dimensions on the surface of a sphere. Then the surface (and universe) is finite with no boundary. The cosmological principle is satisfied. Space is locally flat, and the expansion of the universe maintains both the cosmological principle and the local flatness of space.

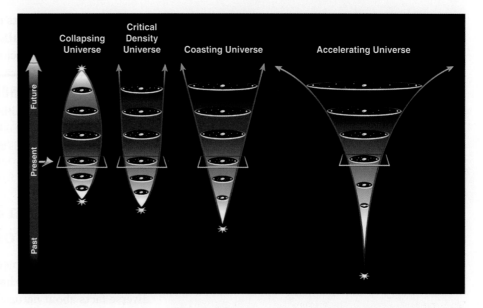

Figure 22.6 Past, present, and future for different scenarios. All models start in the past with a Big Bang (yellow star), expand to the present, and then have different possible futures. The present age of the universe is different in each model, as in Figure 22.5.

Figure 22.7 These two-dimensional representations show possible geometries that space can have in a universe. In a flat universe (a), Euclidean geometry holds: triangles have angles that sum to 180 degrees, and the circumference of a circle equals 2π times the radius. In an open universe (b) or a closed universe (c), those relationships are no longer correct over very large distances.

(a) Flat geometry

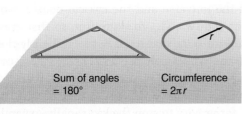

Sum of angles = 180° Circumference = $2\pi r$

If $\Omega_m + \Omega_\Lambda = 1$, the universe is flat.

(b) Open (saddle) geometry

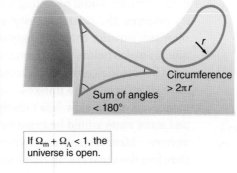

Sum of angles < 180° Circumference > $2\pi r$

If $\Omega_m + \Omega_\Lambda < 1$, the universe is open.

(c) Closed (spherical) geometry

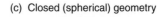
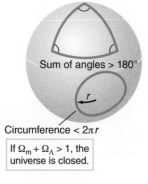

Sum of angles > 180° Circumference < $2\pi r$

If $\Omega_m + \Omega_\Lambda > 1$, the universe is closed.

The measurements to estimate directly which of those shapes describes the universe are difficult. As noted earlier, the data suggest that $\Omega_m = 0.3$ and $\Omega_\Lambda = 0.7$, so $\Omega_m + \Omega_\Lambda$ is close to 1, in which case the universe is very nearly flat.

CHECK YOUR UNDERSTANDING 22.2

Dark energy has been hypothesized to solve which problem? (a) The universe is expanding. (b) The cosmic microwave background radiation is too smooth. (c) The expansion of the universe is accelerating. (d) Stars orbit the centers of galaxies too fast.

22.3 Inflation Solves Several Problems in Cosmology

A century ago, astronomers were struggling to understand the size of the universe. Today scientists have a comprehensive theory that ties together many diverse facts about nature: the constancy of the speed of light, the properties of gravity, the motions of galaxies, and even the origins of the atoms that make up planets and life. The case for the Big Bang is compelling. Even so, improved observations of the cosmic background radiation and measurements of the expansion of the universe have raised some questions about how the universe expanded when it was very young. In this section, we look at those questions and some potential answers.

The Flatness Problem

From the cosmic microwave background radiation (CMB), astronomers have found that the universe is flat—too close to being exactly flat for that to have happened by chance. Any deviation from flatness would grow, so if the universe originally had a value of $\Omega_m + \Omega_\Lambda$ even slightly different from 1, the value would by now be drastically different and easily detectable. For the present-day value of $\Omega_m + \Omega_\Lambda$ to be as close to 1 as it is, when the universe was 1 second old, $\Omega_m + \Omega_\Lambda$ must have been equal to 1 all the way out to at least the tenth decimal place. At even earlier times, it had to be much flatter still. That situation is too special to be the result of chance—a fact referred to in cosmology as the flatness problem: the universe is so flat that something about the early universe must have caused $\Omega_m + \Omega_\Lambda$ to have a value incredibly close to 1.

A universe that contains mass and does not start out perfectly flat has a very different fate. If a universe started out with Ω_m even slightly greater than 1, its expansion would slow more rapidly than that of the flat universe, meaning that less and less density would be required to stop the expansion. Meanwhile, the actual density would be falling less rapidly than in the flat universe. That disparity between the actual density of the universe and the critical density would increase, causing the ratio between the two, Ω_m, to skyrocket, so that gravity quickly overwhelms the expansion. A universe that starts out even slightly closed rapidly becomes obviously closed and would collapse long before stars could form. Conversely, if a universe started with Ω_m even a tiny bit less than 1, the expansion would slow less rapidly than in a flat universe. As time passed, more and more mass would be required for gravity to stop the too-rapidly-expanding universe. Meanwhile, the actual density of the universe would be dropping faster than in a flat universe. In that case, Ω_m would plummet.

Adding Ω_Λ to the picture makes the math more complex but does not change the basic results. Try balancing a razor blade on its edge. If the blade is tipped just a tiny bit in one direction, it quickly falls that way. If the blade is tipped just a tiny bit in the other direction, it quickly falls in the other direction instead. It would seem that the actual universe should be obviously open or obviously closed—analogous to the tipped razor blade. Instead, the universe has $\Omega_m + \Omega_\Lambda$ so close to 1 that telling which way the razor blade is tipped is difficult—if it is tipped at all. Discovering that $\Omega_m + \Omega_\Lambda$ is extremely close to 1 after more than 13 billion years is like balancing a razor blade on its edge and coming back 10 years later to find that it still has not tipped over.

The Horizon Problem

Another problem faced by cosmological models is that the CMB is surprisingly smooth. After its discovery in the 1960s, many observational cosmologists turned their attention to mapping that background radiation. At first, result after result showed that the temperature of the CMB is remarkably constant, with variations of less than one part in 3,000, regardless of where one looks in the sky. Over time, though, that strong confirmation of Big Bang cosmology challenged cosmologists' view of the early universe. Once Earth's motion relative to the CMB is removed from the picture, the CMB is not just smooth—it is *too* smooth.

In Chapter 5, we discussed the bizarre world of quantum mechanics that shapes the world of atoms, light, and elementary particles. When the universe was extremely young, it was so small that quantum mechanical effects played a role in shaping the structure of the universe as a whole. The early universe was subject to the quantum mechanical **uncertainty principle**, which says that as a system is studied at extremely small scales, the properties of that system become less and less well determined. That principle applies to the properties of an electron in an atomic orbital or to the entire universe when it was very young and would have fit within the size of an atom.

Consider a simple analogy of how the uncertainty principle applies to the universe. Imagine sitting on the beach looking out across the ocean. Off in the distance, you see more total ocean and average the surface over larger scales; therefore, the surface of the ocean appears smooth and flat. The horizon looks almost like a geometric straight line. Yet the apparent smoothness of the ocean as a whole hides the tumultuous structure present at smaller scales, where waves and ripples fluctuate dramatically from place to place. Similarly, while the universe on large scales seems steady and smooth, quantum mechanics says that conditions must fluctuate unpredictably at smaller and smaller scales in the universe. In particular, quantum mechanics says that those fluctuations are more dramatic earlier and earlier in the history of the universe. When the universe was young, it could not have been smooth. Dramatic variations ("ripples") must have existed in the density and temperature of the universe from place to place.

If the universe had expanded slowly, those ripples would have smoothed themselves out, but the universe expanded much too rapidly for such smoothing to be possible. After the Big Bang, not enough time was available for a smoothing signal to travel from one region to the other. So when cosmologists look at the universe today, they should see the fingerprint of those early ripples imprinted on the cosmic background radiation—but they do not. The fact that the CMB is so smooth is called the *horizon problem* in cosmology. The **horizon problem**

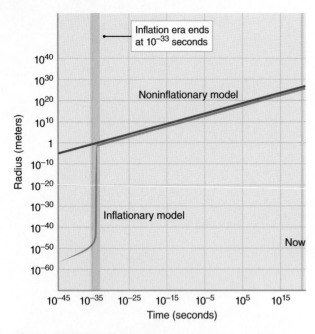

Figure 22.8 This graph illustrates expansion of the observable universe. In an inflationary model, between 10^{-35} and 10^{-33} second after the Big Bang, the universe expanded greatly.

states that different parts of the universe are too much like other parts of the universe that should have been "over their horizon" and beyond the reach of any signals that might have smoothed out the early quantum fluctuations. In essence, the horizon problem is as follows: How can different parts of the universe that underwent different fluctuations and were never able to communicate with one another still show the same temperature in the cosmic background radiation to an accuracy of better than one part in 100,000?

Inflation: Early, Rapid Expansion

In the early 1980s, physicist Alan Guth (1947–) offered a solution to the flatness and horizon problems of cosmology. Guth suggested that the universe has not expanded at a steady pace but that instead it started out much more compact than steady expansion would predict. Then, briefly, the young universe expanded at a rate *far* in excess of the speed of light. That rapid expansion of the universe is called **inflation**. In the first 10^{-33} second of the universe, the distance between points in space increased by a factor of at least 10^{30} and perhaps very much more (**Figure 22.8**). In that incomprehensibly brief instant, the size of the observable universe grew from ten-trillionths the size of the nucleus of an atom to a region about 1 meter across. That is like a grain of very fine sand growing to the size of today's observable universe—all in a billionth of the time that light takes to cross the nucleus of an atom. During inflation, space itself expanded so rapidly that the distances between points in space increased faster than the speed of light. Inflation does *not* violate the rule that no signal can travel through space faster than the speed of light because the space itself was expanding.

To understand how inflation solves the flatness and horizon problems of cosmology, imagine that you are an ant living in the two-dimensional universe defined by the surface of a golf ball (**Figure 22.9**). This universe is positively curved, like the surface of Earth. If you were to walk around the circumference of a circle in your two-dimensional universe and then measure the radius of the circle, you would find the circumference to be less than $2\pi r$. If you were to draw a triangle in your universe, the sum of its angles would be greater than 180 degrees. Another obvious characteristic would be the dimples, approximately a half-millimeter deep, on the surface of the golf ball.

Now imagine that the golf-ball universe suddenly grew to the size of Earth. The curvature of the universe would no longer be apparent. An ant (or person) walking along the surface of the golf-ball universe would think the universe is flat. The circumference of a circle would be $2\pi r$, and a triangle would have 180 degrees. (In fact, it took most of human history for people to realize that Earth is a sphere.) For inflationary cosmology, the universe after inflation would be extraordinarily flat (that is, having $\Omega_m + \Omega_\Lambda$ extraordinarily close to 1) *regardless* of what the geometry of the universe was before inflation. Because the universe was inflated by a factor of at least 10^{30}, $\Omega_m + \Omega_\Lambda$ immediately after inflation must have been equal to 1 within one part in 10^{60}, which is flat enough for $\Omega_m + \Omega_\Lambda$ to remain close to 1 today. If inflation occurred, today's universe is not flat by chance. It is flat because any universe that underwent inflation would become flat.

What about the horizon problem? When the golf-ball universe inflates to the size of Earth, the dimples that covered the surface of the golf ball stretch out as well. Instead of being a half millimeter or so deep and a few millimeters across, those dimples now are only an atom deep but are hundreds of kilometers across. The ant would not detect any dimples at all. For our universe, inflation took the